H.-G. Franck   J.W. Stadelhofer

Industrial Aromatic Chemistry

Raw Materials · Processes · Products

With 206 Figures and 88 Tables
and 720 Structural Formulas

Springer-Verlag
Berlin Heidelberg New York
London Paris Tokyo

Prof. Dr. Dr.-Ing. E.h. Heinz-Gerhard Franck
Dr. Jürgen Walter Stadelhofer
Rütgerswerke AG, Mainzer Landstrasse 217
6000 Frankfurt/M 11, West Germany

Title of the original German edition: Industrielle Aromatenchemie.
© Springer-Verlag Berlin Heidelberg 1987. ISBN 3-540-18146-6

The cover shows the tar refinery of *Rütgerswerke*
in Castrop-Rauxel, West Germany

ISBN-13: 978-3-642-73434-2   e-ISBN-13: 978-3-642-73432-8
DOI: 10.1007/978-3-642-73432-8

Library of Congress Cataloging-in-Publication Data
Franck, H.-G. (Heinz-Gerhard) Industrial aromatic chemistry.
Translation of: Industrielle Aromatenchemie. Bibliography: p. Includes index.
1. Aromatic compounds.   I. Stadelhofer, J.W. (Jürgen Walter), 1949- .   II. Title.
TP248.A7F7313   1988   661'.816   88-4365

This work is subject to copyright. All rights are reserved, whether the whole or part of the material is concerned, specifically the rights of translation, reprinting, reuse of illustrations, recitation, broadcasting, reproduction on microfilms or in other ways, and storage in data banks. Duplication of this publication or parts thereof is only permitted under the provisions of the German Copyright Law of September 9, 1965, in its version of June 24, 1985, and a copyright fee must always be paid. Violations fall under the prosecution act of the German Copyright Law.

© Springer-Verlag Berlin Heidelberg 1988
Softcover reprint of the hardcover 1st edition 1988

The use of registered names, trademarks, etc. in this publication does not imply, even in the absence of a specific statement, that such names are exempt from the relevant protective laws and regulations and therefore free for general use.

Flow sheets: Wolfgang Lücke
Graphic layout: Klaus Langhoff
Media conversion, printing, and bookbinding: G. Appl, Wemding

2152/3140-543210

Preface

Aromatic chemicals represent about 30% of the total of some 8 million known organic compounds; the percentage of aromatic chemicals produced by the entire organic chemical industry is of the same order.

The importance of aromatics in hydrocarbon technology is, however, greater than the percentage figures indicate. Quantitatively, the most important processes in hydrocarbon technology are catalytic reforming to produce gasoline, which has a worldwide capacity of around 350 Mtpa, and the carbonization of hard coal to produce metallurgical coke, on roughly the same scale. A characteristic of both processes is the formation of aromatics. The third most important process in hydrocarbon technology in terms of quantity is catalytic cracking, which is also accompanied by an aromatization, as is the most important petrochemical process, steam cracking of hydrocarbon fractions.

The recovery and further refining of aromatics was the basis of the industrial organic chemistry in the middle of the last century. In the early 1920's, aromatic chemistry was complemented by the chemistry of aliphatics and olefins, which today, in terms of quantity, has surpassed the industrial chemistry of aromatics.

From the beginning of industrial aromatic chemistry there have been fundamental new developments in the production of aromatics. Until the 1920's, coal tar and coke-oven benzole were virtually the sole sources of aromatics available on an industrial scale. Coal tar contains a host of widely used aromatic compounds, such as benzene, toluene, naphthalene, anthracene and pyrene, as well as styrene and indene. In addition coal tar contains some important recoverable aromatic compounds with hetero atoms, such as phenols, anilines, pyridines and quinoline.

As the growing demand for some coal-tar constituents for the development of mass-produced plastics, such as phenolic resins and polystyrene, and the increasing production of explosives could not be met by coal tar alone, new sources of aromatics were developed, starting from petroleum. The development of the production of reformate-gasoline and steam cracking of petroleum fractions has made two further feedstock sources for the production of aromatics available today; renewable raw materials are also used for the manufacture of aromatics, albeit on a much smaller scale.

Processes for the refining of crude aromatics, in common with methods for their further processing, are complicated by the occurrence of by-products; individual aromatics are usually accompanied by associated products, and must be isolated by subsequent separation processes. Aromatic chemistry is characterized by the high reactivity of the π-electron system, which enables substitutions to take place not only at one position in the aromatic ring, but also, especially for polynuclear aromatics, at several different sites, thus leading to isomers as well as multiple substitution. Refining processes have, therefore, to be optimized to produce the desired compounds in pure form from the crude products. Thus, industrial aromatic chemistry involves close collaboration between chemists and process engineers.

The present applications and future developments of aromatic chemicals are characterized by a number of inherent properties. These are in particular:

1. the facile substitution and high reactivity, which can be further increased by the introduction of suitable substituents,
2. the relatively easily activated π-electron system which, coupled with auxochromic groups, is capable of absorbing part of the spectrum of light and is used in the production of dyes and pigments,
3. the high solvent power, especially of alkylated derivatives,
4. the high C/H ratio, which renders polycyclic aromatics particularly suited to the production of high-value industrial carbon products, such as premium coke, graphite and carbon black, and
5. the affinity and tendency for association, which make aromatic molecules suitable mesogens for the formation of liquid-crystalline phases.

Approximately 800,000 tpa of organic dyestuffs (dyes, pigments and optical brighteners) are produced worldwide. Since the beginning of the industrial production of organic dyestuffs, aromatics have been the dominant raw materials for this group of products.

In addition, their versatility regarding substitution and resistance to premature biological degradation have made aromatics essential for the manufacture of plant protection agents. Of the ca. 300 registered plant protection agents in Japan, over one-half are aromatic in nature. A considerable proportion of the 160,000 t of pesticides (also ca. 300) produced in West Germany in 1985 is based on aromatics. In the USA likewise, the most important organic plant protection agents in terms of quantity, such as atrazine, alachlor, trifluralin and metolachlor, are aromatic in nature.

The traditional applications of aromatic chemistry, such as the production of dyestuffs and the manufacture of plant protection agents have constantly expanded since their early days. The latest advances in the production of aromatics have led to new catalytic processes from simple compounds such as methanol and propane/butane. The manufacture of aromatic monomer building blocks for the production of polymers for high-performance engineering plastics is in rapid development. The liquid-crystalline nature of a number of aromatic-based polymers is the key to obtaining high-performance properties in diverse applications such as high-value aramid fibers and carbon artifacts.

Against the background of these developments, the authors of this monograph have put together a review of aromatic chemistry from benzene through the polynuclear aromatics such as naphthalene, anthracene and pyrene, up to industrial graphite products, concentrating on the industrially important raw materials and intermediates as well as the quantitively most important final products. Some particularly interesting compounds of less importance in terms of quantity have been included to illustrate the broad range of aromatic chemistry.

The concentrated survey, complemented by detailed, standardized process flow sheets, provides chemists and process engineers engaged in the production and research, scientists in neighbouring disciplines, and advanced students of chemistry and chemical engineering with a brief insight into industrial aromatic chemistry, thus making possible a comprehensive review.

Thanks are due to numerous colleagues, both at home and abroad, for their valuable suggestions.

Frankfurt/Main (West Germany), *H.-G. Franck*
March 1988 *J. W. Stadelhofer*

Contents

1 History 1

2 The nature of the aromatic character 8

2.1	Molecular considerations	8
2.2	Mechanistic considerations	13
2.2.1	Electrophilic aromatic substitution	13
2.2.1.1	Orientation rules	17
2.2.2	Nucleophilic aromatic substitution	17
2.2.3	Radical reactions	20
2.2.3.1	Pyrolysis processes	20
2.2.3.2	Oxidation reactions	21
2.2.4	Rearrangement reactions	22
2.3	Nomenclature	23

3 Base materials for aromatic chemicals 27

3.1	Origin of fossil raw materials and their composition	27
3.2	Coal	31
3.2.1	Deposits, composition and uses of coal	31
3.2.2	Thermal coal conversion – tar and benzole recovery .	35
3.2.3	Tar refining	37
3.2.4	Aromatics stemming from coal gasification	43
3.2.5	Production of aromatics by coal liquefaction	46
3.2.5.1	Historic development	46
3.2.5.2	Mechanisms of coal liquefaction	47
3.2.5.3	Further developments in coal hydrogenation since 1970	51
3.2.5.4	Hydropyrolysis	55
3.3	Crude oil	55
3.3.1	Reserves and characteristics of crude oil	55
3.3.2	Crude oil refining	60
3.3.2.1	Distillation	60
3.3.2.2	Catalytic cracking	63

3.3.2.3	Catalytic reforming	68
3.3.2.4	Hydrocracking	73
3.3.2.5	Thermal cracking of hydrocarbons	76
3.3.2.5.1	Steam cracking to produce olefins	77
3.3.2.5.2	Production of olefins from crude oil	81
3.3.2.5.3	Other thermal-cracking processes	83
3.4	Production of aromatic hydrocarbons with zeolites	86
3.4.1	Aromatics from alcohols	86
3.4.2	Aromatics from short-chain alkanes	88
3.5	Renewable raw materials	89
3.5.1	Furfural	92
3.5.2	Lignin	93
3.5.3	Vanillin	95
3.6	Summary review of processes for the production of aromatics	96

4 Production of benzene, toluene and xylenes 99

4.1	History	99
4.2	Pre-treatment of mixtures containing crude aromatics	100
4.2.1	Recovery of aromatics	105
4.2.1.1	Liquid-liquid extraction	107
4.2.1.2	Extractive and azeotropic distillation	112
4.3	Separation of mixed aromatics into individual constituents	114
4.4	Dealkylation, isomerization and disproportionation reactions of BTX aromatics	122
4.4.1	Dealkylation of toluene and xylenes to benzene	122
4.4.2	Isomerization of xylenes	125
4.4.3	Disproportionation	126
4.5	Quality standards	128
4.6	Economic data	128
4.7	Process review	130

5 Production and uses of benzene derivatives 132

5.1	Ethylbenzene	133
5.1.1	Recovery of ethylbenzene from aromatic mixtures	133
5.1.2	Synthesis of ethylbenzene	133
5.1.3	Production of styrene	137
5.1.4	Substituted styrenes	144
5.2	Cumene	146
5.3	Phenol	148
5.3.1	Phenol from cumene	149
5.3.2	Alternative methods for the synthesis of phenol	151

5.3.3	Recovery of phenol from the products of coal pyrolysis	155
5.3.4	Phenol derivatives	158
5.3.4.1	Bisphenol A	158
5.3.4.2	Cyclohexanol and cyclohexanone	161
5.3.4.3	Alkylphenols	163
5.3.4.3.1	Cresols	164
5.3.4.3.2	Xylenols	171
5.3.4.3.3	Higher alkyl phenols	174
5.3.4.4	Salicylic acid	175
5.3.4.5	Chlorinated phenols	176
5.3.4.6	Nitrophenols	180
5.3.4.7	Other phenol derivatives	181
5.3.4.8	Polyhydric phenols	183
5.4	Benzene hydrogenation – cyclohexane	191
5.5	Nitrobenzene and aniline	193
5.5.1	Nitrobenzene	194
5.5.2	Aniline	196
5.5.3	Aniline derivatives	199
5.5.3.1	4,4'-Diphenylmethane diisocyanate (MDI)	199
5.5.3.2	Cyclohexylamine and benzothiazole derivatives	200
5.5.3.3	Secondary and tertiary aniline bases	203
5.5.3.4	Other aniline derivatives	205
5.6	Alkylbenzenes and alkylbenzene sulfonates	210
5.7	Maleic anhydride	213
5.8	Chlorobenzenes	218
5.8.1	Chlorobenzene	218
5.8.1.1	Nitrochlorobenzenes	222
5.8.2	Dichlorobenzenes	229
5.8.3	Hexachlorocyclohexane	232
5.9	Process review	234

6 Production and uses of toluene derivatives 236

6.1	Nitro-derivatives of toluene	237
6.1.1	Mononitrotoluene and its products	237
6.1.2	Dinitrotoluenes and their derivatives	242
6.1.3	Trinitrotoluene compounds	247
6.2	Benzoic acid	247
6.3	Chlorine derivatives of toluene	250
6.3.1	Side-chain chlorination of toluene	250
6.3.1.1	Benzyl chloride	251
6.3.1.2	Benzal chloride	256
6.3.1.3	Benzotrichloride	258
6.3.2	Nuclear chlorination of toluene	259
6.4	Sulfonic acid derivatives of toluene	261

6.5	Toluenesulfonyl chloride	263
6.6	Other toluene derivatives	264

7 Production and uses of xylene derivatives 265

7.1	o-Xylene and its derivatives	265
7.1.1	Oxidation of o-xylene to phthalic anhydride	265
7.1.2	Production of phthalic esters	272
7.1.3	Other products from phthalic anhydride	274
7.1.4	Nitration of o-xylene	277
7.2	m-Xylene and its derivatives	279
7.2.1	Production of isophthalic acid	279
7.2.2	Other products from m-xylene	280
7.3	p-Xylene and its derivatives	283
7.3.1	Terephthalic acid	283
7.3.2	Other p-xylene derivatives	290

8 Polyalkylated benzenes – production and uses 291

8.1	Pseudocumene	292
8.2	Mesitylene	294
8.3	Durene	295
8.4	Other cumene derivatives	296
8.4.1	Nitrocumene and isoproturon	296
8.4.2	Cumenesulfonic acid	296
8.5	Indan and indene	297

9 Naphthalene – production and uses 298

9.1	History	298
9.2	Naphthalene recovery	299
9.2.1	Naphthalene from coal tar	299
9.2.2	Naphthalene from petroleum-derived raw materials	305
9.3	Naphthalene derivatives	308
9.3.1	Production of phthalic anhydride (PA) from naphthalene	309
9.3.2	Production and uses of naphthoquinone	310
9.3.3	Production and uses of naphthols	313
9.3.3.1	1-Naphthol and its derivatives	313
9.3.3.2	2-Naphthol and its derivatives	316
9.3.4	Sulfonic acid derivatives of naphthalene	322
9.3.5	Nitro- and aminonaphthalenes	327
9.3.6	Production of tetralin and tetralone	328
9.3.7	Production of isopropylnaphthalene derivatives	329
9.3.8	Other alkylnaphthalenes from naphthalene	331
9.4	Process review	332

10 Alkylnaphthalenes and other bicyclic aromatics -production and uses 334

10.1 Biphenyl 334
10.2 Methylnaphthalenes 336
10.3 Acenaphthene/acenaphthylene 340

11 Anthracene - production and uses 343

11.1 Production of anthracene 343
11.2 Production of anthraquinone 346
11.2.1 Oxidation of anthracene to anthraquinone 346
11.2.2 Synthetic production of anthraquinone 348
11.3 Anthraquinone derivatives 350
11.3.1 Production of anthraquinone derivatives from anthraquinone 350
11.3.2 Synthetic production of anthraquinone derivatives . 354
11.4 Higher condensed dyes from anthraquinone 357
11.5 Anthraquinone as a catalyst in the production of hydrogen peroxide 359
11.6 Wood pulping with anthraquinone 360

12 Additional polynuclear aromatics - production and uses . . . 362

12.1 Phenanthrene 362
12.2 Fluorene 364
12.3 Fluoranthene 365
12.4 Pyrene 365

13 Production and uses of carbon products from mixtures of condensed aromatics 368

13.1 Pyrolysis of aromatic hydrocarbon mixtures in the liquid phase 368
13.1.1 Formation of mesophase 368
13.1.2 Production and uses of coke from aromatic residues by the delayed coking process 375
13.1.3 Pitch coking in horizontal chamber ovens 380
13.1.4 Production of carbon fibers 380
13.2 Pyrolysis of mixtures of aromatics in the gas phase - Carbon black production 382

14 Aromatic heterocyclics – production and uses 387

14.1	Five-membered ring heterocyclics	387
14.1.1	Furan	387
14.1.2	Thiophene	389
14.1.3	Pyrrole	390
14.1.4	Five-membered ring heterocyclics with two or more hetero-atoms	391
14.2	Six-membered ring heterocyclics	394
14.2.1	Pyridine	394
14.2.2	Pyridine derivatives	401
14.2.3	Alkylated pyridines	403
14.3	Pyrimidine	411
14.4	Triazines	412
14.5	Condensed heterocyclics	416
14.5.1	Thianaphthene (2,3-Benzothiophene)	416
14.5.2	Indole	417
14.5.3	Benzothiazole	418
14.5.4	Benzotriazole	418
14.5.5	Quinoline and isoquinoline	419
14.5.6	Carbazole	423
14.5.7	Dibenzofuran (Diphenylene oxide)	425

15 Toxicology / Environmental aspects 426

15.1	Basic toxicological considerations	426
15.2	Aspects of occupational medicine and legislation . .	427
15.3	Environmental aspects and biological degradation of aromatics	444

16 The future of aromatic chemistry 447

Appendix 449

Bibliography 455

Subject index 471

1 History

Aromatic compounds are currently defined as cyclic hydrocarbons in which the carbon skeleton is linked by a specified number of conjugated π-bonds in addition to σ-bonds (Hückel's rule). During the early days of industrial aromatic chemistry in the mid-19th century, the structure of aromatic compounds had not yet been elucidated. The name of this class of compounds is historically-based since the first members were obtained from aromatic, i.e. pleasant-smelling resins, balsams and oils; examples of this are benzoic acid, which was obtained from gum benzoin, toluene from tolu balsam and benzaldehyde from oil of bitter almonds.

The history of aromatic chemistry was, at the outset, closely linked to the development of coal carbonization to produce coke, gas and tar.

Coke was mainly used as a substitute for charcoal in the production of pig iron, coal gas was used for lighting, and coal tar initially replaced wood tar for impregnating timber used in ship-building.

As early as 1584, the Duke of Brunswick recommended the application of 'desulfurized' coal as an alternative to charcoal for the production of salt. The first patent for the production of coke for use in blast furnaces for iron smelting was granted to Dud Dudley in England in 1622.

The first large-scale attempts to manufacture lighting gas were initiated by the French engineer Philippe Lebon in 1790. He degasified wood chippings in an iron retort on the grate of a kitchen stove and fed the resultant gas by pipe to other rooms, where it was burned in lamps. His discovery, which he called 'Thermolamp', aroused great interest but never found practical application.

The real founder of gas engineering was the Scotsman William Murdoch, who carried out experiments in Redruth, Cornwall, in 1792 on the degasification of hard coal. He lit his house with gas which he brought daily from the factory in containers. (This portable gas was one of the major products of the company *Engelhorn & Comp.*, founded in 1848 by Friedrich Engelhorn, later to be one of the co-founders of *BASF*.)

The German philosopher and chemist, Johann Becher, is considered to be the discoverer of coal tar. In 1691, together with the Englishman Henry Serle, he obtained the English patent no. 214 for the production of pitch and tar from coal.

The German privy councilor Winzler, known in England as Windsor, was particularly successful in marketing the idea of using coal gas for lighting and founded a number of gas companies. In London in 1813, Westminster Bridge was lit by gas from the *London & Westminster Chartered Light & Coke Co.*, established by Windsor; in 1819 he introduced gas lighting to Paris. Lighting with coal gas began in Germany in 1824 in Hanover, and 1826 in Berlin, when gas works were set up by the *London Imperial Continental Association*. The first gas works in the USA was already operating in Baltimore in 1802; gas lighting was introduced into New York in 1824.

The first distillation of gas-works tar from coal carbonization was carried out at Leith in Scotland in 1822. The tar oil was used in timber impregnation, while the distillation residue, namely pitch, was employed in coal briquetting.

The key factor in the development of the tar industry was the accelerated growth of the railroad system. Tracks were laid on wooden sleepers, which were impregnated with coal tar oil to preserve them from rapid decay. The first railway lines started operation in England between Stockton and Darlington in 1825, and in Germany between Nuremberg and Fürth in 1835.

In spite of the demand for impregnation oil, by the mid-19th century, owing to the rapid growth in the production of gas for lighting and the tremendous development of the iron and steel industry, there was a considerable over-supply of coal tar. Although some of the tar could be used in the production of roofing tar felts and in carbon black manufacture, these applications were not sufficient to absorb all the tar being produced.

The first estimates of annual tar production from gas works in Europe date from 1884. Great Britain led the way with a production of 450,000 t, followed by Germany (85,000 t), France (75,000 t), Belgium (50,000 t) and the Netherlands (15,000 t).

The parent compound of the aromatics is benzene; it was first discovered by Michael Faraday in 1825 in the condensed part of a lighting gas derived from whale oil and obtained some years later by Eilhard Mitscherlich by decarboxylation of benzoic acid (as calcium benzoate). The occurrence of benzene in coal tar was first described by August Wilhelm v. Hofmann in 1845. John Leigh had already demonstrated to the British Natural Research Conference in 1842, that benzene is present in coal tar; this claim was not immediately published, however. Even before the discovery of benzene, Ferdinand Runge had found aniline and phenol in coal tar in 1834.

The composition of the aromatic mixture, coal tar, was still largely unknown up to the middle of the 19th century. As tar production grew, so analytical investigations increased; August Wilhelm v. Hofmann, a disciple of Justus v. Liebig, was particularly active in this field.

In 1845, Hofmann went to London as Principal of the newly-founded Royal College of Chemistry, to continue his investigations at the original source of coal tar. Hofmann gathered a number of young chemists around him, who concentrated on investigating the reactions of tar components. London thus became the Mecca of aromatic chemistry.

One of the principal objectives of Hofmann's work was to synthesize quinine, at that time the only known agent effective against malaria. William Henry Perkin, one of Hofmann's youngest students, devoted a great deal of imagination to the synthesis of quinine. In 1856, Perkin tried to synthesize quinine by oxidation of N-allyltoluidine, but instead obtained only a red-brown precipitate. As a model reaction Perkin chose to investigate the treatment of aniline sulfate with potassium dichromate.

                    N-Allyltoluidine   Quinine

On working-up the reaction mixture he produced a violet dyestuff; Perkin had synthesized the first tar-derived dyestuff, mauveine. Within barely 18 months, together with his father and brother, he set up a factory in Greenford Green to manufacture aniline dyes, the first time that coal-tar dyestuffs had been produced on an industrial scale. By 1860, there were already five companies in England engaged in the production of synthetic dyes. Apart from Perkin, *Read & Holliday* should be mentioned among the founders of English companies; they established the first subsidiary in the USA in 1861. Synthetic dyestuffs found a ready market, since to satisfy growing demand from the textile industry, 75,000 t of natural dyes were imported annually into Great Britain alone.

Production of coal-tar dyes developed dramatically on the European continent too, following Perkin's discovery. In Lyons in 1859, Francois Emanuel Verguin produced the red-violet fuchsin (Basic Violet 14) by oxidation of technical aniline, a mixture of aniline and toluidines. This dyestuff provided the basis for the production of coal-tar dyestuffs in France; it is still important today. At the World Exhibition in London in 1862, the coal-tar dyestuffs industry celebrated great triumphs. The thirteen prize winners were almost exclusively English and French dye manufacturers.

          Mauveine (Main component)   Fuchsin

Following the pattern of England and France, new manufacturing facilities were set up in Germany for production of synthetic dyes. The first were established by existing natural-dye traders such as Rudolf *Knosp* and Gustav *Siegle* in

Stuttgart; these companies later became part of the *Badische Anilin- und Sodafabrik* (*BASF*). In 1860, the dyestuff merchant Friedrich Bayer set up a fuchsin factory in Elberfeld. In 1863, Meister, Lucius and Brüning also began the production of fuchsin in Höchst (*Hoechst*).

In the same year, Paul Wilhelm *Kalle* founded a dyeworks in Biebrich (Wiesbaden). In 1865, the *Badische Anilin- und Sodafabrik* was established in Mannheim; its founder company, *Sonntag, Engelhorn & Clemm* had already begun production of coal-tar dyes in 1861. In 1865, Carl Alexander Martius, a disciple of Hofmann, returned from England to Germany and became a co-founder of the *Aktiengesellschaft für Anilinfarben* (*Agfa*). In 1870, the dyestuffs company *Leopold Cassella & Companie* was founded in Mainkur (Frankfurt).

The growing demand from the dyestuffs industry for aromatics was met by the rapidly developing tar industry. In Germany, the companies *Rütgerswerke*, founded initially for timber impregnation by Julius Rütgers in 1849, and the *Gesellschaft für Teerverwertung* (*GfT*), founded by August Thyssen, are worth mentioning. The basis of tar refining was fractional distillation, first used by Charles Blachford Mansfield in 1847 for the production of benzene from coal tar on a large scale.

Lively competition developed among the European dyestuff producers. The race to produce alizarin was particularly dramatic. In 1867, Adolphe Wurtz and August Kekulé had discovered that the sulfonic acid group in aromatics could be replaced by a hydroxyl group by alkali fusion. Perkin used this process to manufacture alizarin from anthraquinone, which up to this time was produced from madder. When he applied for a patent on his process, he learned that Carl Graebe and Carl Liebermann had already submitted an application for this product on 25th June 1859, just one day earlier than himself. However, Perkin did not give up, but developed a new process, which subsequently led to an exchange of licenses. In 1873, he produced 435 t of alizarin, while German alizarin production had already reached 1,000 t.

Alizarin

Just as madder extracts were used for red coloring, so indigo had been used since olden times as a blue coloring agent, especially in India. The source of indigo was woad, which was cultivated in Germany, mainly in Thuringia. In the 16th and 17th centuries, there were large woad plantations found in the area around Gotha, Erfurt and Weimar, which were largely abandoned when imports of indigo from India took over. Production of natural indigo was around 8,200 tpa in 1885, with over half this dyestuff being produced in Bengal by biochemical decomposition from Indigofera tinctoria.

Synthesis of indigo was a particular challenge for dyestuff chemists in the second half on the 19th century. Adolf von Baeyer first successfully synthesized indigo in 1869, starting from o-nitrocinnamic acid.

o-Nitrocinnamic acid   Indigo

Since Baeyer's work was more directed to the elucidation of the structure of indigo, the synthesis he had discovered could not be applied economically on a large scale. Karl Heumann, at the Federal Polytechnic in Zurich, discovered a method of synthesis which was based on phenylglycine, which can be produced from aniline and chloroacetic acid. The yield from this process, however, was still unsatisfactory. Heumann's second proposal used phenylglycine-o-carboxylic acid, obtained by first oxidizing naphthalene to phthalic anhydride.

Naphthalene   Phthalic anhydride   Phenylglycine-o-carboxylic acid

The oxidation of naphthalene was carried out with chromic acid and chromates, regenerated by electrochemical reoxidation. This method was first used by the *Farbwerke Hoechst*. In 1891, it was accidentally discovered at *BASF* that naphthalene could be oxidized by concentrated sulfuric acid in the presence of mercury. In 1897, *BASF* introduced the first synthetic indigo to the market, followed shortly thereafter by *Farbwerke Hoechst*. Synthetic indigo rapidly replaced the natural product in the market in spite of harsh competition of the producers of natural indigo, especially of the Provence/France.

As a result of intensive research in the dyestuff sector by the turn of the century, around 15,000 dyes had already been patented in Germany. The largest portion were azo-dyes, obtained by coupling diazotized amines with suitable organic compounds.

Peter Griess, a student of August Wilhelm v. Hofmann, living in England, discovered diazotization in 1857. In 1861, the first azo-dye, aniline yellow (Solvent Yellow 1) appeared on the market in England; a further milestone in the development of azo-dyes was Congo red, a substantive dye discovered by Paul Böttiger in 1884.

Aniline yellow

Congo red

Production of dyestuffs initially followed empirical methods, but academic research, especially that of August Wilhelm v. Hofmann, Adolf v. Baeyer, Carl Graebe, Carl Liebermann, Emil Fischer and Heinrich Caro established the scientific bases which gave a powerful impetus to the German dyestuffs industry. Already by 1880, the German share in world dyestuffs production had reached one-half and in 1900 it exceeded 80%.

Concurrent with the rather haphazard discovery of the first coal-tar dyes, important scientific knowledge was being accumulated, which significantly advanced the understanding of the chemical reactions involved in the production of dyestuffs. After Friedrich August Kekulé, a Professor at Bonn University, had postulated the tetravalent bonding of carbon in 1857, he proposed, in 1865, the ring formula for benzene, which provided the basis for understanding the essentials of aromatic chemistry.

On the basis of this progress in fundamental research in organic chemistry, the German dyestuff industry also enjoyed great success in the new field of synthetically-produced pharmaceuticals. Products discovered at the beginning of this development, and worthy of mention, are the antipyretics antipyrine (1-phenyl-2,3-dimethylpyrazolin-5-one, 1883) and phenacetin (p-ethoxyacetanilide, 1888), as well as aspirin (acetylsalicylic acid, 1899).

Antipyrine

Phenacetin

Aspirin

Initially, coal-derived raw materials were the almost exclusive source of aromatics. However, beginning in the 1930's, the growth of the automobile industry brought petroleum to the fore in ever-increasing quantities as a source of raw materials for monocyclic aromatics.

Alongside pyrolytic processes for the production of aromatics, this new development was accompanied by the introduction of catalysis; even today, catalytic and purely thermal processes still complement each other in the production of aromatics.

2 The nature of the aromatic character

2.1 Molecular considerations

Since the middle of the last century, chemists have devoted considerable effort trying to explain the chemical bond. The knowledge hitherto accumulated has also been enormously beneficial in expanding the understanding of the nature of aromatic compounds.

Originally the classification 'aromatic' included all substances with an aromatic odor which were obtained from balsams, resins and other renewable raw materials before the beginning of industrial aromatic chemistry.

After Michael Faraday discovered benzene in the lighting gas from whale oil in 1825, Eilhard Mitscherlich established in 1834 that the summation formula for benzene, which he had produced by decarboxylation of benzoic acid, is C_6H_6.

In 1865, August Kekulé postulated the ring formula for benzene in a publication entitled 'Investigations into aromatic compounds', which he summarized in the sentence: 'These facts clearly lead to the conclusion that in all aromatic substances one and the same atomic group or, if one prefers, a common core can be found, which consists of six carbon atoms'.

Kekulé extended the static bond concept by the oscillation hypothesis, according to which the two benzene formulae constantly interchange with each other.

In 1866, Emil Erlenmeyer explained aromaticity on the basis of specific reactivity; he also suggested the structural formula for naphthalene, which was confirmed by Carl Graebe in 1868.

Investigations by Arthur Lapworth and Christopher K. Ingold into the mechanisms of electrophilic aromatic substitution provided an additional fundamental extension of the understanding of aromatics. Since olefins and conjugated polyenes principally undergo electrophilic addition reactions with polar reagents, it was surprising to discover that conjugated cyclic polyenes are electrophilically substituted with the same reagents. It was reasonable to accommodate this special behavior of cyclic polyenes by a single concept; they were thus designated aromatic compounds. Their special chemical behavior was considered to be a characteristic of the presence of aromatic properties.

It was later recognized that the characteristic chemical behavior of certain planar cyclic polyenes is the result of increased thermodynamic stability caused by the delocalized π-electron system. Thus, a thermodynamic criterion for distinguishing between aromatic and non-aromatic compounds was created.

The gain in stability of benzene through delocalization of the π-electrons (resonance) can be seen in the hydrogenation energy diagram (Figure 2.1).

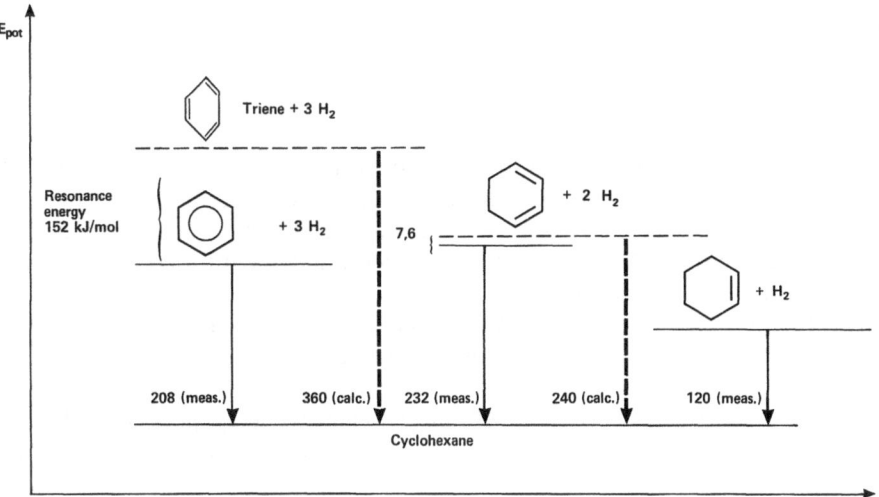

Figure 2.1: Energy diagram for the hydrogenation of benzene

Hydrogenation of cyclohexene is an exothermic reaction in which 120 kJ/mol are released; hydrogenation of 1,3-cyclohexadiene produces 232 kJ/mol in exothermal reaction, that is 8 kJ/mol less than would be expected based on the cyclohexene. The diene is thus more stable than two corresponding localized $C=C$ bonds; the resonance energy of 1,3-cyclohexadiene is therefore 8 kJ/mol.

The heat of hydrogenation of benzene is only 208 kJ/mol and is thus considerably below the theoretical heat of hydrogenation of the hypothetical cyclohexatriene, which should be 360 kJ/mol. A cyclohexatriene with the known resonance energy of cyclohexadiene must produce a value of $360-3\times 8$, i.e. 336 kJ/mol, which is also clearly in excess of the heat of hydrogenation of benzene of 208 kJ/mol. Consequently, benzene is fundamentally more stable than a normal, although hypothetical, 1,3,5-cyclohexatriene. Similar considerations apply for naphthalene, anthracene and other polynuclear aromatics.

Table 2.1 summarizes the resonance energies of some aromatic hydrocarbons and heterocyclics.

Table 2.1: Resonance energies of aromatic hydrocarbons (kJ/mol)

Compound	Resonance energy
Benzene	152
Biphenyl	296
Naphthalene	257
Azulene	139
Anthracene	353
Phenanthrene	388
Naphthacene	544
Chrysene	563
[18]Annulene	500
Pyridine	95
Pyrrole	89
Furan	66
Thiophene	120

In 1931, Erich Hückel deduced from quantum-mechanical calculations, that the resonance energy gain typical of aromatic compounds is possible when a monocyclic compound has $(4n+2)$ π-electrons $(n=0, 1, 2, 3)$ which can be delocalized across the entire ring. The delocalization reaches a maximum for compounds of a planar configuration (Hückel's rule).

The energy states of the π-orbitals can be determined graphically for simple cyclic π-systems using the scheme developed by Arthur A. Frost and Boris Musulin. The polygon is drawn standing on a vertex in a circle, the radius of which, by definition, corresponds to double of the resonance energy β. One molecular orbital is allocated to each corner of the polygon. The vertical distance from the

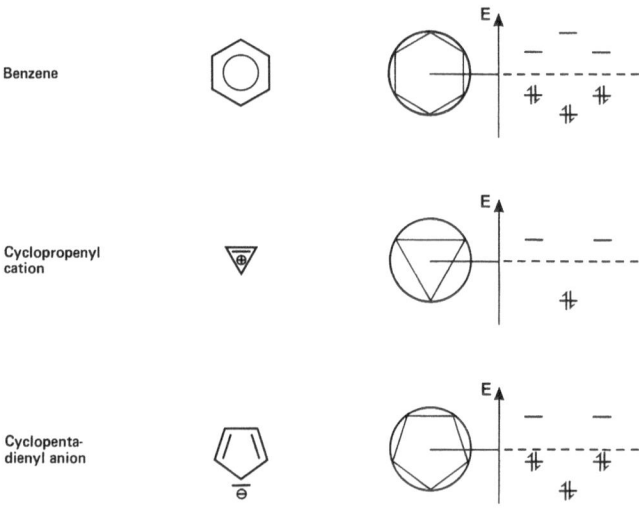

horizontal drawn through the center of the circle represents the π-energy of the orbital. Each electron, which can be allocated to a molecular orbital below the horizontal, causes a gain in stabilization energy.

In benzene, the maximum possible stabilization is achieved (and a total π-energy of 8 β), since all the binding orbitals are occupied with a total of 6 π-electrons. It can also be deduced from Hückel's rule, that the cyclopropenyl cation and the cyclopentadienyl anion exhibit aromatic character; this is also true for azulene, which has been found in coal tar.

Azulene

After World War II, Michael J. S. Dewar, in particular, developed Hückel's rule further by explaining the α-tropolone structure in 1945.

α-Tropolone

More advanced developments of Hückel's theory, such as the Pariser, Parr, Pople Self-Consistent Field Theory, which takes into account the interaction of the electrons, have led to a deeper understanding of the aromatic concept and are widely applied to determine physical properties of aromatics, particularly when describing dyestuffs.

Nevertheless, the debate on aromaticity is still continuing. In recent times, it has been suggested that the term 'aromatic' should no longer be used, and its place taken by, for example, the terms 'regenerative' (tendency to retain type) or 'meneid' (of constant form).

Whereas aromatic systems are defined by a positive resonance energy, anti-aromatic systems are characterized by a negative resonance energy. As a rule, anti-aromatic compounds are unstable and contain $4n$ π-electrons in a cyclic planar, completely conjugated arrangement. Cyclobutadiene belongs to this category and is stable only in a solid matrix at very low temperatures (20 K).

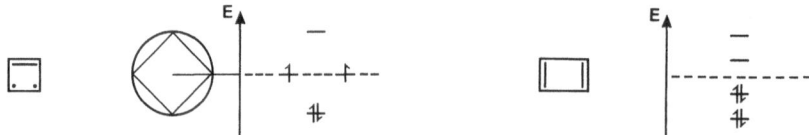

Figure 2.2: Energy diagram for the anti-aromatic compound cyclobutadiene (formulated as a biradical and diene)

A further criterion for determining the aromatic character of organic compounds is provided by nuclear resonance spectroscopy. If an aromatic or anti-aromatic molecule is put into a strong magnetic field, the π-electrons form a cyclic electrical loop under the influence of the magnetic field. In $(4n+2)$ π-electron systems, a diamagnetic ring current is induced, which weakens the magnetic field inside, above and below the electrical loop, and strengthens it on the periphery of the molecule. An example of these compounds, defined as diatropes, is [18]annulene, with 18 π-electrons; the resonance positions of the outer protons lie at the low field end of the spectrum at 9.28 ppm, whereas the inner protons experience a high field shielding to -2.99 ppm.

In anti-aromatic systems, on the other hand, a paramagnetic ring current is produced, in which the magnetic field strengthens the outer magnetic field inside, above and below the loop, and weakens it on the periphery. [16]Annulene, an example of an antiaromatic substance, therefore displays absorption signals at 5.33 ppm for the outer protons, and 9.44 ppm for the inner protons.

[16]Annulene   [18]Annulene

In spite of frequent discussions on the validity of the concept of aromaticity, the term 'aromatic compound' is still widely used. For the wide range of industrially produced aromatics, the definition of aromaticity based on Hückel's rule and the associated reaction mechanisms is perfectly adequate.

(In order to simplify presentation, in accordance with other authors, this monograph 'loosely' uses the ring method of portraying naphthalene and higher condensed aromatics, for each ring of the π-electron system, without the criterion of Hückel's rule being fulfilled for all the rings.)

In industrial hydrocarbon technology, the term 'aromaticity' is also used to characterize mixtures of aromatic hydrocarbons. In this case, aromaticity is defined as the relationship of the number of aromatic carbon atoms to the total number of carbon atoms in the 'average' molecule. This statistical quantity is mainly determined with the aid of NMR spectroscopy (see Chapter 3.2.1).

2.2 Mechanistic considerations

In the early days of industrial aromatic chemistry, reaction mechanisms were completely unknown. Their elucidation, especially in the 1930's and 1940's together with considerable improvements in analytical techniques for identifying intermediates and by-products, was enormously important for the industrial aromatic chemistry. It is hardly possible to design and equip modern processes without in-depth knowledge of the reaction mechanisms.

One of the most important reactions in the production of industrial aromatics is electrophilic aromatic substitution; another prominent type of reaction is nucleophilic substitution, which is favored for aromatics with electron-withdrawing groups. Free radical reactions, which occur especially in thermal pyrolysis processes and in side-chain oxidation and chlorination reactions, are even more important, in quantitative terms, than electrophilic and nucleophilic substitution reactions. Typical examples are thermal cracking of naphtha and gas-oil fractions, the oxidation of naphthalene to phthalic anhydride, and the side-chain chlorination of toluene. Rearrangement reactions are less significant.

2.2.1 Electrophilic aromatic substitution

Because of their high negative charge density, aromatics are generally attacked by electrophiles. Although electrophilic mono-substitution of benzene yields only a single product, with substituted aromatics the electrophile can often be introduced at one of several positions. As a rule secondary substitution is influenced by the existing substituents and occurs in the o- and p- or m-position; attack on the existing substituent position (ipso-reaction) is virtually unknown in industrial aromatic chemistry.

Figure 2.3 shows a reaction-energy diagram for electrophilic aromatic substitution, occurring in four stages (see page 14).

In the first stage, the aromatic molecule is attacked by the electrophile and a π-complex is formed, which retains the aromatic state. The π-complex is converted in a second reaction stage into the σ-complex. The high stability, which the aromatic electron sextet gives to the hydrocarbons, is lost in this process. The σ-complex is a reactive intermediate, which can be isolated through substitution with suitable electron donors.

The third stage in the reaction is the formation of the second π-complex by separation of a proton; complete detachment of the proton to yield the end product occurs in the fourth stage, by the action of a proton acceptor (base).

The relative rate of the four reaction stages depends on the electrophilic agent and the aromatic substrate. Formation of the first π-complex, as for example in the nitration of benzene, can be rate-determining. One example of a reaction in which the formation of the σ-complex determines the rate, is the reaction of benzene with bromine in acetic acid. Reaction of the σ-complex to the second π-complex generally does not govern the reaction rate.

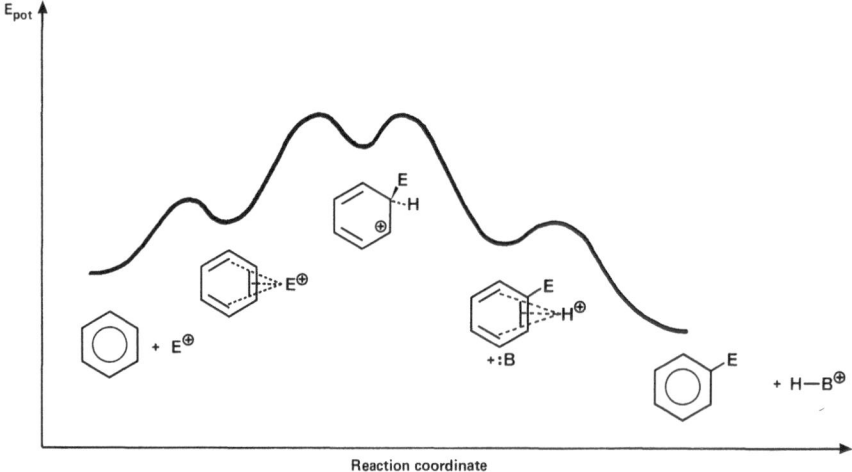

Figure 2.3: Reaction-energy diagram of electrophilic aromatic substitution

Alkylation is the paramount electrophilic substitution reaction in industrial aromatic chemistry, for example, in the production of ethylbenzene, cumene, diisopropylbenzenes and diisopropylnaphthalenes. A carbonium ion generally acts as the electrophilic agent and is produced by reaction of a Lewis acid with an olefin. The most stable of the possible carbonium ions normally predominates in the reaction; nevertheless, attention must also be paid to the formation of isomers.

Care should be taken in Friedel Crafts alkylation since the alkylaromatic produced displays increased reactivity in comparison with the original unsubstituted aromatic, because of the activating effect of the alkyl group, so that the formation of by-products is unavoidable. This formation of by-products is frequently countered by working at low conversion rates or by subsequent transalkylation.

Friedel-Crafts acylation is related to Friedel-Crafts alkylation, with an acylium cation acting as the electrophile. However, in industrial aromatic chemistry, because of the high consumption of catalyst, this reaction is of much less importance than Friedel-Crafts alkylation. Nonetheless, it has been used, for example, in the manufacture of anthraquinone from phthalic anhydride and benzene.

Nitration is another technically important example of electrophilic aromatic substitution. The nitronium ion $NO_2^{\oplus}$ acts as the electrophilic agent and is produced from sulfuric acid and nitric acid.

$$2\ H_2SO_4\ +\ HNO_3\ \rightleftharpoons\ NO_2^{\oplus}\ +\ H_3O^{\oplus}\ +\ 2\ HSO_4^{\ominus}$$

Table 2.2 summarizes the relative nitration rates for benzene derivatives; electron-attracting moieties, such as nitro groups, have a strongly deactivating effect.

Table 2.2: Relative reaction rates in nitration of benzene derivatives

C_6H_5-	relative rate
OH	1000
CH_3	25
$CH_2COOCH_2CH_3$	3.8
H	1
CH_2Cl	0.71
CH_2CN	0.35
Cl	0.033
NO_2	$6 \cdot 10^{-8}$

Historically, sulfonation has been one of the most important electrophilic aromatic substitutions, particularly in the production of 1- and 2-naphthol, as well as alizarin. Unlike the previously mentioned electrophilic reactions, it is frequently reversible. SO_3, which occurs in low concentration in sulfuric acid, acts as the electrophilic agent.

Sulfonation is carried out on an industrial scale in particular on benzene, toluene, cumene, naphthalene and anthracene.

Sulfonation of naphthalene with concentrated sulfuric acid under kinetically controlled reaction conditions (at low temperature) produces predominantly naphthalene-1-sulfonic acid. Increasing the temperature to over 160 °C greatly augments the proportion of naphthalene-2-sulfonic acid.

Azo-coupling, in which the diazonium cation functions as an electrophile, is also of great industrial significance, particularly in the production of dyes. Because of the weak electrophilic character of the diazonium cation, only reactive aromatic systems such as phenols and amines are normally attacked. Azo-coupling is performed in weakly acidic solutions, to ensure high concentrations of the diazonium cation.

Halogen groups can also be introduced by electrophilic aromatic substitution. A Lewis acid is then needed to polarize the halogen molecule.

$$Cl_2 \xrightarrow{FeCl_3} |\overset{\delta\oplus}{\underline{\underline{Cl}}} - \overset{\delta\ominus}{\underline{\underline{Cl}}}| \cdots FeCl_3$$

Electrophilic halogenation is applied on a large scale, for example, in the production of chlorobenzenes.

2.2.1.1 Orientation rules

The entry of additional substituents into aromatic molecules, which have already been substituted, is affected by the existing substituents. Thus electron donors such as methoxy-, amino- or alkyl groups lead mainly to increased o-/p-substitution. In this substitution mode, which is governed by the stability of the π-complex, particular attention should be given to steric and rearrangement effects, which can cause marked variations within the orientation rule.

Figure 2.4 (see overleaf) shows the directing effect of functional groups on the nitration of benzene derivatives.

2.2.2 Nucleophilic aromatic substitution

If an aromatic molecule is substituted with strong electron acceptors, such as nitro groups, a nucleophilic attack is possible on the aromatic system. Substitution generally follows an addition-elimination mechanism. In the first stage, the nucleophile is added to the aromatic system with the formation of a relatively stable intermediate. The negative charge is thus delocalized over the entire conjugated π-electron system. With suitable substitution these intermediate stages are stabilized and can be isolated; they are known as Meisenheimer complexes after Jacob Meisenheimer, who discovered them in 1902.

Meisenheimer complex

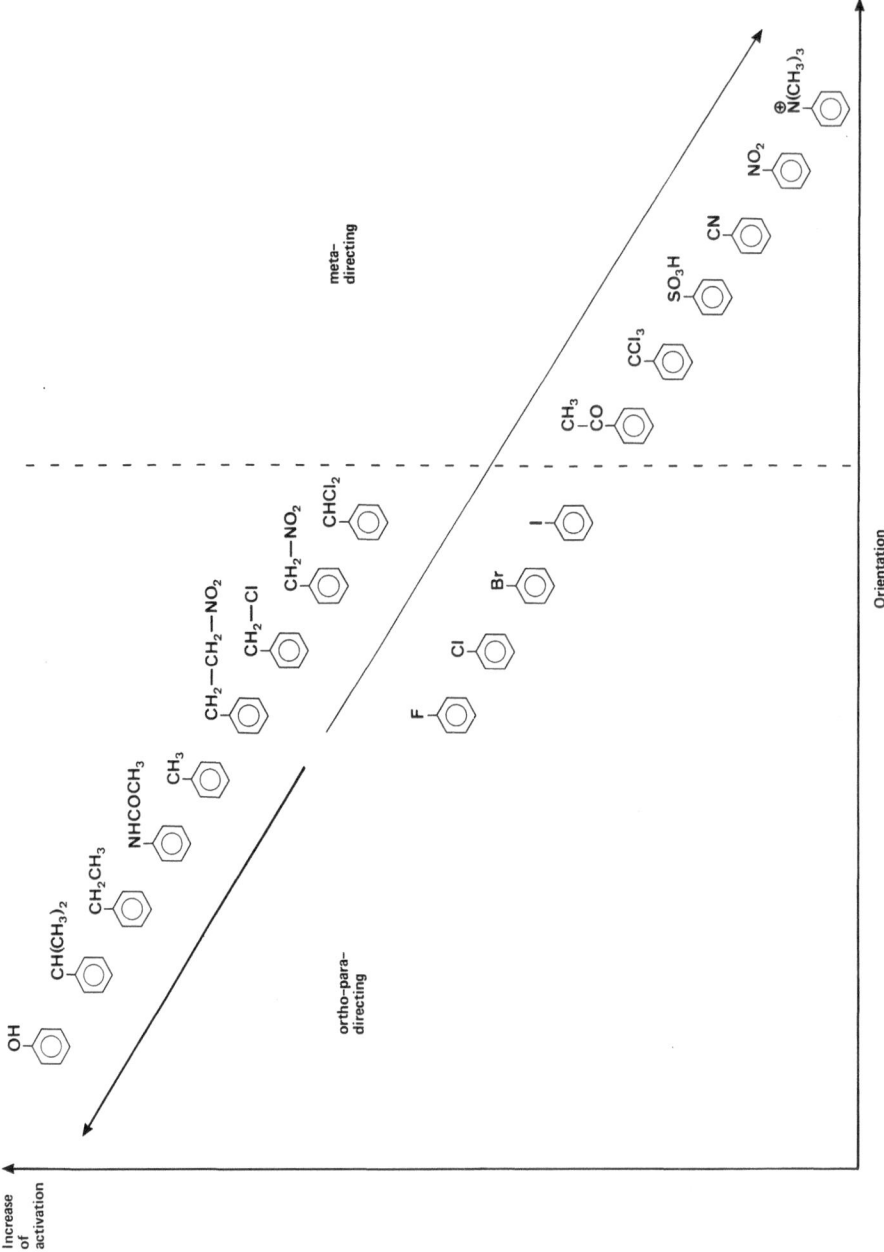

Figure 2.4: Directing influence of substituents on the nitration of aromatics

In the second stage, the substituent is separated, together with its electron pair.

The reaction is a combination of two equilibrium reactions. It therefore occurs more easily, the more nucleophilic the attacking group and the more stable the leaving nucleofuge. Halogens in the form of halide ions, diazonium ions as nitrogen, sulfonic acid groups as sulfite anions (for example in NaOH fusion of sulfonates to produce phenols) and hydrogen, if the energy-rich departing hydride ion is oxidized, are all suitable leaving groups.

In addition to nucleophilic substitution of aromatics by the addition-elimination mechanism, nucleophilic substitution is also possible through the effect of a strong base ($B|^{\ominus}$) via an elimination-addition mechanism. Arynes occur as intermediates in such a process.

This type of reaction occurs, for example, in the production of m-cresol from o- or p-chlorotoluene. The highly reactive aryne can be attacked in two positions, producing a mixture of isomers.

Proof of the occurrence of an 'aryne' intermediate can be found, for example, by intercepting the dehydrobenzene with a diene component, such as anthracene; this produces triptycene.

Triptycene

20   The nature of the aromatic character

Analogous in overall formula, but following a radical mechanism, is the Sandmeyer reaction, in which replacement of diazonium groups by $CN^\ominus$, $Cl^\ominus$, $Br^\ominus$ or $N_3^\ominus$ occurs in the presence of $Cu^\oplus$ salts.

$$Ar-\overset{\oplus}{N}\equiv N\; Cl^\ominus + CuCl \longrightarrow Ar\cdot + N_2 + CuCl_2$$

$$Ar\cdot + CuCl_2 \longrightarrow Ar-Cl + CuCl$$

2.2.3 Radical reactions

2.2.3.1 Pyrolysis processes

Radical reactions are of paramount importance in industrial aromatic chemistry, especially in the production of aromatics from coal and suitable petroleum fractions, as well as in the pyrolysis of aromatic mixtures to produce coke. The reaction temperature of such processes is generally well above 500 °C, so that a wide spectrum of products results from pyrolysis.

The thermal dealkylation of toluene (or methylnaphthalenes) also follows a radical reaction mechanism.

Kinetic studies have shown that thermal hydro-dealkylation is a first-order reaction with respect to the aromatics, and a half-order reaction in relation to hydrogen.

$$-\frac{dc_T}{dt} = k \cdot c_{H_2}^{0,5} \cdot c_T$$

Here, c_T is the toluene concentration, c_{H_2} the hydrogen concentration, and k the reaction rate constant. The activation energy of the reaction is 220 kJ/mol. Figure 2.5 shows the reaction diagram for the dealkylation of toluene.

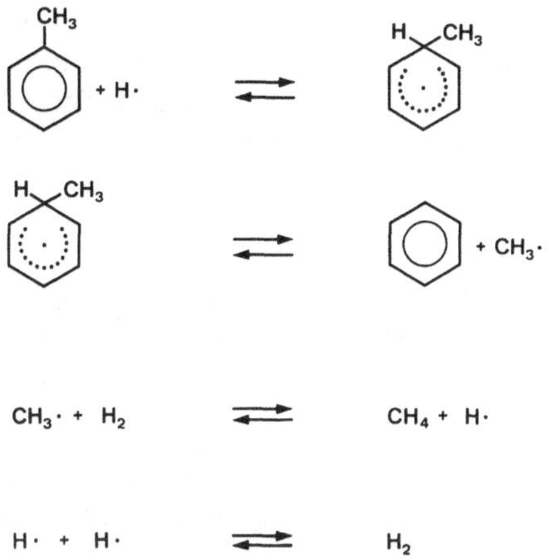

Figure 2.5: Reaction diagram for the dealkylation of toluene

A further important process which follows a radical mechanism is the catalytic dehydrogenation of ethylbenzene for the production of styrene (see chapter 5.1.3).

2.2.3.2 Oxidation reactions

The most prominent oxidation reactions in industrial aromatic chemistry are those performed in the gas phase, particularly on V_2O_5 catalysts, along with liquid-phase oxidation in acetic acid/Mn-/Co-salts system; both types are radical reactions.

In gas-phase oxidation processes at temperatures above 300 °C, as used for the production of phthalic anhydride, maleic anhydride and naphthalic anhydride, careful selection of a suitable catalyst ensures that the smallest possible amount of by-product arises.

The kinetics of naphthalene oxidation can be represented by the following equation, with constant oxygen partial pressure:

$$-\frac{dp_N}{dt} = \frac{k_N \cdot p_N}{1 + C \cdot k_N \cdot p_N}$$

Here, P_N represents the naphthalene partial pressure, k_N the rate constant, and C a proportionality constant.

With low naphthalene partial pressure, oxidation follows first-order reaction kinetics, while with relatively high partial pressures, as used in industrial applications, the reaction becomes zero-order.

At lower temperatures in the liquid phase, industrial cumene oxidation produces cumene hydroperoxide. In this case, the oxygen bi-radical does not attack the aromatic ring, but rather the activated CH group. The following equation represents the kinetics of the auto-catalytic reaction:

$$-\frac{dc_{RH}}{dt} = -\frac{dc_{O_2}}{dt} = \sqrt{\frac{2 k_i}{k_t}} \cdot k_p \cdot c_{ROOH}^{0,5} \cdot c_{RH}$$

In this equation, k_i is the rate constant in the initiating stage, k_p the constant of the chain reaction, k_t the constant of the radical termination, RH cumene, and ROOH the hydroperoxide.

Liquid-phase oxidation is carried out predominantly in the acetic acid/$Co^{2\oplus}$/$Br^{\ominus}$ system; chromium oxidation is also used. In the predominant reaction of p-xylene to terephthalic acid, which takes place in two stages via methylbenzaldehyde, the following rate equation is valid for the first stage of the process:

$$r_1 = k_1 \cdot c_{Co^{2\oplus}} \cdot c_{Br^{\ominus}} \cdot c_{Xyl}.$$

The equation below applies to the rate of transformation of p-methylbenzoic acid to terephthalic acid:

$$r_2 = k_2 \cdot c_{Co^{2\oplus}} \cdot c_{O_2}^{0,5}$$

Here, r_1 and r_2 represent the relevant reaction rates, k_1 and k_2 the rate constants and c the concentration of various reactants.

These rate equations corroborate the conclusion that radical reactions take part in the production of terephthalic acid.

Radical reactions also play a dominant role in the side-chain chlorination of alkylated aromatics, such as toluene.

2.2.4 Rearrangement reactions

The most important rearrangements in industrial aromatic chemistry are those of alkylaromatics (transalkylations), which occur under the influence of Lewis acids. They are particularly significant in the production of benzene from xylenes, and in optimizing the production of alkylaromatics.

Rearrangement reactions occurring in the production of phenol from cumene via cumene hydroperoxide are described in Chapter 5.3.1.

The benzilic acid and benzidine rearrangements are of less industrial importance. The benzilic acid rearrangement is used in the production of fluorene-9-hydroxy-9-carboxylic acid from phenanthrenequinone.

Benzidine rearrangement of hydrazobenzene is used to obtain 4,4'-diphenyldiamines from nitrobenzenes. This reaction occurs intra-molecularly, by way of a p-quinonoid intermediate.

Presently, the reaction is restricted virtually exclusively to the production of substituted benzidines.

Before the introduction of p-xylene oxidation on a large scale, the production of terephthalic acid by the Henkel rearrangement from benzoic acid and phthalic acid (see Chapter 7.3.1), at 400 to 420 °C and 10 bar, was of considerable industrial importance, especially in Japan.

2.3 Nomenclature

The more important benzene derivatives are given trivial names: benzene, toluene, o-xylene, m-xylene, p-xylene; the residue groups are characterized by the final syllable -yl.

24   The nature of the aromatic character

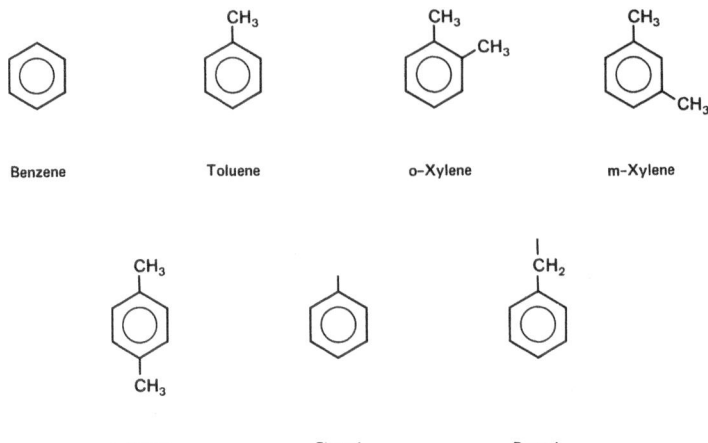

Condensation of several six-membered rings leads to polycyclic aromatic compounds; in these, the C atoms are indicated by numbers, the bonds by letters.

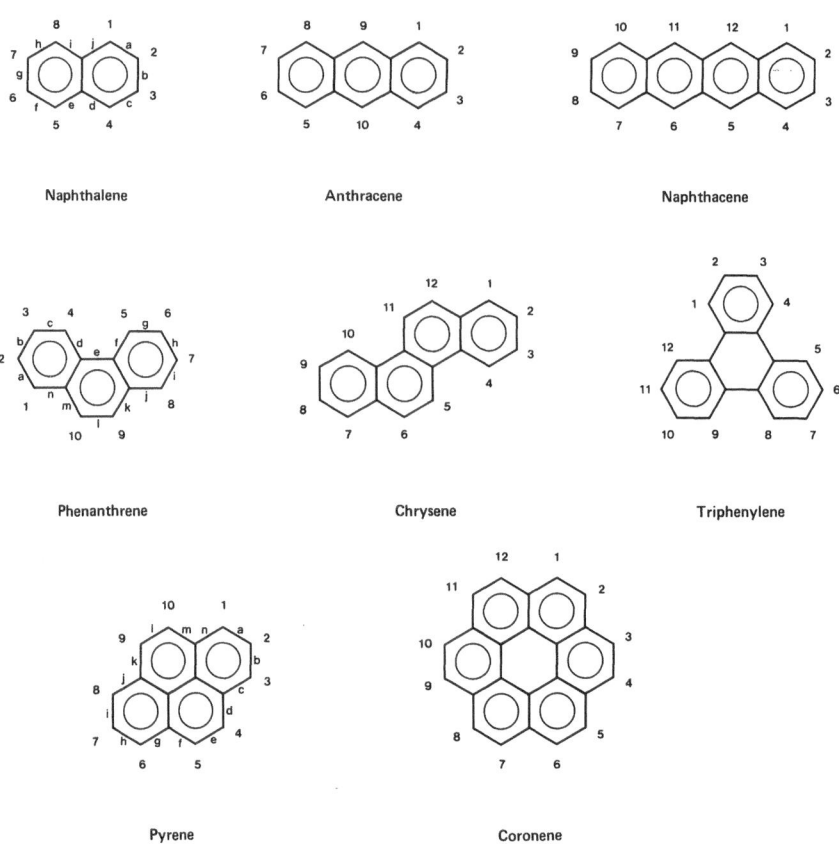

In naming substituted aromatics, the same procedure is used as for aliphatic compounds; the substituents are placed in alphabetical order, in the smallest possible numbered positions.

The naphthalene derivative depicted above is therefore called 2-chloro-1-methyl-5-phenylnaphthalene, (and not 6-chloro-5-methyl-1-phenylnaphthalene).

Polynuclear aromatics are considered to be composed from simple aromatics. A condensed benzene ring is indicated by benzo-, an added naphthalene ring by naphtho-.

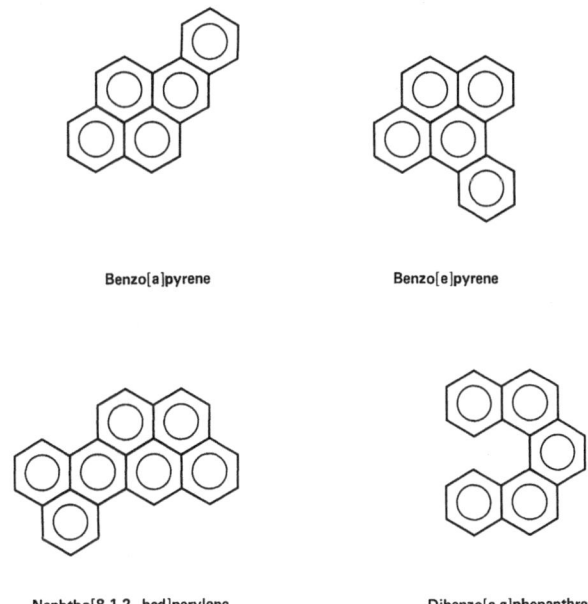

Benzo[a]pyrene

Benzo[e]pyrene

Naphtho[8,1,2−bcd]perylene

Dibenzo[c,g]phenanthrene

26   The nature of the aromatic character

In practice, the most common compounds containing heteroatoms are also given trivial names.

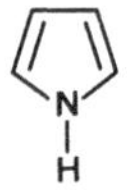

Pyrrole

Furan

Thiophene

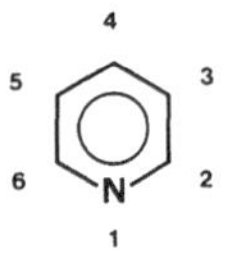

Pyridine

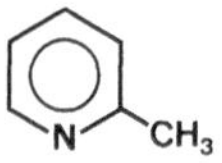

α-Picoline

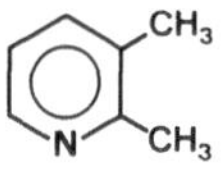

2,3-Lutidine

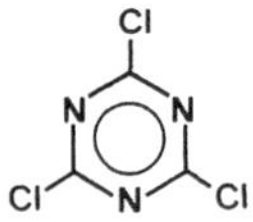

Cyanuric chloride

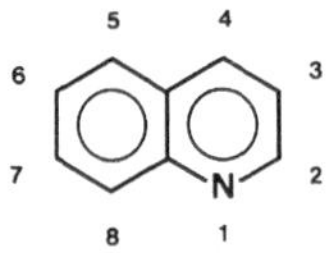

Quinoline

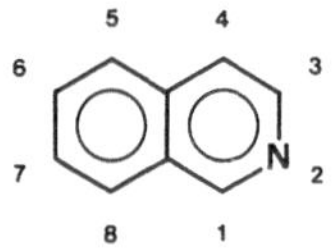

Isoquinoline

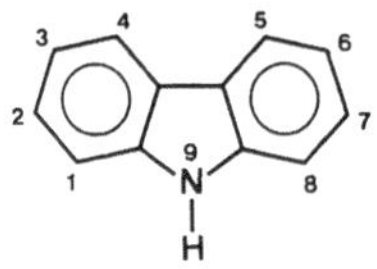

Carbazole

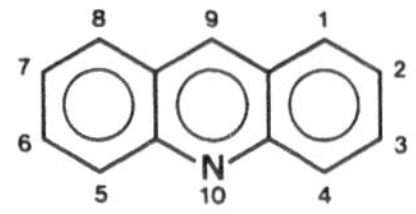

Acridine

3 Base materials for aromatic chemicals

The chemical industry meets its demand for carbon-containing feedstocks for the production of organic compounds from fossil raw materials - coal, oil and natural gas - as well as from renewable raw materials.

In the selection of a raw material, availability and chemical nature are deciding factors. Olefinic and aliphatic chemicals such as ethylene, propylene and methanol are therefore produced from crude oil fractions and suitable natural gas, whereas polynuclear aromatics such as naphthalene, anthracene and pyrene are recovered almost exclusively from coal-derived raw materials. Mononuclear aromatics such as benzene, toluene and xylene occupy a medial position, being obtainable from both crude oil and coal feedstocks. Renewable raw materials are, owing to their chemical structure, particularly suitable for the production of compounds containing oxygen.

The worldwide demand of the chemical industry for carbon containing feedstocks is around 245 Mt oil equivalent (1985). The main source (Figure 3.1) is crude oil (135 Mt), followed by natural gas (65 Mt), coal (25 Mt) and renewable raw materials (20 Mt).

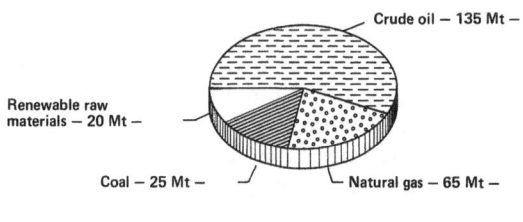

Figure 3.1: Origin of raw materials in the organic chemical industry

3.1 Origin of fossil raw materials and their composition

The origin of the fossil raw materials crude oil, coal and natural gas is rooted in photosynthesis, in which plants convert carbon dioxide and water into carbohydrates, i.e., organic material, using the energy of sunlight. Present-day knowledge confirms that coal and crude oil are of organic origin, but, in earlier times, a non-organic genesis of fossil raw materials was considered possible. Crude oil could thus have been formed from CO and H_2, analogously to Fischer-Tropsch synthesis, while carbides were considered as intermediates in the genesis of coal.

The formation of crude oil began 500 to 600 million years ago, while the origin of coal dates back 300 million years. The origins of coal can be traced to land plants, whereas the organisms from which oil was formed were marine in nature; these are phytoplankton, zooplankton, higher plants and bacteria. These starting materials are relatively rich in hydrogen, whereas terrestrial material, especially wood is characterized by a low content of hydrogen and displays, to some extent, a basic aromatic character.

The formation of organic material in the sea takes place almost exclusively in the upper layer, known as the euophotic zone, penetrated by sunlight to a depth of around 200 m. Almost all the synthesized organic substances are again oxidized bacterially to CO_2 in this zone, and are thus returned into the carbon cycle (1st cycle). Only a small proportion, generally 0.1% (at most 4%), of the organic substance escapes oxidation and sinks to the sea-bed with the sediment (clay-muds) brought from dry land. In the absence of oxygen, and in the presence of anaerobic bacteria, a sediment containing a sludge of decaying organisms builds up (2nd cycle). The sediment consists of around 45% cellulose, 45% proteins and 5 to 10% fats.

This composition indicates that a large proportion of petroleum-forming microorganisms belongs to the phytoplankton class; the presence of chlorophyll corroborates this assumption.

In the course of time, the deposition of further strata over the layer containing organic residues compacted and condensed the organic sediment. The upper layers gradually became solid, so that the parent rock of crude oil was eventually formed. The carbohydrates, proteins and fats of the organic sludge were transformed into a finely-divided heavy oil-type asphalt substance, i.e. kerogen, as a result of increased pressure and temperatures between 160 and 180 °C. As the depth of deposition increased, gas was the main product from the crude oil parent rock in the so-called metagenesis phase, at depths from 3000 to 4000 m, whereas in the catagenesis phase at depths of 2000 to 3000 m a wide range of long-chained hydrocarbons was formed in addition to gas.

A small proportion of the hydrocarbons, with carbon structures characteristic of biologically-derived molecules, stems directly from dead organisms in virtually unchanged form, without passing through the kerogen phase. These geochemical markers include, e.g., cholesterol and cholestane, together with the isoprene compounds such as phytane.

$$CH_3-\underset{\underset{CH_3}{|}}{CH}-(CH_2)_3-\underset{\underset{CH_3}{|}}{CH}-(CH_2)_3-\underset{\underset{CH_3}{|}}{CH}-(CH_2)_3-\underset{\underset{CH_3}{|}}{CH}-CH_2-CH_3$$

C_{20} – Phytane

Base materials for aromatic chemicals 29

Cholesterol

Cholestane

All organic substances are unstable under geological conditions and become modified by the pressure and temperature conditions as the sediment rock subsides. Because of this, both the insoluble kerogen in the parent rocks and the aliphatic hydrocarbon mixtures released during catagenesis and metagenesis undergo changes. The metastable kerogen becomes richer in carbon and poorer in hydrogen during this 'maturing process', with the formation of more basic aromatic structures. The hydrocarbon mixtures become depleted of complex molecules and therefore of geochemical markers, so that the average molecular weight is steadily reduced until finally only methane remains. In the n-alkanes, the tendency towards odd-numbered molecules disappears; as Figure 3.2 shows, these predominate in the original aquatic sediments.

The presence of the various geochemical markers can be used in crude-oil exploration to assess the maturity of the oil.

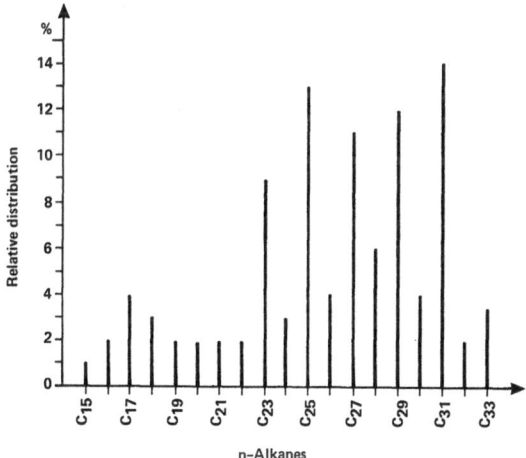

Figure 3.2: Distribution of n-alkanes in aquatic sediments

The hydrocarbons which are formed pass up through the parent rock into higher layers, until diffusion is blocked by impermeable strata and the crude-oil storing rock strata are formed. This crude-oil formation process explains why crude oil deposits are found predominantly in areas which were formerly marine sedimentary basins, or in undisturbed continental coastal regions, viz. the shelf areas.

In contrast to crude-oil formation, modification of the hydrocarbon skeleton is not the most important reaction in the formation of coal. Coal is formed by conversion of land-based plants into peat, lignite, hard coal, anthracite and graphite.

The main components of terrestrial plants consist of cellulose and other polymeric carbohydrates. In wood, the cellulose is wrapped up by aromatic lignin.

If land plants are covered with rock layers and thereby cut off from the air, anaerobic decomposition of their organic substances, such as cellulose, lignin and other biopolymers, commences. Transformation into coal begins, i.e. the conversion of plant residues via an increasing aromatization into elementary carbon. The cellulose is then largely 'fermented' by the effects of bacteria and fungi; carbon dioxide and methane are released. Lignin is transformed into humic acid. In the first stage of transformation into coal, fibrous peat is produced. During the conversion of peat into lignite, the humic acids lose part of their water content. The transformation to hard coal brings about further changes in the humic acids, with carboxyl groups in particular being lost. The final stages of coal formation are the geochemical processes, which take place over long periods of time at temperatures up to 200 °C. The process of coal formation is depicted in the H/C and O/C diagram in Figure 3.3.

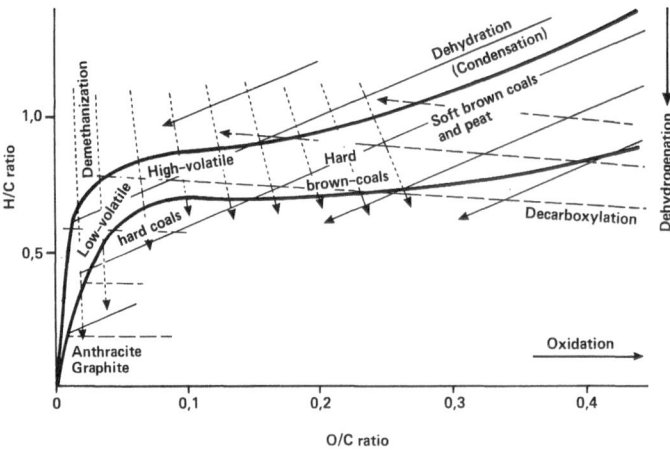

Figure 3.3: The process of coal transformation

The difference in aromaticity between coal and crude oil is, therefore, predominantly a result of the base materials: marine organic matter (like fats and amino acids), as a consequence of the base structures present, are rich in hydrogen and form crude oil containing materials of predominantly aliphatic structure; terrestrial plants, characterized by a higher carbon content and greater aromaticity, lead to coals with a high proportion of aromatics.

3.2 Coal

3.2.1 Deposits, composition and uses of coal

Hard coal, representing over 50% of the fossil raw materials, is the carbonaceous material with by far the greatest availability. Confirmed world resources of hard coal are around 6,900 billion t; of these, some 550 billion t can be recovered by current mining techniques. Brown coal deposits total some 6,500 billion t, of which 430 billion t are recoverable. In comparison with coal reserves, economically recoverable oil reserves are only 95 billion t, and natural gas deposits 90,000 billion m^3 (around 72 billion t).

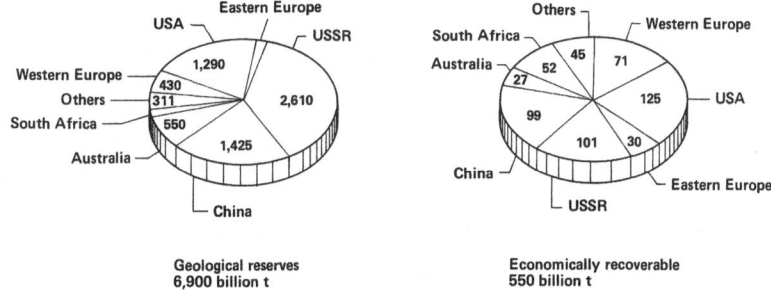

Figure 3.4: World coal reserves (1985)

The geographical location of coal deposits (Figure 3.4) is generally more balanced than the distribution of oil reserves, which are predominantly concentrated in the Middle East.

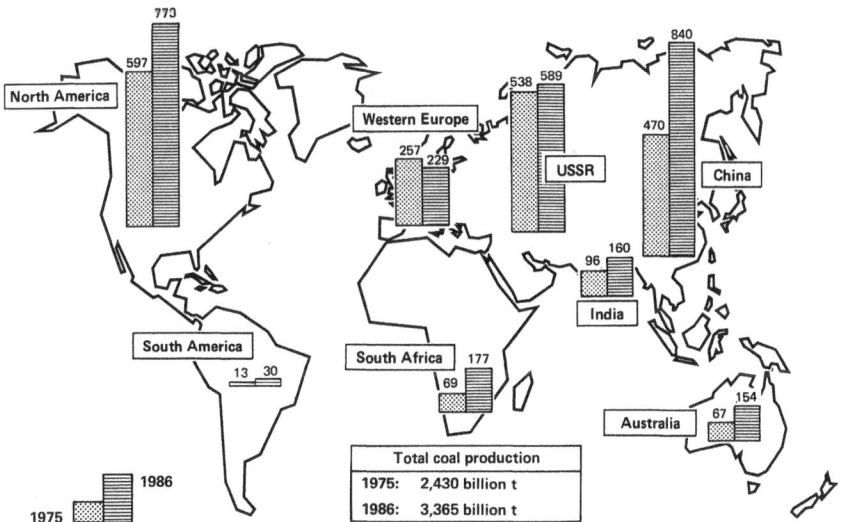

Figure 3.5: Coal production 1975/1986

Current world coal supplies are around 3.3 billion tpa (Figure 3.5). The largest coal producers are China (840 Mt), the Soviet Union (589 Mt) and the USA (742 Mt). Coal production in Western Europe is around 230 Mt. Whereas coal production in Western Europe has stagnated and even fallen in recent years, an increase in coal supply is noticeable particularly in the developing countries.

An understanding of the chemical structure of the different types of coal is of fundamental importance for the recovery of aromatics. Coal chemists have been occupied with the vexing task of elucidating the structure of coal for over 80 years. The statistical distribution of the carbon atoms of the aliphatic and aromatic parts of the macromolecules of which coal is composed presented a particular challenge. However with the development of NMR spectroscopy, especially ^{13}C-NMR-spectroscopy since the mid-1970's, assessment of the distribution and the direct determination of the aromaticity, defined as the ratio of aromatic carbon to total carbon in a molecule, is now possible. New methods of determining molecular weight have shown that coals are composed of macromolecules with molecular weights up to 100,000. The predominantly aromatic and hydroaromatic structures of the macromolecules are linked together by methylene bridges or longer aliphatic chains. The degree of aromatization of coals is higher as the geological age of the coal increases.

A typical model for hard coal was proposed by William R. Ladner (Figure 3.6).

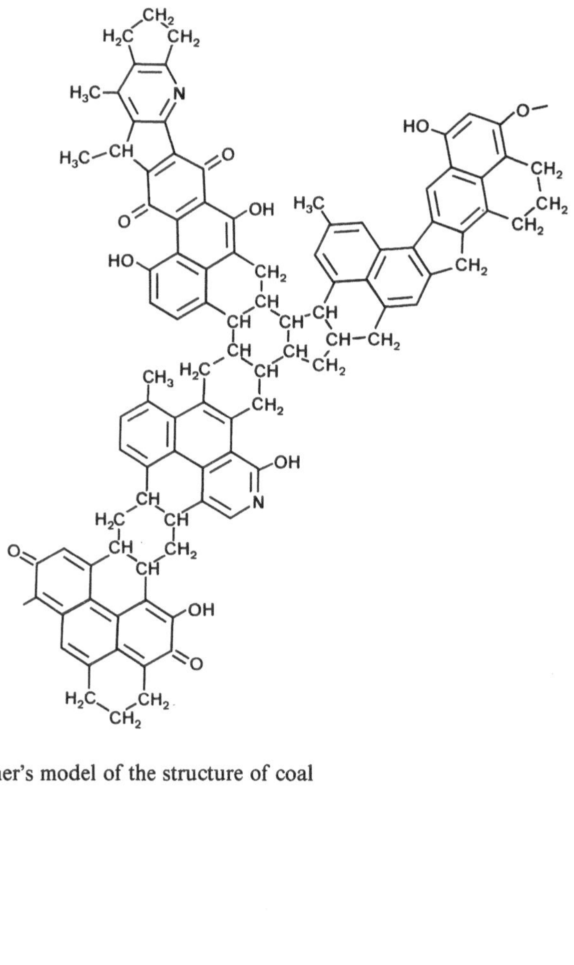

Figure 3.6: Ladner's model of the structure of coal

It largely corresponds to the aromatic structures of compounds which, among others, can be extracted from coal (Figure 3.7).

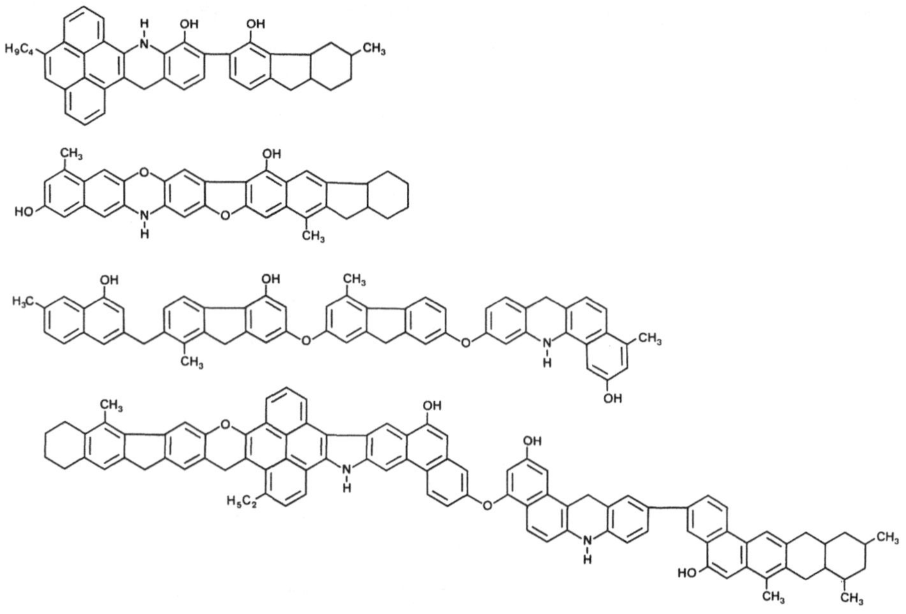

Figure 3.7: Aromatic extracts from coal

In brown coal, the proportion of aromatic structures is considerably lower than in hard coal. Anthracite exhibits a particularly high degree of aromatization, because of its great geological age.

Apart from the aromatic model of the structure of coal discussed thusfar, the polyadamantane structures attracted particular interest in the mid-1970's. Figure 3.8 shows the structure of adamantane and a hypothetical polyadamantane model of the composition $C_{66}H_{59}$.

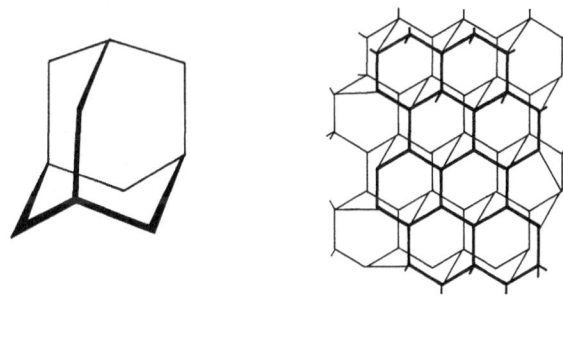

Adamantane   Polyadamantane

Figure 3.8: Structure of adamantane and a polyadamantane

With the aid of ^{13}C-NMR-spectroscopy, which enables a direct insight into the carbon skeleton, the proposed adamantane structure model of coal could be largely dismissed. Figure 3.9 shows the solid-state ^{13}C-NMR-spectra of coals with varying degrees of coalification. The resonance intensities in the aliphatic region at around 40 to 50 ppm decline markedly as coalification increases; anthracite is formed practically exclusively from sp^2-hybridized carbon, whereas the sub-bituminous coal still has a large proportion of aliphatics.

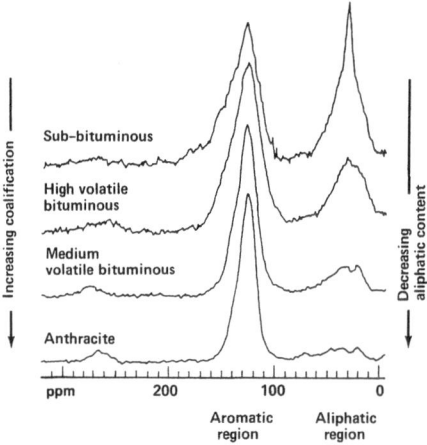

Figure 3.9: Solid-state ^{13}C-NMR-spectra of different coals

Next to combustion, which is mainly for the generation of electricity, the most important use for coal consists in carbonization to make metallurgical coke for the production of pig-iron. Worldwide, some 500 Mt are carbonized per year; concurrently, the two important aromatic chemical raw materials, i.e. coal tar (ca. 16 Mt) and benzole (ca. 5 Mt) are formed as by-products. Complementary processes for upgrading coal are coal hydrogenation and coal gasification. Coal hydrogenation, which was developed to industrial maturity in Germany in the 1930's, is aimed at producing oil from coal, i.e., the production of a synthetic crude oil (syncrude) from which fuels, heating oils and chemicals can be recovered. At current crude oil price levels, coal hydrogenation is not economic. However, since crude oil reserves are sufficient only for a limited time, developmental work is continued on optimizing coal hydrogenation processes. Coal gasification to produce synthesis gas is in a better position, with low coal costs, to compete with crude oil and natural gas. Thus, gas has been produced from coal for many years in countries which have no indigenous crude oil or natural gas reserves, but which have access to inexpensive coal. This applies particularly to India, South Africa, Finland and Zambia.

3.2.2 Thermal coal conversion – tar and benzole recovery

The pyrolytic conversion of coal into coke, gas and aromatic liquid products is the oldest and, in quantitative terms, most important coal-refining process. In the absence of air, carbonization processes are considered to occur in stages; up to 150 °C, carbon dioxide, water and volatile C_2 to C_4 hydrocarbons are evolved. At pyrolysis temperatures above 180 °C the volatile components also contain aromatics. At temperatures in excess of 350 °C, rapid degasification occurs, which continues to around 550 °C, leading to semi-coke. The rate of degasification approximately follows a reaction of the 1st order, which can be explained by the rupture of the bonds of the macromolecules in the coal. In the secondary degasification of the semi-coke (600 to 800 °C) hydrogen and methane are the main products.

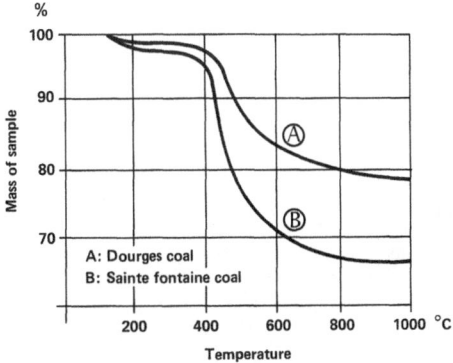

Figure 3.10: Hard coal degasification process

Figure 3.10 shows the degasification process for hard coal belonging to the middle coal rank (medium volatile bituminous (A), high volatile bituminous (B)) at a heating rate of 2 K/min.

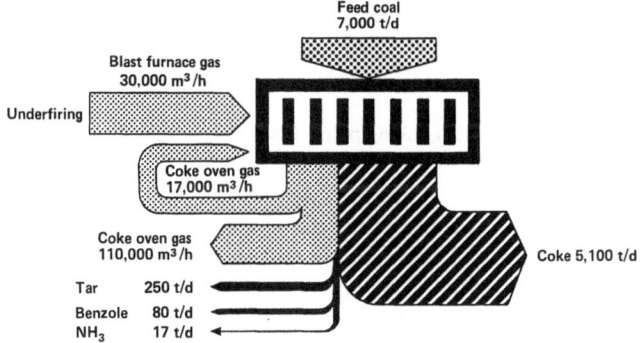

Figure 3.11: Flow chart of a medium sized cokery

36 Base materials for aromatic chemicals

Large-scale carbonization of hard coal is performed at temperatures between 1,000 and 1,200 °C. The production of blast-furnace coke takes 14 to 20 hours. Each ton of coal yields 750 kg of coke, 370 m^3 coke-oven gas, 35 kg of crude tar, 11 kg benzole, 2.4 kg ammonia and 150 kg water. Figure 3.11 shows the quantitative flow chart for a coke plant with a daily coal throughput of 7,000 t. The blast-furnace gas is supplied by the blast furnaces which are linked for energy supply (underfiring) to the coke ovens.

In recent times, large-chamber ovens 450 to 550 mm wide, 7 to 8 m high and 16 to 18 m long have been used for the production of coke. A large number of oven chambers are set up side by side to form a battery. Very large modern cokeries have a total capacity of 3 to 4 Mtpa of coke. Figure 3.12 shows the pushing side of the Zollverein coking plant (*Krupp-Koppers* construction) operated by *Ruhrkohle*, which produced around 2.3 Mt of metallurgical coke in 1986.

Figure 3.12: *Ruhrkohle*'s Zollverein coking plant, Essen/West Germany

The 750 to 850 °C hot gaseous products which are released during carbonization are taken to the gas-collection main by ascension pipes. The crude gas is quenched to around 80 to 100 °C by ammonia liquor, removing 60 to 70% of the crude tar. The aqueous phenolic condensate which arises at the same time undergoes extractive dephenolation. The phenols can be processed together with the phenolic products of the distillation of tar (see Chapter 3.2.3). Cooling the crude gas in the gas pre-cooler to around 25 °C brings about the separation of the remaining 30 to 40% of tar. After passing through the electrostatic precipitator, ammonia, hydrogen sulfide and benzole are washed out of the cooled gas (Figure 3.13). The crude benzole is extracted from the coke-oven gas counter-currently in towers, with benzole-absorbing oil, a tar-oil fraction boiling between 220 to 300 °C. The benzole is separated from the saturated wash oil in a refining plant (see Chapter 4.2).

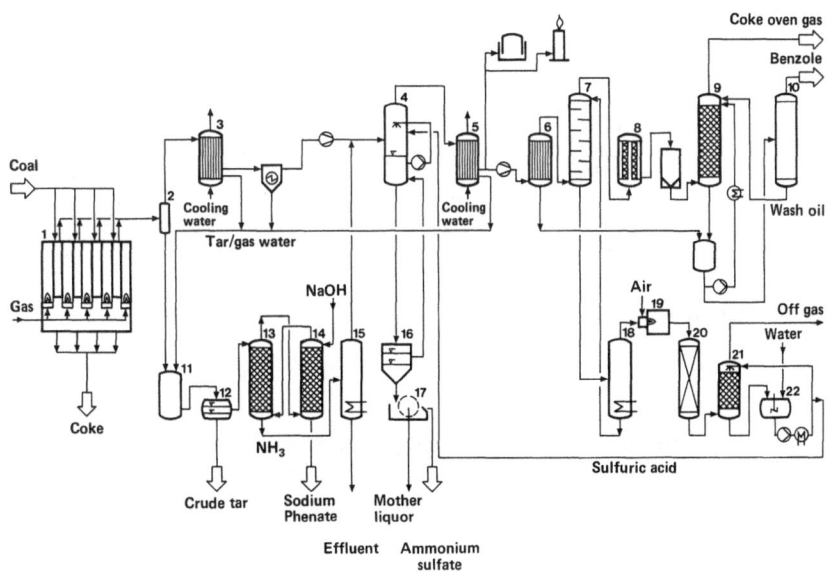

1 Coke oven; 2 Gas separator; 3 Pre-cooler; 4 NH_3-spray saturator; 5 Final cooler; 6 Cooler; 7 H_2S scrubber; 8 Gas purification; 9 Benzole cold scrubber; 10 Benzole column; 11 Underground condensate tank; 12 Tar separator; 13 Phenol extraction; 14 Benzole regeneration; 15 NH_3 separator; 16 Ammonium sulfate slurry container; 17 Filter; 18 H_2S separator; 19 H_2S combustion; 20 SO_2 oxidation; 21 Absorption column; 22 Dilution vessel

Figure 3.13: Flow diagram for the purification of coke oven gas

3.2.3 Tar refining

Some 16 Mtpa of tar is obtained worldwide as a by-product of coke production. The major tar-producing countries are the USSR, Japan, the USA, China, West Germany, Poland and France. With the development of the iron and steel industry, tar production is also growing in the South-East Asian countries, particularly Korea (Figure 3.14).

38 Base materials for aromatic chemicals

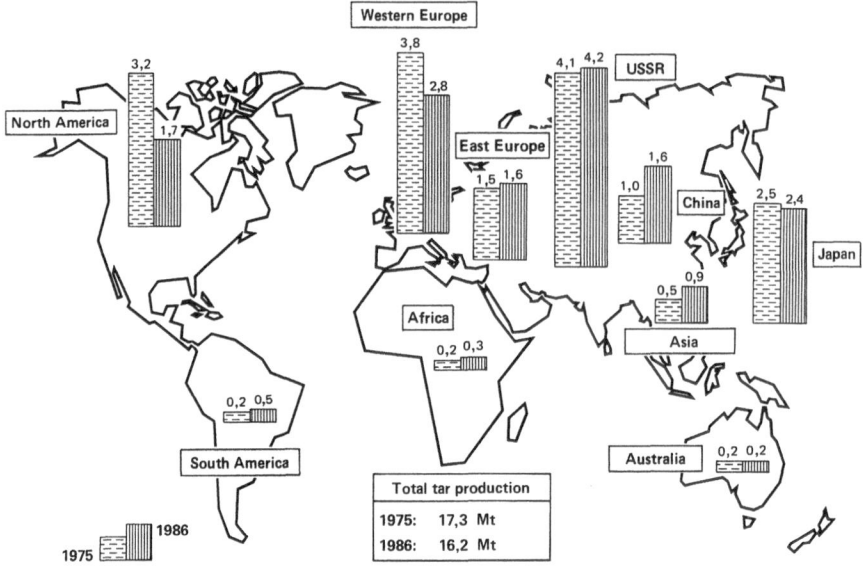

Figure 3.14: Tar production 1975/1986

Coal tar is a complex mixture consisting almost exclusively of aromatic compounds. The main components of coal tar are naphthalene, phenanthrene, fluoranthene, pyrene, acenaphthene, anthracene, the heterocyclics carbazole, quinoline and isoquinoline, phenol and benzofuran-derivatives, as well as sulfur compounds such as thianaphthene. Olefinic compounds are also present in coal tar (Table 3.1). The total number of constituents is estimated at 10,000.

Table 3.1: Constituents of coal tar

Compound	B.pt. in °C at 1013 mbar	M.pt. °C	Average content %
Hydrocarbons:			
Naphthalene	217.95	80.29	10.0
Phenanthrene	338.4	100.5	4.5
Fluoranthene	383.5	111.0	3.0
Pyrene	393.5	150.0	2.0
Acenaphthylene	270.0	93.0	2.5
Fluorene	298.0	115.0	1.8
Chrysene	441.0	256.0	1.0
Anthracene	339.9	218.0	1.3
Indene	182.8	− 1.8	1.0
2-Methylnaphthalene	241.1	34.6	1.5
1-Methylnaphthalene	244.4	− 30.5	0.7
Biphenyl	255.9	71.0	0.4
Acenaphthene	277.2	95.3	0.2

Table 3.1 (continued)

Compound	B.pt. in °C at 1013 mbar	M.pt. °C	Average content
Heterocyclics:			
Carbazole	354.75	245.0	0.9
Dibenzofuran	287.0	83.0	1.3
Acridine	243.9	111.0	0.1
Quinoline	273.1	− 15.0	0.3
Dibenzothiophene	331.4	97.0	0.4
Thianaphthene	219.9	31.3	0.3
Isoquinoline	243.25	26.5	0.1
Quinaldine	246.6	− 2.0	0.1
Phenanthridine	349.5	107.0	0.1
7,8-Benzoquinoline	340.2	52.0	0.2
2,3-Benzodiphenylene oxide	394.5	208.0	0.2
Indole	254.7	52.5	0.2
Pyridine	115.26	− 41.8	0.03
2-Methylpyridine	129.41	− 66.7	0.02
Phenolics:			
Phenol	181.87	40.89	0.5
m-Cresol	202.23	12.22	0.4
o-Cresol	191.00	30.99	0.2
p-Cresol	201.94	34.69	0.2
3,5-Dimethylphenol	221.96	63.27	0.1
2,4-Dimethylphenol	210.93	24.54	0.1

The number of possible structures (Table 3.2) of polycyclic aromatic hydrocarbons increases with the total number of six-membered rings. For a molecule like pyrene, which is formed from four rings, six other ring arrangements are possible (Figure 3.15), leading to the compounds naphthacene, benz[a]anthracene, chrysene, benzo[c]phenanthrene, triphenylene, pyrene and benzo[a]phenalenyl.

Table 3.2: Number of equiannular polycyclic aromatic hydrocarbons in relation to the ring number n

n	kata-condensed PAH	peri-condensed PAH	Total
1	1	0	1
2	1	0	1
3	2	1	3
4	5	2	7
5	12	10	22
6	37	45	82
7	123	210	333
8	446	1002	1448

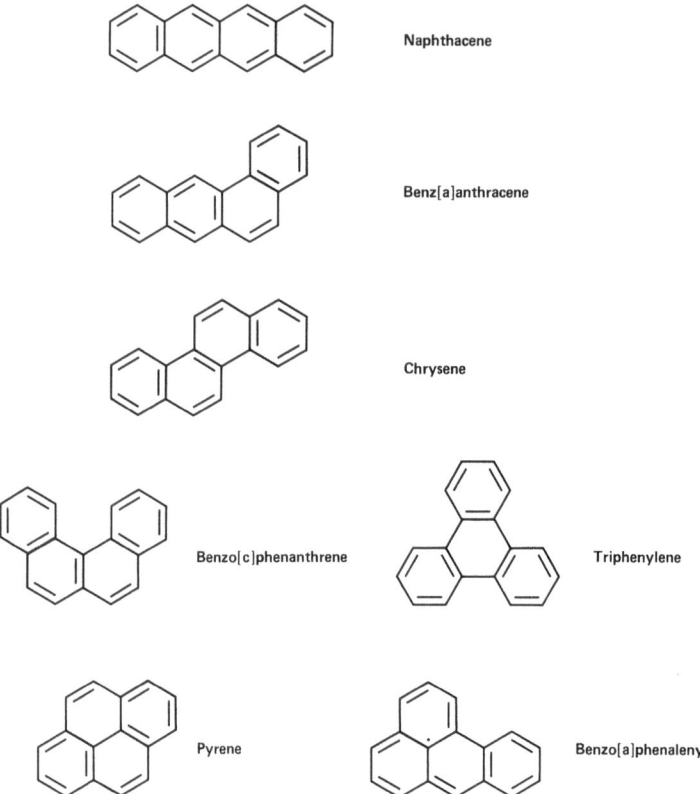

Figure 3.15: Condensed aromatics consisting of four hexagons

Because of its origins, coal tar also contains high molecular weight toluene-insoluble components, which are collectively known as 'Toluene Insolubles' (TI) together with soot-like components of a size of 100–1000 nm. These particles are referred to as 'Quinoline Insolubles' (QI), because of their insolubility in quinoline. A proportion of the mineral- and volatile inorganic co-compounds of coal, e.g. zinc, are found in tar, predominantly in the ash. Table 3.3 shows the typical composition of a tar from the Ruhr region.

Table 3.3: Typical composition of a coke oven tar from the Ruhr region (West Germany)

Density	g/cm^3	1.175
Water	%	2.5
Toluene Insolubles	%	5.50
Quinoline Insolubles	%	2.0
Carbonization residue		
(by Muck)	%	14.6
Carbon (waf)*	%	91.39
Hydrogen (waf)	%	5.25

Table 3.3 (continued)

Nitrogen (waf)	%	0.86
Oxygen (waf)	%	1.75
Sulfur	%	0.75
Chlorine	%	0.03
Ash	%	0.15
Zinc	%	0.04
Naphthalene	%	10.0
Distillation range (DIN 1995):		
up to 180 °C Water	%	2.5
Light oil	%	0.9
180–230 °C	%	7.5
230–270 °C	%	9.8
270–300 °C	%	4.3
300 °C– to pitch	%	20.1
Pitch**	%	54.5
Distillation loss	%	0.5

* waf = water and ash free
** softening point (K&S) 67 °C

Coal tar refining is nowadays predominantly carried out in central distillation plants. There are, at present, more than 100 tar refineries in operation in the world, with capacities up to 750,000 tpa. Coal tar refining at the beginning of this century provided the pattern for the subsequent refining of crude oil and has contributed to the development of the pipe still (Borrmann system). Coal tar processing differs from crude oil distillation however, in that the coal tar refining process is predominantly directed to the recovery of technically pure aromatic base chemicals, while crude oil refining is geared exclusively to the recovery of distillate fractions.

Crude tar leaves the coke oven with a water content of 2 to 10%; it is stored in tanks to further remove some water before the distillation process. The electrostatic coalescence of water droplets and subsequent removal of water, which is common practice in crude oil refining, is not possible in tar distillation, because of the different density ratios. Since crude tar usually contains chloride, neutralization with soda or sodium hydroxide is required to avoid damage by corrosion. Various schemes are in operation for the distillation of crude tar. Irrespective of the scheme, the first stage of the distillation process is dewatering. The sensible heat of the distillates is used to pre-heat the crude tar. Of the numerous processes for tar distillation, vacuum rectification with bottom pump-around has shown particular merit. In this process (Figure 3.16) the dewatered tar is fed into a main column with around 60 trays after heating up in a gas- or oil-fired pipe still, then distilled into 4 to 5 fractions, together with the pitch residue. Further concentration of the tar constituents, such as naphthalene and anthracene then takes place in side columns.

42   Base materials for aromatic chemicals

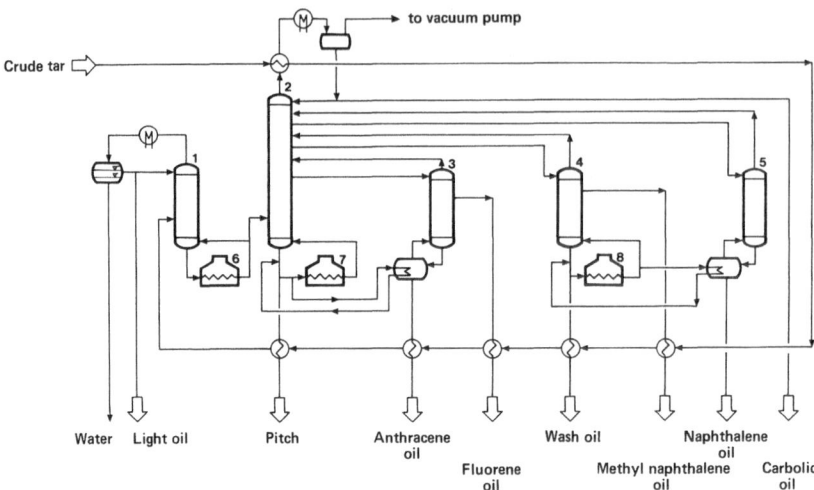

1 Dewatering column; 2 Main column; 3 Anthracene-oil side column; 4 Wash-oil side column; 5 Naphthalene-oil side column; 6-8 Tubular furnaces

Figure 3.16: Flow diagram for the distillation of coal tar with bottom pump-around

The advantage of this process consists in the short residence time of the pitch at high temperature which can be advantageous for certain applications, such as the production of binders for carbon electrodes.

The boiling ranges of typical tar fractions are given in Figure 3.17.

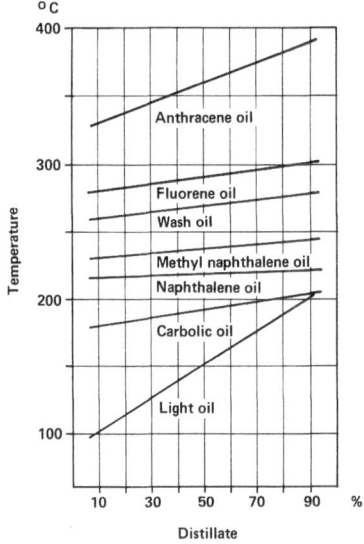

Figure 3.17: Boiling ranges of tar distillation fractions

In addition to pitch, which at 50 to 55% forms the largest proportion of coal tar, proportions of the various distillates are typically 0.5 to 1% light oil, 2 to 3% carbolic oil, 10 to 12% naphthalene oil, 2 to 3% methyl naphthalene oil, 7 to 8% wash oil, 2 to 3% fluorene oil and 20 to 30% anthracene oil.

The crude tar is not completely inert during distillation, and some aromatics such as acenapthylene, anthracene and indene are partially converted into the corresponding hydroaromatics by hydrogen transfer from the pitch.

Products recovered from light oil and carbolic oil include phenols and pyridines (see Chapters 5.3.3 and 14.2.1), from naphthalene oil, naphthalene (see Chapter 9.2.1), from wash oil, acenaphthene (see Chapter 10.3) and from anthracene oil, anthracene, phenanthrene and carbazole (see Chapters 11.1, 12.1, 14.5.5). The pitch residue is further processed to yield industrial carbon products (see Chapter 13).

3.2.4 Aromatics stemming from coal gasification

Coal is gasified on an industrial scale to produce hydrogen, synthesis gas (CO/H_2) and gas for heating. Current coal-gasification plants are predominantly operated to generate hydrogen for the production of ammonia for fertilizers. In addition, in South Africa, the *South African Oil & Gas Corporation (SASOL)* uses coal gasification to produce fuels via synthesis gas by the Fischer-Tropsch method in the *SASOL* I, II and III plants. The process converts carbon monoxide and hydrogen into hydrocarbons at temperatures from 200 to 350 °C and at 10 to 25 bar on iron or cobalt catalysts.

The Fischer-Tropsch synthesis leads almost exclusively to aliphatic hydrocarbons and oxygen compounds; the proportion of aromatics depends on the type of reactor system used. The aromatics content of the gasoline and diesel oil produced by the fixed-bed process in the *SASOL* I plant is negligible, while the aromatics content of gasoline produced by the fluidized-bed method by *SASOL* II and III is around 7%, and the level in diesel oil 10%. The increase in aromatics with the fluidized-bed reactor is a result of the reaction temperature, which is approximately 100 °C higher. Since the concentration is rather low, it is not economical to recover aromatics from mixtures of hydrocarbons produced by the Fischer-Tropsch method; however, synthesis gas and its product, methanol, can be transformed into aromatics by the *Mobil* process (see Chapter 3.4).

There are currently three industrially-proven methods for coal gasification, namely fixed-bed gasification (e.g. *Lurgi*), fluidized bed reactors (e.g. Fritz Winkler (*BASF*)) and entrained bed processes (e.g. *Texaco, Koppers*-Totzek). Other processes are being developed, e.g. coal gasification in an iron bath.

Table 3.4 shows the operating conditions for the three technically proven gasification processes.

Table 3.4: Operating conditions for coal gasification processes

Reactor type	Fixed bed reactor	Fluidized bed reactor	Entrained bed reactor
Physical form	Bulk material	Fluidized bed	Flowing
Coal size (mm)	10 to 30	1 to 10	less than 0.1
Steam/O_2 ratio (kg/Nm^3)	9:1 to 5:1	2.5:1 to 1:1	0.5:1 to 0.02:1
Movement of fuel	Counter current	Turbulant co-current	Co-current
Fuel contact time	60 to 90 min	15 to 60 min	less than 1 sec
Fuel specification	should not cake or disintegrate	highly reactive; should not disintegrate	ash melting point less than 1450 °C
Max. gas exit temperature (°C)	370 to 600	800 to 950	1400 to 1600
Pressure (bar)	20 to 30	1.03	1 to 30
Composition of crude gas (vol%)			
CO + H_2	62	84	85
CH_4	12	2	0.1
Organic by-products	Tar, oil, phenols, gasoline, effluent	none	none

Aromatic hydrocarbons occur as by-products only in fixed-bed processes, since the coal does not undergo a carbonization stage in fluidized or entrained bed gasification processes. As a result of the high reaction temperatures, all volatile hydrocarbons are rapidly converted to gas.

The tar produced by fixed-bed gasification with counter-current feed of the two reactants, coal and air/steam, is basically dependent on the geological age of the coal, i.e, its content of volatiles, which ranges from 20 kg to 90 kg per t of coal. The level of crude phenol varies correspondingly between 3 and 10 kg per t of coal.

As a result of the relatively low temperatures in the fixed-bed gasification process, which is generally carried out using the *Lurgi* method, the tar produced has a composition comparable to low-temperature carbonization tar, i.e., the aromaticity is appreciably lower than that of high-temperature coal tar. Table 3.5 shows a comparison of the composition of fixed-bed gasification tar, produced by the *Lurgi* process, and a typical high-temperature coal tar.

Table 3.5: Constituents of high-temperature tar and tar arising from coal gasification

Compound	Coal tar (%)	Gasification tar (Lurgi) (%)
Naphthalene	10.0	2.5
1-Methylnaphthalene	0.7	0.7
2-Methylnaphthalene	1.5	1.0
Dimethylnaphthalene	<0.1	1.8
Biphenyl	0.4	0.1
Acenaphthene	0.2	0.9
Fluorene	1.8	1.1
Phenanthrene	4.5	1.6
Anthracene	1.3	0.7
Carbazole	0.9	0.3
Fluoranthene	3.0	0.4
Pyrene	2.0	0.3
n-Alkanes	<0.1	3.3

The following fractions are recovered from tars produced by *Lurgi* fixed-bed gasification at the *SASOL* plant: light naphtha (5%), heavy naphtha (7%), middle creosote oil (25%), heavy creosote oil (20%), pitch distillate (13%) and pitch (30%).

Figure 3.18 shows the flow diagram for refining *SASOL* tar.

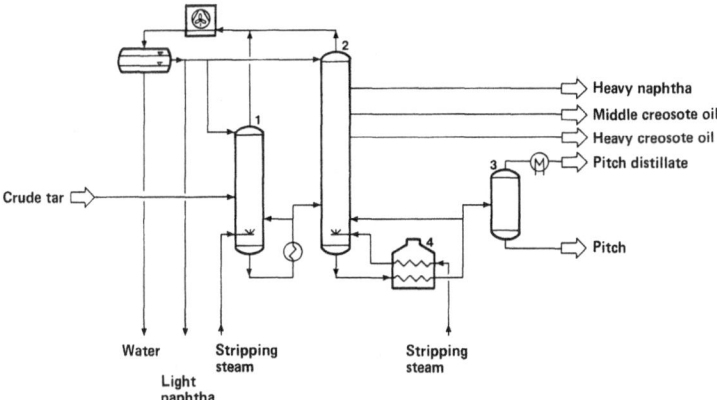

1 Dewatering column; **2** Distillation column; **3** Pitch flash column; **4** Tubular furnaces

Figure 3.18: Diagram of the *SASOL* tar distillation

The naphtha streams are further processed to produce gasoline by low-pressure hydrogenation (50 bar/370 °C). The middle and heavy creosote oils are also hydrogenated under pressure, but under more severe conditions, i.e., higher pressure and higher temperatures (180 bar/400 °C), to produce diesel fuel. The crude phenol can be refined by methods used to recover phenols from coal tar. The pitch is not suitable as a binder for anodes, due to its ash content and low aromaticity, but it can be used as a road binder.

3.2.5 Production of aromatics by coal liquefaction

3.2.5.1 Historic development

In simple terms, the carbon/hydrogen ratio of coal is reduced in coal liquefaction by hydrogenation to the extent that coal is transformed into a 'synthetic crude oil' (syncrude), while largely retaining its aromatic character.

There are two types of processes available for coal liquefaction, namely direct conversion of coal with hydrogen, invented by Friedrich Bergius in 1913, and coal extraction with hydrogenating solvents, first tested in a pilot plant by Alfred Pott and Hans Broche in 1935. The products from both processes can be further hydrogenated in a second stage to obtain fuels.

Hydrogenation of coal has been developed particularly in Germany and Great Britain, to reduce the dependence on imported crude oil. In both countries, large-scale coal hydrogenation plants were operated up to the end of World War II (although in the UK only creosote was hydrogenated during the War) having a total capacity of around 4.5 Mtpa of fuels. Table 3.6 gives details of the German and British hydrogenation plants and their operating conditions.

Table 3.6: German and British coal hydrogenation works

Starting date	Works site	Raw materials	Pressure (bar) Liquid phase	Gas phase	Capacity (tpa) 1943/1944
1927	Leuna	Brown coal, tar	200	200	650,000
1933	Billingham	Hard coal (75 %), tar oil (25 %)	300	300	150,000
1936	Böhlen	Brown coal tar	300	300	250,000
1936	Magdeburg	Brown coal tar	300	300	220,000
1936	Scholven	Hard coal	300	300	280,000
1937	Welheim	Pitch	700	700	130,000
1939	Gelsenberg	Hard coal	700	300	400,000
1939	Zeitz	Brown coal tar	300	300	280,000
1940	Lützkendorf	Tar, oil	500	500	50,000
1940	Pölitz	Hard coal, oil	700	300	700,000
1941	Wesseling	Brown coal	700	300	250,000
1942	Brüx	Brown coal tar	300	300	600,000
1943	Blechhammer	Hard coal, tar	700	300	420,000

Figure 3.19 shows the flow diagram for a coal-hydrogenation plant using an advanced version of the Bergius process (*IG*-hydrogenation process).

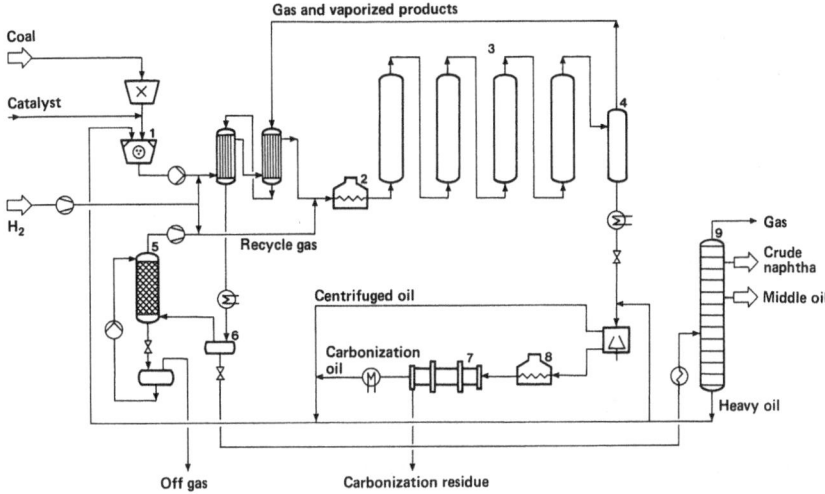

1 Coal-slurry production; **2** Tubular furnace; **3** Hydrogenation reactors; **4** Hot separator;
5 Recycle-gas scrubber; **6** Cold separator; **7** Rotary kiln; **8** Tubular furnace; **9** Distillation

Figure 3.19: Flow diagram of Bergius' coal hydrogenation (*IG*-process)

In the Pott-Broche process, tetralin was used as a solvent for coal extraction in a mixture with cresols (80/20). This method was used to produce 30,000 tpa of coal extract in a plant operated by *Ruhröl* in Welheim (Bottrop), Germany between 1938 and 1944. Extraction was carried out under a pressure of around 100 bar at a temperature of 415 to 435 °C. The coal extract was used as a low-sulfur fuel, for the production of electrode coke, or refined by hydrogenation to yield benzole, middle oil and heavy oil. The solvent had to be regenerated, i.e. hydrogenated, before each extraction.

After World War II, some of the German hydrogenation plants were converted to hydrogenate petroleum-derived oil residues, and were in use until the early 1960's. These plants were the forerunners of the hydrocrackers.

Coal liquefaction received a further innovative impetus, particularly in the USA, West Germany, Great Britain and Japan, with the beginning of the first world 'oil crisis' in 1973.

3.2.5.2 Mechanisms of coal liquefaction

Whereas gasification causes a drastic decomposition of coal into low-molecular weight compounds (mainly CO, CH_4 and H_2), the basic aromatic skeleton of coal is retained in coal liquefaction by hydrogenation, so that the coal oil is suitable not only for the production of heating oils and fuels, but also for the production of aromatics. Brown coals (lignites) and, among hard coals, high-volatile bituminous coals are particularly suitable for hydrogenation. Low ash content, a content of volatiles between 25 and 40% and a high level of vitrinite of 60 to 80% are prereq-

48 Base materials for aromatic chemicals

uisites for a coal with good hydrogenation properties. Table 3.7 shows the composition of a bituminous coal from the Ruhr district, and for two hydrogenation coals from the USA.

Table 3.7: Composition of hydrogenation coals (in %)

	Ruhr coal	Indiana V	Illinois No. 6
Volatile components (waf)	37.9	39.7	42.8
Ash	4.6	10.6	10.2
Carbon	82.8	77.3	77.4
Hydrogen	5.2	5.3	5.2
Sulfur	1.0	3.7	3.2
Nitrogen	0.5	1.6	1.7
Oxygen	10.5	11.1	12.4

Coal liquefaction basically follows the stages described below, regardless of the type of process:

1. Mixing finely-crushed coal with a suitable solvent,
2. Hydrocracking the hot reaction mixture under pressure with molecular or transferable hydrogen (from donor solvents) using catalysts
3. Depressurising the reaction product and separating the residue from the distillable portion and
4. Further processing the coal-oil distillate.

In the first stage of coal liquefaction, finely-crushed coal (particle size <0.1 mm) is slurried with a solvent, to render the coal flowable and pumpable. The choice of solvent is particularly important, since it must be suitable to stabilize the coal fragments and to dissolve the smaller disintegrated molecular moieties. Due to their similarity in chemical nature, coal-derived oils are particularly efficient. Anthracene oil from coal-tar processing was therefore preferred as a solvent, when coal hydrogenation was being developed.

During the heating of the coal-oil mixture, decomposition of the coal occurs, as evidenced by an increase in the proportion of quinoline-insoluble material.

Table 3.8: Solvent power of aromatics and hydroaromatics for coal

Compound	Standard extraction yield
Acenaphthene	85.2
Anthracene	32.4
Carbazole	87.3
Dibenzofuran	65.9
9,10-Dihydroanthracene	76.9
Biphenyl	13.0

Table 3.8 (continued)

Compound	Standard extraction yield
Fluoranthene	80.1
Fluorene	66.0
2,3-Dihydroindole	96.0
1-Methylnaphthalene	48.1
2-Methylnaphthalene	50.4
Naphthalene	14.0
Phenanthrene	54.8
Pyrene	83.0
1,2,3,4-Tetrahydroquinoline	94.7
Tetralin	85.5

Table 3.8, which shows the solvent power of polynuclear aromatics and hydroaromatics for coal, illustrates that hydroaromatics, which also occur in anthracene oil, display particularly high solvent powers.

Figure 3.20 shows a comparison of the solubility of coal in aromatic, hydroaromatic and aliphatic solvents at 400 °C.

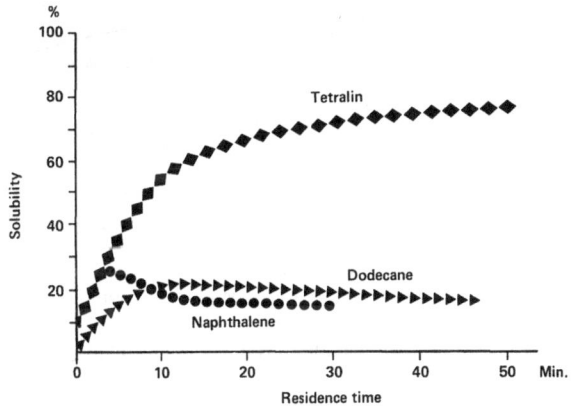

Figure 3.20: Solubility of coal at 400 °C in C_{10}-hydrocarbons

To achieve high solubility, it is necessary to saturate the coal fragments with hydrogen, for which only the hydroaromatic tetralin is suitable, in contrast to naphthalene and dodecane. As an example, the possible reaction of 1-phenanthrene-2-pyrene-ethane, a coal fragment model, with tetralin is shown overleaf.

Hydrogen donors labelled with deuterium, such as tetralin-d_{12} and naphthalene-d_8, have also been used to study the reaction mechanism of the hydrogen transfer. Tetralin-d_{12} is rapidly substituted in the 1-position by hydrogen from coal. At a reaction temperature of 400 °C the deuterium content of the 1-position remains, after a reaction time of 15 min, relatively constant at 66%. The 2-position, however, participates much less in the exchange reaction. On the other hand, the deu-

50 Base materials for aromatic chemicals

[Scheme: Coal → (400 °C) → Tetralin intermediates]

terium atoms of the aromatic ring, are hardly involved in the hydrogen exchange. Naphthalene-d_8 reacts likewise predominantly at the 1-position. In interpreting the experiments with deuterium labelled solvents one has to have in mind, however, that smaller coal fragments are also in a position to participate in hydrogen exchange reactions (hydrogen shuttle).

The oil yield from coal liquefaction by reaction with hydrogen or with a hydrogenating solvent is particularly dependent on the level of hydrogen consumption (Figure 3.21).

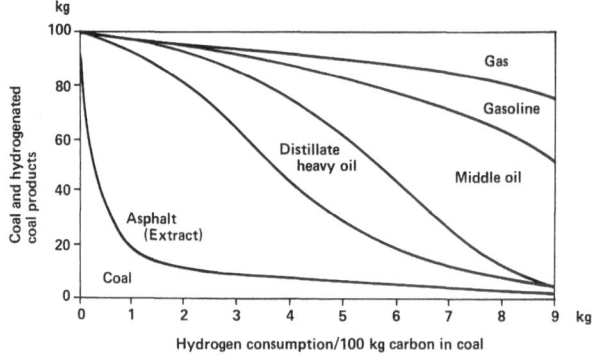

Figure 3.21: Distribution of coal-hydrogenation products versus hydrogen consumption

The reaction conditions used in the first phase (sump phase) are generally temperatures of 400 to 500 °C and pressures ranging from 100 to 700 bar. Molybdenum and tungsten oxides are commonly used as catalysts, together with iron compounds. In the *IG*-hydrogenation process, Bayer-mass ('red mud'), a by-product of bauxite processing, was used as an iron catalyst. Coals which contain mineral compounds with the necessary catalytic activity can be hydrogenated without the addition of catalysts.

The first large-scale coal-hydrogenation plants, which started operation in Billingham, England in 1933 and in Scholven, Germany in 1936, were designed to operate at a pressure of 250 or 300 bar. At these relatively low pressures, 0.06% of tin oxalate (based on dry coal) was used as catalyst. Later coal-hydrogenation operations used higher pressures (700 bar), resulting in higher coal conversions and enabling the use of Bayer-mass and weakly-acidic iron sulfates as catalysts.

The reaction mixture formed during hydrogenation is generally taken to a hot separator to separate gases, vapor, and liquid-solid mixtures. The hydrogenation residue can be refined by filtration, centrifugation or distillation. The coal-oil products are further processed to gasoline and diesel oil by methods commonly used in the petroleum industry.

Tables 3.9 and 3.10 summarize typical characteristics of a sump-phase gasoline and a sump-phase middle oil.

Table 3.9: Composition and characteristic data for a sump-phase gasoline

Phenols in crude gasoline %	18.0
Dephenolated gasoline	
density at 15 °C (g/ml)	0.795
boiling range up to 100 °C %	23.0
final boiling point °C	185
paraffins %	32.5
naphthenes %	35.0
aromatics %	22.5
olefins %	10.0
octane rating RON	80

Table 3.10: Characteristics of a sump-phase middle oil

Density at 20 °C (g/ml)	0.970
Aniline point °C	-16
Cetane no.	6
Pour point °C	-44
Initial boiling point °C	195
Boiling range	
to 250 °C %	50
Final boiling point °C	348
Hydrogen content %	9.5
Phenol content %	16

The hydrogenation plants operated before 1945 had to cope with a number of fundamental difficulties. Since coal swells when mixed with the solvent, heat transfer deteriorated during pre-heating before hydrogenation. Furthermore, coke deposits formed on the heater walls, which also hampered heat transfer. In the hydrogenation reactors, the prevention of coke build-up through overheating was a major task. There was also a risk of coking in the hot separator. The plants therefore had a maximum operating time of only some 200 days. In the *IG*-hydrogenation process, the residue was refined by centrifugation and subsequent carbonization of the slurry. If the asphaltene content of the centrifuge overflow was too high, its tendency to carbonize was increased when it was fed back into the preheater and the reactor.

3.2.5.3 Further developments in coal hydrogenation since 1970

Renewed efforts to improve coal liquefaction by hydrogenation and extraction, particularly in the early 1970's, were aimed at overcoming these difficulties and obtaining the highest possible oil yields at relatively low pressures and reasonable hydrogen consumption.

The SRC (solvent refined coal) process of *Pittsburgh & Midway Coal Mining* (*Gulf Oil*) is an advanced version of the Pott-Broche extraction process. The purpose of the SRC method is to desulfurize coal to yield a solid, low-sulfur fuel.

In the SRC process, the crushed coal is mixed with process-derived oil to produce a 35% suspension. (Figure 3.22). The suspension is fed into the pressure reactor after heating and addition of hydrogen (70 bar, 425 °C).

The process operates without catalysts, since the catalytic activity of the finely divided pyrites in the coal feedstock is sufficient. Hydrogen consumption is around 2%. The undissolved residue is separated by pressure filtration. A pilot plant was started up at Fort Louis in Tacoma/Wash., USA in 1974, with a throughput of 50 t/d of coal. By increasing the severity of cracking, this method can also be used to produce heavy heating oil (SRC-II process).

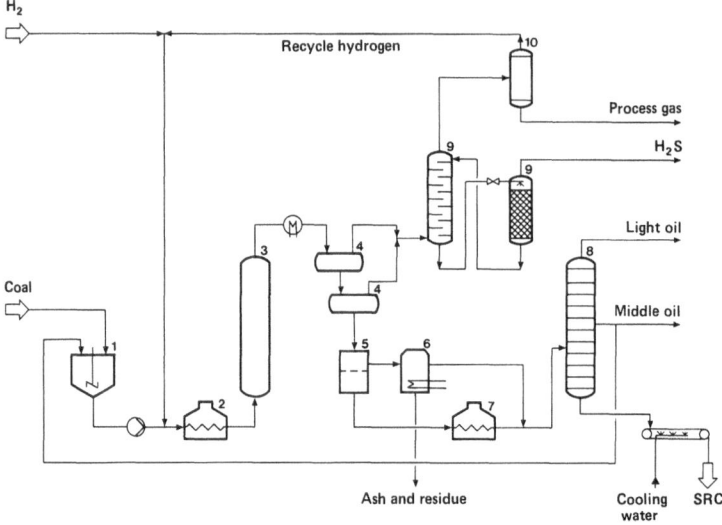

1 Coal-slurry production; **2** Tubular furnace; **3** Hydrogenation reactor; **4** Gas separator; **5** Pressure filter; **6** Solvent recovery; **7** Tubular furnace; **8** Vacuum distillation; **9** Gas purification; **10** Gas separation

Figure 3.22: Process scheme of coal hydrogenation using the SRC process

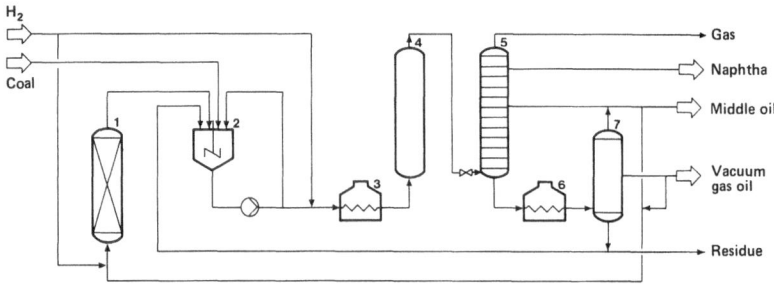

1 Solvent hydrogenation: **2** Coal-slurry production; **3** Tubular furnace; **4** Hydrogenation reactor; **5** Distillation; **6** Tubular furnace; **7** Vacuum distillation

Figure 3.23: Process scheme of coal hydrogenation using the *Exxon*-Donor-Solvent process

Similar to the SRC version is the *Exxon*-Donor-Solvent-Coal-Liquefaction process (EDS) developed by *Exxon*. Rehydrogenation of the hydrogen-donating solvent is achieved in a special process stage through a fixed-bed catalyst (Figure 3.23).

A large-scale pilot plant with a throughput of 250 t/d of coal was operated at the *Exxon* refinery in Baytown, Texas from June 1980 to August 1982.

Further developments of the *IG*-hydrogenation process include the *Ruhrkohle/ VEBA Oel*-hydrogenation process, and *Hydrocarbon Research's* H-Coal process. In the latter method, crushed and dried coal is mixed with a process-derived oil, the mixture saturated with hydrogen and hydro-cracked after preheating at 455 °C and 180 to 210 bar. Cobalt and molybdenum oxides are used as catalysts on an aluminum oxide carrier. The H-Coal process is a variant of the H-Oil process, in which heavy hydrocarbons, e.g. crude oil residues, tar or shale oil, are cracked in a catalytic fluidized bed. The catalyst is added in the form of extruded pellets. A particular advantage of the H-Coal process is the thorough mixing of the coal suspension with the catalyst, together with the uniform reactor temperature. Hydrogen consumption is over 4% (based on waf coal). Figure 3.24 shows a flow chart for the H-Coal process and Table 3.11 gives the process conditions and characteristic properties of the reaction products.

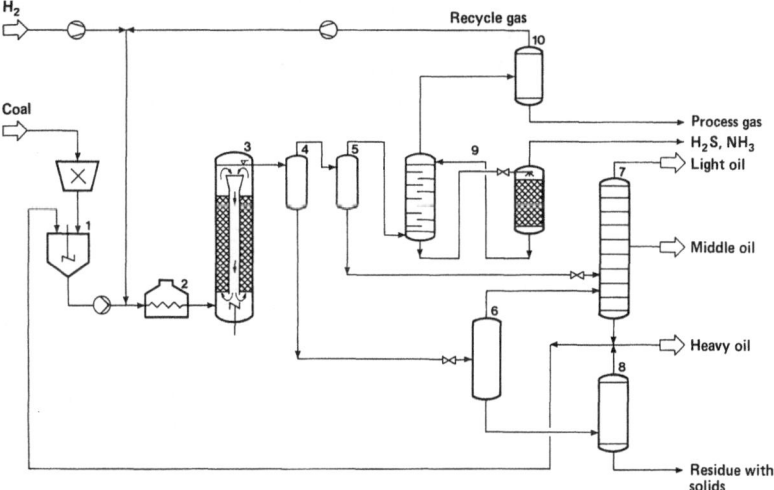

1 Coal-slurry production; **2** Tubular furnace; **3** Hydrogenation reactor; **4** Hot separator; **5** Cold separator; **6** Flash chamber; **7** Atmospheric distillation; **8** Vacuum distillation; **9** Gas purification; **10** Gas separation

Figure 3.24: Process scheme of coal hydrogenation using the H-Coal process

Table 3.11: Process conditions and product yields for coal hydrogenation using the H-Coal process

Operating conditions		
Pressure	bar	190
Temperature	°C	460
Coal throughput	kg h^{-1} l^{-1}	0.490
Coal feedstock		
Volatiles	% wf	37.8
Ash	% wf	11.8
Sulfur	% wf	3.63
Products		
C_1 to C_3	%	8.3
C_4 to 204 °C	%	22.3
204 to 343 °C	%	11.4
343 to 524 °C	%	11.7
Residue > 524 °C	%	15.6
unconverted coal, af	%	10.6
Ash	%	11.7
H_2O, NH_3, CO_2	%	9.8
H_2S	%	2.8

The *Ruhrkohle/VEBA Oel* coal-liquefaction plant came into operation in Bottrop, West Germany in 1981. Today, it is the largest pilot plant in the world for direct coal liquefaction, with a coal throughput of 200 t/d yielding 18 t/d of liquefied gas, 29 t/d of light oil and 69 t/d middle oil.

Finely-crushed coal and an iron catalyst (2% based on waf coal) are mixed together with process oil and, after addition of hydrogen, heated in a tubular fur-

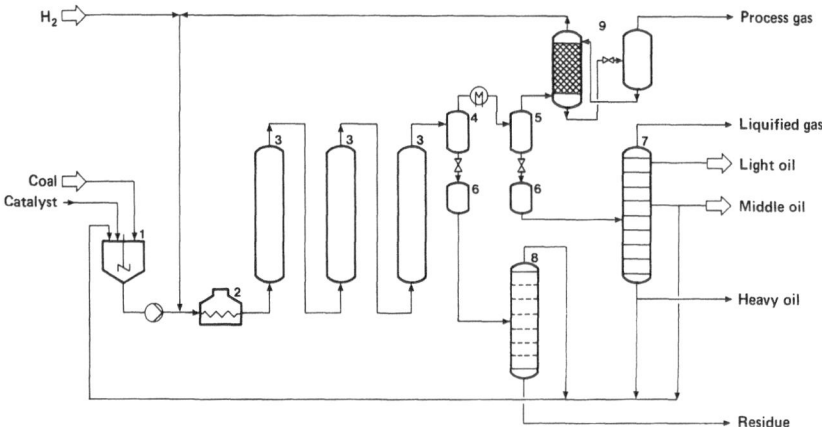

1 Mixing vessel; **2** Pre-heater; **3** Hydrogenation reactors; **4** Hot separator; **5** Cold separator; **6** Flash vessel; **7** Atmospheric distillation; **8** Vacuum distillation; **9** Gas purification (scrubber)

Figure 3.25: Flow diagram of the *Ruhrkohle/VEBA Oel* coal-liquefaction plant in Bottrop

nace (Figure 3.25). The hydrocracking reaction is carried out in a large-chamber hydrogenation reactor or in three reactor trains at a pressure of 300 bar and a temperature of 485 °C. The reaction product is separated into an overhead and bottom product in the hot separator. The bottom product, which contains the high-boiling and all the solid materials (i.e. unconverted coal, ash and catalyst) is distilled in a vacuum column. The vacuum overhead product is recycled to the process, while the residue, containing ash and unreacted coal, is gasified to produce hydrogen. The overhead product from the hot separator is further cooled. In the cold separator, additional separation into gas and liquid hydrocarbons takes place. The gas is fed back into the liquid-phase hydrogenation after being purified by oil scrubbing. The condensate from the cold separator is split by atmospheric distillation into light oil (≤ 185 °C), middle oil (185 to 325 °C) and a higher-boiling bottom product. The bottom product from the atmospheric column is likewise returned to the process as mixing oil. The coal oil distillates, light and middle oil, correspond to the overall oil production in the liquid-phase hydrogenation. They can be further refined to gasoline, diesel oil and home heating oil. The thermal efficiency of this coal-oil production is in excess of 80%. The oil yield, based on water and ash-free coal feedstock, is over 50%. Further improvements to the process are being developed.

3.2.5.4 Hydropyrolysis

Hydropyrolysis occupies a position midway between coal pyrolysis and coal liquefaction. This process, which is currently being developed to complement coal liquefaction, operates at pressures up to 100 bar and temperatures up to 900 °C; the residence time of the coal in the pyrolysis zone is around 4 to 5 sec.

Hydropyrolysis is characterized by the fact that the yield of coke, which can reach 80% in purely pyrolytic processes, is significantly reduced and more gaseous hydrocarbons and around 20% of tar are produced. The aromaticity of the tar generated by hydropyrolysis at 530 °C is 0.80, rising to 0.95 at a reaction temperature of 950 °C.

The hydropyrolysis of coal is related to the hydropyrolysis of naphtha; the latter was operated in a pilot plant belonging to *Naphtha Chimie/Pierrefitte-Auby* to produce ethylene, in 1975.

3.3 Crude oil

3.3.1 Reserves and characteristics of crude oil

By far the largest petroleum reserves are found in the Middle East, particularly in Saudi Arabia. The proportion of the industrial nations in the list of world oil reserves is relatively small. The USA owns 6%, the USSR 10%. The share held by West European states is extremely small; Great Britain, the leading West European oil producer, owns just over 2% of world oil reserves.

Figure 3.26 and Figure 3.27 depict reserves and production rates of the major world petroleum producers.

56 Base materials for aromatic chemicals

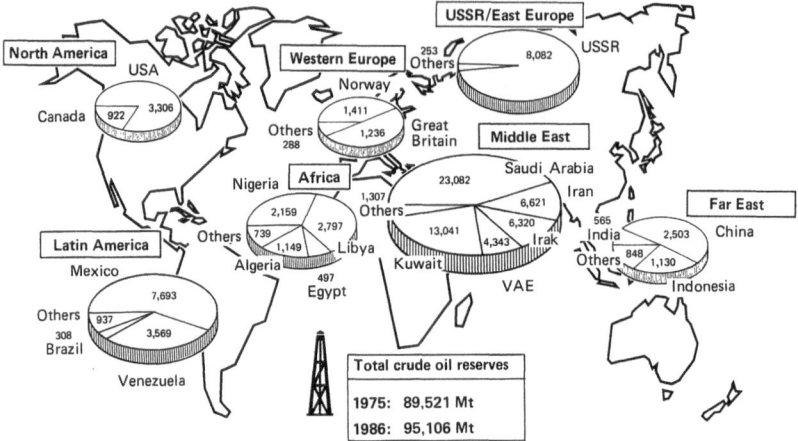

Figure 3.26: Distribution of crude oil reserves (1986)

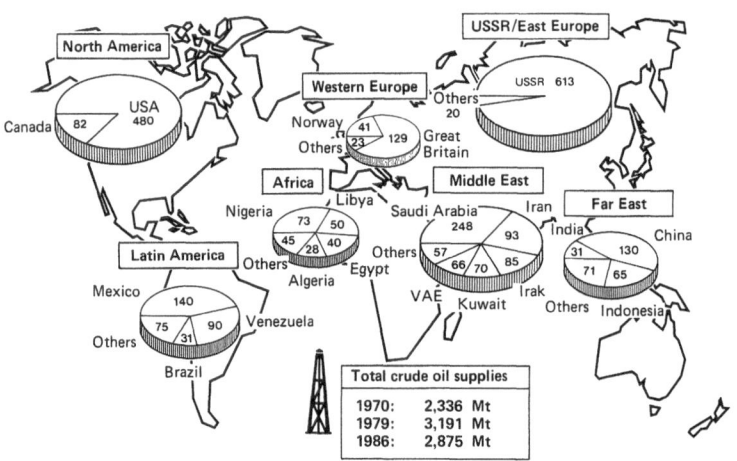

Figure 3.27: World oil supplies (1986)

As a result of intensive exploration, proven crude oil reserves have tripled in the last 30 years (Figure 3.28) so that the expected life span calculated as the quotient of reserves and consumption today is around 35 years.

Base materials for aromatic chemicals 57

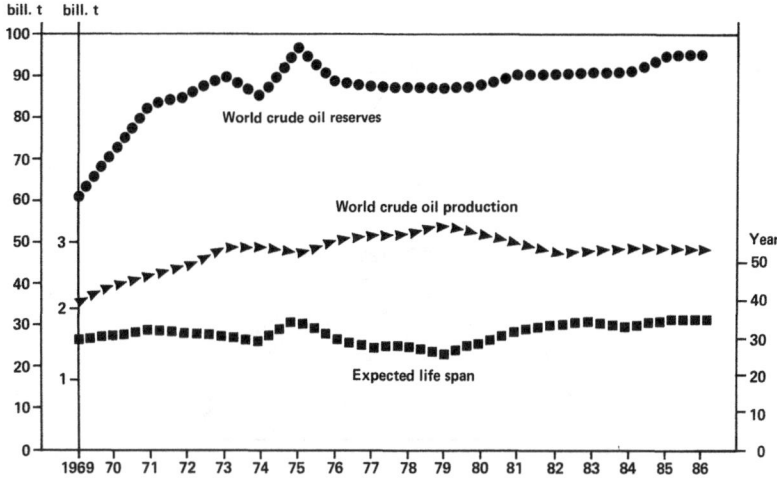

Figure 3.28: Development of crude oil reserves and their expected life span

The carbon content of crude oil generally ranges from 85% to 90%, hydrogen content from 10 to 14%; vanadium is particularly common as a trace element (Table 3.12).

Table 3.12: Basic composition of crude oil

Carbon	85–90 %
Hydrogen	10–14 %
Sulfur	0.2–3.0 % (max. 7 %)
Nitrogen	0.1–0.5 % (max. 2 %)
Oxygen	0–1.5 %
Vanadium	ca. 30 ppm
Nickel	ca. 10 ppm

Crude oils are classified as paraffin-based, mixed-based, naphthene-based and the rather rare aromatic oils, depending on their paraffin, naphthene and aromatic hydrocarbon content (Figure 3.29).

58 Base materials for aromatic chemicals

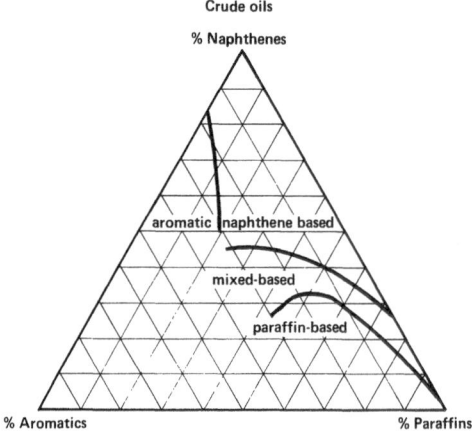

Figure 3.29: Crude oil classification diagram

The density of crude oil rangesfrom the extremes of 0.75 to 1.0 g/ml, but is generally between 0.8 and 0.9 g/ml. The API-gravity introduced by the American Petroleum Institute, is determined by the following formula:

$$\text{API}° = \frac{141.5}{\text{sp. gravity } 15.5/15.5\,°C} - 131.5$$

Table 3.13 summarizes characteristic properties of various crude oils.

Table 3.13: Characteristic properties of various crude oils

Origin of crude	Middle East Kuwait	N. Africa Libya	Venezuela Tia Juana M.	North Sea Forties Field	Alaska Prudhoe Bay
Total crude					
Density at 15 °C (g/ml)	0.869	0.824	0.897	0.835	0.893
API gravity (° API)	31.3	37.6	26.5	38.0	27.0
Sulfur (wt%)	2.5	0.14	1.55	0.29	0.82
Viscosity at 38 °C (mm^2/s)	9.6	9.0	25	4.2	18.0
Pour point (°C)	−4	+24	−35	0	−10
Vanadium content (mg/kg)	27	<0.5	170	10	25
Distillation analysis, yields:					
C$_4$ and lighter (vol%)	2.52	2.3	0.70	4.00	0
Naphtha fraction, C$_5$ – 150 °C (vol%)	16.65	17.3	13.70	18.75	11.8
Middle distillate, 150 – 370 °C (vol%)	35.15	37.4	32.65	39.80	38.4
Residue >370 °C (vol%)	45.75	43.0	52.95	36.00	49.8
Sulfur content (wt%)	4.16	0.15	2.35	0.65	1.55

Table 3.13 (continued)

Origin of crude	Middle East Kuwait	N. Africa Libya	Venezuela Tia Juana M.	North Sea Forties Field	Alaska Prudhoe Bay
Vacuum distillate, 370–525 °C (vol%)	19.85	22.9	23.10	20.40	(49.8)
Vacuum residue, > 525 °C (vol%)	25.90	21.1	29.85	15.60	

Whereas coal and the tar which arises during carbonization are composed predominantly of aromatic constituents, petroleum contains mainly aliphatic hydrocarbons. The proportion of aromatics is generally small.

Table 3.14 shows the contents of benzene, toluene and C_8 aromatics of some crude oils.

Table 3.14: Aromatics content of crude oils (in wt%)

	Libya	Louisiana Gulf	West Texas	Venezuela	Nigeria	Iran
Benzene	0.07	0.15	0.18	0.15	0.11	0.19
Toluene	0.37	0.45	0.51	0.60	0.92	0.56
C_8 aromatics	0.56	0.50	1.10	1.10	1.47	1.05
Total aromatics	1.00	1.10	1.79	1.85	2.50	1.80

Since the concentration of aromatics in crude oils is low, production of aromatics from crude oil generally necessitates additional aromatization of naphthenic and aliphatic hydrocarbons by dehydrogenation and cyclization. Before the development of modern crude oil refining technologies, crude oils rich in aromatics, mainly from Southeast Asia, were commonly resorted to, especially for the production of toluene.

The rare crudes with a high proportion of aromatics, particularly in the form of polycyclic aromatics, are found principally in California, Texas, Burma, Mexico, and in the Urals. In some cases, the aromatics content of these crudes is over 35%.

In general, the greater the geological age of the crude oil, the higher is its aromatics content (Figure 3.30). Crude oils from the Tertiary, Silurian and Cambrian periods have especially high levels of aromatics, while Permian, Carbonaceous and Devonian oils are distinguished by very low levels of aromatics.

Crude oil is split by refining into a large number of fractions to recover marketable products. Only a small proportion was, a number of years ago, used in pyrolysis processes as a feedstock for olefins and aromatics production without previous refining (e.g. *Kureha/Union Carbide* process, see Chapter 3.3.2.5.2).

60   Base materials for aromatic chemicals

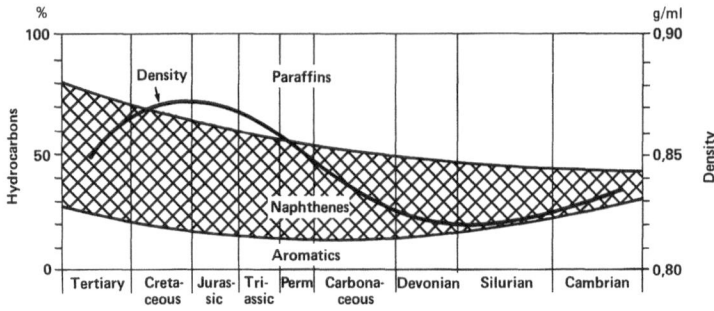

Figure 3.30: Aromatics content of mineral oil in relation to its geological age

3.3.2 Crude oil refining

3.3.2.1 Distillation

There are approximately 730 petroleum refineries operating worldwide, around 190 of which are in the USA alone. Distillation capacity at the end of 1986 was

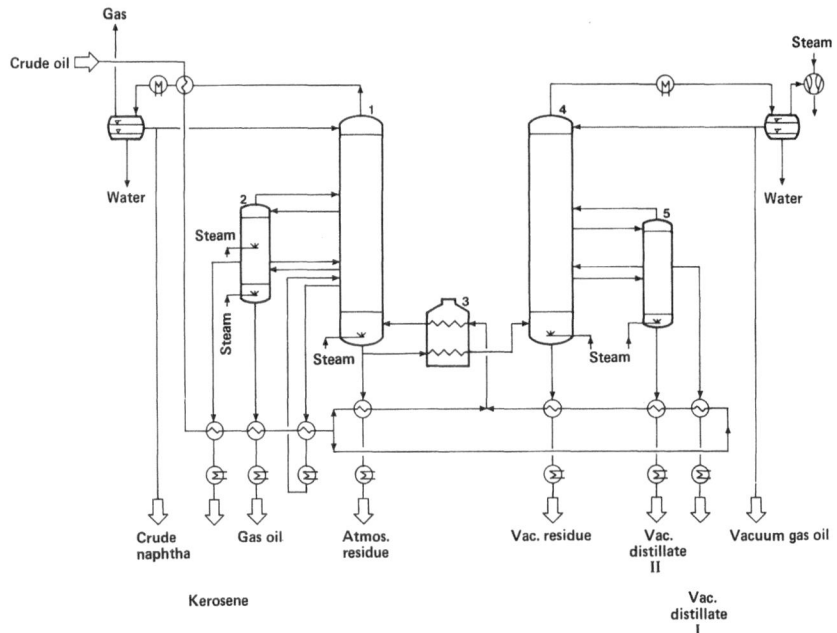

1 Atmospheric distillation; **2** Side column; **3** Tubular furnace; **4** Vacuum distillation; **5** Side column

Figure 3.31: Scheme of crude oil distillation with atmospheric section and vacuum section

3.6 billion tons; a medium-sized refinery has a throughput of 6 to 7 Mtpa of crude oil.

When the crude has been delivered by ship or pipeline, the crude oil is stored in large tanks, capable of holding up to 100,000 m^3. The first stage in refining is dewatering and desalination, normally by electrostatic methods. This is followed by atmospheric distillation, frequently complemented by vacuum distillation (Figure 3.31). Atmospheric distillation is carried out with the injection of steam; gases, naphtha fractions, kerosene and light gas oil are recovered. The topped residue is reheated in a tubular furnace and split by flash distillation in the vacuum column into vacuum gas oils and the vacuum residue. Since hydrocarbons have a tendency to carbonize, the flash temperature is generally below 400 °C.

The most important properties of the main distillate streams for light and heavy Arabian crude oil are summarized in Table 3.15.

Table 3.15: Characteristic data for two Arabian crudes

	Arabian heavy	Arabian Light
Crude oil		
Density (g/ml)	0.886	0.851
Sulfur, wt%	2.84	1.70
Pour point, °C	−34	−34
Viscosity, mm^2/s		
at 21 °C	35.8	8.2
at 38 °C	18.9	5.4
Light naphtha		
Boiling range, °C	20–100	20–100
Yield, vol%	7.9	10.5
Density (g/ml)	0.669	0.677
Sulfur, wt%	0.0028	0.055
Paraffins, vol%	89.6	87.4
Naphthenes, vol%	9.5	10.7
Aromatics, vol%	0.9	1.9
RON (Research Octane Number)	59.7	54.7
Heavy naphtha		
Boiling range, °C	100–150	100–150
Yield, vol%	6.8	9.4
Density (g/ml)	0.737	0.744
Sulfur, wt%	0.018	0.057
Paraffins, vol%	70.3	66.3
Naphthenes, vol%	21.4	20.0
Aromatics, vol%	8.3	13.7
Kerosene		
Boiling range, °C	150–235	150–235
Yield, vol%	12.5	18.4
Density (g/ml)	0.787	0.788
Sulfur, wt%	0.19	0.092
Paraffins, vol%	58.0	58.9
Naphthenes, vol%	23.7	20.5

Table 3.15 (continued)

	Arabian heavy	Arabian light
Kerosene		
Aromatics, vol%	18.3	20.6
Crystallisation point, °C	−53	−55
Viscosity, mm^2/s		
at 34 °C	4.74	5.09
at 38 °C	1.12	1.13
Light gasoil		
Boiling range, °C	235–343	235–343
Yield, vol%	16.4	21.1
Density (g/ml)	0.846	0.838
Sulfur, wt%	1.38	0.81
Viscosity, mm^2/s		
at 38 °C	3.65	3.34
at 99 °C	1.40	1.32
Heavy gasoil		
Boiling range, °C	343–565	343–565
Yield, vol%	26.3	30.6
Density (g/ml)	0.923	0.905
Viscosity, mm^2/s		
at 38 °C	62.5	49.0
at 99 °C	7.05	6.65
Residual oil (I)		
Boiling range, °C	>343	>343
Yield, vol%	53.1	38.0
Density (g/ml)	0.984	0.924
Sulfur, wt%	4.35	2.04
Conradson Carbon, wt%	13.2	4.5
Viscosity, mm^2/s		
at 38 °C	5,400	146
at 99 °C	106	12,4
Residual oil (II)		
Boiling range, °C	>565	>565
Yield, vol%	26.8	7.4
Density (g/ml)	1.004	0.990
Sulfur, wt%	5.60	3.0
Pour point, °C	49	27
Conradson Carbon, wt%	24.4	19.0
Viscosity, mm^2/s		
at 99 °C	13,400	393
Metal content, ppm		
Vanadium	171	12
Nickel	53	7
Iron	28	36

As the recovery of gasoline and light fuel oil by simple distillation of crude oil does not produce sufficient quantities to meet market demand, conversion processes have been developed since the beginning of this century to achieve higher yields. In thermal- and catalytic-cracking processes, high-molecular weight units are split into low-molecular weight hydrocarbons, with simultaneous disproportionation into hydrogen-rich and hydrogen-poor products. The resulting product range depends on temperature conditions, reaction time and the catalyst used. The aromatics content of cracker products can be varied by altering the reaction conditions.

A major factor in petroleum refining is the production of high-octane gasoline. Since aromatics show particularly high octane ratings, aromatization is an especially important facet of gasoline production.

3.3.2.2 Catalytic cracking

Catalytic cracking is the most important process for converting heavy hydrocarbons into high-value gasoline and light fuel oil components. The reaction is carried out using a catalyst to produce hydrocarbons with high octane ratings, i.e. olefins, isoparaffins and aromatics. In the USA, with a production of 200 Mtpa, over 50% of all gasoline is produced in catalytic cracking plants (cat-crackers); in Europe, cat crackers are less common.

Amorphous aluminum silicates, with an aluminum oxide content between 10 and 15%, have been used as catalysts. In the last 15 years, zeolite catalysts have gained importance in catalytic cracking. Zeolite catalysts make conversion rates of around 85% possible, whereas conversion rates using amorphous SiO_2 catalysts are only 70 to 75%; furthermore, the aromatics content of the gasoline is higher. Because of their acidity the suitability of zeolites as effective cracker catalysts was recognized early, but the first attempts to use them were abortive, since the crystallinity of the zeolites was lost during regeneration due to the high partial steam pressures. It was then found that, as a result of ion-exchange of the alkali metal with rare earths, e.g. lanthanum, the crystallinity of the zeolites was retained. The catalyst used in catalytic cracking acts as a Lewis acid. This leads to the formation of carbonium ions as intermediates, which are converted into branched paraffins, which have a high octane quality.

Although catalytic cracking was originally carried out in fixed beds, nowadays practically all processes use a fluidized bed. Finely-crushed catalyst (2 to 200 μm diameter) is suspended in the ascending hydrocarbon gas stream (riser). The feedstock, e.g., a vacuum distillate or heavy gas oil from atmospheric distillation, is fed with steam to the base of the catalyst riser, where it is contacted by a hot stream of the regenerated catalyst. The resultant hydrocarbon vapor rises with the catalyst at a rate of 10 to 20 m/s to the top of the reactor, where separation of the catalyst and cracked products occurs. The cracker products leave the reactor via cyclones, which retain the catalyst powder. The catalyst sinks to the base of the reactor and is stripped with steam, to remove any adsorbed hydrocarbons, before it passes via an inclined pipe into the regenerator. In the regenerator, the coke which has formed on the catalyst surface is burned off, and the catalyst returned to the base of the riser, to be brought again into contact with fresh feedstock material. Figure 3.32 shows a diagram of an *Exxon* Fluid Cat-Cracker unit (FCC).

64   Base materials for aromatic chemicals

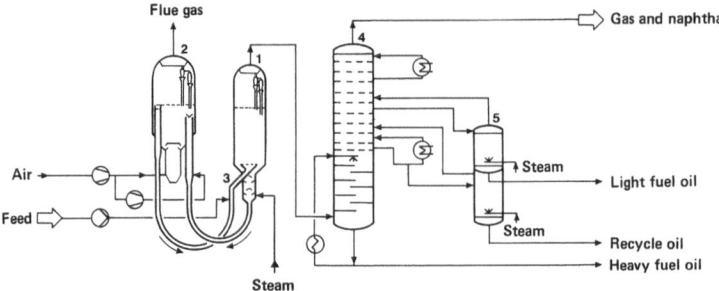

1 Reactor; 2 Regenerator; 3 Riser; 4 Fractionation column; 5 Side columns

Figure 3.32: Flow diagram of Fluid Catalytic Cracking

Reaction temperatures in catalytic cracking range generally between 480 and 530 °C; regeneration of the catalyst occurs at temperatures between 550 and 700 °C. The heat required for regeneration is produced by burning the coke deposits on the catalyst. The cracked gases leaving the reactor are separated in a fractionation column into a cycle stock, heavy and light gas oil, together with an overhead product.

This overhead product is condensed by a compressor and separated into light gases and gasoline.

Table 3.16: Raw materials and composition of products of catalytic cracking

FCC feedstock (380–580 °C)	Brega	Arabian light	Arabian heavy
Naphthenes wt%	42.2	30.5	29.7
Paraffines wt%	18.6	14.7	13.6
Aromatics wt%	31.5	49.6	53.9
Polars wt%	7.7	5.2	2.8
Conradson Carbon wt%	0.51	0.96	1.53
Heavy metals Ni (ppm)	0.16	0.17	0.39
V (ppm)	0.11	0.17	0.07
Fe (ppm)	0.61	1.21	1.05
Conversion (%)	70	62	52
Product composition			
$C_1 + C_2$ wt%	2.7	2.4	3.6
$C_3 + C_4$ unsaturated wt%	5.9	5.4	4.9
$C_3 + C_4$ saturated wt%	4.4	3.3	2.1
C_5 unsaturated wt%	5.7	5.8	4.6
C_5 saturated wt%	6.4	4.4	2.4
C_6 to 221 °C wt%	38.7	35.7	28.4
Naphtha composition			
FIA aromatics vol%	25.8	26.2	24.6
saturated compounds vol%	41.9	27.1	6.0
olefins vol%	32.3	46.7	69.3

Table 3.16 shows the characteristic properties of some cat-cracker feedstocks and their products, together with the product distribution. Where the level of aromatics is high, conversion is lowest, and the olefins content of the gasoline increases noticeably; the octane number of the gasoline produced is around 92.

Figure 3.33 shows the different behavior of hydrocarbons during catalytic cracking (480 °C, Al_2O_3/SiO_2-catalyst) and illustrates how the yield of gasoline, with a final boiling point of 220 °C, is dependent on the severity of cracking. In model tests, the ratio of the surface area of the catalyst to the weight of feedstock serves as an indicator of the cracking severity. Commercially and synthetically produced gas oil fractions with an average boiling range of 370 °C were chosen as model mixes.

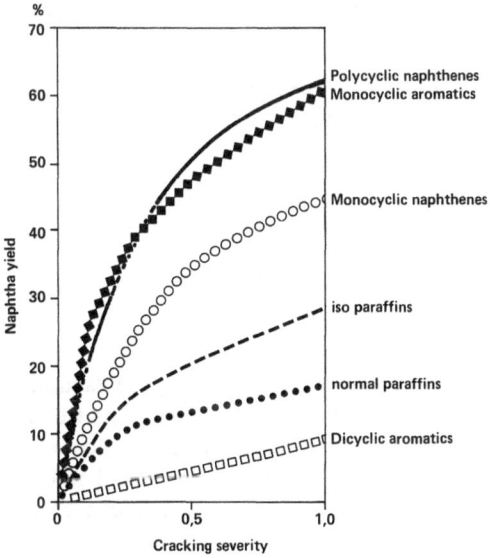

Figure 3.33: Naphtha yield from catalytic cracking of various hydrocarbon mixtures

The highest yield is obtained from polycyclic naphthenes, alkylsubstituted monocyclic aromatics and monocyclic naphthenes. Isoparaffins and normal paraffins give medium yields, whereas the gasoline yield from naphthalene derivatives is relatively low by comparison.

The tendency toward coke build-up during catalytic cracking of various gas oil fractions with an average boiling point of 370 °C has also been investigated. Coke formation increases considerably as the degree of aromaticity grows (Figure 3.34).

66   Base materials for aromatic chemicals

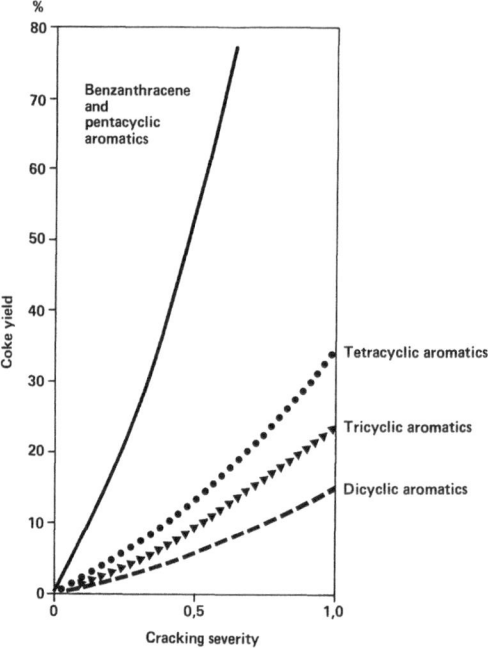

Figure 3.34: Coking of aromatics during catalytic cracking

Gasoline yield can be increased by pretreating the vacuum distillates which are the usual feedstock for cat crackers. The objective of pretreatment is the hydrogenation of polycyclic aromatics, which make virtually no contribution to the gasoline yield in catalytic cracking but are rapidly convertred into coke. Hydrogenation transforms aromatics into hydroaromatics, which can then be broken up into

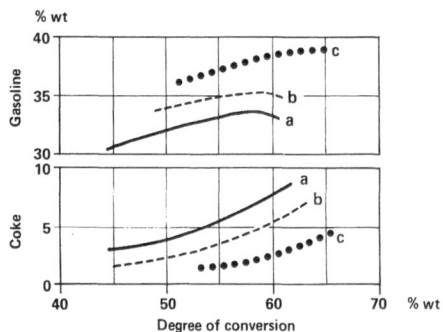

a: untreated feedstock; **b:** hydrogenated with 80 Nm³ H_2 per t of feedstock; **c:** hydrogenated with 120 Nm³ H_2 per t of feedstock

Figure 3.35: Effect of hydrogenation on product yield in catalytic cracking

smaller alkyl aromatics and reappear in the gasoline as high-value octane boosters. Coke formation can also be greatly reduced by hydrogenation. Figure 3.35 shows the influence of hydrogenation on the yield of gasoline and on coke formation in cat cracking.

The gas oil fractions obtained from the fractionation of the cat-cracker reaction product are rich in aromatics. They are therefore suitable as a basis for the recovery of polycyclic aromatics, e.g. naphthalene. Because of the high C/H ratio, they can also be used for the production of high-value carbon in the form of premium coke and as feedstock for the manufacture of carbon black (see Chapter 13).

Table 3.17 gives the composition of aromatics in a cycle stock oil (boiling range 340 to 470 °C).

Table 3.17: Distribution of aromatics in a cycle stock oil (%)

Hydrocarbon groups:	
Benzene and benzene-homologous	5.3
Naphthalene and naphthalene-homologous	13.2
Phenanthrene and anthracene	41.7
Pyrenes	18.9
Chrysene, benzanthracene and benzophenanthrene	12.6
Five-ring aromatics	8.3
Total	100.0

The aromatics content of the cat-cracker gasoline, which increases the octane rating, can also be influenced by the choice of catalyst. Table 3.18 shows the composition of a gasoline recovered by cat-cracking gas oil on a silica/alumina catalyst and a zeolite. With the zeolite, the formation of aromatics increases at the expense of olefin formation.

Table 3.18: Composition of a gasoline fraction from catalytic cracking on aluminum oxide and zeolite catalysts

Feedstock	Calif. straight-run gas oil		Calif. coker-gas oil		Gachsaran gas oil	
Catalyst	Zeolite	Al_2O_3/SiO_2	Zeolite	Al_2O_3/SiO_2	Zeolite	Al_2O_3/SiO_2
Paraffins (%)	21.0	8.7	21.8	12.0	31.9	21.2
Naphthenes (%)	19.3	10.4	13.4	9.5	14.3	15.7
Olefins (%)	14.6	43.7	19.0	42.8	16.3	30.2
Aromatics (%)	45.0	37.3	45.9	35.8	37.4	33.1

3.3.2.3 Catalytic reforming

Catalytic reforming, developed by Vladimir Haensel in 1949 at *Universal Oil Products (UOP)*, is the second most important process for converting hydrocarbons in petroleum refining, after catalytic cracking. Following the introduction of catalytic reforming in 1950 as an improvement to the traditional thermal reforming, only catalysts based on platinum and a nonferrous metal oxide were used at first. However, since 1967, bimetallic catalysts have been mainly employed, consisting of a second noble metal component, usually iridium or rhenium in addition to platinum. Bimetallic catalysts enable to operate the reformer with increased stability with regard to yield and reaction time.

The purpose of reforming, by which in the USA alone some 70 Mt annually of gasoline are produced, is to convert paraffinic hydrocarbons into aromatics or branched alkanes, in order to increase the octane rating. Since the compression ratios of Otto-engines rose between 1915 and the mid-1960's from 4 to 7-9 in Europe, and to 9-10 in the USA, the octane rating of gasoline had also to be steadily increased during the same period; catalytic reforming therefore gained considerable importance, especially in Europe and Japan.

Table 3.19 gives the mixed octane rating (RON) of typical gasoline hydrocarbons.

Table 3.19: Octane numbers of hydrocarbons

Hydrocarbon	Octane No.	Hydrocarbon	Octane No.
Paraffins			
Isobutane	122	n-Heptane	0
n-Pentane	62	2-Methylhexane	41
Isopentane	99	3-Methylhexane	56
n-Hexane	19	2,2-Dimethylpentane	89
2-Methylpentane	83	2,3-Dimethylpentane	87
3-Methylpentane	86	2,4-Dimethylpentane	77
2,3-Dimethylbutane	96	2,2,3-Trimethylbutane	113
		2,2,4-Trimethylpentane (Isooctane)	100
Naphthenes			
Methylcyclopentane	107	1,2-Dimethylcyclohexane	85
cis-1,3-Dimethylcyclopentane	98	cis-1,3-Dimethylcyclohexane	67
trans-1,3-Dimethylcyclopentane	91	1,1,3-Trimethylcyclohexane	85
Cyclohexane	110	cis-1,3,5-Trimethylcyclohexane	60
Methylcyclohexane	104		
Ethylcyclohexane	43	Isopropylcyclohexane	62
Aromatics			
Benzene	99	Ethylbenzene	124
Toluene	124	Isopropylbenzene	132
o-Xylene	120	1-Methyl-2-ethylbenzene	125
m-Xylene	145	1-Methyl-3-ethylbenzene	162
p-Xylene	146	1-Methyl-4-ethylbenzene	155
		1,2,4-Trimethylbenzene	171

Table 3.19 (continued)

Hydrocarbon	Octane No.	Hydrocarbon	Octane No.
Olefins			
1-Pentene	91	trans-3-Octene	73
1-Octene	29	4-Methyl-1-pentene	96

The processes occurring during reforming are:

1. Dehydrogenation of cycloalkanes to aromatics,
2. Isomerization of n-alkanes to branched alkanes,
3. Ring extension and dehydrogenation of alkylcyclopentanes to benzene derivatives,
4. Dehydrogenation and cyclization of alkanes to aromatics,
5. Hydrocracking of alkanes and cycloalkanes, together with hydrogenation of the cracker products to hydrocarbons with low molecular weight.

Figure 3.36 shows the equilibrium diagram for the n-hexane/benzene/hydrogen system at pressures of 17.5 and 35 bar with different hydrogen/hydrocarbon ratios.

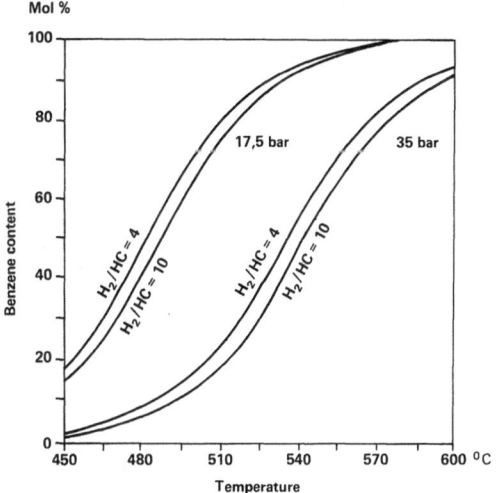

Figure 3.36: Equilibrium diagram for the n-hexane/benzene/ hydrogen system

Thermodynamic data for C_6 hydrocarbon reactions under catalytic reforming conditions are summarized in Table 3.20.

Table 3.20: Thermodynamic data for C_6 hydrocarbon reactions

Reaction	k (at 500 °C)	ΔH_R/kJ mol^{-1}
cyclohexane → benzene + 3 H_2	6×10^5	221
methylcyclopentane → cyclohexane	0.086	− 16.0
n-hexane → benzene + 4 H_2	0.78×10^5	266
n-hexane → 2-methylpentane	1.1	− 5.9
n-hexane → 3-methylpentane	0.76	− 4.6
n-hexane → 1-hexene + H_2	0.037	130

When converting paraffins and naphthenes by catalytic reforming into aromatics, hydrogen is released in relatively pure form. Reforming is therefore an important source of hydrogen for the refinery and the associated petrochemical plants.

The formation of aromatics is facilitated by a low hydrogen pressure and high temperature. Nevertheless, care should be taken since the hydrocarbons carbonize rapidly on the catalyst as the temperature rises.

Apart from the various types of catalyst, the reforming processes differ in the method of regenerating the catalyst.

Until the early 1970's, two methods were widely used, namely semi-regenerative and cyclical regenerative processes. The *UOP* semi-regenerative process (*UOP* platform process) is particularly common; the *Exxon* Powerforming process and the *Atlantic Richfield/Engelhard* Magnaforming process operate in a similar way.

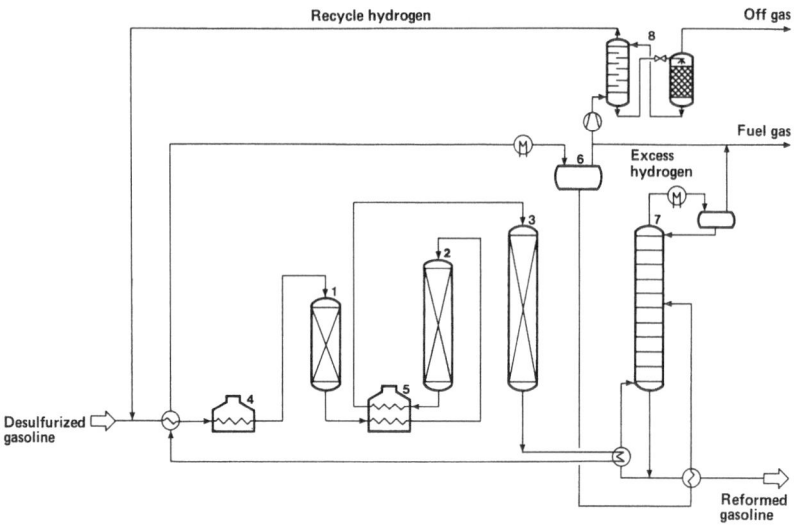

1 Reactor I; **2** Reactor II; **3** Reactor III; **4** and **5** Tubular furnaces; **6** Gas separator; **7** Stabilizer; **8** Gas purification

Figure 3.37: Flow diagram for three-stage reforming

The feedstock for the platformer is desulfurized straight run gasoline, which must have particularly low concentrations of metal impurities (e.g. Pb, As, Cu etc.). The desulfurized gasoline is heated with recirculated hydrogen and fed into the first reactor, where mainly the dehydrogenation of the naphthenes takes place. Before entering the second reactor, intermediate heating occurs in the tubular furnace; in the second reactor, slow isomerization of the C_5-naphthenes to cyclohexane homologs and their dehydrogenation takes place. In the third reactor, mild hydrocracking occurs, leading to a slight increase in temperature.

Reactor pressure is around 15 to 25 bar, reaction temperature around 510 to 540 °C; reforming is carried out with a molar hydrogen excess of 5-6:1. Pt/Re catalysts are used with a platinum content of ca. 0.5%; flow through the reactor bed occurs radially.

The catalyst is regenerated every 4 to 8 months by burning off the carbon.

Figure 3.37 shows the flow diagram for the semi-regenerative, three stage process.

Table 3.21 shows the composition of the feedstock gasoline and the reformate gasoline.

Table 3.21: Typical data for catalytic reforming feedstock and products

Characteristics of raw material	
Density g/cm^3	0.767
Octane no. (unleaded)	55.4
Paraffins vol%	38.1
Naphthenes vol%	42.6
Aromatics vol%	19.3
ASTM distillation °C	
Initial boiling point	99
10 %	115
50 %	134
90 %	157
Final boiling point	177
Product octane no. (unleaded)	95.8
Yields	
Hydrogen m^3/m^3	184
C_1-C_4 fraction, wt%	5.66
Benzene vol%	2.3
Toluene vol%	11.8
C_8 fraction vol%	21.2
C_8^+ aromatic fraction vol%	49.0
Total gasoline production	86.9

The octane rating can be increased from 35-60 to over 100 (unleaded).

In cyclical regeneration, a fourth reactor is used, which is switched into service during catalyst regeneration (swing reactor).

As a result of the necessary increase in severity of reforming when reforming gasoline fractions from conversion plants (FC-cracker, hydrocracker, coker), a

72 Base materials for aromatic chemicals

reforming system was developed in the mid 1970's with continuous catalyst regeneration. In the *UOP* process, the three reactors are stacked in a tower; the catalyst bed moves from top to bottom through all three reactors in plug flow.

Continuous reforming is characterized by high hydrogen yield and a high aromatic content in the gasoline; the octane number can be as high as 105.

Figure 3.38 shows a part of the continuous reformer of *Wintershall* (*BASF*), Lingen/West Germany with a capacity of ca. 650,000 tpa.

Figure 3.38: Catalytic reformer (*UOP*) of *Wintershall*, Lingen/West Germany

3.3.2.4 Hydrocracking

Hydrocracking processes date back to the work of Friedrich Bergius in 1910, which was originally aimed at converting heavy crude oil residues into distillates and later led to the development of coal hydrogenation. Whereas cat-cracking is aimed at splitting long-chain high-boiling hydrocarbons and catalytic reforming is used to convert hydrocarbons to produce high proportions of gasoline with a high aromatics content, hydrocracking is applied to break down high-molecular weight aliphatic and aromatic hydrocarbons. Modern hydrocracking processes operate with bi-functional catalysts, which on the one hand accelerate cracking and on the other, hydrogenation. Current hydrocracking plant capacity worldwide is over 60 Mtpa.

Special hydrocracking variants can be used as required:

1. to produce base oils for high-viscosity lubricants,
2. to convert vacuum gas oils into feedstocks for steam crackers to produce olefins,
3. to produce isobutenes for alkylation plants,
4. to produce liquid gas,
5. for selective cracking of n-alkanes using small-pored zeolites such as erionite and mordenite,
6. for hydro-dewaxing.

Ideally, in the hydrocracking of long alkanes, practically no small fragments, such as methane and ethane, are formed; the molar distribution of the cracking products is therefore symmetrical. Figure 3.39 shows the molar distribution of cracking products from 'ideal hydrocracking' of dodecane on a platinum/zeolite catalyst, together with the effect of increased temperature on the formation of smaller fragments.

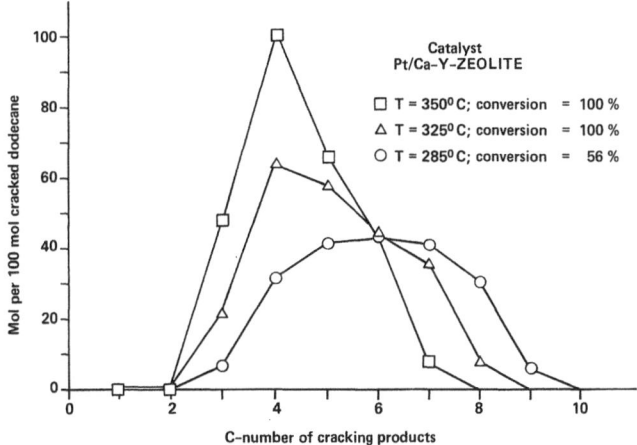

Figure 3.39: Molar distribution of cracking products from ideal hydrocracking of dodecane

In practical hydrocracking, a catalytic cracking reaction occurs in addition to the ideal hydrocracking, so that small fragments of the long-chain hydrocarbons are formed.

The aromatics pass through a series of intermediate stages and are finally converted into paraffins (Figure 3.40). Partial hydrogenation of the aromatics occurs relatively rapidly, whereas further hydrogenation to fully-hydrogenated cyclohexane-type compounds is comparatively slow.

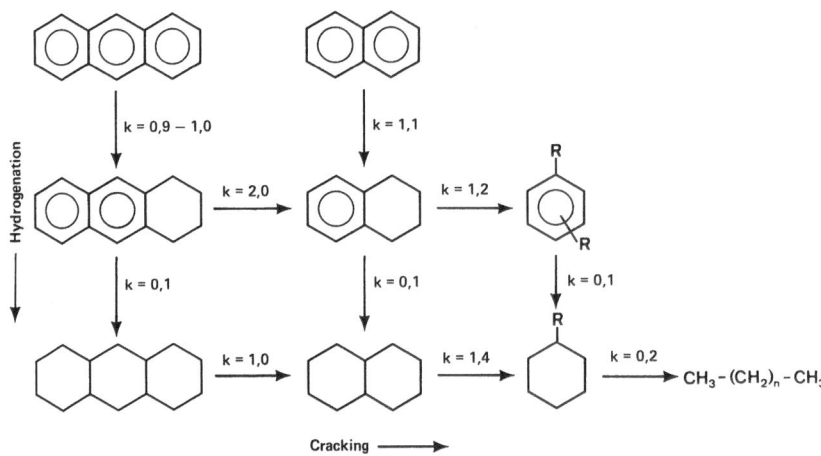

Figure 3.40: Relative reaction rates for hydrocracking of aromatics

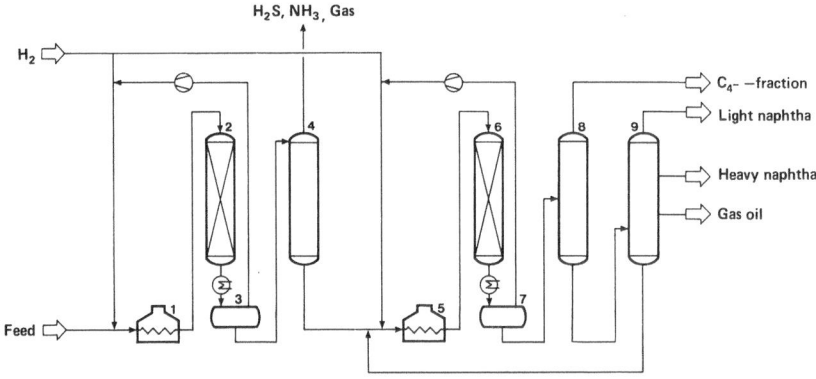

1 Tubular furnace; 2 1st-stage reactor; 3 Gas separator; 4 Stripping column; 5 Tubular furnace; 6 2nd-stage reactor; 7 Gas separator; 8 Stabilizer; 9 Fractionation column

Figure 3.41: Flow diagram for a two-stage hydrocracker

In two-stage hydrocracking, the first stage comprises the removal of sulfur and nitrogen compounds by hydrogenation to produce hydrogen sulfide and ammonia. This is effected over nitrogen- and sulfur-resistant catalysts based on cobalt/molybdenum oxides on alumina carriers. After the separation of the hydrogen sulfide and ammonia, the actual high-pressure cracking reaction is performed in the second reactor, e.g. on a highly-active bi-functional zeolite catalyst. The level of cracking conversion is between 40 and 70%. The reactor can, in principle, be operated in two modes, namely as a fixed-bed or fluidized-bed. Most hydrocrackers operate with fixed-bed catalysts (e.g. the *UOP* and *Chevron* Isomax process). Figure 3.41 shows the flow diagram for a two-stage hydrocracker.

The *Hydrocarbon Research* process, the H-Oil process, operates with a fluidized catalyst bed (Figure 3.42).

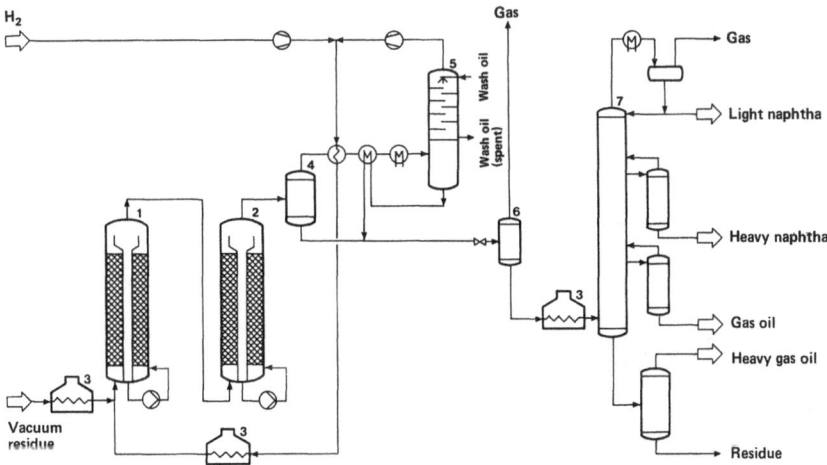

1 and **2** Reactors with fluidized catalyst bed; **3** Tubular furnace; **4** Hot separator; **5** Cold separator with gas scrubber; **6** Flash vessel; **7** Tubular furnace; **8** Fractionation column with side columns

Figure 3.42: Flow diagram for the H-Oil hydrocracker

Table 3.22 shows the yields obtained by hydrocracking a Texan vacuum residue by the H-Oil process. The aromatics content of the gasoline is relatively low in comparison with that of cat-cracker gasoline and the octane rating of the gasoline is correspondingly moderate. The octane number can be raised by adding octane boosters, or by further reforming.

The hydrocracker is distinguished from the cat cracker by increased process flexiblity. Yields of gasoline, kerosene and middle distillates can be varied within wide ranges from processing vacuum gas oil and other heavy distillate feedstocks. A disadvantage is the high investment and operating costs; to date, only a small number of hydrocrackers have been built in Europe.

Table 3.22: Yields and product specifications from hydrocracking a Texan vacuum residue using the H-Oil process

Crude material: Vacuum residue (West Texas)
Relative density 15/15 °C 0.981
Sulfur, wt% 2.95

	Desulfurization	Cracking	
		mild	severe
Hydrogen consumption, m^3/m^3	110	143	223
Yields			
H_2S and NH_3, wt%	2.6	2.3	2.5
C_1-C_3 fraction, wt%	1.0	3.7	4.8
C_4 fraction, vol%	0.7	2.2	2.9
C_5 fraction, vol%			
(Final boiling point 82 °C)	1.4	3.1	4.4
Boiling range			
82–177 °C, vol%	4.0	9.3	12.7
Boiling range			
177–343 °C, vol%	9.7	22.3	28.5
Heavy gasoil, vol%	30.9	34.0	35.2
Residue, vol%	57.0	32.0	20.0
Product specifications			
C_5 fraction			
(Final boiling point 82 °C)			
Relative density 15/15 °C	0.682	0.682	0.685
Research octane no.	73	..	71
Boiling range 82–177 °C			
Relative density 15/15 °C	0.751	0.755	0.759
Composition of hydrocarbons:			
Paraffins, vol%	52	..	45
Olefins, vol%	10	..	10
Naphthenes, vol%	28	..	35
Aromatics, vol%	10	..	10
Boiling range 177–343 °C			
Relative density 15/15 °C	0.850	0.860	0.865
Sulfur, wt%	0.1	0.2	0.2
Heavy gasoil			
Relative density 15/15 °C	0.916	0.922	0.928
Sulfur, wt%	0.3	0.7	0.7
Residue			
Relative density 15/15 °C	0.960	1.014	1.060
Sulfur, wt%	1.1	2.0	2.1

3.3.2.5 Thermal cracking of hydrocarbons

Thermal cracking to cleave hydrocarbons may be divided into high-temperature processes, operating at temperatures above 750 °C, and processes operating in the medium-temperature range, at around 450 to 550 °C. Olefin production predominates in high-temperature processes, with simultaneous formation of aromatics. Thermal cracking in the middle-temperature range is used to produce gasoline.

Base materials for aromatic chemicals   77

Again, partial aromatization of hydrocarbons occurs, but to a lesser extent than in high-temperature pyrolysis.

Table 3.23 shows the free bonding enthalpies of various hydrocarbon classes, which are causal for the high energy consumption of thermal cracking processes.

Table 3.23: Free bond enthalpies of various hydrocarbon types

		(kJ/mol)
C–C	aliphatic	297
C–C	aliphatic to aromatic nucleus	335
C–C	aromatic	401
C–H	aliphatic	389
C–H	aromatic	426
C=C	olefinic	513
C≡C	triple bond	686

3.3.2.5.1 Steam cracking to produce olefins

The most important petrochemical process is steam cracking of hydrocarbons to produce ethylene, propylene, C_4 olefins and higher unsaturated compounds. Ethylene is the most important basic organic chemical in terms of quantity; production worldwide was 40 Mt in 1985.

When hydrocarbon chains are heated to temperatures of around 800 °C, C-C and C-H bonds are split, producing unstable radicals. These radicals can react further to produce saturated or unsaturated molecules. Furthermore, diolefins which arise can cyclize with other olefins to yield aromatics so that aromatics are also formed by high-temperature pyrolysis. Apart from the bond-fission reaction, therefore, other reactions occur in steam cracking which lead to larger molecules. Polymerization of olefins can give high-molecular weight compounds, which are deposited on the heater wall and can eventually be converted into coke by loss of hydrogen. Undesirable polymerization and coke formation are facilitated by high concentrations of hydrocarbons, whereas the required cracking reaction is favored by low hydrocarbon concentration levels. For this reason, the hydrocarbons used as feedstock in steam cracking are diluted with steam. Table 3.24 shows typical feedstocks, together with the final product spectrum.

Table 3.24: Composition of pyrolysis products from steam cracking various feedstock materials (%)

Feed material	Ethane	Propane	n-Butane	i-Butane	Light naphtha	Full range naphtha	Raffinate after aromatics extraction	Kerosene	Light gas oil	Heavy gas oil	Heavy vacuum gas oil
Pyrolysis products											
Hydrogen	3.7	1.31	0.9	1.25	0.98	0.86	0.9	0.65	0.6	0.51	0.43
Methane	2.8	25.2	20.9	22.6	17.4	15.3	16.5	12.2	10.6	8.82	7.7
Acetylene	0.26	0.65	0.55	0.6	0.95	0.75	0.85	0.35	0.4	0.21	0.16
Ethylene	50.5	38.9	37.3	10.7	32.3	29.8	28.4	25.0	24.0	21.36	17.5

Table 3.24 (continued)

Feed material	Ethane	Propane	n-Butane	i-Butane	Light naphtha	Full range naphtha	Raffinate after aromatics extraction	Kerosene	Light gas oil	Heavy gas oil	Heavy vacuum gas oil
Ethane	40.0	3.7	4.5	0.6	3.95	3.75	3.9	3.7	3.1	4.54	2.8
Methylacetylene	0.03	0.6	0.8	3.0	1.25	1.15	1.1	0.75	1.05	0.19	0.45
Propylene	0.8	11.5	16.4	21.2	15.0	14.3	14.5	14.5	14.7	13.25	13.4
Propane	0.16	7.0	0.15	0.3	0.33	0.27	0.3	0.4	0.45	0.86	0.35
Butadiene-1,3	0.85	3.55	3.85	2.15	4.75	4.9	4.8	4.4	4.8	6.15	5.46
Butene	0.2	0.95	1.8	17.5	4.55	4.15	5.7	4.2	4.4	5.54	6.29
Butane	0.23	0.1	5.0	8.0	0.1	0.22	0.25	0.1	0.1	0.05	0.11
C_5 fraction	0.22	1.6	1.6	2.0	3.85	2.35	3.5	2.0	3.3	2.18	5.5
C_6-C_8 non-aromatics					2.02	2.05	3.8	1.55	1.5	2.51	4.7
Benzene	0.2	2.2	2.0	3.06	5.6	6.0	6.1	6.2	5.7	5.43	2.95
Toluene	0.05	0.4	0.9	1.4	1.65	4.6	2.35	2.9	3.0	3.27	2.7
C_8 fraction			0.35	0.4	0.72	1.65	1.8	1.2	1.2	0.74	1.3
Styrene					0.65	0.85	0.95	0.7	0.7	0.50	0.4
C_9 fraction (Final boiling point 200 °C)		1.0	1.3	3.25	0.65	3.1	0.9	3.1	2.3	3.08	3.0
Pyrolysis tar		1.34	1.7	1.99	3.3	3.95	3.4	16.1	18.1	20.81	24.8
Total	100	100	100	100	100	100	100	100	100	100	100

In the USA, and to some extent in Great Britain and Norway, ethane is the dominant feedstock for steam cracking. It is recovered from wet natural gas and gives high yields of ethylene, hydrogen and methane. From naphtha, the preferred feedstock in Europe and Japan, additional principal products are propylene, C_4 hydrocarbons and pyrolysis naphtha as well as highly aromatic pyrolysis tar.

In industrial steam cracking (Figure 3.43), naphtha passes through coiled tubes, first through the convection zone of the furnace where it is vaporized by the heat from hot flue gases. At the same time, approximately half a ton of steam is added for each ton of naphtha. The mixture, preheated to around 600 °C, then passes into the cracking tubes, which are arranged vertically in the furnace. These pyrolysis tubes (diameter ca. 60 to 100 mm) of a single-train furnace with an ethylene production capacity of around 40,000 tpa are between 60 and 80 m long. The heat flux in short contact time cracking furnaces is around 300,000 kJ/m^2 h.

The hydrocarbons are cleaved at a pyrolysis temperature from 800 to 900 °C in 0.1 to 0.5 sec. The cracked gases, which leave the furnace at a temperature of around 850 °C, must be cooled very rapidly ('quenched') to stop the reaction and to avoid undesirable secondary reactions; this is carried out in quenchers. The heat from the gases is recovered to produce high-pressure steam, which is used to drive the cracker-gas compressors, having been superheated to around 500 °C. The gases are further cooled by oil spraying and freed from high-boiling components by washing oil in a main column. The high-boiling compounds are removed from the column as pyrolysis tar. Water and gasoline are taken off in a quench column (water quench).

The cracked gases are compressed to 30 to 40 bar, freed from carbon dioxide and hydrogen sulfide and dried. Low temperature distillation at temperatures of −30 to 140 °C effects separation into ethylene, ethane, propylene, propane, mixtures of C_4- and C_5-hydrocarbons and light pyrolysis gasoline. This refining pro-

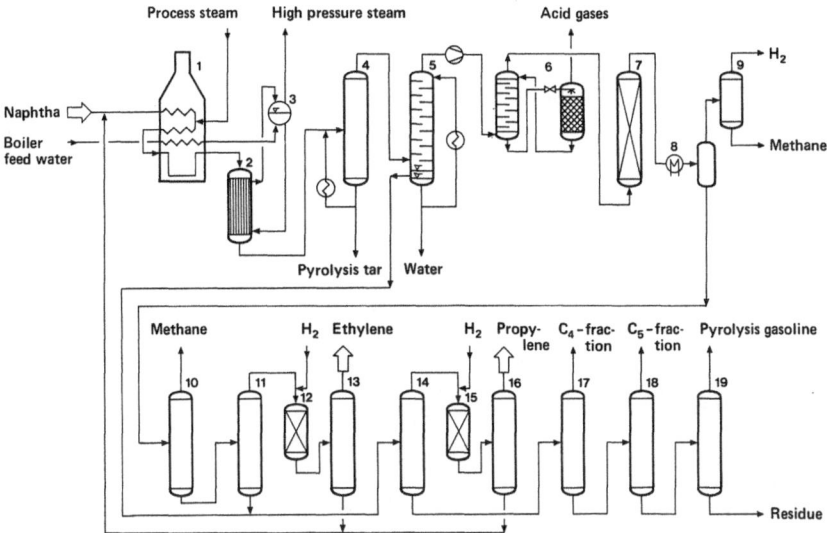

1 Cracking furnace; 2 Quenching cooler; 3 Steam generator; 4 Primary fractionation; 5 Quench column; 6 Gas purification; 7 Drier; 8 Low-temperature cooler; 9 Hydrogen/methane separation; 10 Demethanization column; 11 Deethanization column; 12 Acetylene hydrogenation; 13 Ethylene column; 14 Depropanization column; 15 Methylacetylene hydrogenation; 16 Propylene column; 17 Debutanization column; 18 Depentanization column; 19 Residue column

Figure 3.43: Flow diagram for steam cracking of naphtha

cess yields ethylene and propylene purities in excess of 99.9%. Hydrogen and methane are split at even lower temperatures (Joule-Thomson effect).

The upper cracking temperature is limited by the construction material of the cracker tubes. At final exit temperatures of 850 °C, temperatures in excess of 1,000 °C occur at the outer tube wall, at which steels tend to take up carbon from the reaction medium and thus become brittle. For this reason, centrifugally cast iron pipes containing over 25% chromium and 20 to 35% nickel have been introduced; these have a long service life, even at temperatures above 1,000 °C.

Figure 3.44 shows the series of cracking furnaces at the *Shell Nederland Chemie* ethylene plant at Moerdijk/Netherlands, which has a production capacity of 500,000 tpa of ethylene.

Figure 3.44: Cracking furnace at *Shell Nederland Chemie* ethylene plant, Moerdijk/Netherlands

The two most important sources of aromatics from steam cracking are pyrolysis gasoline and pyrolysis tar. The composition of typical pyrolysis gasoline from cracking is shown in Table 3.25.

Table 3.25: Characteristics of pyrolysis gasoline from different levels of cracking

Characteristics	Pyrolysis gasoline from	
	mild cracking	severe cracking
Density at 20 °C (g/ml)	0.790	0.835
Initial boiling point °C	40	44
50 vol% at °C	95	96
Final boiling point °C	200	182
Distillate to 70 °C, vol%	35	16
Distillate to 100 °C, vol%	65	55
Bromine number g/100 g	60	63
Diene content wt%	12	15
RON, unleaded	97	103.5
Composition		
Non-aromatics wt%	41.0	27.4
Benzene wt%	22.0	33.8
Toluene wt%	17.5	19.4
o-Xylene wt%	2.3	1.1
m-, p-Xylene and ethylbenzene	6.0	6.6
Styrene	3.2	5.3

Benzene, toluene and xylene can be recovered from pyrolysis gasoline as required. After hydrogenation, principally to saturate diolefins, pyrolysis gasoline is used as a motor fuel and is noted for its high octane rating.

The composition of pyrolysis tar exhibits some similarity to high temperature-coal tar, although its aromaticity is lower (Table 3.26).

Table 3.26: Composition of pyrolysis tar from naphtha steam cracking

Indene	1.4%
Naphthalene	12.8%
2-Methylnaphthalene	4.8%
1-Methylnaphthalene	4.0%
Biphenyl	2.2%
Dimethylnaphthalenes	2.6%
Fluorene	0.9%
Phenanthrene/Anthracene	1.3%
Fluoranthene	0.1%
Pyrene	0.2%
C/H ratio	1.1

Some of the constituents of pyrolysis tar, such as the unsaturated C_9 compounds and naphthalene, can be used as chemical raw materials. Because of its high C/H ratio, pyrolysis tar is also suitable for use as a raw material in the production of carbon black and as a feedstock for coking to premium coke (see Chapter 13.1.2).

3.3.2.5.2 Production of olefins from crude oil

In West Germany and Japan especially, processes have been developed to produce ethylene by cracking crude oil, to counter the dependence of ethylene production on the availability and price fluctuations of naphtha. Owing to the temperatures, which are needed in such processes, ethylene formation is accompanied by an aromatization, so that aromatics can also be obtained from these processes.

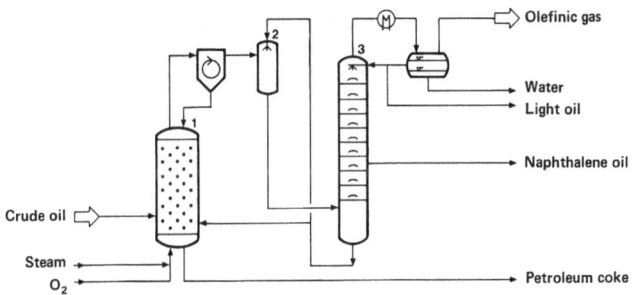

1 Fluidized-bed reactor; **2** Quencher; **3** Fractionation column

Figure 3.45: Flow diagram for crude oil cracking according to the *BASF* process

In Germany in the early 1960's, *BASF* and *Hoechst* carried out trials to crack crude oil, although these processes were rapidly replaced by steam cracking. The *BASF* method provides for crude-oil cracking in a fluidized bed of coke granules at 700 to 750 °C. In addition to a gas containing olefins, light oil, a naphthalene fraction and petroleum coke are recovered (Figure 3.45).

The light oil consists principally of benzene, toluene and xylenes; in terms of its composition, it resembles coke-oven benzole. The naphthalene fraction, recovered as a 4% yield of the crude oil, contains 42% naphthalene along with mainly methylnaphthalenes.

Table 3.27: Quantities produced by the *BASF* fluidized-bed cracker process using 1,000 kg of crude oil and 400 kg (280 Nm^3) of oxygen

Product	kg/t crude oil
Ethylene	230
Propylene	125
C$_4$ fraction	60
Light oil (45 % benzene)	140
Naphthalene oil (42 % naphthalene)	40
Hydrogen	8
Methane	120
Ethane	45
Propane	7
Acetylene	1
Carbon dioxide	395
Carbon monoxide	100
Coke	45
Water	64
Loss	20

The *Kureha/Union Carbide* process operates at a higher temperature than the *BASF* method (Figure 3.46). It uses steam as a heat carrier, at a temperature of 2,000 °C. Preheated crude oil is fed into the superheated steam in an adiabatic reactor, from which 30% ethylene is recovered. Reaction time is only 15 to 20 msec. This process also produces an aromatics-rich oil of relatively high naphthalene content which boils above the pyrolysis gasoline. The aromatic pyrolysis tar concurrently produced can be used to manufacture medium-modulus carbon fibers.

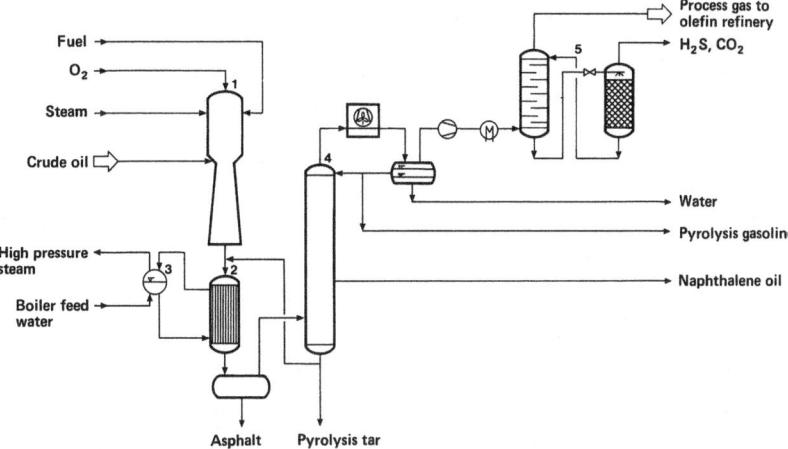

1 Reactor; **2** Quenching cooler; **3** Steam generator; **4** Fractionation column; **5** Gas purification

Figure 3.46: *Kureha/Union Carbide* process for crude oil cracking

3.3.2.5.3 Other thermal-cracking processes

In crude oil refining, the visbreaker process, the delayed coking process (see Chapter 13.1.2) and thermal cracking are used in the middle-temperature range to convert heavy petroleum residues into lighter gasoline fractions and middle distillates. The aromaticity of the fractions recovered, however, is relatively low.

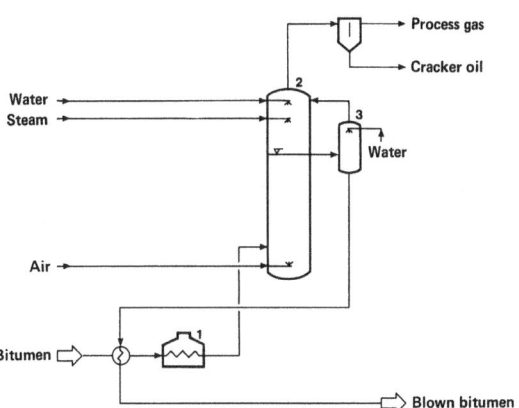

1 Tubular furnace; **2** Blowing tower; **3** Hot bitumen receiver

Figure 3.47: Flow diagram for the production of blown bitumen

The production of blown bitumen from straight-run bitumen also involves thermal cracking and aromatization; although the temperature for blowing bitumen with air is only 280 to 300 °C, the use of air leads to an increase in aromaticity, from around 0.35 to 0.45 after blowing, by dehydrogenation of naphthenes. Blown bitumen is not used for the recovery of aromatics, however, because of its relatively low aromaticity. Figure 3.47 (see page 83) shows the flow diagram for a continuous blowing plant.

The feedstock for the visbreaker process is generally an atmospheric residue or a vacuum residue from crude oil distillation. The residue is heated to 450 to 480 °C in a tubular furnace, which leads to thermal cracking reactions. The reaction is carried out under a pressure of 5 to 50 bar. In order to terminate the cracking reaction, the cracking products leaving the furnace are quenched with cold gas oil. The reaction product is then broken down into the required distillates and a heavy fuel-oil residue in the fractionation column. The alkylaromatics which are formed during aromatization, characterized by a high solvent power, bring about a reduction in the viscosity of the cracked bitumen in comparison with the original raw material. The flow diagram of a visbreaker and typical yields from a feedstock of a Louisiana residue are shown in Figure 3.48 and Table 3.28.

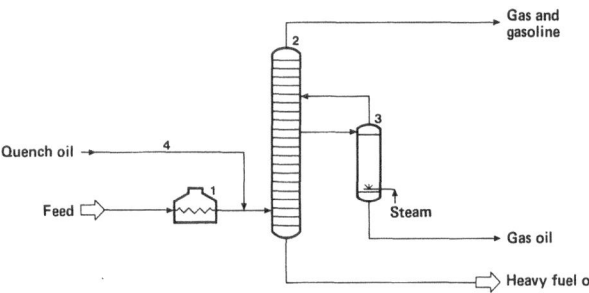

1 Cracking tubular furnace; **2** Fractionation column; **3** Side column; **4** Gas oil quenching

Figure 3.48: Flow diagram for a visbreaker

Table 3.28: Characteristic properties of feedstock and products from visbreaking a South Louisiana residue

Raw material: Louisiana crude oil residue	
Relative density 15/15 °C	0.987
Conradson carbon wt%	10.6
Viscosity at 98.9 °C, mm^2/s	107
Sulfur wt%	0.61
Operating conditions:	
Cracking temperature, °C	480
Pressure, bar	7
Yield:	
C_1-C_3 fraction, wt%	0.6
C_4 fraction, vol%	1.0

Table 3.28 (continued)

	Gasoline	Gas oil	Residue
Gasoline (Final boiling point 220 °C), vol%			6.2
Gas oil (220–340 °C), vol%			6.3
Residue, vol%			88.4
Product specifications:			
Relative density, 15/15 °C	0.748	0.851	0.990
Sulfur, wt%	0.26	0.33	0.58
Conradson carbon, wt%	0.0	0.01	15.0
RON (unleaded)	66.8	..	..
Viscosity at 98.9 °C, mm^2/s	..	..	34

Wide variations in pressure, temperature and residence time, as well as in feedstock material, are possible with visbreaking by thermal cracking.

The classic thermal cracking of heavy gas oil fractions occurs at higher temperatures. It is operated at 500 °C and 50 bar. Figure 3.49 shows the flow diagram of the thermal-cracking process, originally designed by Petroleum Carbon Dubbs (*UOP*) for cracking wax distillates.

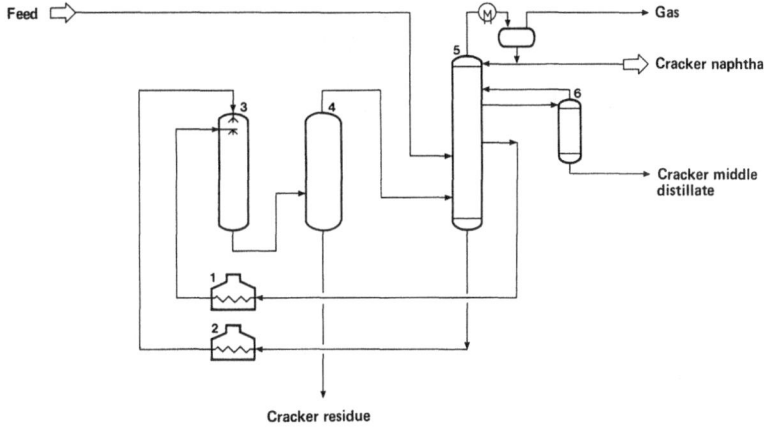

1 Light-oil tubular furnace; **2** Heavy-oil tubular furnace; **3** Reaction chamber; **4** Flash chamber; **5** Distillation column; **6** Side column

Figure 3.49: Flow diagram for the Dubbs process

The cracker tar which is generated by this process is suitable as feedstock for production of premium coke in the delayed coker; the gasoline quality is somewhat better than that of coker naphtha.

3.4 Production of aromatic hydrocarbons with zeolites

3.4.1 Aromatics from alcohols

Methanol is the principal feedstock for the production of aromatic gasolines with zeolite catalysts.

Zeolites are crystalline alumino-silicates, which contain a three-dimensional network of SiO_4- and AlO_4-tetrahedral voids of a specific size. These voids have different diameters, depending on the structure. Figure 3.50 shows zeolite structures with voids of varying diameters.

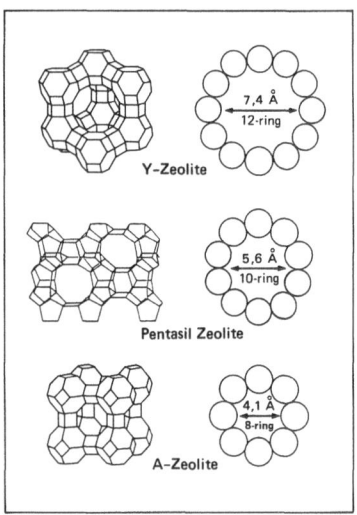

Figure 3.50: Zeolite structures with various diameters

A distinction is made between wide, medium and narrow-pored zeolites. In wide-pored Y-zeolites, the structures leading to the large voids are formed by 12 SiO_4 or AlO_4 tetrahedra. These apertures have a diameter of 7.4 Å; the largest voids have a diameter of around 13 Å.

In the narrow-pored A-zeolites, 8 tetrahedra produce an aperture of 4.1 Å. Between these ranges lie the pentasil zeolites, with a slightly elliptical pore section. These pores are formed from 10 tetrahedra; the pore size is approx. 5.6 Å.

Zeolites are inorganic ion-exchangers; their catalytic activity is basically a result of acid groups on the intracrystalline surface. The zeolites per se can be used as catalysts, but also serve as carriers for active components. The composition of a typical molecular-sieve zeolite is shown in Table 3.29.

Table 3.29: Composition of a mordenite-zeolite catalyst (wf, as %)

SiO_2	74–81
Al_2O_3	10–13
Fe_2O_3	0.5–1.0
TiO_2	0.6
CaO	0.6
MgO	0.6
Na_2O	5–8.5
K_2O	0.6

The specific surface of the zeolite catalysts varies between 800 and 1,000 m²/g.

Against the background to reduce the dependence of motor fuel on crude oil, *Mobil Oil* developed the Methanol to Gasoline (MTG) process to convert methanol to gasoline. Since methanol can be produced not only from crude oil but also from natural gas and coal by currently available technology, the MTG process is not dependent on crude oil as a raw material.

Figure 3.51 shows the flow diagram for the two-stage *Mobil Oil* fixed-bed process.

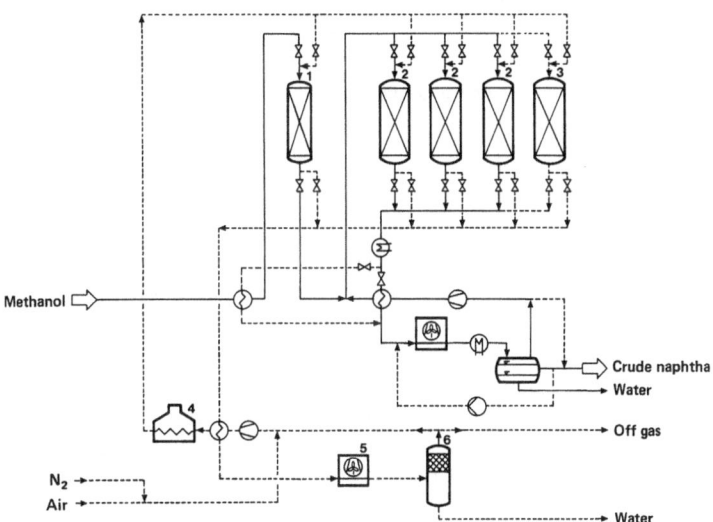

1 Dimethyl ether reactor; 2 Conversion reactors; 3 Stand-by reactor; 4 Start-up and regeneration gas furnace; 5 Regeneration gas cooler; 6 Water separator

Figure 3.51: Flow diagram for the *Mobil Oil* fixed-bed process to convert methanol into gasoline

The reaction is carried out at a pressure of 20 bar and a temperature of 380 °C. The process produces 44% of hydrocarbons along with 56% of water. In terms of its composition and characteristics, such as octane rating, boiling range and other specifications, the gasoline corresponds to currently-used motor gasolines.

Figure 3.52 depicts the dependence of methanol conversion and product composition on contact time; as contact time increases, (reciprocal of the catalyst load), the yield of gasoline and aromatics increases. (In English terminology, the catalyst load is defined as (LHSV) 'liquid hourly space velocity').

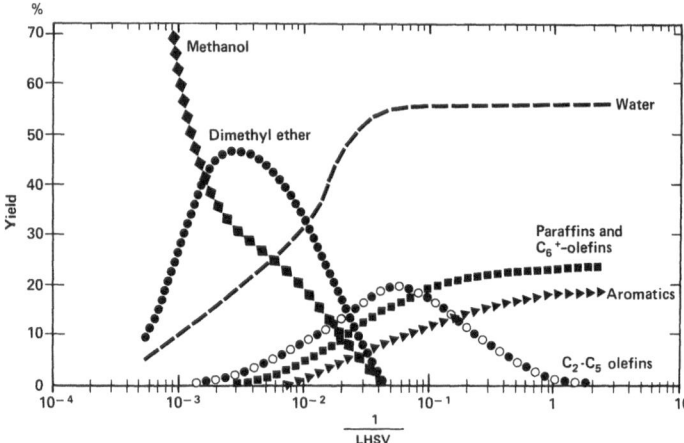

Figure 3.52: Dependence of methanol conversion and product composition on contact time in the *Mobil Oil* process

Pentasil zeolites are used as catalysts, since their shape-selectivity leads to the formation of hydrocarbons with a carbon numer below 11; in terms of boiling point, this corresponds to conventional gasoline.

The first part of a plant, which uses the MTG process and has a planned final capacity of 800,000 tpa of gasoline with an octane rating of 92 to 94, is already in operation in New Zealand; the methanol is produced from natural gas.

Until 1986, *UK-Wesseling,* in collaboration with *Mobil Oil* and *Uhde* operated a pilot plant with a capacity of 25 t/d in West Germany, to optimize the fluidized-bed process.

3.4.2 Aromatics from short-chain alkanes

UOP/BP have recently developed a process related to reforming to produce aromatics-rich gasoline from aliphatic hydrocarbons. In this 'Cyclar' process, the low-molecular weight alkanes, propane and butane, are used instead of naphtha. The flow diagram for the Cyclar process, which operates with zeolite catalysts, is shown in Figure 3.53.

Base materials for aromatic chemicals   89

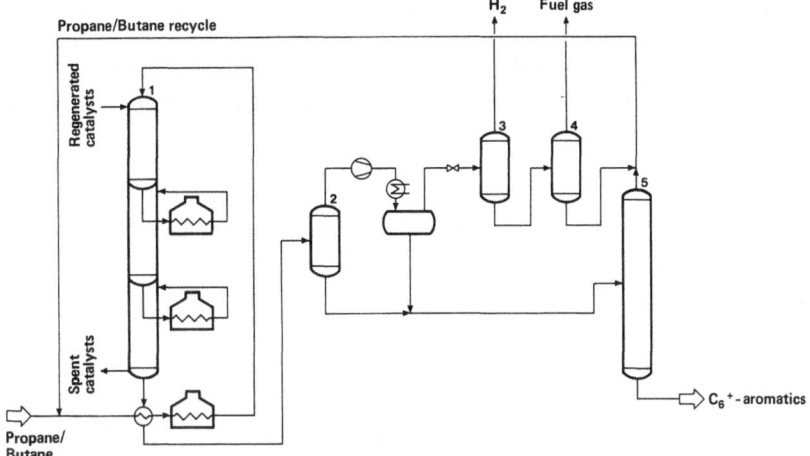

1 Fixed-bed reactors; **2** Separator for light-boiling compounds; **3** Hydrogen recovery; **4** Propane/butane recovery; **5** Stripping column

Figure 3.53: Flow diagram of the Cyclar process for aromatization of propane/butane

A yield of 62 to 67% of aromatic gasoline (dependent on the propane/butane ratio) can be produced by cyclization. The benzene and toluene content of gasoline produced by the Cyclar process is higher than the level of xylenes and C_9-aromatics from conventional platforming processes (Figure 3.54). The construction of a 500 bblpd plant at Grangemouth/UK has been announced.

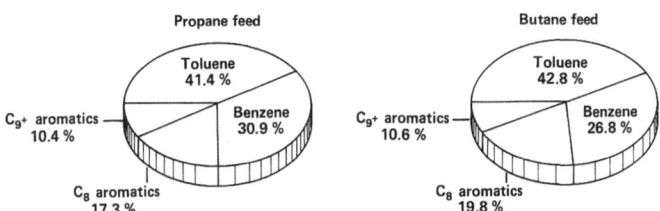

Figure 3.54: Yield structure of the Cyclar process for propane and butane feed

3.5 Renewable raw materials

Apart from fossil raw materials, atmospheric carbon dioxide provides a further source of chemical carbon, via photosynthesis, i.e. the formation of carbohydrates from carbon dioxide and water using energy from the sun. Photosynthesis produces around 2×10^{11} t of biomass annually, with an energy content of 3×10^{21} Joule.

Some 25% of this arises in swamps, meadows and tundra, over 25% in forests and 10% in cultivated ground. Around one third of the biomass is produced in the sea. Table 3.30 compares the chemical composition of some renewable raw materials with corresponding values for hard coal. The greatest differences are in oxygen content, which decreases as coalification increases.

Table 3.30: Elementary composition of renewable raw materials in comparison with coal

Elementary analysis (%)	Cellulose	Pinewood	Seaweed	Hard coal
Carbon	44.44	51.8	27.65	69.0
Hydrogen	6.22	6.3	3.73	5.4
Oxygen	49.34	41.3	28.16	14.3
Nitrogen		0.1	1.22	1.6
Sulfur		0.0	0.34	1.0

Only around 1% of the biomass is used as foodstuffs; a further 2% is used as a raw material for energy and fiber production. The vast majority of the biomass rots away unused; it is generally decomposed by bacteria into carbon dioxide and water and returned to the carbon cycle.

Chemical exploitation of renewable raw materials has a long tradition. Before the utilization of the fossil carbon resources coal and petroleum, they were the exclusive source of organic raw materials. Even today, a wide range of chemicals is produced on this basis, e.g. natural rubber, cellulose, fatty acids, ethanol and essential oils, citric acid, enzymes and antibiotics. In terms of quantity, around 8% of organic chemicals are recovered from renewable raw materials. Of the 20 Mt of renewable raw materials used annually, oils and fats have the largest share, amounting to some 40% (Figure 3.55).

This is followed by plant secretions and extracts at 25%, wood derivatives at 20% and carbohydrates at 15%.

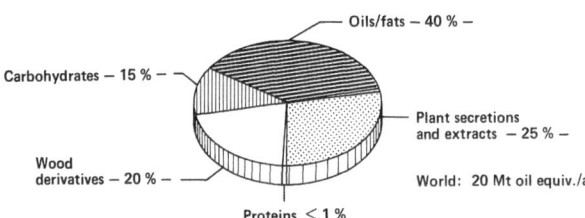

Figure 3.55: Fields of applications of renewable chemical raw materials (1985)

The most important process for transforming biomass into chemicals is pyrolysis, used e.g. with wood, to produce acetic acid and charcoal. Other conversion processes which are also used include partial oxidation of waste containing wood or cellulose to produce synthesis gas, hydrolysis of wood to sugar, together with fermentation, by which plant materials containing sugar or starch are transformed microbially into ethanol and other products such as glycerine, acetone, fumaric acid and methane.

The most important renewable raw material is wood. World production of wood is around 1.5 billion tpa. Some 15% of wood production is used in the paper industry, where the wood is broken down into its constituents - cellulose, hemicellulose and lignin. Since wood consists of around 50% cellulose, cellulose is the most common organic substance on Earth.

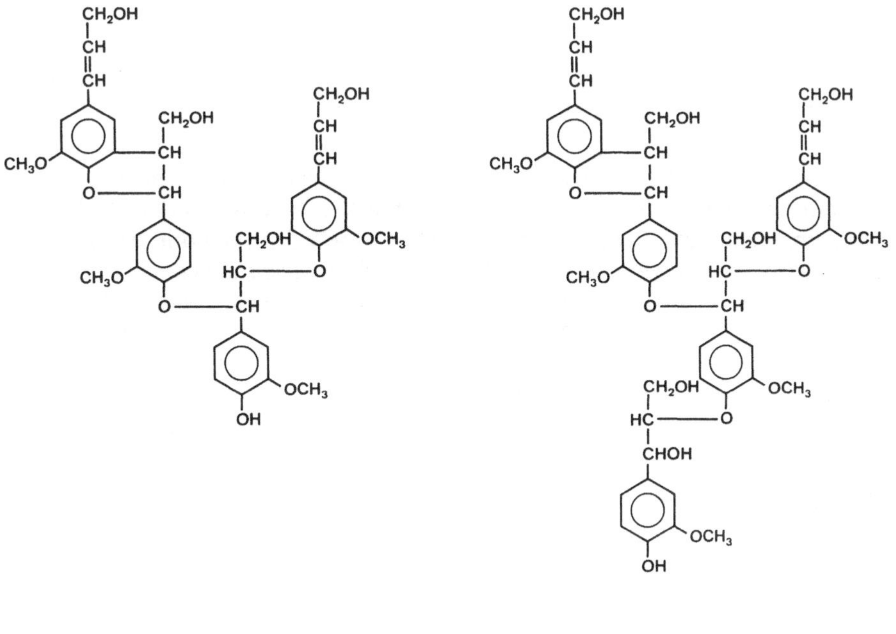

Tetralignol Pentalignol

Figure 3.56: Structural model for lignin

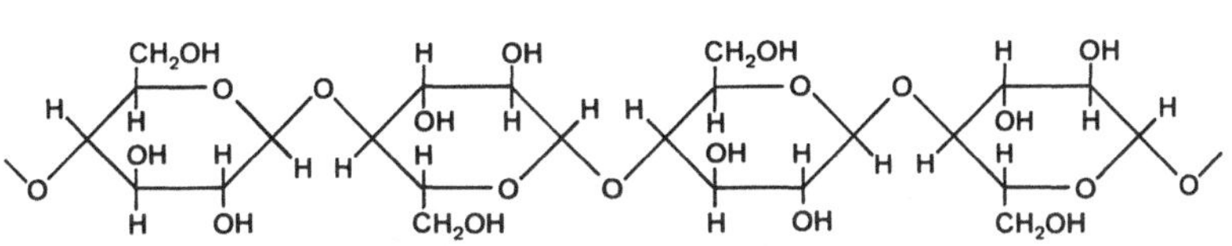

Figure 3.57: Structural model for cellulose

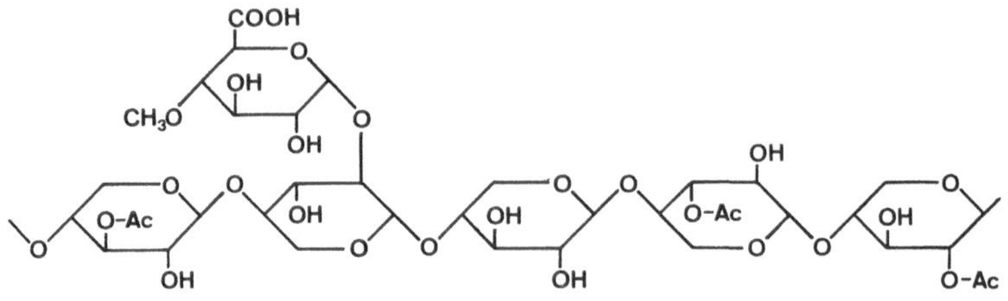

Figure 3.58: Structural model for hemicellulose

While cellulose is mainly used to produce aliphatic products, such as cellulose chemical fibers and cellulose ethers, hemicellulose and lignin are used to produce chemicals exhibiting aromatic character.

3.5.1 Furfural

The production of aromatic chemicals from renewable materials is mainly concentrated on the exploitation of wood products together with recovery of furfural by converting pentosans. Pentosans are almost as widely distributed in nature as cellulose, although in lower quantities. Pentosan-rich materials are commonly used in the production of furfural, particularly waste from corncobs, grain chaff and cotton pod cases.

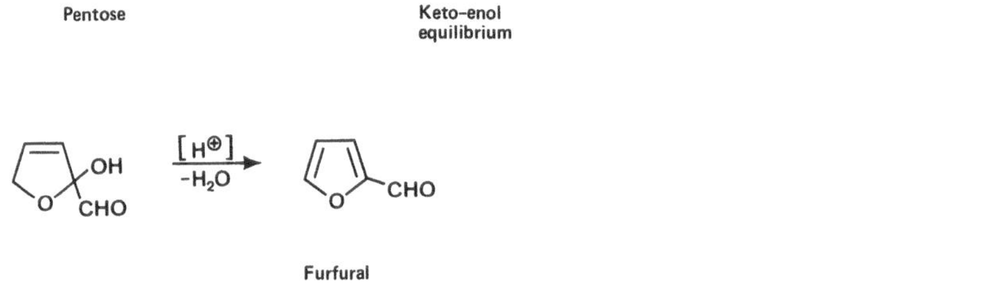

In furfural production, the collected feedstock is treated with sulfuric acid and superheated steam in rotating drums under pressure. Furfural and water evaporate and are subsequently separated by distillation (Figure 3.59).

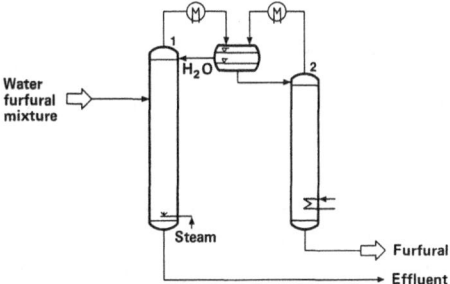

1 Stripping column; **2** Dewatering column

Figure 3.59: Flow diagram for furfural refining

Current world production of furfural is around 150,000 tpa; the largest producer is *Quaker Oats Chem.* (USA). The main areas of application are the production of the solvents furfuryl alcohol and tetrahydrofurfuryl alcohol. Furfural itself is used as a solvent, especially for separating unsaturated and saturated compounds in petroleum refining.

Furfuryl alcohol is the base material for ranitidine, one of the best selling drugs (see Chapter 14.1.1).

Special resins are produced by condensing furfuryl alcohol and furfural with urea and formaldehyde (e.g. foundry resins).

3.5.2 Lignin

The most important class of compounds among renewable raw materials with a marked aromatic character is lignin. This is produced during the processing of wood to recover pulp for the paper industry.

Table 3.31 shows the composition of some types of wood.

Table 3.31: Composition of various species of wood

Wood species	Cellulose	Hemi-cellulose	Lignin	Pentosan
Pine (Pinus radiata)	45.5	16.3	26.8	9.3
Fir (Abies balsamea)	49.4	15.4	27.7	7.0
Cedar (Thuja plicata)	47.5	14.7	32.5	8.1
Walnut (Carya tomentosa)	56.2	–	23.4	18.8
Oak (Quercus spec.)	40.5	23.3	22.2	17.5
Birch (Betula papyrifera)	48.5	–	19.4	25.1
Teak (Tectona grandis)	39.1	–	13.0	13.0

In wood processing, in addition to mechanical methods, sulfite and sulfate processes are used. In the sulfite process, used for around 10% of wood pulping, gaseous SO_2 reacts with calcium carbonate to produce calcium hydrogen sulfite. Wood chips are boiled in this solution at a pH-value of 3 to 5 at 150 to 170 °C. This causes the lignin and hemicellulose to dissolve, while the cellulose remains as undissolved pulp. In the sulfate process (Kraft process), by which some 75% of wood is processed, the wood is treated with sodium hydroxide/sodium sulfate to separate cellulose from lignin and hemicellulose.

Lignin, which is produced wordwide in quantities of around 50 Mtpa, has only limited application at the present time as an aromatic polymer, in the form of lignin sulfonates, which are used as surfactants, e.g. in tertiary crude-oil recovery and as a concrete plasticizer. Lignin can be broken down into its monomeric components by pyrolysis or hydrocracking.

Figure 3.60 shows the hydrocracking process developed by *Hydrocarbon Research* to produce phenols from lignin.

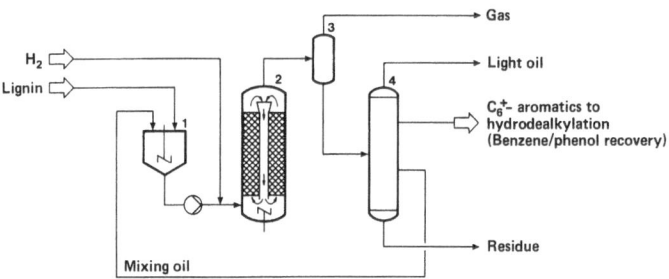

1 Mixing vessel; **2** Hydrogenation reactor; **3** Gas separator; **4** Distillation column

Figure 3.60: Flow diagram for recovering phenol from lignin (*HRI*)

The phenol yield from this process, developed along the lines of the H-Coal and H-Oil processes, is around 50% (Table 3.32). This method could gain industrial importance if prices of petrochemicals rise considerably.

Table 3.32: Pattern of yields from lignin hydrogenation using the *HRI* process

Product	Recovery in relation to organic lignin substance
H_2O	17.9 %
C_1–C_5	25.2 %
C_6–150 °C neutral oils	8.3 %
150–240 °C neutral oils	5.7 %
150–240 °C phenols	37.5 %
240–260 °C	8.7 %
Residue	2.4 %
Hydrogen consumption	−5.7 %
	100.0 %

3.5.3 Vanillin

Next to furfural, vanillin is one of the most important aromatic compounds based on renewable raw materials. Current world production of vanillin is around 8,000 tpa.

There are three industrially proven methods for recovering vanillin, the first two being based on renewable raw materials.

1. Extraction from the waste sulfite lye from pulp production.
2. Oxidation of isoeugenol, which is itself obtainable from clove oil.
3. Synthetic production from guaiacol and glyoxylic acid.

To recover vanillin by the Howard Smith method from waste sulfite lye, the lye is mixed in a pressure reactor with equal amounts of nitrobenzene (oxidizing agent) and sodium hydroxide (Figure 3.61). The mixture is then heated to 200 °C and boiled for 1 to 2 hours. After depressurizing, the reaction product is 'sprung' with CO_2. In general, only a small quantity of sulfuric acid is added for full acidification. The liberated vanillin is extracted from the mixture in an extractor with benzene or butanol; afterwards the solvent is distilled off.

The vanillin is purified by steam distillation followed by crystallization.

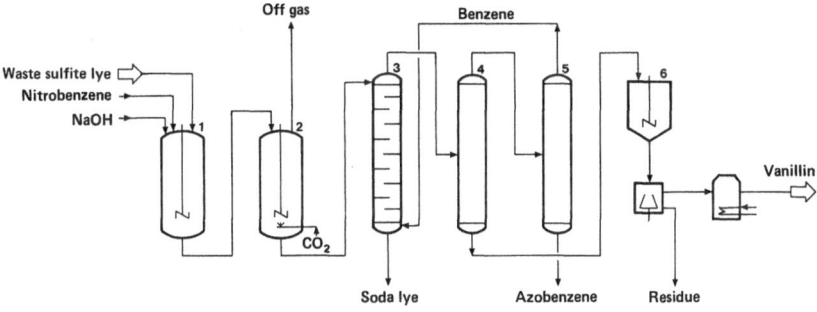

1 Pressure reactor; 2 Neutralization vessel; 3 Extraction column; 4 Distillation column;
5 Solvent recovery; 6 Crystallization vessel

Figure 3.61: Flow diagram for the production of vanillin from sulfite lye

Ontario Paper has developed a process which uses $Ca(OH)_2/ Na_2CO_3$, to avoid the necessary recovery of sodium hydroxide. The calcium carbonate which arises during the process is calcined to CaO and can be returned to the cycle.

Oxidation of isoeugenol, which is obtained from the eugenol of clove oil, is no longer practiced on a large scale for the production of vanillin.

Synthetic production of vanillin from guaiacol and glyoxylic acid takes place via the intermediate hydroxyacetic acid; this method is presently by far the most important source for vanillin.

Vanillin is mainly used in the manufacture of perfumes and as a raw material for pharmaceuticals (e.g. methyldopa).

3.6 Summary review of processes for the production of aromatics

The main methods of producing aromatics involve thermal and catalytic processes. In thermal processes, the hydrocarbons are treated in a temperature range between 400 and 2,000 °C, to bring about rearrangement of the carbon skeleton.

Naturally, the degree of aromatization of the rearranged products depends greatly on the structure of the hydrocarbon feedstock. Nevertheless, it is possible to deduce a qualitative relationship (Figure 3.62) when cracking severity, which is a function of reaction temperature (T) and residence time (τ) of the hydrocarbon mixture in the thermal reaction, is plotted versus the aromaticity of the liquids produced.

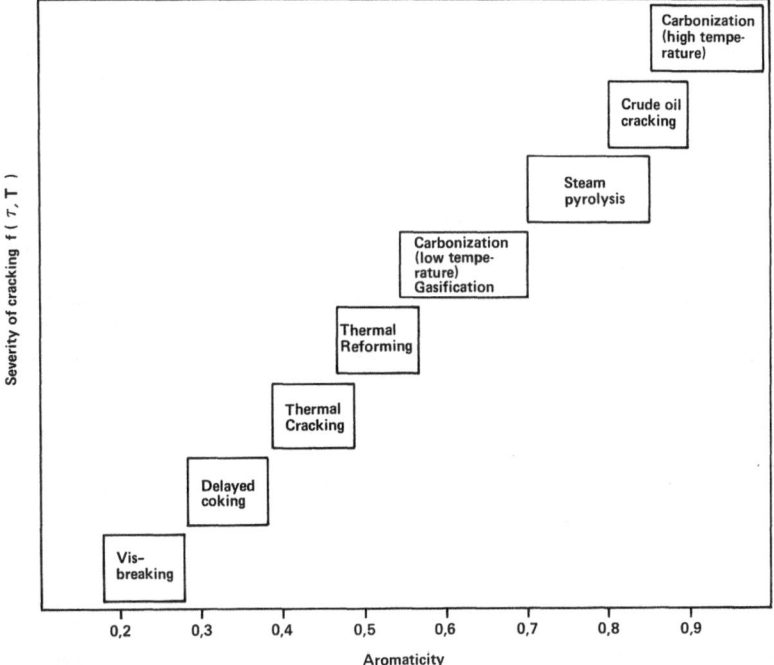

Figure 3.62: Formation of aromatics during thermal transformation of hydrocarbons

The mildest thermal cracking process is visbreaking, in which aromatization is relatively low. An intermediate position in aromatization is occupied by thermal cracking, thermal reforming (previously used to produce gasoline), together with low-temperature carbonization and gasification of coal. Aromatization increases with processes which have greater cracking severity, such as steam cracking, crude oil cracking and high-temperature carbonization of coal.

Among the catalytic processes, a distinction should be made between hydrogenation processes on the one hand, and reforming and catalytic cracking processes on the other.

Hydrogenation processes, whether they are applied to coal or petroleum residues, aim at increasing the hydrogen content, particularly of the distillates; the aromaticity of the distillable part of the product therefore decreases.

In reforming and catalytic cracking processes, in contrast, there is an increase in aromatization, since isomerization reactions occur under the reaction conditions and cycloaliphatic hydrocarbons are converted into aromatics.

Base materials for aromatic chemicals

Because of these differences, it is less expedient to derive a dependence of product aromaticity on reaction conditions in catalytic processes than in purely thermal processes for producing aromatics.

Table 3.33 reproduces the main processing data for thermal and catalytic processes for converting hydrocarbons and producing aromatics.

Table 3.33: Review of major processes and typical process conditions for converting hydrocarbons and producing and transforming aromatics

Process	Objective of process	Process conditions				Other characteristics
		Pressure (bar)	Temperature (°C)	Catalyst	Reaction components	
1. Thermal processes:						
Visbreaking	Reducing viscosity of vacuum residues, light conversion	5–18	450–480	–	–	simple conversion method; low investment costs
Delayed Coking	Producing gasoline and middle distillates	5	480	–	–	petroleum coke by-product
Thermal cracking	Producing gasoline and middle distillates from heavy gas oil	50	500	–	–	only used in isolated cases
Thermal reforming	Increasing octane rating of gasoline	40	520	–	–	obsolete; replaced by catalytic reforming
Steam cracking	Olefins production	atmos.	850–900	–	H_2O	co-production of aromatic-rich pyrolysis gasoline and pyrolysis tar
High-temperature carbonization	Production of metallurgical coke	atmos.	1200	–	–	co-production of the aromatic feedstocks tar and benzole
Crude oil pyrolysis (Kureha/UCC)	Production of acetylene and olefins	atmos.	2000	–	H_2O	co-production of naphthalene-rich gas oil
Bitumen oxidation	Increasing plasticity of bitumen	atmos.	280–300	–	O_2	cont. process; also used for pitch blowing
Coal gasification	Production of synthesis gas	20–30	max. 1000	–	O_2, H_2O	Aromatics arise only with counter-current flow of coal and air/steam reactants
2. Catalytic processes:						
Hydrocracking	Conversion of heavy oil distillates into gasoline and middle distillate	70–150	350–450	Mo, W	H_2	very flexible conversion process; high investment cost, originally developed for coal hydrogenation
Reforming	Increasing octane rating of straight-run gasoline	20	500	Pt, Re, Ir	–	most important source of aromatics in USA; source of hydrogen
Catalytic cracking	Converting heavy-oil distillates into gasoline and middle distillates	0.5–1	500	Zeolite	–	important for gasoline production, esp. in the USA

4 Production of benzene, toluene and xylenes

Since the BTX-aromatics benzene, toluene and xylene are co-generated, alongside other aromatics and non-aromatics during coal conversion and petroleum refining, their concurrent recovery is described jointly in this chapter.

4.1 History

Benzene was discovered in 1825 by Michael Faraday during the pyrolysis of whale oil. Although benzene aromatics were being used as solvents for rubber, e.g. in the production of waterproof Mackintosh raincoats, as early as the 19th century, the development of the industrial use of benzene began when Michael Faraday treated benzene with nitric acid and produced nitrobenzene, a base material for aniline-derived dyestuffs such as Perkin's mauveine. A further milestone in the history of industrial applications of benzene was the production of phenol, which was used during World War I as a feedstock for picric acid, an explosive. However, picric acid is now of little commercial importance.

Of far-reaching and more lasting importance for the industrial use of benzene was the commencement of the production of styrene by *IG Farbenindustrie* in 1929, and the hydrogenation of benzene to cyclohexane as a feedstock for nylon production, after the discovery of nylon synthesis from hexamethylenediamine and adipic acid by Wallace H. Carothers of *Du Pont* in 1935.

Like benzene, toluene was also discovered in the pyrolysis of a renewable raw material, by Pierre J. Pelletier and Philippe Walter in 1837, during investigations into the by-products from the manufacture of illumination gas from pine resin. The name is derived from the small harbor-town of Tolu in Columbia, where Tolu balsam is produced. Henri Saint-Claire Deville was the first to produce toluene by destructive distillation of this renewable raw material in 1838.

Although Charles B. Mansfield, a disciple of August W. von Hofmann, detected the presence of toluene in coal tar in 1849, it found only limited application at first as a chemical raw material. However, this changed in World War I, when toluene was used in the production of the explosive trinitrotoluene (TNT). Up to the turn of the century, coal tar and coke-oven benzole remained the only source of toluene, but during the World War I it was also produced by fractional distillation of aromatic crude oils from the Far East (e.g. Borneo, Java).

Xylenes were also first discovered during pyrolysis of renewable raw materials, namely by Auguste Cahours in crude wood spirit, in 1850; the name was therefore taken from the Greek word for wood. In 1855, H. Ritthausen and A. H. Church found xylenes in coal tar.

The importance of xylenes as major industrial chemicals began with the rise of the plastics industry in the 1920's and 30's, when polyesters were developed from p-xylene; from the mid-1950's, o-xylene also gained prominence as a raw material in the production of phthalic anhydride, alongside coal-derived naphthalene.

Before these developments, xylenes were mainly used as solvents or as components of fuels.

4.2 Pre-treatment of mixtures containing crude aromatics

The most important sources for BTX aromatics are reformer gasoline, pyrolysis gasoline and coke oven benzole. Reformer gasoline is the major source in countries where the production of ethylene is primarily based on gas, so that production of pyrolysis gasoline is relatively low; this applies particularly to the United States, where around two-thirds of its ethylene is produced from wet natural gases. With a share of around 75%, reformer gasoline is therefore the major raw material for BTX aromatics in the USA.

The composition of reformer gasoline is naturally dependent on the original feedstock and on reforming conditions. Table 4.1 shows the composition of a typical reformer gasoline.

Table 4.1: Composition of a typical reformer gasoline

Benzene	5%
Toluene	24%
Ethylbenzene	4%
p-Xylene	4%
m-Xylene	9%
o-Xylene	5%
C_9 and C_{10} aromatics	4%
Non-aromatics	45%

Owing to the relatively low benzene content of reformer gasoline, in the USA in particular, toluene and xylene are converted to benzene, to meet the demand.

The ratio of aromatics to each other is less dependent on processing conditions than is the proportion of non-aromatics, which drops as the severity of reforming increases. In contrast to benzole from coal carbonization or pyrolysis gasoline from ethylene production, the sulfur content of reformer gasoline is low (ca. 0.5 to 1 ppm). Prior to catalytic reforming, the sulfur content of the straight run gasoline, which ranges from 200 to 1,500 ppm, is reduced to well below 5 ppm, by hydrodesulfurization at 350 to 380 °C and a partial hydrogen pressure from 15 to 35 bar over a cobalt/molybdenum catalyst with aluminum oxide as a carrier. The nitrogen content of reformer gasoline is also extremely low (<1 ppm), since the nitrogen compounds in the crude gasoline (ca.1 to 40 ppm) are also broken down under the conditions of hydrogenation. If small quantities of unsaturated compounds are present in the reformer feedstock, these are converted to saturated compounds under the high partial hydrogen pressure, so that reformer gasoline requires only a distillative treatment before being used for the production of aromatics.

In Japan and Western Europe, pyrolysis gasoline is by far the most important source of aromatics, since large quantities of aromatic-rich gasoline arise during steam cracking of naphtha. The composition of pyrolysis gasoline varies according to cracking conditions. As cracking intensity increases, the proportion of benzene rises considerably (Table 4.2).

Table 4.2: Properties of pyrolysis gasolines in relation to type of cracking

Characteristics	Pyrolysis gasolines from	
	mild cracking	severe cracking
Density, g/ml	0.790	0.835
Initial boiling point, °C	40	44
50 vol% at °C	95	96
Final boiling point, °C	200	182
Bromine no., g/100 g	60	63
Diene content, %	12	15
Composition:		
Non-aromatics, %	41.0	27.4
Benzene, %	22.0	33.8
Toluene, %	17.5	19.4
o-Xylene, %	2.3	1.1
m-, p-Xylene and Ethylbenzene, %	6.0	6.6
Styrene, %	3.2	5.3

Unlike reformer gasoline, pyrolysis gasoline cannot be used directly for the production of aromatics. It must first undergo a hydrogenation process to remove olefins and diolefins, together with sulfur compounds, so that aromatics can be produced to specified purity.

The desulfurization of pyrolysis gasoline and hydrogenation of olefins and diolefins takes place in a two-stage process (Figure 4.1), since the reaction conditions for desulfurization and olefin hydrogenation differ considerably from conditions for diolefin hydrogenation. In the first reactor, the diolefins are hydrogenated, while in the second reactor, olefins and sulfur compounds are converted after the reaction mixture has been re-heated. There are two main variants of the diolefin hydrogenation process, which differ in the catalyst and the temperature and pressure conditions used. Mild diolefin hydrogenation operates with palladium catalysts at temperatures between 80 and 160 °C, at pressures of from 20 to 30 bar, and with a LHSV of 3-8 h^{-1}. If a nickel/alumina catalyst is used, the reaction is performed at a temperature of 120 to 160 °C with a pressure of between 20 and 60 bar in a trickle-bed reactor. The LHSV is only one-third of that for the palladium process.

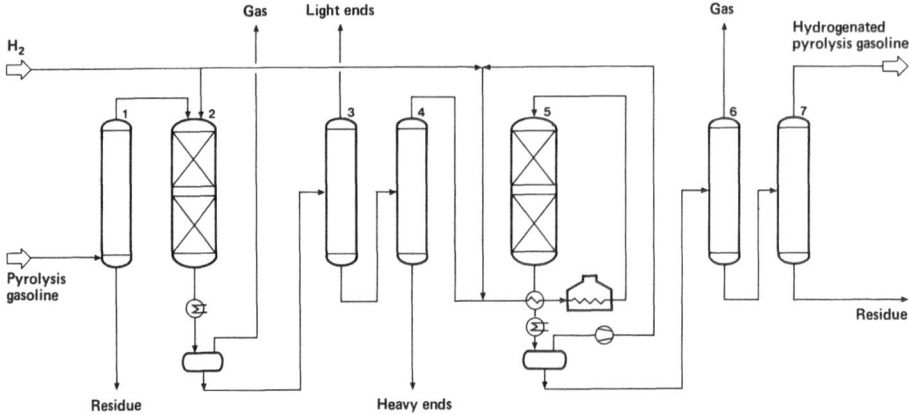

1 Predistillation; 2 Diolefin-hydrogenation reactor; 3 and 4 Intermediate distillation;
5 Mono-olefin hydrogenation reactor; 6 Stabilizer; 7 Distillation column

Figure 4.1: Flow diagram for two-stage hydrogenation of pyrolysis gasoline

In the second reaction stage, where hydrogenation of the mono-olefins and desulfurization are carried out, molybdenum and cobalt catalyst on aluminum oxide carriers are used. The reaction temperature in this hydrogenation stage is between 280 and 350 °C, with a partial hydrogen pressure of 15 to 25 bar, and a total pressure of around 45 to 65 bar.

Table 4.3 shows the composition of a typical pyrolysis gasoline fraction after hydrogenation.

Table 4.3: Typical data for hydrogenated pyrolysis naphtha

Density at 20 °C, g/ml	0.828
Bromine no., g/100 g	0.1
Mono-olefin content, %	<0.1
Diolefin content, %	≤0.1
Initial boiling point, °C	40
Final boiling point, °C	180
RON (unleaded)	101.2
MON (unleaded)	87.3
Benzene, %	39.2
Toluene, %	21.6
Sulfur, ppm	0.5

Figure 4.2 shows the benzene plant operated by *Mitsubishi Petrochemical Co.* in Kashima/Japan, which refines 175,000 tpa of pyrolysis gasoline by hydrogenation and extraction using the Sulfolane process (see Chapter 4.2.1.1).

Figure 4.2: Benzene plant operated by *Mitsubishi Petrochemical Co.* in Kashima, Japan

The third source of BTX aromatics is benzole, which is formed during coking of hard coal. Because of the high temperature of carbonization, the benzene content of benzole (Table 4.4) is noticeably higher than that of pyrolysis and reformer gasolines. The composition of benzole, similarly to coal tar, is less variable than that of corresponding petroleum-based fractions.

Table 4.4: Typical composition of benzole from coal carbonization

Benzene	67 %
Toluene	16 %
Xylenes and ethylbenzene	6 %
Styrene	1.3 %
C_9^+ aromatics	7 %
Non-aromatics	2 %
Sulfur	4,000 ppm
Nitrogen	1,800 ppm
Oxygen	100 ppm

Figure 4.3 shows the flow diagram of the recovery of benzole from coke oven gas by distillation.

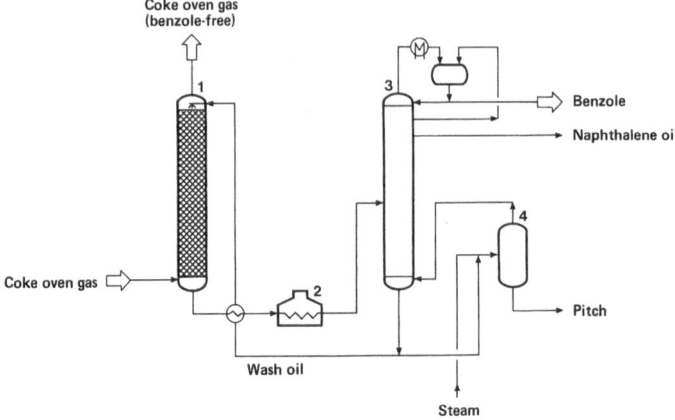

1 Benzole scrubber; 2 Tubular furnace; 3 Benzole stripper; 4 Pitch separator

Figure 4.3: Flow diagram of the recovery of benzole from coke oven gas

As with pyrolysis gasoline, the level of olefins, particularly diolefins such as cyclopentadiene, and of sulfur in benzole stemming from carbonization of coal is too high to produce pure BTX aromatics without pretreatment.

Two methods are used to refine benzole, namely:

1. treatment with sulfuric acid,
2. hydrorefining under pressure.

Sulfuric acid refining nowadays has only limited application. An advantage in comparison with hydrogenation consists in the fact that reactive constituents such as cyclopentadiene and indene which would otherwise be hydrogenated, are unaffected and may be extracted for further use. Furthermore, unlike hydrogenation, no additional azeotropic components of benzene such as heptane or cyclohexane are formed which hamper distillative refining of the refined product. A major disadvantage of sulfuric acid refining is the loss of the products through formation of acid sludge. In addition, it is not economically possible to reach the required sulfur content of the purified benzene below 1 ppm by simple sulfuric acid refining. However, the addition of unsaturated compounds such as styrene permits the resinification of the benzene impurity thiophene, so that with multi-stage treatment, a sulfur content of around 1 ppm in the refined benzene can be achieved.

Hydrogenation is currently the most important method for pretreatment of coke oven benzole. The *BASF/Scholven* process has proved particularly effective, operating at temperatures between 300 and 400 °C with molybdenum or cobalt/molybdenum catalysts and treating the benzole with various gases containing hydrogen, under pressure (20 to 50 bar). By this process the sulfur content of the refined product is reduced to below 0.5 ppm.

Table 4.5 compares the basic data for benzole and refined benzene produced by hydrogenation with coke oven gas.

Table 4.5: Typical data for benzole and refined benzene before and after hydrogenation

	Benzole	after hydrogenation with coke oven gas
Density at 15 °C, g/ml	0.879	0.876
Bromine no., g/100 cm^3	13	0.06
Non-aromatics, %	1.5	2.0
Total sulfur, ppm	3,800	<1

Figure 4.4 shows the flow diagram for this classic benzole refining process, which also provided the basis for development of the hydrogenation process of pyrolysis gasoline.

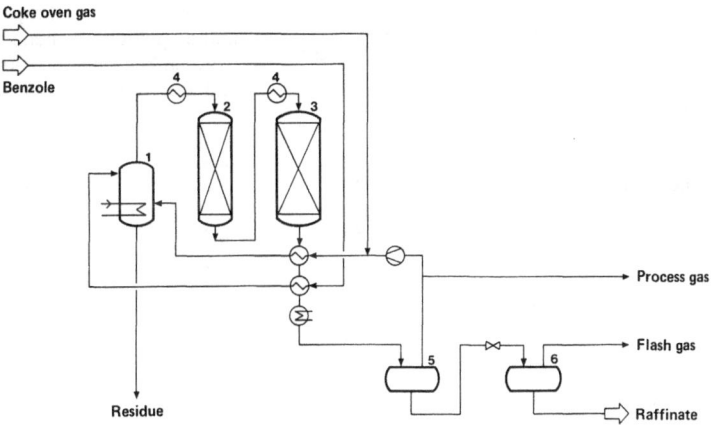

1 Vaporizer; **2** 1st stage reactor; **3** Main reactor; **4** Superheater; **5** Gas separator; **6** Flash vessel

Figure 4.4: Flow diagram for the *BASF/Scholven* process for the pressure hydrogenation of crude benzole

4.2.1 Recovery of aromatics

Production of individual BTX aromatics from the refined raw material depends on the different boiling and melting points (Table 4.6), and on differences in solubility in selective solvents.

Table 4.6: Physical data of pure benzene aromatics

	Melting point (°C)	Boiling point (°C)	Density (20 °C) (g/cm³)
Benzene	+ 5.53	80.10	0.8790
Toluene	−94.99	110.63	0.8669
Ethylbenzene	−94.98	136.19	0.8670
m-Xylene	−47.87	139.10	0.8642
p-Xylene	+13.26	138.35	0.8610
o-Xylene	−25.18	144.41	0.8802

Catalytic reformate, hydrogenated pyrolysis gasoline and hydrogenated coke oven benzole consist of a wide range of paraffinic, naphthenic and aromatic hydrocarbons with boiling points which in some cases lie very close together. Many hydrocarbons in the gasoline boiling range form azeotropic mixtures (Table 4.7).

Table 4.7: Boiling point of some non-aromatics and their azeotropes with benzene

	Boiling point, °C		Difference in boiling point between benzene and azeotrope	Proportion of benzene in azeotrope (%wt)
	Hydrocarbon	Azeotrope		
Benzene	80.1	–	–	–
Cyclohexane	80.6	77.7	2.4	51.8
Methylcyclopentane	71.8	71.5	8.6	9.4
Hexane	69.0	68.5	11.6	9.7
2,2-Dimethylpentane	79.1	75.9	4.2	46.3
2,3-Dimethylpentane	89.8	79.2	0.9	79.5
2,4-Dimethylpentane	80.8	75.2	4.9	48.3
n-Heptane	98.4	80.1	0	99.3
2,2,3-Trimethyl-butane (979 mbar)	79.9	75.6	4.5	50.5
2,2,4-Trimethyl-pentane	99.2	80.1	0	97.7

It is generally impracticable to produce pure aromatics from the refined feedstock material exclusively by distillation. The separation of the accompanying paraffins and naphthenes can be achieved by one of three methods:

1. Liquid-liquid extraction
2. Extractive distillation
3. Azeotropic distillation

Liquid-liquid extraction and extractive distillation are most commonly used to separate non-aromatics from coke-oven benzole, pyrolysis gasoline and reformer gasoline.

The advantage of liquid-liquid extraction is that a single process stage can produce an aromatics fraction of a broad boiling range containing benzene, toluene as well as C_8- and C_9-aromatics; the aromatics are then separated in a subsequent stage. A drawback of the liquid-liquid extraction process consists in relatively high investment and operating costs.

With extractive distillation, only the required aromatics are isolated, usually benzene. This requires a pre-distillation with its associated energy costs; a comparable separation step is carried out later in liquid-liquid extraction.

Azeotropic distillation can only be used for aromatics fractions with high concentrations of benzene. The boiling point of the polar solvent should be appreciably lower than the boiling range of the aromatics, so that the solvent can be taken off at the top of the column. The disadvantage of this is given by the fact that the azeotropic medium must be vaporized, which involves high energy consumption.

4.2.1.1 Liquid-liquid extraction

The principle of liquid-liquid extraction is based on the differences in solubilities of aromatics and non-aromatics in polar solvents (Figure 4.5); the selective solvent for aromatics should ideally have a boiling point 40 °C higher than the aromatics, and a miscibility gap in the phase diagram; additionally, low viscosity and relatively high density are also important criteria.

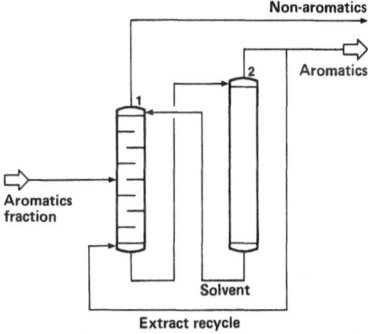

1 Extraction column; **2** Distillation column

Figure 4.5: General flow diagram for the extraction of aromatics

Usually, the crude aromatics stream is fed below the middle of the extractor. The solvent, which flows down from the top, dissolves the aromatics, which are removed from the extractor bottom together with the solvent and separated by subsequent distillation. Non-aromatics leave the extractor through the top.

The most traditional method of this type is the Udex process, developed by *Dow* and *UOP* and first used on a large scale in 1951/52. Diethylene glycol and triethylene glycol, sometimes with the addition of water, are used as solvents. Figure 4.6 shows the flow diagram for the Udex process, which operates at 140 to 150 °C and a pressure of around 9 bar.

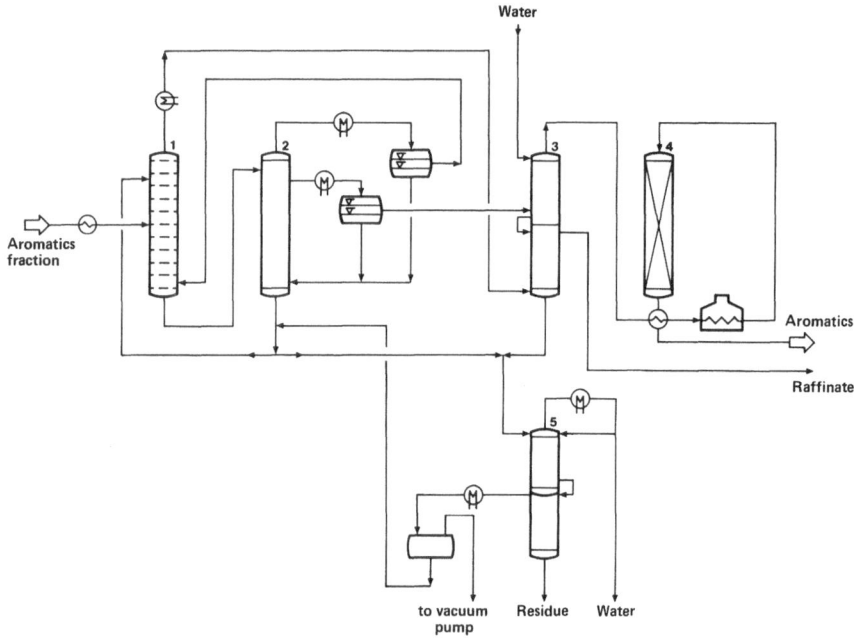

1 Extraction column; 2 Stripping column; 3 Water scrubber; 4 Active-earth treatment;
5 Water and solvent distillation

Figure 4.6: Flow diagram of the Udex process for the extraction of aromatics

Owing to the low solvent capacity of diethylene glycol for aromatics, a high proportion of solvent is needed in relation to the crude aromatic material, the ratio being typically 6-8:1, and in individual cases up to 20:1.

Sulfolane has a markedly better capacity than diethylene glycol for dissolving aromatics. In the Sulfolane process, first carried out technically by *Shell* and *UOP* in 1961, the mass ratio of solvent to feedstock is 3-6:1. A large number of plants which used to apply the Udex process have now been revamped to sulfolane use (Figure 4.7).

The raw material which is pre-heated to around 115 °C, is fed at atmospheric pressure to the extraction column (e.g. rotating disc contactor, sieve tray extractor), where the aromatics are preferentially dissolved by sulfolane. The solvent is separated from the aromatics mixture at a temperature of around 190 °C and a pressure of just 1 bar. The sulfolane which boils at 287 °C, remains at the base of the column and is recycled to the top of the extraction column. The aliphatics are recovered by counter-current washing with sulfolane and subsequent distillation.

Production of benzene, toluene and xylenes 109

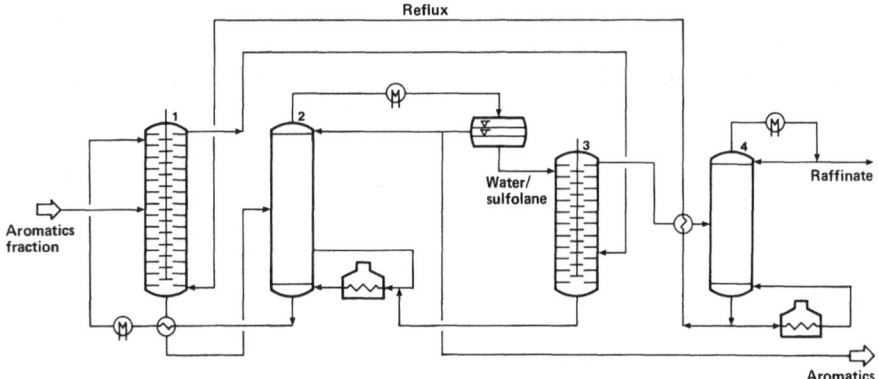

1 Extraction column (rotating disc contactor); **2** Extract-stripper; **3** Washing column (rotating disc contactor); **4** Rectification column

Figure 4.7: Flow diagram for the extraction of aromatics using the Sulfolane process

The *Lurgi* Arosolvan process for extracting aromatics uses N-methylpyrrolidone (NMP) as a solvent, with the addition of 12 to 14% water. Its first large-scale application was in 1963.

Figure 4.8 shows an example of the phase diagram for the model mixture toluene/n-heptane/NMP, illustrating the increase in solvent power of the NMP through the addition of water.

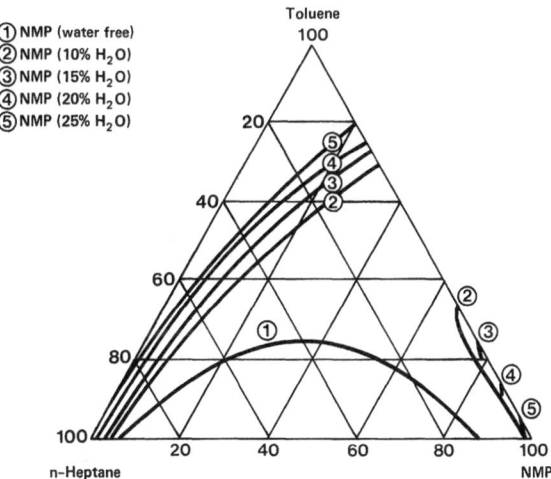

Figure 4.8: Phase diagram for the toluene/n-heptane/N-methylpyrrolidone system in relation to water content

N-methylpyrrolidone has a low viscosity and suitable solvent power for aromatics, so that extraction can be carried out at low temperatures (20 to 40 °C) under normal pressure. Since the boiling point of N-methylpyrrolidone at 206 °C is lower than that of sulfolane and diethylene glycol, the extract phase must be refined in two stages. In the first stripper, the low-boiling non-aromatics and part of the benzene are distilled off. The bottom product and distilled water are taken to a second stripper, where the aromatics-free solvent, containing water, is recovered as a bottom product and fed back into the extractor. An azeotropic mixture of pure aromatics and water is taken off as the top product. In the pentane column, the lower-boiling non-aromatics are removed from the raffinate phase as an overhead fraction and used together with the distillate from the first stripper as reflux for the extractor (Figure 4.9).

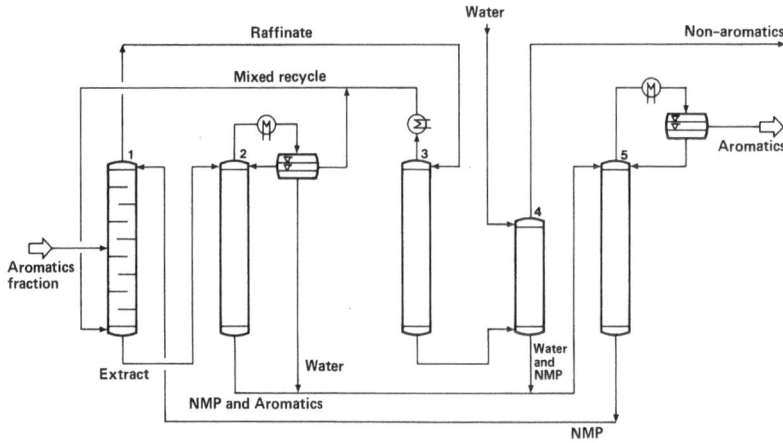

1 Extraction column; **2** Stripping column; **3** Pentane column; **4** NMP-recovery; **5** Stripping column

Figure 4.9: Flow diagram for the extraction of aromatics with N-methylpyrrolidone

Another method for the extraction of aromatics is the DMSO process developed by the *Institut Francais du Pétrole* (*IFP*), which uses dimethyl sulfoxide (DMSO) as a solvent with the addition of around 10% water, in two extractors (Figure 4.10). The extraction of aromatics takes place in the first extractor, while the second extractor is used to recover the solvent by extraction with a low-boiling, paraffinic auxiliary solvent, such as butane.

N-Formylmorpholine is also suitable as a solvent; it was first used on a large scale in 1972 in the highly energy-efficient Morphylex process devised by *Krupp-Koppers*.

The yield of aromatics from the various extraction processes is over 99% for benzene, up to 99.5% for toluene and up to 97% for xylenes.

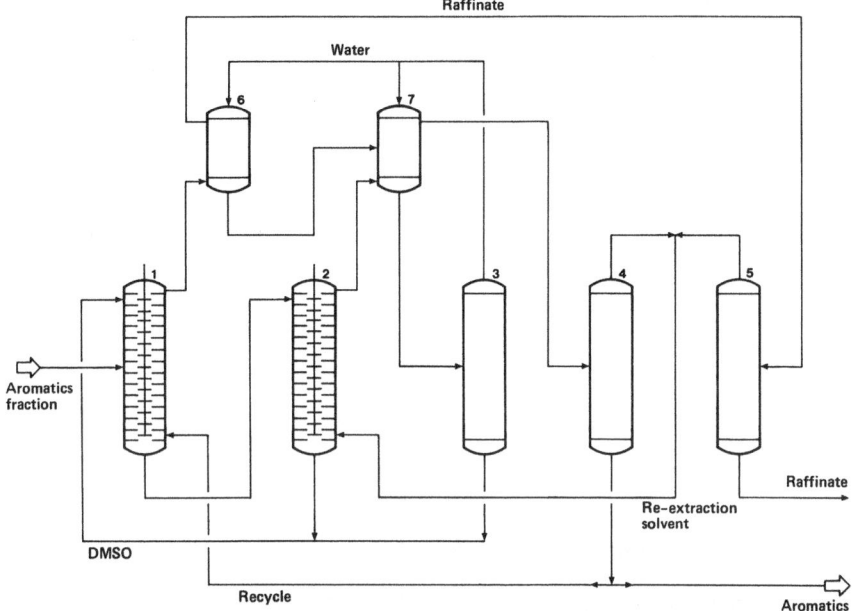

1 Extraction column I (rotating disc contactor); **2** Extraction column 2 (rotating disc contactor); **3** Solvent recovery; **4** Aromatics column; **5** Non-aromatics column; **6** Raffinate washer; **7** Extract washer

Figure 4.10: Flow diagram for extracting aromatics with dimethyl sulfoxide

Table 4.8 shows a comparison of basic parameters for the major aromatics extraction processes.

Table 4.8: Process conditions for the extraction of aromatics

Process	Solvent	Boiling point of solvent (wf) in °C (normal pressure)	Extraction conditions	Ratio of solvent to raw material
Udex (UOP-Dow)	Diethylene glycol	245	130–150 °C, 5–8 bar	6–8 : 1
Sulfolane (Shell-UOP)	Sulfolane	287	100 °C, 2 bar	3–6 : 1
Arosolvan (Lurgi)	N-Methylpyrrolidone	206	20–40 °C, 1 bar	4–5 : 1
IFP (IFP)	Dimethyl sulfoxide	189	20–30 °C, 1 bar	3–5 : 1
Morphylex (Krupp-Koppers)	N-Formylmorpholine	244	180–200 °C, 1 bar	5–6 : 1

HO—CH₂—CH₂—O—CH₂—CH₂—OH

Diethylene glycol

CH₃—S—CH₃
‖
O

Dimethyl sulfoxide

Sulfolane

N-methylpyrrolidone

N-formylmorpholine

4.2.1.2 Extractive and azeotropic distillation

Extractive distillation and azeotropic distillation share as a common characteristic, in that an auxiliary solvent is added to the crude aromatics fraction to achieve better separation by distillation. Extractive distillation takes place in the presence of an extractive material with high solvent power for aromatics, which has relatively low volatility compared with the compounds which are to be separated, and is constantly added at the top of the fractionation column. The purpose of the auxiliary solvent is to change the vapor pressures of the hydrocarbon components in such a way that they can be more easily separated by distillation; e.g. the vapor pressure of benzene is lowered to the point when the accompanying non-aromatics can be distilled off as an overhead fraction.

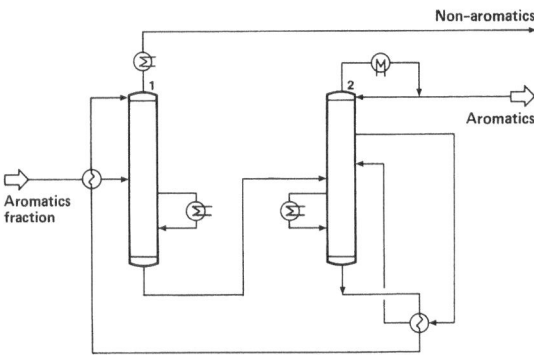

1 Extractive distillation; **2** Stripping column

Figure 4.11: Flow diagram for production of aromatics by extractive distillation

In azeotropic distillation, on the other hand, the additive and the component to be separated form an azeotrope i.e. a mixture boiling at a given temperature and with a constant composition. Azeotropic distillation can only be used to refine highly-enriched mixtures of aromatics, such as occur in coke-oven benzole, whereas extractive distillation can also be used to separate aromatics which are present in low concentrations. As early as World War I toluene used in the production of explosives was obtained by extractive distillation, using phenol as the extractive material.

The main methods of extractive distillation employed today to recover aromatics (Figure 4.11) use dimethylformamide, N-formylmorpholine (*Krupp-Koppers*), N-methylpyrrolidone (Distapex/*Lurgi*) and sulfolane (*Shell/UOP*) as extractive agents.

The Morphylane process, which operates with N-formylmorpholine, can also be run in conjunction with benzene predistillation, allowing low energy consumption rates to be achieved through optimized heat recovery. The largest plant of this kind is operated by *Redestillationsgemeinschaft* in Gelsenkirchen, West-Germany and has a capacity of 336,000 tpa of hydrorefined benzole (Figure 4.12).

Operating the rectification columns under pressures up to 18 bar gives the overhead vapors a greater heat content; these are used to heat the distillation feed. The yield of benzene is over 99%.

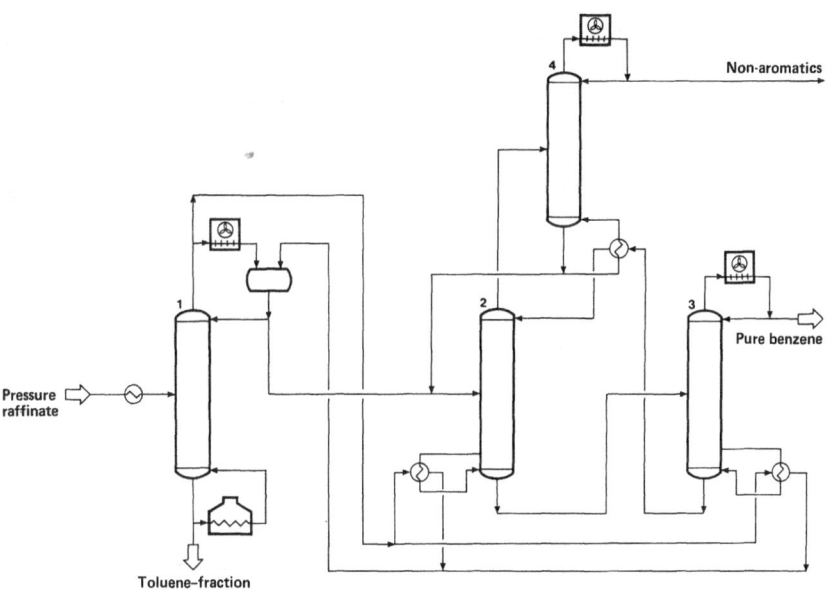

1 Predistillation column; **2** Extractive distillation column; **3** Stripping column; **4** Raffinate column

Figure 4.12: Flow diagram for production of aromatics by the Morphylane process with optimum heat economy

Azeotropic distillation is only of minor importance in the production of BTX aromatics; the most important azeotropic entrainers are methyl ethyl ketone and methanol. Figure 4.13 shows the production of toluene by azeotropic distillation with methanol.

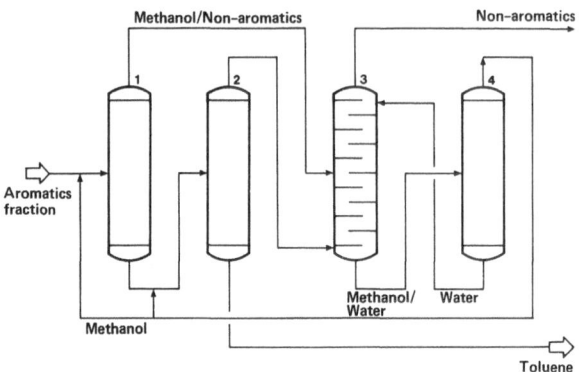

1 and **2** Azeotropic distillation columns; **3** Extraction column; **4** Methanol column

Figure 4.13: Flow diagram for azeotropic distillation for the production of toluene

4.3 Separation of mixed aromatics into individual constituents

Benzene and toluene are produced from the purified aromatic hydrocarbon mixture, obtained by extraction or extractive distillation, by means of distillation, which is relatively simple since the two compounds have well-separated boiling points (Figure 4.14). Rectification columns with around 60 theoretical trays are used.

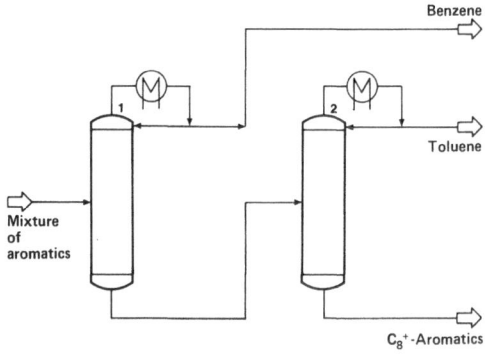

1 Benzene column; **2** Toluene column

Figure 4.14: Flow diagram for the production of benzene and toluene from mixed of aromatics

It is also possible to produce benzene by crystallization (*Newton-Chambers* method); a pre-requisite is the absence of thiophene, since this sulfur compound, unlike the co-boiling paraffins which are present, cannot economically be separated by crystallization. However, in contrast to the recovery of other aromatics by crystallization, there is presently no large-scale application of benzene crystallization.

Since the boiling points of the C_8-isomers lie close together, their separation is more difficult than for benzene and toluene. The separation of ethylbenzene by distillation alone requires a system of columns with some 300 theoretical trays, with reflux ratios of around 100:1. The yield of ethylbenzene from this superfractionation is over 95%, with 99.8% purity.

Since it is basically more economical to synthesize ethylbenzene from benzene and ethylene (see Chapter 5.1.2), distillation is used only in isolated cases, particularly in the USA (*Cosden*).

Table 4.9 summarizes the properties of C_8-aromatics, on which industrial separations are based.

Table 4.9: Specific properties for the separation of C_8- aromatics

	Ethylbenzene	p-Xylene	m-Xylene	o-Xylene
Crystallisation point, °C	−95.0	+13.3	−47.9	−25.2
relative stability of complexes with HF/BF$_3$		1	20	2
relative enrichment factor by adsorption	0.5	1.0	0.3	0.2
Boiling point, °C	136.2	138.3	139.1	144.4

The crystallization point of p-xylene is markedly higher than that of the other C_8-aromatics; it is therefore possible to separate p-xylene by crystallization. Since p-xylene can be better adsorbed than the accompanying C_8-aromatics, adsorption can also be used in its recovery.

The stability of the complexes of m-xylene with HF/BF$_3$ is appreciably higher than for corresponding complexes of other aromatics; m-xylene can therefore be separated as a HF/BF$_3$ complex.

The boiling point of o-xylene is around 5 °C higher than that of m-xylene, its nearest co-boiling compound. o-Xylene is therefore recovered by fractional distillation in columns with 100 to 150 trays, which requires a reflux ratio of 8-10:1. The high-boiling aromatics are separated from the o-xylene by distillation in the o-xylene column with 40 to 60 trays and a reflux ratio of around 1:1.

Figure 4.15 shows a typical flow diagram for the separation of o-xylene by distillation, yielding a product of over 95% purity; this concentration is sufficient for the main use of o-xylene, the production of phthalic anhydride.

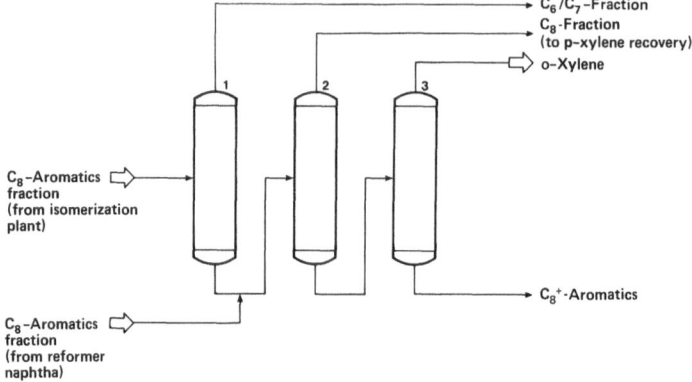

1 Benzene/toluene column; 2 C_8-column; 3 o-Xylene column

Figure 4.15: Flow diagram for production of o-xylene from a C_8-aromatics fraction by distillation

p-Xylene cannot be separated from m-xylene by fractional distillation, since it boils only 0.653 °C below m-xylene. Crystallization was initially the most important method of producing p-xylene.

Production of p-xylene by melt-crystallization is hampered, in spite of the relatively high melting point, by the formation of binary and ternary eutectic mixtures with the accompanying C_8-aromatics. The eutectic of m-xylene and p-xylene alone limits the technically achievable yield of p-xylene to around 65% (Tables 4.10 and 4.11).

Table 4.10: Composition of binary eutectic systems of C_8– aromatics

System	Crystallization point (°C)	Composition in mol%	
pX–oX	−34.93	23.88 pX	76.12 oX
pX–mX	−52.55	12.18 pX	87.82 mX
pX–Etb	−94.8	1.37 pX	98.63 Etb
mX–oX	−61.20	68.14 mX	31.86 oX
mX–Etb	−99.44	16.30 mX	83.70 Etb
oX–Etb	−96.37	6.61 oX	93.39 Etb

oX: o-xylene; pX: p-xylene; mX: m-xylene; Etb: Ethylbenzene

Table 4.11: Composition of ternary eutectic systems of C_8– aromatics

System	Crystallisation point (°C)	Composition in mol%
pX–oX–mX	− 63.55	7.47 pX–29.13 oX -63.36 mX
pX–oX–Etb	− 96.7	1.21 pX– 6.30 oX -92.29 Etb
pX–mX–Etb	− 99.71	1.00 pX–16.10 mX–82.90 Etb
oX–mX–Etb	−100.84	5.18 pX–15.29 mX–79.53 Etb

oX: o-xylene; pX: p-xylene; mX: m-xylene; Etb: Ethylbenzene

Figure 4.16 shows the phase diagram for m/p-xylene with the eutectic point at −52.55 °C. At this temperature, m- and p-xylene crystallize out simultaneously, so that the solution and the crystal mixture have the same composition.

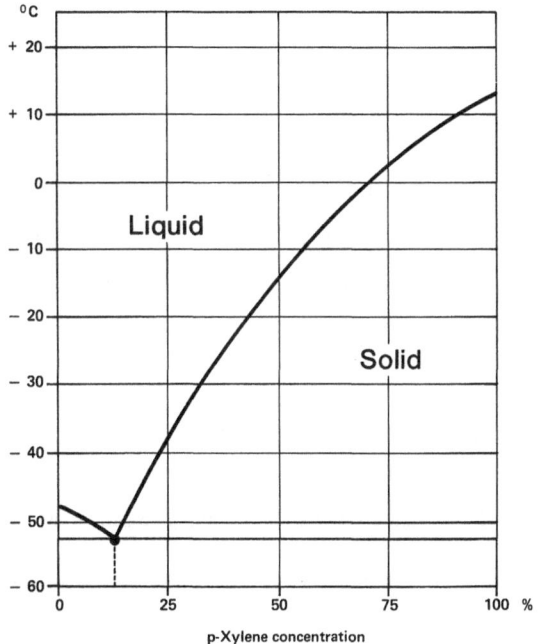

Figure 4.16: Phase diagram of m/p-xylene

The crystallization processes used to produce p-xylene are generally carried out in two stages. The feedstock material is first dried to a residual water content of around 10 ppm, to avoid the formation of ice, then cooled to around -55 to -70 °C, and separated from the mother liquor in centrifuges or rotating vacuum drum filters. The mother liquor from the first stage is transferred to an isomerization unit. The crystallizate of the first stage with a p-xylene content of around 90% is melted, then cooled again and separated from the mother liquor. The purity of the produced p-xylene surpasses 99%.

The differences of the industrial crystallization processes consist in the arrangement of equipment in the cooling cycle and the second refining stage. The feedstock may be cooled either directly or indirectly. With direct cooling, the coolant, e.g. liquid ethylene or liquid CO_2, is fed directly into the crystallizer. The feedstock is cooled by vaporization of the cooling agent, which is then condensed by compression and recycled into cooling system. Indirect cooling is carried out in pipe or scraped-wall chillers. Methanol or freon 13 is used as the coolant.

The *Chevron* process has proved particularly effective in producing p-xylene by crystallization. Crystallization is carried out by direct cooling with CO_2. Distillation follows the crystallization process, removing toluene and light hydrocarbons from the mother liquor. The distillation residue is water-free and is returned to the crystallization plant. In the *Chevron* process, relatively large crystals are produced, which can easily be separated from the mother liquor in centrifuges. The flow diagram is depicted in Figure 4.17.

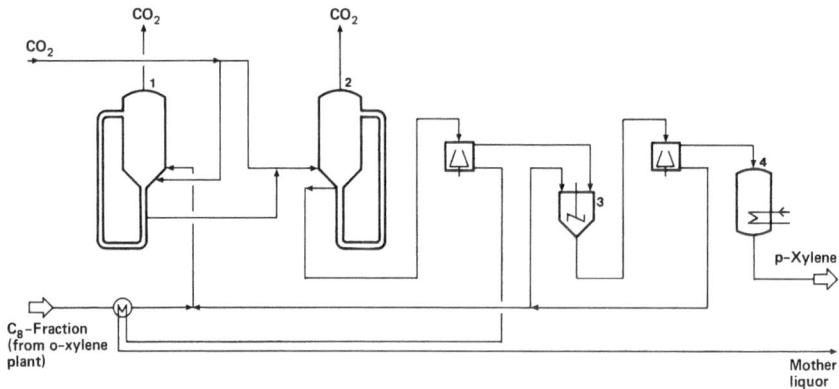

1 Pressure crystallizer; **2** Vacuum crystallizer; **3** Mixing vessel; **4** Melting vessel

Figure 4.17: Flow diagram for p-xylene production by crystallization using the *Chevron* process

A crystallization process for the recovery of p-xylene which operates with indirect cooling is the *Amoco* process, in which ethylene is generally used in the first stage and propane in the second stage as coolant (Figure 4.18).

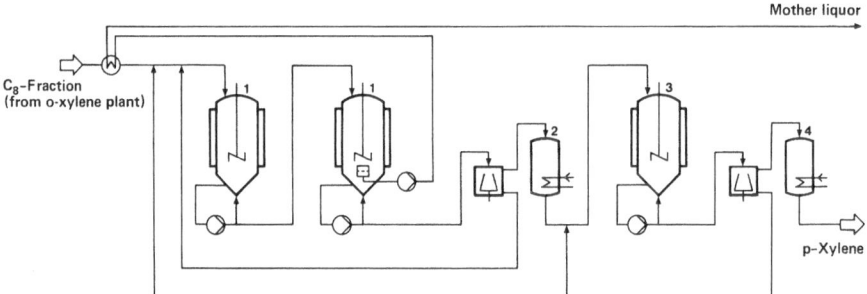

1 First stage crystallizers; **2** Melting vessel; **3** Second stage crystallizer; **4** Melting vessel

Figure 4.18: Flow diagram for p-xylene production by crystallization using the *Amoco* process

Figure 4.19 shows parts of the p-xylene plant of *Amoco*, Decatur, USA with a total capacity of 450,000 tpa.

Figure 4.19: Parts of *Amoco*'s p-xylene plant, Decatur/USA

Other crystallization methods for producing p-xylene include the *Maruzen* process, which uses ethylene as a direct coolant, the *Arco* process, which is very similar to the *Amoco* process, and the three-stage *Krupp-Koppers* process.

In the *Phillips* process, which nowadays is applied in only few cases, the crystallizate is refined by counter current crystallization in a pulse column through the ascending melt (Figure 4.20); this process was introduced in 1957.

The high cost of equipment, low temperature levels, high energy consumption and limited yield of the crystallization processes has led to an increase in importance for adsorptive p-xylene production in recent years. The Parex process, first introduced by *UOP* in 1971, is currently the most common method of this type (Figure 4.21).

In the Parex process, the p-xylene is separated at 120 to 175 °C by selective adsorption. Separation of the C_8-aromatics is effected by an adsorbing solid material (adsorbant) and a suitable liquid. The process is based on the fact that the various components, adsorbed to a different degree, are recovered from the active surface of the adsorbant. A synthetic zeolite is used as the adsorption agent, the active centres of which are formed by cations from the first and second group (K, Ca) of the periodic table. A hydrocarbon with a low adsorption capacity, such as toluene or p-diethylbenzene is used for desorption, and can easily be separated from p-xylene by distillation. In this process, the liquid phases are fed through a

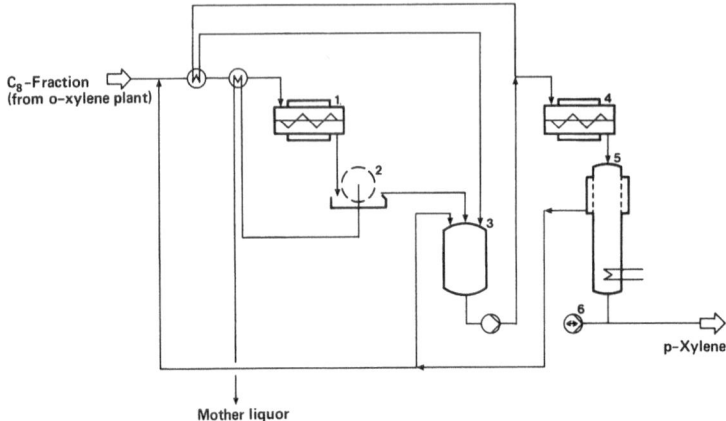

1 Scraped wall chiller (1st stage); 2 Rotating drum filter; 3 Melting vessel; 4 Scraped wall chiller (2nd stage); 5 Pulse column with wall filter; 6 Pulsation pump

Figure 4.20: Flow diagram of p-xylene production by countercurrent crystallization using the *Phillips* process

column consisting of around 12 chambers. These chambers are filled with molecular sieves; the flow of the liquid streams is controlled by a rotating valve. The p-xylene produced is 99.5% pure. It contains traces of ethylbenzene in particular, together with m- and o-xylene. The relatively high concentration of ethylbenzene is due to the high adsorption affinity of this hydrocarbon. Ethylbenzene can be detrimental in the production of terephthalic acid as benzoic acid is the end-product of its oxidation.

The Aromax process developed by *Toray* (Japan) to recover p-xylene operates in a similar way to the Parex process; the process is carried out at around 180 °C and a pressure of 20 bar. The yield of p-xylene is over 90%.

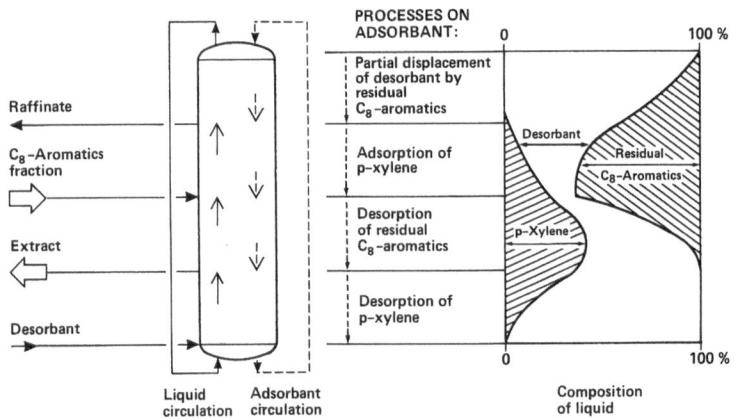

Figure 4.21: Concentration profile of adsorptive recovery of p-xylene by the Parex process

Production of pure m-xylene is predominantly carried out by a process developed by *Mitsubishi Gas Chemical* (*MGCC*), based on the formation of complexes with HF/BF_3. When a C_8-aromatic stream is treated with HF/BF_3, two phases form, with m-xylene being selectively concentrated in the HF phase. The efficiency of m-xylene separation is increased by the addition of a diluent, generally a C_6-paraffin. Due to the high basicity of m-xylene, its HF/BF_3 complex is particularly stable. Pure m-xylene (over 99%) is recovered by thermal decomposition of the complex; HF and BF_3 are then recirculated. The flow diagram for the *MGCC* process is shown in Figure 4.23.

The *Mitsubishi Gas Chemical* process can be used not only for production of pure m-xylene, but also as an isomerization process.

Figure 4.22 shows the 300,000 tpa xylene plant operated by *Mitsubishi Gas Chemical Co.* in Mizushima/Japan.

Figure 4.22: Xylene plant operated by *Mitsubishi Gas Chemical Co.* in Mizushima/Japan

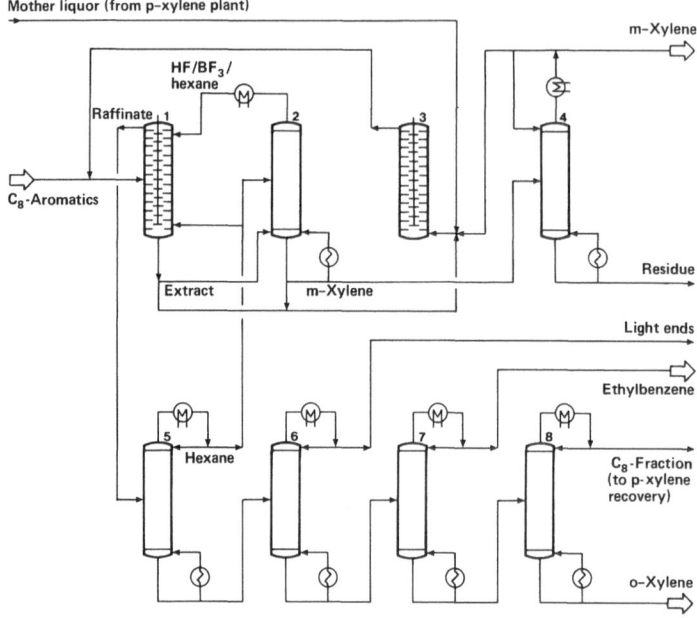

1 Extractor; 2 Decomposer; 3 Isomerization column; 4 Heavy ends column; 5 Raffinate-stripping column; 6 Light ends column; 7 Ethylbenzene column; 8 o-Xylene column

Figure 4.23: Flow diagram for production of m- and o-xylene using *MGCC's* HF process

4.4 Dealkylation, isomerization and disproportionation reactions of BTX aromatics

The most important benzene aromatics in terms of quantity are benzene and p-xylene. Since the production from reformer gasoline, pyrolysis gasoline and coke-oven benzole is frequently inadequate to meet the demand, isomerization, disproportionation and dealkylation methods have been developed to complement the direct production from aromatics mixtures.

4.4.1 Dealkylation of toluene and xylenes to benzene

The dealkylation of toluene and xylenes to benzene is carried out not only catalytically but also as a purely thermal reaction in the presence of hydrogen. In the catalytic process, pressures ranges from 35 to 70 bar and temperatures from 550 to 650 °C; chromium oxides on alumina are used as catalysts. Thermal dealkylation takes place at temperatures up to 750 °C and pressures of around 45 bar.

Figure 4.24 shows the flow diagram for the *Houdry* Litol process for refining and dealkylating benzole.

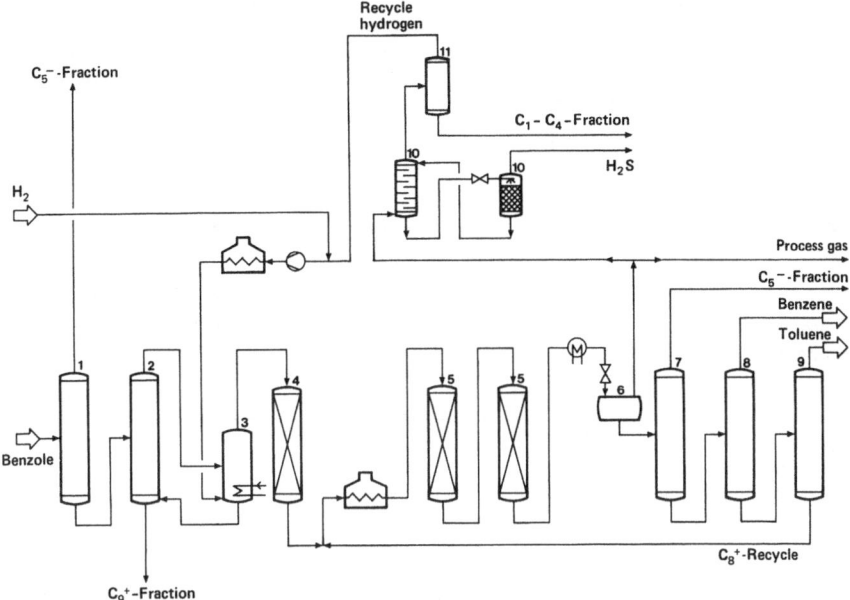

1 and 2 Prefractionation column; 3 Vaporizer; 4 Prehydrogenation reactor; 5 Litol reactors; 6 Flash vessel; 7 Stabilizer; 8 Benzene column; 9 Toluene column; 10 Gas purification; 11 Low temperature gas separation

Figure 4.24: Flow diagram for refining benzole using the *Houdry* Litol process

The *Houdry* Litol process is widely used to produce benzene by dealkylation; the following processes occur concurrently:

1. Desulfurization
2. Removal of paraffins and naphthenes
3. Saturation of unsaturated compounds
4. Dealkylation

Hydrogenation is carried out in two stages. After preliminary hydrogenation, the benzole is heated to 600 °C and fed through a system of fixed bed reactors, in which sulfur compounds are converted into hydrogen sulfide. The non-aromatics are then subjected to a hydro-cracking reaction and transformed into gas; the alkylaromatics are dealkylated. When the reaction mixture has been stabilized by removing the low-boiling compounds, benzene is recovered by distillation.

Yields with and without toluene recycling are shown in Table 4.12.

After a long period of operation, carbon deposited on the surface of the catalyst reduces its effectiveness. The carbon is burned off with oxygen, after the system has been purged with nitrogen.

The Hydeal process developed by *UOP*, the *Houdry* Detol process and the *BASF* process for the dealkylation of toluene operate in a similar way to the

Table 4.12: Yields from benzole refining by hydrogenation using the *Houdry* Litol process

Products	Feedstock (wt%)*	Product** (wt%)*	Feedstock (wt%)*	Product*** (wt%)*
Hydrogen	0.95	0.10	1.13	0.13
H_2S	-	0.41	-	0.41
C_1-C_5 fraction	0.53	6.80	0.59	8.13
C_6-C_8 non-aromatics	0.89	0.02	0.89	0.02
Benzene	74.09	86.66	74.09	92.88
Toluene	19.22	7.36	19.22	0.02
C_8 fraction	3.34	-	3.34	-
Styrene	0.62	-	0.62	-
C_9^+ aromatics	0.70	0.01	0.70	0.01
Thiophene	1.02	-	1.02	-
Total	101.36	101.36	101.60	101.60

* based on hydrocarbon feed
** without toluene recycle
*** with toluene recycle

Houdry Litol process. The theoretical yield of benzene is 84.8 wt%, based on the toluene feedstock; in practice, a yield of around 83 wt% is achieved. Since dealkylation is strongly exothermic, cold hydrogen is generally fed into the reactor system for quenching. The molar ratio of hydrogen to toluene is up to 8:1.

Among non-catalytic processes, a technology developed by *Hydrocarbon Research* is particularly widely used. The molar hydrogen/toluene ratio is around 4:1; the molar yield is around 97 to 99%, i.e., of the same order as with catalytic dealkylation.

To avoid using expensive hydrogen, a process has been developed to convert toluene into benzene with steam at a temperature of around 430 °C and pressures of from 5 to 15 bar on rhenium/alumina catalysts. This produces carbon monoxide, carbon dioxide and hydrogen. Benzene selectivity of this process, which has not yet been applied on a large scale, is 87%, with 46% toluene conversion.

$$\text{C}_6\text{H}_5\text{CH}_3 \xrightarrow[\begin{array}{c}-\text{CO}\\-\text{CO}_2\\-\text{H}_2\end{array}]{+\text{H}_2\text{O}} \text{C}_6\text{H}_6$$

4.4.2 Isomerization of xylenes

Although the demand for p-xylene to produce terephthalic acid for fibers, and for o-xylene to obtain phthalic anhydride for the manufacture of plasticizers has risen considerably in recent decades, demand for m-xylene for the production of isophthalic acid is relatively small. m-Xylene and ethylbenzene are therefore isomerized to o-and p-xylene.

Figure 4.25 shows the equilibrium concentrations (at 1 bar) of C_8-aromatics versus temperature. While the proportion of m-xylene falls as temperature increases, the proportion of ethylbenzene in particular rises.

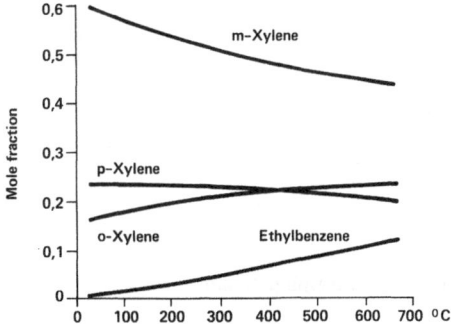

Figure 4.25: Equilibrium concentrations of C_8-aromatics

Although xylenes can be converted through an acid-catalyzed carbonium ion mechanism, the presence of hydrogen is necessary to convert ethylbenzene into xylenes.

One of the most important processes for the isomerization of m-xylene to produce large quantities of o- and p-xylene is the Octafining process, developed by *Atlantic Richfield* in 1960. In this method isomerization takes place on a platinum/ aluminum silicate catalyst in a hydrogen atmosphere. The platinum content of the catalyst is around 0.5%. The catalyst is similar in composition to reforming catalysts, since transalkylation (i.e. isomerization) also occurs during gasoline reforming. The reaction is carried out at a pressure of 10 to 30 bar. The hydrogen/hydrocarbon ratio is around 4–6:1; the initial temperature is ca. 425 °C which is slowly increased to 480 °C. The catalyst can last in service up to 5 years.

In the Octafining process (Figure 4.26) only a small amount of ethylbenzene is converted into xylenes, since the original concentration of ethylbenzene in the feedstock is generally well below the thermodynamic equilibrium level.

When catalysts other than noble metals are used, such as $AlCl_3$ (Isomar, *UOP*) and AlF_3 (Isarom, *IFP*), their service lives are shorter; reaction temperatures are usually between 370 to 540 °C, with pressures from 15 to 30 bar.

Figure 4.27 shows the reactions which are possible during the isomerization (rearrangement, ring contraction, ring expansion, dealkylation). Isomerization follows a 1,2-rearrangement, i.e., o-xylene is not converted directly into p-xylene.

In the early 1970's *Mobil Oil* developed zeolite catalysts for the isomerization of xylene without the use of hydrogen (LTI method). This process is carried out in the liquid phase at a pressure of 30 bar and temperatures of from 200 to 260 °C. The service life of the catalyst is generally over 2 years. Regeneration (coke removal) is carried out from a reaction temperature of around 275 °C.

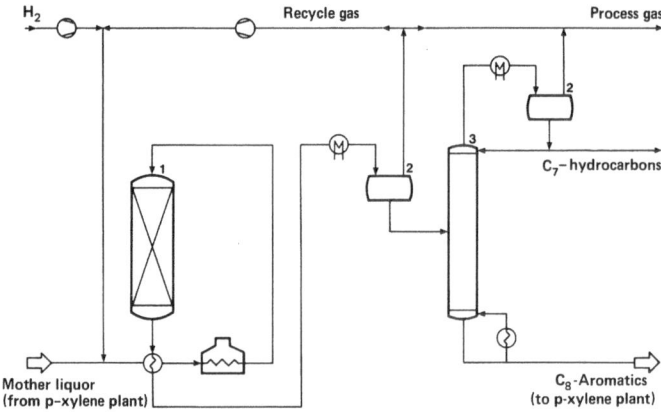

1 Reactor; 2 Gas separator; 3 Stripping column

Figure 4.26: Flow diagram for Octafining process for the isomerization of C_8-aromatics

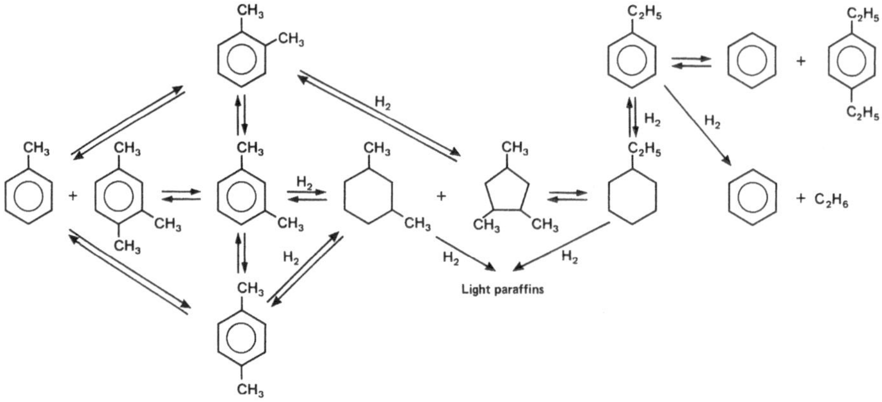

Figure 4.27: Reaction diagram for the interconversion of C_8-aromatics

4.4.3 Disproportionation

By disproportionation toluene is converted into benzene and xylenes. It is also possible to transfer alkyl groups from higher alkylated benzenes to benzene or lower-alkylated benzenes. The catalytic LTD process developed by *Mobil Oil* (Figure 4.28) to transform toluene into benzene and xylenes operates at a pressure of around 46 bar and temperatures of from 260 to 320 °C in the liquid phase; the service life of the catalyst is around 1.5 years.

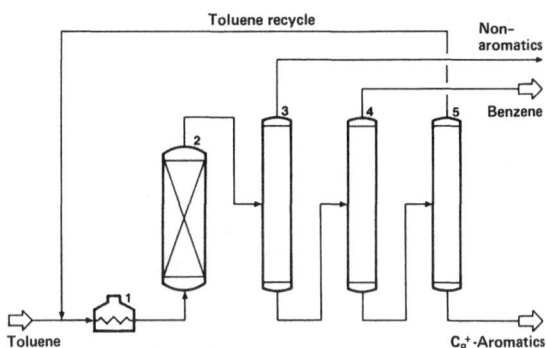

1 Tubular furnace; **2** Reactor; **3** Non-aromatics column; **4** Benzene column; **5** Toluene column

Figure 4.28: Flow diagram of toluene disproportionation using the *Mobil* LTD process

Methyl group transfer can also be effected in the gas phase by the Tatoray process, developed by *UOP* and *Toray* (Figure 4.29).

The reaction occurs on a zeolite catalyst, which is doped with transition metals, at 30 to 40 bar and temperatures of from 410 to 470 °C; the molar hydrogen/hydrocarbon ratio reaches 20:1. The yield of benzene and xylenes is around 95%.

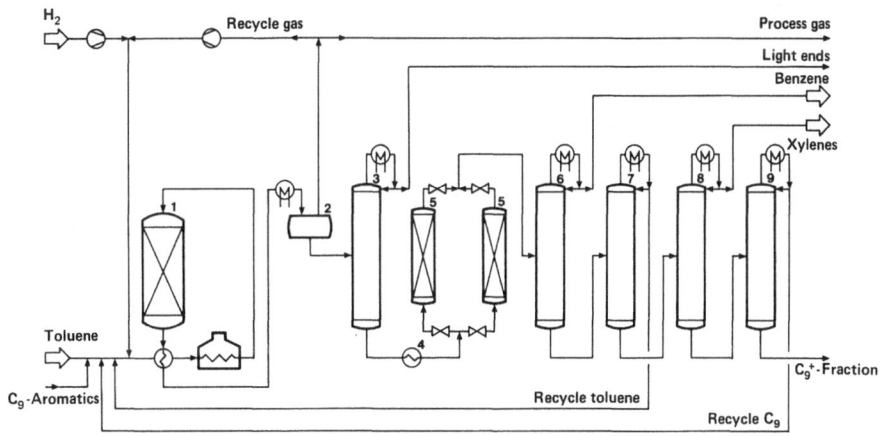

1 Reactor; **2** Gas separator; **3** Stabilizer; **4** Heater; **5** Clay treatment; **6** Benzene column; **7** Toluene column; **8** Xylene column; **9** C_9 column

Figure 4.29: Flow diagram of the Tatoray process for producing aromatics from toluene

The Xylene-Plus process developed by *Atlantic Richfield* also operates in the gas phase, but at low pressures (2 bar) and without hydrogen, and using Al_2O_3/SiO_2 catalysts.

4.5 Quality standards

Quality standards of benzene hydrocarbons are generally determined by their applications.

The criteria for benzene purity are mainly set by its use for cyclohexane production by hydrogenation. For this application, a low sulfur content is especially important, because of the possibility of catalyst poisoning.

The purity requirements for p-xylene are especially stringent, because of its use in the production of terephthalic acid and its esters, for the manufacture of polyester fibers.

The main product of m-xylene, isophthalic acid, can be produced from comparatively low-quality m-xylene by a process developed by *Amoco,* since the phthalic acid isomers can be relatively easily separated from each other. The *Mitsubishi Gas Chemical* process produces m-xylene of high purity.

o-Xylene is supplied in concentrations of at least 95% for the manufacture of its principal derivative, phthalic anhydride.

The major important quality criteria for benzene, toluene and xylenes are summarized in Table 4.13. The specifications in various countries are covered by comparable national standards (ASTM, DIN, BSS etc.).

Table 4.13: Quality criteria for technically pure benzene aromatics

	Benzene	Toluene	o-Xylene	p-Xylene	m-Xylene
Purity, %	–	–	99.0	99.4	95,4
Distillation interval, °C	0.6–0.8	0.6–0.8	1	1	1
Crystallisation point, °C	5.40 (min)	–	–25.5	13.1	–
Density (15.5/15.5 °C)	0.883–0.886	0.869–0.872	0.882	0.865	0.869
Sulfur content (max. ppm)	1	4	10–15	10–15	10–15
Non-aromatics (max. wt%)	0.1	0.2	0.1	0.05	0.2

4.6 Economic data

Tables 4.14, 4.15 and 4.16 show the regional distribution of the production of benzene, toluene and xylene.

Table 4.14: Benzene production, 1985

	(1,000 t)
USA	4,460
Canada	635
Brazil	500
West Germany	1,580
Great Britain	740
France	630
Italy	490
Benelux	1,100
USSR	2,000
Poland	250
German Democratic Republic	300
Japan	2,280
India	100
China	450
Other countries	1,985
Total production	17,500

The proportion of coal-derived benzene in some countries with high levels of coke production, such as Japan or West Germany is between 10 and 15%.

Table 4.15: Toluene production, 1985

	(1,000 t)
USA	2,250
Canada	310
Mexico	220
West Germany	370
Great Britain	80
France	40
Italy	120
Benelux	140
USSR	950
German Democratic Republic	150
Japan	830
China	200
Other countries	840
Total	6,500

Table 4.16: Production of o- and p-xylene, 1985

	(1,000 t)	
	o-Xylene	p-Xylene
USA	310	1,940
Brazil	80	75
Mexico	45	110
West Germany	210	230
Great Britain	–	270

Table 4.16 (continued)

	(1,000 t)	
	o-Xylene	p-Xylene
France	60	50
Italy	100	235
Benelux	50	110
Portugal	30	70
Spain	30	20
Yugoslavia	30	60
Taiwan	50	150
Japan	180	720
Other countries	125	260
Total production, Western world	1,300	4,300

Production of m-xylene is confined to a small number of countries; capacity in the USA is 110,000 tpa (*Amoco*), in Italy 30,000 tpa (*Nurachem*) and in Japan around 60,000 tpa (*Mitsubishi Gas Chemical*).

4.7 Process review

Table 4.17 summarizes the most important processes for the production of aromatics from coal and petroleum feedstocks.

Table 4.17: Summary of the most important processes for the production of aromatics

Process	Aim of process	Process conditions				Other characteristics
		Pressure (bar)	Temperature (°C)	Catalyst	Reaction component	
1. Refining processes:						
Hydrogenation of pyrolysis gasoline	Hydrogenation of diolefins and desulfurization	40–60	200–250	Co, Mo, Ni, Pd	H_2	Two-stage process
Benzole pressure refining	Hydrogenation of coke oven benzole	20–50	350	Co, Mo	H_2	Reduction of sulfur to below 0.5 ppm; removal of unsaturated hydrocarbons, which hamper the production of benzene by distillation

Table 4.17 (continued)

Process	Aim of process	Process conditions				Other characteristics
		Pressure (bar)	Temperature (°C)	Catalyst	Reaction component	
2. Dealkylation processes:						
Houdry Litol	Production of benzene from toluene	50	600	Co, Mo	H_2	Hydrogenation of unsaturated compounds; hydrocracking of non-aromatics; desulfurization, dealkylation and dehydrogenation of naphthenes lead to higher benzene yields
Houdry dealkylation (HDA)	Production of benzene from toluene	45	max. 750	–	H_2	Benzene yield up to 99 %
3. Isomerization processes:						
Octafining	Increasing prop. of p-xylene	10–30	425–480	Pt/Zeolite	H_2	Comparable with Isomar (*UOP*), Isoforming (*Exxon*) and Isarom (*IFP*) processes
4. Transalkylation:						
Arco	Production of benzene and C_8 aromatics from toluene	2	480–520	Al_2O_3/SiO_2	–	Fluidized-bed process in the gas phase
Tatoray	Production of benzene and C_8 aromatics from toluene	10–50	350–530	Zeolite	H_2	adiabatic process
Mobil LTD	Production of benzene and C_8 aromatics from toluene	46	260–315	Zeolite	–	Service life of catalyst around 1.5 years
5. Extraction processes:						
Udex process	Extraction of aromatics	5–8	130–150	Diethylene glycol*		6–8 : 1**
Sulfolane process	Extraction of aromatics	2	100	Sulfolane*		3–6 : 1
Arosolvan process	Extraction of aromatics	1	20–40	N-methyl-pyrrolidone*		4–5 : 1
IFP process	Extraction of aromatics	1	20–30	Dimethyl-sulfoxide*		3–5 : 1
Morphylex process	Extraction of aromatics	1	180–200	N-formyl-morpholine*		5–6 : 1
6. Extractive distillation:						
Distapex process	Production of aromatics	1	≤ 170	N-methyl-pyrrolidone*		2,5–4 : 1
Morphylane process	Production of aromatics	2	180–200	N-formyl-morpholine*		3 : 1
7. Crystallisation processes:						
Amoco process	Production of p-xylene	atmos.	−55 to −65	–	–	two-stage melt crystallisation

* Solvent
** Solvent ratio

5 Production and uses of benzene derivatives

Benzene is not only the most important aromatic raw material in terms of quantity, but it is also the most versatile from the viewpoint of its uses.

The major industrial products from benzene are alkylated derivatives such as ethylbenzene and cumene, which are used as basic materials for the production of styrene and phenol, and long-chain alkylbenzenes, which are used as feedstocks in the manufacture of surfactants.

Other significant processes for the conversion of benzene include hydrogenation to produce cyclohexane, oxidation to manufacture maleic anhydride, nitration to obtain nitrobenzene as an intermediate in the production of aniline, and halogenation to obtain chlorobenzene derivatives.

Figure 5.1 compares the main uses of benzene in Western Europe, the USA and Japan.

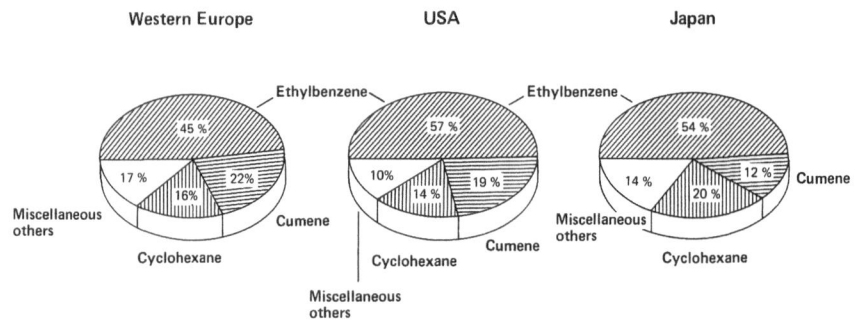

Figure 5.1: Main uses of benzene (1985)

This overview of applications for benzene shows that the three most important products from benzene are used in the manufacture of plastics in the form of polystyrene, phenolic resins and polyamide fibers. Other uses for benzene, of lesser importance in terms of quantity but still extremely versatile, are halogen and nitrogen derivatives for the chemistry of dyestuffs, production of plant protection agents and additives for rubber and plastics processing, as well as for the manufacture of pharmaceuticals.

5.1 Ethylbenzene

The synthesis of ethylbenzene from benzene and ethylene was discovered in 1879 by M. Balsohn, who introduced ethylene gas into a mixture of benzene and aluminum chloride, and with heating to 70 to 80 °C obtained ethylbenzene and higher alkylated benzenes. In 1891, ethylbenzene was found in the xylene fraction of coal tar.

Ethylbenzene is produced either by the alkylation of benzene with a C_2 component, or by isolation from aromatic fractions derived from coal or petroleum refining.

5.1.1 Recovery of ethylbenzene from aromatic mixtures

The ethylbenzene content of the xylene fraction from reformer gasoline is generally around 15%. Separation of ethylbenzene and p-xylene requires high efficiency fractionation (see Chapter 4.3) owing to their boiling points, which lie closely together (difference of 2.2 °C). In practice, columns with 300 theoretical trays and reflux ratios of approx. 100 are used. However, because of the high cost of this energy-intensive distillation, the recovery of ethylbenzene from petroleum-derived fractions has noticeably diminished in recent years and is now of minor importance, in comparison with synthesis.

5.1.2 Synthesis of ethylbenzene

The synthesis of ethylbenzene is carried out by Friedel-Crafts alkylation of benzene with ethylene.

$$C_6H_6 + CH_2{=}CH_2 \longrightarrow C_6H_5{-}CH_2{-}CH_3 \qquad \Delta H_{298} = -114 \text{ kJ/Mol}$$

A fundamental problem in the alkylation of aromatics is the likelihood of polyalkylation of the initially produced reactive alkyl aromatics. Therefore, to obtain high yields, alkylation is operated with low primary conversion and recycle of polyalkylated aromatics. In the ethylation of benzene, the primary conversion rate is kept to around 40%, and a total yield of 98 mol% of ethylbenzene is achieved after transalkylation.

Figure 5.2 shows the thermodynamic equilibrium diagram for the production of ethylbenzene.

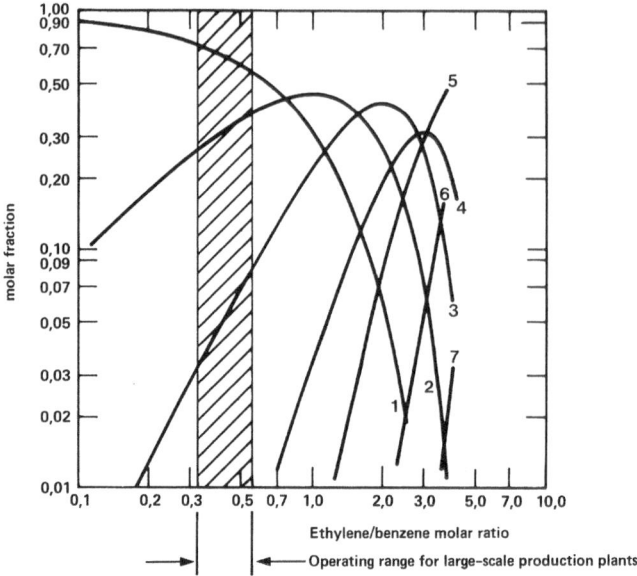

1 Benzene; **2** Ethylbenzene; **3** Diethylbenzene; **4** Triethylbenzene; **5** Tetraethylbenzene; **6** Pentaethylbenzene; **7** Hexaethylbenzene

Figure 5.2: Thermodynamic equilibrium of ethylbenzenes

The optimum molar ratio of ethylene to benzene is around 0.9; as ethylene concentration increases, so the level of polyalkylated benzene derivatives rises. In order to assure high selectivity with satisfactory conversion rates, the synthesis of ethylbenzene is carried out industrially with molar ratios of ethylene to benzene of only 0.35 to 0.55.

On an industrial scale, the reaction takes place either in the gas or the liquid phase. The application of a fixed bed reaction system requires higher temperatures than the liquid phase system.

Friedel-Crafts alkylation with aluminum chloride in the liquid phase, using co-catalysts, is of major technical importance. Because of the corrosive effect of the catalyst, the traditional process is carried out in enamelled, glass-lined or ceramic tile-clad tower reactors, with a capacity of around 80 m^3, at temperatures of 125 to 140 °C and a pressure of 2 to 4 bar; more modern processes operate (for better energy efficiency) at pressures of 7 to 10 bar and temperatures up to 180 °C. The average residence time of the reaction mixture in the bubble column reactor, into which the ethylene is sprayed and the higher alkylated benzenes are recirculated, is around 30 to 60 min. Ethyl chloride, BF_3, $FeCl_3$, $ZrCl_4$, $SnCl_4$, H_3PO_4 and alkaline-earth phosphates as Lewis acids, are used as co-catalysts. The catalyst is normally activated by HCl. Figure 5.3 shows a typical flow diagram (*Union Carbide/ Badger*) for the synthesis of ethylbenzene.

The benzene used in the reaction is dried to a residual water content of around 30 ppm (distillation, molecular sieve). The ethylene which is fed into the lower section of the reactor is 99.9% pure. The reaction product leaving the reactor is

Production and uses of benzene derivatives 135

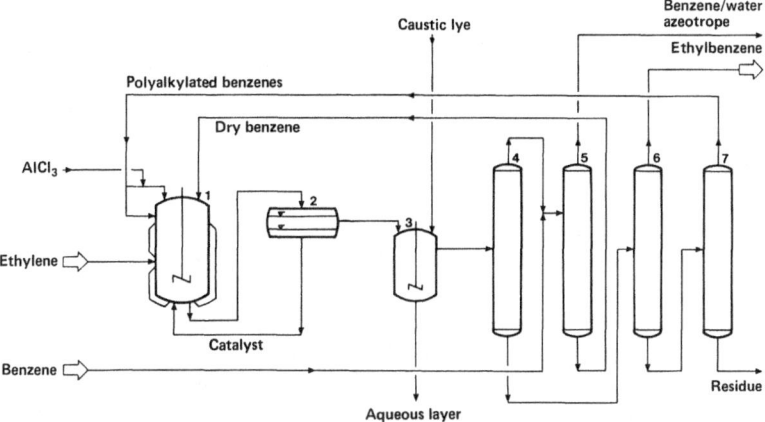

1 Alkylation reactor; **2** Catalyst separator; **3** Neutralization of catalyst residues; **4** Benzene column; **5** Benzene dewatering column; **6** Ethylbenzene column; **7** Column for polyalkylated benzenes

Figure 5.3: Flow diagram for Friedel-Crafts alkylation of benzene with ethylene in the liquid phase

Figure 5.4: Parts of the ethylbenzene unit of *Polysar*, Sarnia/Canada

cooled and taken to the settling vessel, in which the organic phase is separated from the red-brown catalyst complex. The catalyst is returned to the reactor. The organic phase is washed with water and sodium hydroxide, to remove any remaining traces of catalyst. After neutralization and separation of the aqueous phase, the crude ethylbenzene is purified by distillation; the benzene and ethylbenzene columns are each fitted with around 45 trays for this purpose.

On the basis of the ethylbenzene produced, 0.2 to 0.4% catalyst, together with 0.1% hydrogen chloride and 0.5% caustic solution, are needed.

Figure 5.4 shows parts of the ethylbenzene unit of *Polysar,* Sarnia/Canada with a total capacity of 315,000 tpa.

The Alkar process, introduced by *UOP* in 1958 and using BF_3/Al_2O_3 as a catalyst, is likewise operated in the liquid phase. Alkylation in this process occurs at 120 to 150 °C, while transalkylation is carried out in an additional reactor at temperatures of from 170 to 180 °C and a pressure of from 35 to over 100 bar. For this process, the water content of the feed benzene must be reduced to 2 to 3 ppm, since the water reacts with BF_3 to form volatile boroxyfluorides, which are concentrated in the recycled benzene.

The alkylation of benzene with ethylene in the gas phase was first achieved in 1942. Initially, Al_2O_3/SiO_2 was used as a catalyst, but this was replaced later by P_4O_{10}/SiO_2. However, transalkylation is not complete when these catalysts are applied and the yield is unsatisfactory. Since 1980, zeolite catalysts have found application in gas-phase production of ethylbenzene. The *Mobil-Badger* process uses a ZSM-5-zeolite catalyst at a temperature of 420 to 430 °C and a pressure of

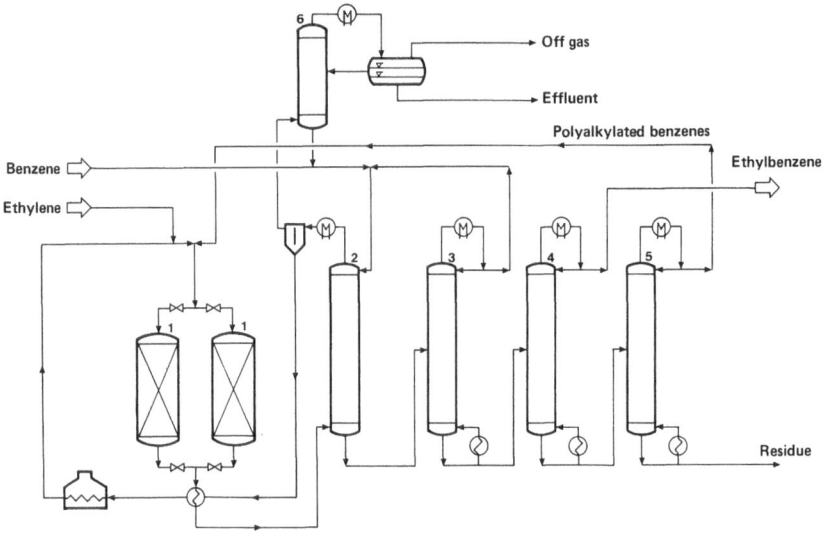

1 Alkylation reactors; **2** Prefractionation column; **3** Benzene column; **4** Ethylbenzene column; **5** Column for polyalkylated benzenes; **6** Waste-gas scrubber

Figure 5.5: Process flow diagram for the alkylation of benzene by the *Mobil-Badger* process

15 to 20 bar. The advantage of this process consists in the non-corrosiveness of the catalyst; since it can be regenerated, there are no problems of deposition arising from the spent catalyst.

In continuous plant operation, the coke produced by carbonization is burned off every 2 to 3 weeks; a switch is then made over to a stand-by reactor (swing reactor).

Figure 5.5 shows the flow diagram for the *Mobil-Badger* ethylbenzene process.

Production methods for ethylbenzene are very sophisticated from the technical point of view, as indicated by current high yields of around 98%.

Since ethylbenzene is used almost exclusively for the production of styrene, the quality of technical ethylbenzene, of around 98% purity, corresponds to the needs of the styrene process. The boiling range is between 134 and 137 °C, density from 0.8676 to 0.8684 and benzene content 1% maximum.

Table 5.1 summarizes the capacities of the major ethylbenzene producer-countries.

Table 5.1: Ethylbenzene production capacities (1985)

	(1,000 t)
USA	4,200
Canada	750
Brazil	320
France	590
West Germany	1,100
Netherlands	1,300
Great Britain	300
Spain	120
USSR	1,280
Saudi Arabia	350
Japan	1,500
Korea (South)	100
Australia	140
Taiwan	400
Others	1,550
Total capacity	14,000

Consumption of ethylbenzene in 1985 was 3.9 Mt in the USA, 800,000 t in the USSR, 1.0 Mt in the Netherlands, 1.9 Mt in the Far East (Japan, Taiwan, Korea) and 1.03 Mt in West Germany. The largest producers are *Dow, Sterling, Cos-Mar* (USA) and *BASF* (West Germany).

5.1.3 Production of styrene

Next to ethylene and vinyl chloride, styrene is the most important monomer building block in the production of plastic materials; in addition, it is used to make synthetic rubber, such as styrene-butadiene copolymer (SBR) and other polymers. The first mention of styrene appeared in the Dictionary of Practical and Theoretical Chemistry, published by William Nicholson in 1786. H. Bonastre succeeded in isolating styrene from the essential oils of the liquid amber orientalis tree in 1831. The Berlin pharmacist, Eduard Simon, gave the name styrene to the oil which he obtained from steam distillation of storax in 1839. Pierre E. M. Berthelot later found styrene in the xylene fraction of coal tar.

Eduard Simon established that when styrene was left to stand for a period of time, it changed into a viscous fluid, which eventually became solid; he erroneously took polystyrene to be an oxidation product of styrene. In 1848, at the Royal College of Chemistry in London, August Wilhelm v. Hofmann and John Blythe established that this product had the same composition as styrene; they called the polymer 'mety-styrene'. In 1920, Hermann Staudinger determined the exact nature of the styrene polymer and named it polystyrene.

$$\left[CH_2 - CH\underset{\displaystyle\bigcirc}{} \right]_n$$

Polystyrene

In 1925, the *Naugatuck Chemical Co.* (*Uniroyal*) in the USA initiated production of polystyrene (PS); in 1930, *Dow Chemical Co.* and *IG Farbenindustrie* in Ludwigshafen started production. Styrene gained more importance during World War II, particularly in the production of rubber-like copolymers with butadiene.

The commercial production of styrene nowadays is carried out almost exclusively by catalytic dehydrogenation of ethylbenzene. *Toray* has developed a process for recovery from pyrolysis gasoline, which contains 3 to 5% styrene. The method involves hydrogenation of the aliphatic diene components of a close-cut pyrolysis gasoline (130 to 140 °C) followed by extractive distillation with dimethylacetamide.

$$CH_3-\underset{\underset{\displaystyle N(CH_3)_2}{\|}}{C}$$

Dimethylacetamide

Isolation of styrene from coke-oven benzole or coal tar light oil is no longer of any practical significance; styrene, however, has to be separated from these coal-derived C₉-fractions (indene) for the manufacture of indene/coumarone resins as styrene brings about a deterioration of compatibility.

$$\underset{}{C_6H_5-CH_2-CH_3} \longrightarrow \underset{}{C_6H_5-CH=CH_2} + H_2 \quad \Delta H_{873} = 125 \text{ kJ/Mol}$$

The dehydrogenation of ethylbenzene is carried out at temperatures up to approx. 600 °C. At higher temperatures, the thermal decomposition of styrene and ethylbenzene increases and reduces the yield. Since dehydrogenation is endothermic, high temperatures and low pressure (less than 1 bar in modern units) favor the reaction (Le Cateliers's principle). In order to lower the partial pressure and prevent carbon deposits or the need for the frequent regeneration of the catalyst, steam is injected into the reactor. In industrial practice, the primary conversion is restricted to limit the formation of undesirable by-products; the unconverted ethylbenzene is recycled, to achieve a high styrene yield; thus only small amounts

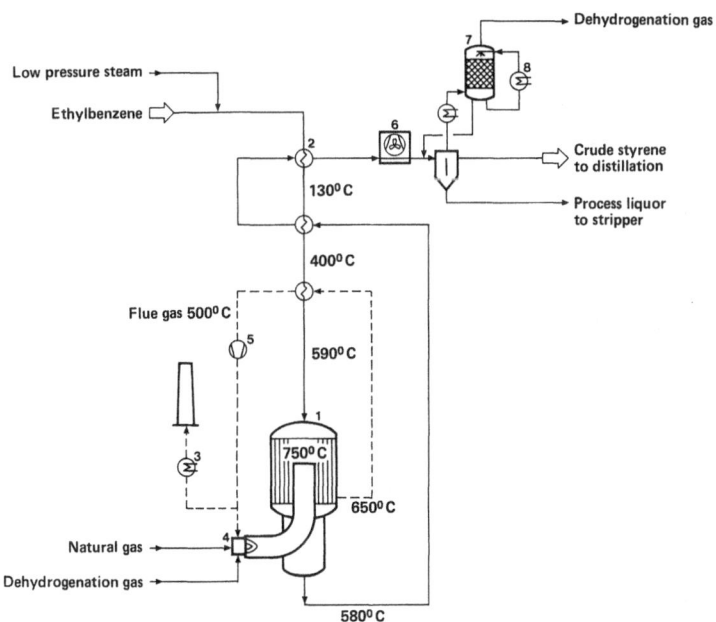

1 Reactor; **2** Ethylbenzene-vaporizer; **3** High-pressure steam generator; **4** Combustion chamber; **5** Flue gas blower; **6** Air condenser; **7** Scrubber; **8** Brine cooler

Figure 5.6: Flow diagram for the isothermal dehydrogenation of ethylbenzene

of benzene, toluene and higher boiling aromatics are produced. Iron oxides are generally used as catalysts (active components); they are modified with chromium oxide (stabilization component), potassium oxide (coking inhibitor) as well as a promoter (CuO, V_2O_5). Conversion of ethylbenzene with the newer catalyst types is 60 to 70%, selectivity is 85 to 90 %. Catalyst life ranges typically from 2 to 4 years.

Both the isothermal and the adiabatic dehydrogenation processes are applied in industrial styrene production. The flow diagram for the *BASF* dehydrogenation process is shown in Figure 5.6.

Ethylbenzene and low-pressure steam (steam ratio approx. 1.2) are fed to a vaporizer, heated in heat exchangers to over 590 °C and then passed into a reactor. A multi-tubular reactor is used (tube diameter 100 to 200 mm, length 2.5 to 4.0m), and is heated by flue gas at 720 to 750 °C. The reaction product, leaving the reactor at a temperature of around 580 °C, is cooled in heat exchangers and the ethylbenzene vaporizer to 160 °C and condensed in an air cooler.

The crude styrene contains 60 to 65% styrene and higher boiling aromatics, along with 30 to 35% ethylbenzene; the proportion of benzene and toluene is around 5%. The reaction gas consists of approx. 90 to 95% hydrogen and also contains some each 5% of CO_2 and C_1-C_2 components. Crude styrene is purified by distillation in a series of distillation trains.

The adiabatic dehydrogenation process, which is operated in single-train plants with a capacity of up to 500,000 tpa, has been developed by *Dow;* Figure 5.7 shows the flow diagram for the process.

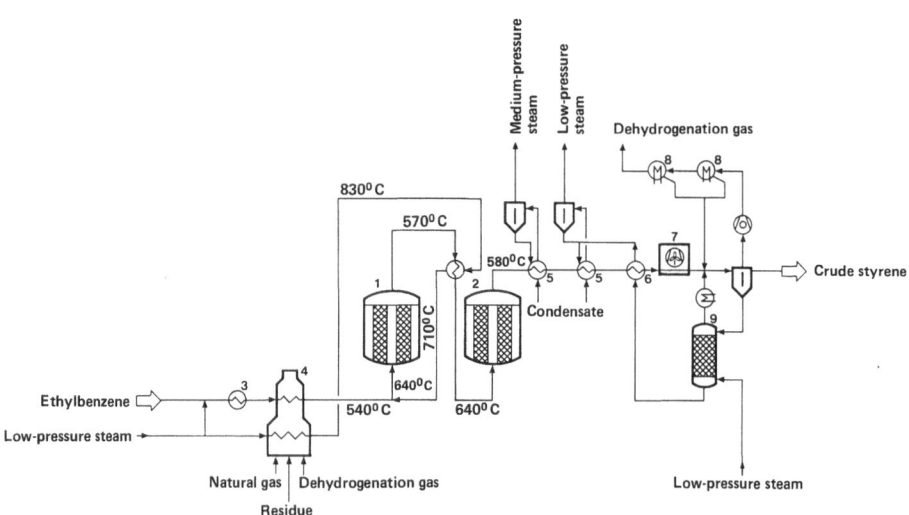

1 and 2 Reactor; 3 Vaporizer; 4 Tubular furnace; 5 Steam generator; 6 Preheater; 7 Air condenser; 8 Brine cooler; 9 Stripper

Figure 5.7: Flow diagram for adiabatic dehydrogenation of ethylbenzene

The dehydrogenation is carried out in two stages. In the first, ethylbenzene and steam (approx. 10 to 15% of the total dilution steam) are heated to around 550 °C. Additional superheated steam at 710 °C is fed in, so that the mixture enters the first reactor at a temperature of 630 to 640 °C, and some 35% of the ethylbenzene is converted. In the second stage, further dehydrogenation occurs, following additional superheating to 640 °C. The styrene content of the reaction product is 60 to 65%.

Purification of the crude styrene must be carried out at the lowest possible temperature by vacuum distillation, to avoid polymerization; as a further measure to suppress the tendency to polymerize, polymerization inhibitors, such as 4-tert-butylcatechol, are added.

The separation of ethylbenzene and styrene is relatively intricate, because of the close boiling points (136.2 °C, and 145.2 °C resp.). To avoid overheating the bottoms, highly-efficient trays with low pressure drop are used.

Figure 5.8 shows a flow diagram for the distillation of crude styrene.

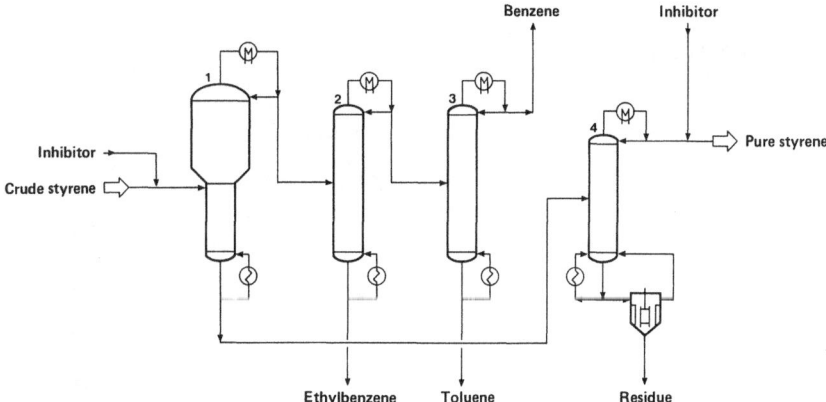

1 Ethylbenzene column; **2** Light ends column; **3** Benzene/toluene column; **4** Styrene column

Figure 5.8: Flow diagram for the distillation of styrene

Figure 5.9 shows part of the distillation section of the *Dow* 700,000 tpa styrene plant at Terneuzen/Netherlands.

A comparison of adiabatic and isothermal processes, with regard to investment costs, comes down in favor of the adiabatic process, since in the isothermal process the reactors have an individual capacity of 100,000 tpa maximum, whereas the adiabatic process can use significantly larger reactor units. The total economics of styrene production, however, depend to a great extent on local energy and raw material costs.

Figure 5.9: Part of the styrene distillation plant operated by *Dow*, Terneuzen/Netherlands

The specifications for purified styrene are based on its use in polymerization. The concentration is generally 99.9 wt%, the content of benzaldehyde below 50 ppm and the content of C_8- and C_9-aromatics between 500 and 1,000 ppm.

Other processes for the production of styrene, apart from dehydrogenation of ethylbenzene have been developed, but these have only limited industrial importance.

The production of styrene by dehydrochlorination of chloroethylbenzene, which is produced by side-chain chlorination of ethylbenzene, has the disadvantage that nuclear-chlorinated aromatics arise during the process and can only be separated with great difficulty by up-stream refining.

[Reaction scheme: ethylbenzene + Cl₂, −HCl → 1-chloro-1-phenylethane (PhCHCl–CH₃) and 2-chloro-1-phenylethane (PhCH₂–CH₂Cl); −HCl → styrene (PhCH=CH₂)]

Reduction of acetophenone to methylphenylcarbinol and its dehydration to styrene is also of no industrial significance at the present time.

[Reaction scheme: acetophenone (PhCO–CH₃) + H₂ → methylphenylcarbinol (PhCH(OH)–CH₃) → −H₂O → styrene (PhCH=CH₂)]

Methylphenylcarbinol is also an intermediate in the *Halcon* process, in which ethylbenzene is oxidized to a hydroperoxide at around 130 °C with air, then converted with propylene into propylene oxide and carbinol. The carbinol is subsequently dehydrated on a titanium catalyst at 180 to 280 °C to styrene. This process, first commercialized by *Atlantic Richfield,* has found large-scale application in a few isolated cases (e.g. *Shell* (Netherlands), *Alcudia* (Spain) and *Nihon Oxirane* (Japan)); it is only viable if there is sufficient demand for propylene oxide.

[Reaction scheme: ethylbenzene + O₂ → ethylbenzene hydroperoxide (PhCH(OOH)–CH₃); + CH₃–CH=CH₂, − CH₃–CH–CH₂ (propylene oxide) → methylphenylcarbinol (PhCH(OH)–CH₃); −H₂O → styrene (PhCH=CH₂)]

Toluene is used as a raw material in the oxidative conversion of toluene into stilbene, which can be transformed in a subsequent reaction (metathesis) with ethylene into two molecules of styrene.

[Reaction scheme: 2 toluene → stilbene (PhCH=CHPh); + CH₂=CH₂ → 2 styrene (PhCH=CH₂)]

The production of styrene from butadiene has also been examined. The dimerization of butadiene in a Diels-Alder reaction yields 4-vinylcyclohexene, which can be dehydrogenated oxidatively into styrene. This process has not yet achieved any commercial importance.

Table 5.2 summarizes the production of styrene in the most important producer countries; the largest producers are *Dow, Cos-Mar* (USA) and *BASF* (West Germany).

Table 5.2: Production of styrene (1985)

	(1,000 t)
USA	3,400
Brazil	240
Canada	500
West Germany	1,000
Netherlands	870
Italy	290
Great Britain	230
Spain	90
Japan	1,400
Korea (South)	80
USSR	940
Taiwan	240
Australia	110
Others	410
Total production	9,800

5.1.4 Substituted styrenes

Among the substituted styrenes, divinylbenzene, vinyltoluene and α-methylstyrene are of importance as building blocks for polymers.

o-Vinyltoluene   m-Vinyltoluene   p-Vinyltoluene

Production and uses of benzene derivatives   145

Divinylbenzene

α-Methylstyrene
(Isopropenylbenzene)

Divinylbenzene is obtained from diethylbenzene by dehydrogenation, at temperatures in excess of 600 °C with superheated steam (Figure 5.10). o-Diethylbenzene is converted into naphthalene by this process, and this must be separated during subsequent refining of the reaction product, since it contains no polymerizable functional groups and, moreover, it has a certain plasticizing effect. The main application is the manufacture of ion-exchange resins, by co-polymerization with styrene.

1 Reactor; 2 Crude divinylbenzene column; 3 Light ends column; 4 Pure divinylbenzene column

Figure 5.10: Flow diagram for the production of divinylbenzene

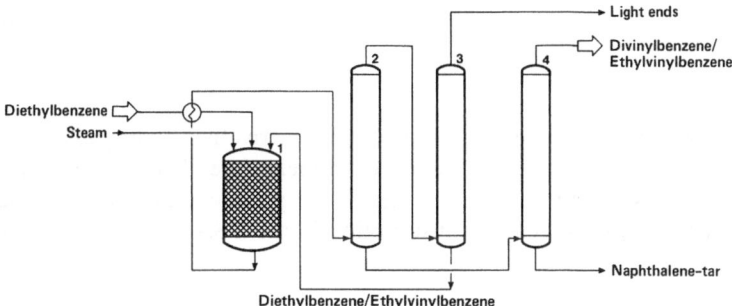

The demand for vinyltoluene is considerably higher than that for divinylbenzene, reaching around 30,000 to 35,000 tpa. Ethyltoluene is used as a raw material. It is produced by Friedel-Crafts alkylation of toluene with ethylene in the presence of $AlCl_3$ at 80 °C in the liquid phase, or in the gas phase, using zeolite catalysts. During the refining operation, care must be taken to ensure that the o-ethyltoluene is separated from the isomer mixture, since it forms indene during the adiabatic steam pyrolysis carried out at 450 to 500 °C; indene has a marked effect on the properties of the polymers. The commercial product consists of around 60% m- and 40% p-vinyltoluene. The most important use for vinyltoluene is its application as a co-monomer in combination with styrene, particularly for the production of copolymers with high heat resistance and good flow properties, and for modifying drying oils and alkyd resins. The presence of the methyl group in the aromatic ring increases the solubility in aliphatic solvents.

A new method of synthesizing p-vinyltoluene (p-methylstyrene) is based on the reaction between toluene and acetaldehyde, whereby p-methylstyrene can be produced in over 90% yield at high conversion rates from the intermediate product 1,1-ditolylethane (DTE).

α-Methylstyrene is another important styrene derivative, arising as a by-product in the synthesis of phenol by oxidation of cumene (see Chapter 5.3.1); it is used predominantly as a reactive co-monomer in the production of acrylonitrile/butadiene/styrene polymers (ABS). Consumption in the USA in 1985 was 22,000 t.

5.2 Cumene

After ethylbenzene, cumene (isopropylbenzene) is the most important industrial derivative of benzene. Cumene is also produced by benzene alkylation; recovery from the corresponding coal or petrochemical streams is uneconomical, owing to the low levels of cumene present.

$\Delta H_{298} = -112$ kJ/Mol

Cumene gained particular technical importance following the discovery of the phenol synthesis via the oxidation of cumene by Heinrich Hock and Shon Lang in Clausthal/Germany in 1944. Cumene had already been obtained by Cornelius Radziewanowski in 1895, by reaction of benzene and isopropyl chloride in the presence of aluminum filings and dry hydrogen chloride. A 66 % yield was achieved. Large-scale production began in the USA in the 1940's, when cumene was needed as an additional component in high-value aircraft fuels.

The reaction of benzene and propylene is carried out in a similar way to the reaction of benzene and ethylene in the production of ethylbenzene. An excess of benzene is used and, in special processes, transalkylation and separation of the reaction products by distillation and subsequent recycle are included to reduce the formation of highly alkylated products.

As in ethylbenzene production, alkylation can be performed either in the gas or in the liquid phase. Propylation of benzene in the liquid phase is achieved with sulfuric acid or aluminum chloride as catalyst at 30 to 40 °C, or with hydrogen fluoride at 50 to 70 °C, and propylene pressures up to 7 bar. The propylene used must be largely free from other olefins. A propylene/propane mixture, such as occurs in refinery gases, can be used for the reaction, since propane is not converted and can readily be stripped from the reaction products.

Gas-phase alkylation with a benzene/propylene ratio of 5:1 takes place using phosphoric acid silicate catalysts at temperatures of 250 to 350 °C and pressures of 30 to 45 bar. The reaction is carried out in the presence of steam to achieve better control of the exothermic conversion, and at the same time to fix the phosphoric acid better to the carrier through the formation of hydrates. In order to produce one part by weight of cumene, 0.67 parts of benzene and 0.38 parts of propylene are converted. Around 0.05 parts by weight of heavy aromatics (tar) arise through polyalkylation and condensation; selectivity is 96 to 97 mol %.

Figure 5.11 shows the dominant gas phase process (*UOP*) carried out in single-train plants with a capacity of up to 300,000 tpa.

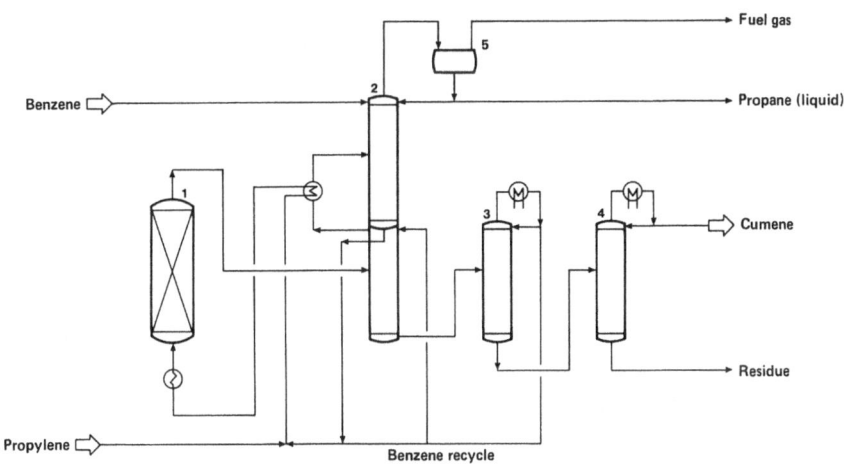

1 Reactor; 2 Fractionation column; 3 Benzene column; 4 Cumene column;
5 Gas separator

Figure 5.11: Flow diagram for the synthesis of cumene

148   Production and uses of benzene derivatives

There is no transalkylation in this process, since the high excess of benzene keeps the proportion of polyalkylated benzenes low; in addition, the phosphoric acid catalyst does not have a transalkylating effect.

The specifications for pure cumene are generally based on its use in the production of phenol. Generally, the degree of purity is 99.9%. The content of butylbenzenes and propylbenzenes, together with ethylbenzene, is set at 500 ppm maximum. The total sulfur content should not exceed 2 ppm and the content of olefins is limited to 200 to 700 ppm.

Table 5.3 summarizes the production figures for the major cumene producer-countries.

Table 5.3: Production of cumene (1985)

	(1,000 t)
USA	1,500
Brazil	150
France	230
West Germany	430
Italy	420
Netherlands	240
Great Britain	180
Spain	100
Soviet Union	585
Roumania	145
Japan	370
Others	150
Total production	4,500

Production capacity is around 1.8 Mt in the USA, 550,000 t in Japan and around 1.8 Mt in Western Europe. The largest producers are *Chevron, Shell* (USA), *Veba Oel* (West Germany) and *Mitsui Petrochemical* (Japan).

Apart from the small quantities used in the direct production of α-methylstyrene, cumene is used virtually exclusively in the production of phenol; additionally, small amounts are also used in the manufacture of p-nitrocumene (see Chapter 8.4).

5.3 Phenol

Phenol was discovered in coal tar as early as 1834, by Friedlieb Ferdinand Runge. Coal tar remains an important source of phenol and its alkyl derivatives, even today. For a long time, phenol derived from coal tar was available in sufficient quantities to meet demand. However, requirements for phenol rose sharply with the use of picric acid as an explosive during the Boer War (1899-1902) and World War I, so that new means of production were developed.

The first large-scale synthesis of phenol was based on benzene sulfonation and subsequent alkali fusion; other methods of phenol synthesis were developed during the first half of the present century.

5.3.1 Phenol from cumene

The consumption of phenol rose noticeably after the 1920's with the increase in its importance as a raw material for phenolic resins, for the manufacture of ε-caprolactam and the production of bisphenol A. The oxidative conversion of cumene into phenol and acetone has proved to be an environmentally safe and economical process; thus it gained an edge over the other processes and is presently the only phenol process operated on a worldwide scale.

Conversion of cumene occurs via the cumene hydroperoxide; protonation and subsequent transformation of the phenyl group produces a carbonium ion which is stabilized by the intermediate phenyl ether, and reacts further to phenol and acetone. In a secondary reaction, α-methylstyrene is formed by a radical-initiated mechanism (see Chapter 2.2.3.2).

Cumene conversion was developed to the production stage by *BP Chemicals* and *Hercules* in the early 1950's, and modified further by *Phenolchemie*.

Figure 5.12 shows the flow diagram for phenol production from cumene.

Cumene is first oxidized at a temperature of 90 to 100 °C and at a pressure of 6 bar by air, with the addition of soda solution (pH value 7 to 8), to yield the hydroperoxide; oxidation is carried out to a cumene hydroperoxide content of around 30%. Unconverted cumene is then separated from the hydroperoxide by distillation and the hydroperoxide is concentrated to 65 to 90%. Catalytic cleavage of cumene hydroperoxide occurs with the addition of 0.1 to 2% sulfuric acid (40%) at boiling temperature, whereby the reaction mixture is cooled by vaporization of acetone. To avoid increased formation of by-products, the residence time is restricted to 45 to 60 seconds.

The reaction product is neutralized with dilute caustic solution or phenolate lye and, after the separation of the catalyst, split into its constituents by distillation.

Crude acetone is distilled off as the first fraction and is refined by an alkali wash followed by distillation to yield the pure product. From the bottom product of the crude acetone column, firstly cumene is recovered and fed back into the process, followed by the recovery of α-methylstyrene. If the α-methylstyrene cannot be directly marketed, it is transformed into cumene by hydrogenation.

The crude phenol, which was distilled off in the crude phenol column is refined by extractive distillation with water; subsequent redistillation yields 99.9% pure phenol. The high molecular weight tar which remains as a distillation residue contains acetophenone and cumylphenol, among other compounds, and can be thermally cracked in the liquid phase into phenol, α-methylstyrene and cumene; tar oils boiling over 300 °C are used as a reaction medium.

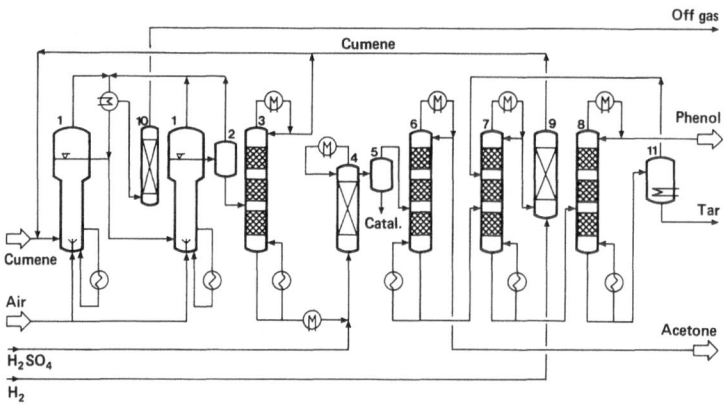

1 Bubble column reactor; 2 Gas separator; 3 Cumene column; 4 Cleavage reactor; 5 Catalyst separation vessel; 6 Acetone column; 7 Cumene/methylstyrene column; 8 Phenol column; 9 Hydrogenation reactor; 10 Waste gas purification; 11 Cracker

Figure 5.12: Flow diagram for the synthesis of phenol from cumene

The acetophenone is made to react with formaldehyde to produce light-resistant resins with a softening point between 75 and 80 °C, which are used, in particular, as additives in nitrocellulose paints.

Acetophenone   Acetophenone synthetic resins

The production of phenol by cumene oxidation is carried out in plants with a total capacity in excess of 400,000 tpa. The process is superior to the methods described below, despite the co-production of acetone, since the cycle of inorganic materials is very small and the carbon skeleton of the aromatic feed is fully utilized, unlike in toluene oxidation. Figure 5.13 shows the oxidation reactors of the 400,000 tpa phenol plant operated by *Phenolchemie*, Gladbeck/West Germany.

Figure 5.13: Oxidation reactors for the production of phenol from cumene, *Phenolchemie*, Gladbeck/West Germany

5.3.2 Alternative methods for the synthesis of phenol

The oldest process for phenol production is based on the sulfonation of benzene. This route goes back to Michael Faraday, who recognized as early as 1825 that phenol could be obtained from benzenesulfonic acid by alkali fusion.

Charles Adolphe Wurtz and August Kekulé further improved the process in 1867. The sulfonation route was developed to technical maturity in particular by *Bayer* and *Monsanto*. The first stage in the synthesis is the sulfonation of benzene with sulfuric acid, generally using a 100% excess of sulfuric acid. In the second stage, the sulfonation product is neutralized with sodium hydroxide or sodium sulfite. The resulting benzenesulfonate, in solid form or as a concentrated aqueous solution, is heated with sodium hydroxide at temperatures of 320 to 340 °C, in cast-iron pans. The alkaline sodium phenolate solution is neutralized ('saturated') with CO_2 or, in the *Monsanto* process, with SO_2. When the water has been distilled off substantially pure phenol with a crystallizing point of 40.5 °C is obtained.

The disadvantages of the benzenesulfonic acid process are the production of large amounts of sodium sulfite and the highly corrosive reaction conditions of the alkali fusion; in addition, the process is only carried out semi-continuously.

Figure 5.14 shows the flow diagram for the *Monsanto* process for the production of phenol.

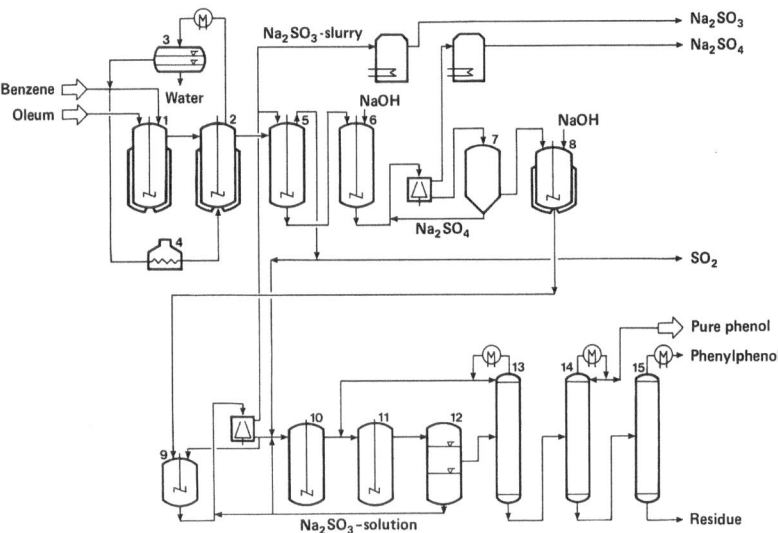

1 and 2 Sulfonation reactors; 3 Water separator; 4 Benzene vaporizer; 5 Pre-neutralization vessel; 6 Final neutralization vessel; 7 Concentrator; 8 Fusion vessel; 9 Dissolving tank; 10 and 11 Neutralization vessel (crude phenol precipitation); 12 Crude phenol separator; 13 Dewatering column; 14 Phenol column; 15 Residue column

Figure 5.14: Flow diagram for synthesis of phenol by benzene sulfonation and alkali fusion

In the early 1930's a process was developed by *Raschig* for the synthesis of phenol, the principle of which had already been found in England in the very early days of aromatic chemistry, namely, oxychlorination of benzene with subsequent alkaline hydrolysis. This route had been originally discovered by L. Dusart and Ch. Bardy in 1872.

In the first stage of this process, benzene is transformed into chlorobenzene with air and hydrochloric acid in the presence of a copper catalyst on an alumina base at 275 °C. To avoid the formation of polychlorinated benzenes as far as possible, conversion is limited to 15% and the process is operated with a large excess of benzene, so that only 5 to 8% dichlorobenzenes result. After the reaction mixture has been separated, the unconverted benzene is recycled; the polychlorinated benzenes are refined to recover o-/p-dichlorobenzenes (see Chapter 5.8.2).

In the second stage, chlorobenzene is converted with steam at 450 to 500 °C on a calcium phosphate apatite catalyst. Conversion is generally 10 to 12%. By recycling the unconverted chlorobenzene, a gross conversion rate for chlorobenzene to phenol of 90 to 93% is achieved.

Figure 5.15 shows the flow diagram for the *Raschig* process.

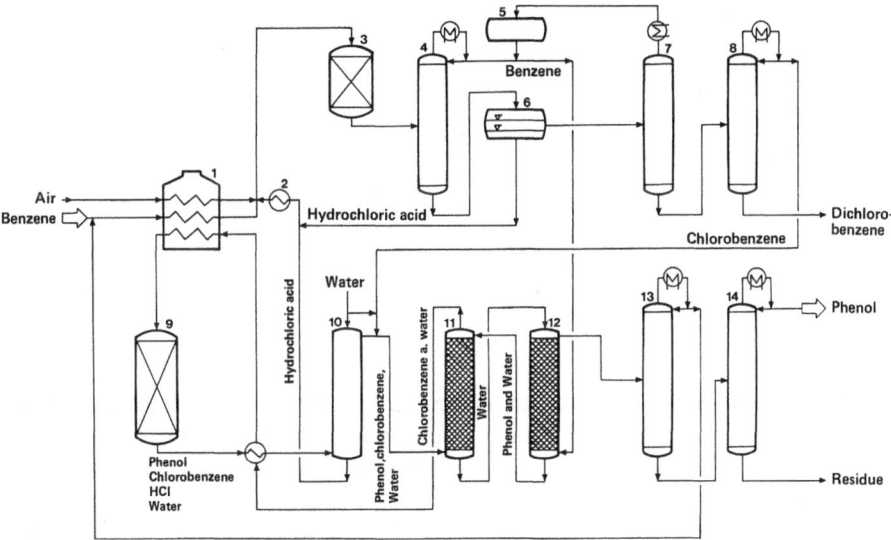

1 Vaporizer/Heater; 2 Hydrochloric acid vaporizer; 3 Chlorination reactor; 4 Pre-column; 5 Benzene storage tank; 6 Chlorobenzene separator; 7 Benzene column; 8 Chlorobenzene column; 9 Hydrolysis reactor; 10 Hydrochloric acid separator; 11 and 12 Extraction columns; 13 Benzene column; 14 Phenol column

Figure 5.15: Flow diagram for the synthesis of phenol by benzene chlorination

A modified method for the hydrolysis of chlorobenzene, operated at around 30 bar and 290 °C, was developed by *Dow*. The reaction was carried out in long tubular reactors, which were built from mild steel. The conversion of chlorobenzene to phenol was around 85%. Diphenyl ether, produced as a by-product, was used as a heat transfer oil.

The chlorobenzene processes for the production of phenol have lost their importance since the 1970's. Occasionally, the toluene oxidation process, also developed by *Dow* is still used. In the first stage of this process, toluene is oxidized to benzoic acid with air in the liquid phase at 150 to 170 °C and 5 to 10 bar, in the presence of cobalt salts, with 90% selectivity. By-products are methylbiphenyls, benzyl alcohol, benzaldehyde and esters. Following the purification of the crude product by distillation or crystallization, the benzoic acid is transformed into phenol in the presence of copper (II) salts with air and steam at 230 to 250 °C, and 2 to 10 bar, by way of the intermediate compounds copper benzoate, benzoylsalicylic acid and phenyl benzoate. The recovered crude phenol is refined by distillation. The molar yield of phenol is around 85 to 90%.

The economics of this process are determined by the price of the toluene. When there is high demand for aromatics, e.g. for gasoline to increase the octane rating, the consequent increase in the value of toluene makes the process uneconomical in comparison with the cumene route.

Figure 5.16 shows the flow diagram for the production of phenol by toluene oxidation.

Production and uses of benzene derivatives   155

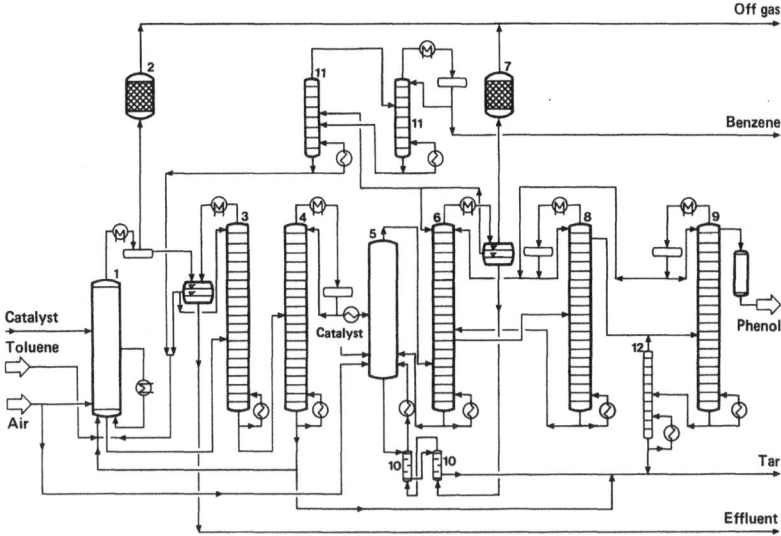

1 Toluene oxidation reactor; **2** Waste-gas purification; **3** Toluene stripping column; **4** Benzoic acid column; **5** Benzoic acid oxidation reactor; **6** Water-HC stripping column; **7** Waste-gas purification; **8** Crude phenol column; **9** Phenol column; **10** Residue extraction; **11** Benzene columns; **12** Phenol bottoms column

Figure 5.16: Flow diagram for the synthesis of phenol from toluene

The phenol process based on the oxidation of cyclohexane has been operated for a short time by *Monsanto* in Australia and is of less importance. In this process, a mixture of cyclohexanone and cyclohexanol is dehydrogenated to phenol at 400 °C, using platinum/activated carbon or nickel/cobalt catalysts. The degree of conversion can reach 90±5%. The crude phenol is refined by distillation. A particular disadvantage of this process lies in the difficulty in refining the crude oxidation mixture from cyclohexane oxidation.

5.3.3 Recovery of phenol from the products of coal pyrolysis

The oldest commercial process for the manufacture of phenol is its extraction from tar fractions. Since oxygen is contained in the macromolecules of coal, phenols are formed during their pyrolytic decomposition. The conversion of coal-oxygen into phenolic-oxygen is dependent on temperature. At the high carbonization temperatures (1,200 °C) required for the production of blast-furnace coke, part of the oxygen is converted into water so that the proportion of phenols in high temperature tar is relatively low. Tars from low-temperature carbonization (400 to 700 °C) have a particularly high content of phenols. Such tars arise during the manufacture of smokeless solid fuels and during *Lurgi* coal gasification.

In the tar refining process, phenol is recovered from the carbolic oil fraction after primary distillation (see Chapter 3.2.3).

Table 5.4 shows the composition of a carbolic oil obtained by refining of high-temperature coal tar; its processing scheme is depicted in Figure 5.17.

Table 5.4: Composition of carbolic oil

Phenols		25%
Phenol	44%	
o-Cresol	15%	
m-Cresol	22%	
p-Cresol	11%	
Xylenols	8%	
Neutral oils		72%
Bases		3%
Total fraction		100%

Carbolic oil, with a phenol content of around 25%, is extracted in the first stage with caustic soda. The bases and neutral oils are driven off from the crude phenolate lye by steam distillation ('steam stripping'). In the next step of the process, a two-stage neutralization ('springing') of the refined phenolate lye with CO_2 is carried out, to recover the crude phenol. The crude phenol can be further refined by extraction with diisopropyl ether, and then worked up to phenol and alkylphenols by distillation and crystallization.

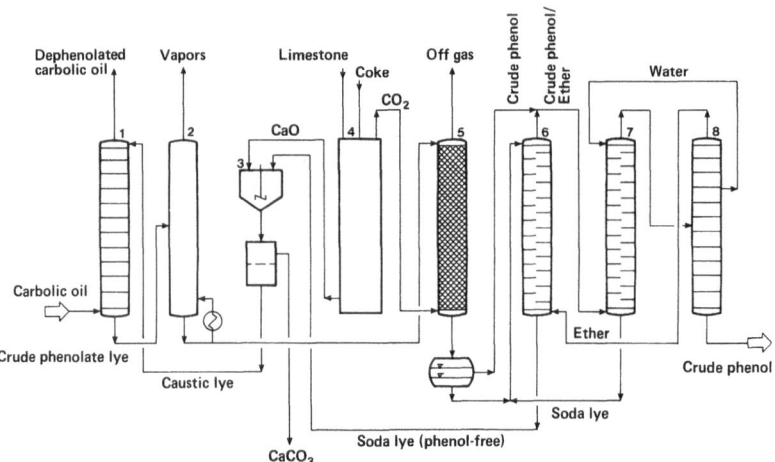

1 Carbolic oil extraction; 2 Steam stripping; 3 Causticization; 4 Lime kiln; 5 Acidification (liberation of crude phenols); 6 Soda lye extraction; 7 Crude phenol extraction; 8 Distillation

Figure 5.17: Flow diagram for the recovery of crude-phenol from carbolic oil

In addition to refining crude phenol by distillation, it is possible to carry out the adsorptive separation of the components by the Sorbex process (*UOP*).

Production and uses of benzene derivatives 157

The specifications for the quality of phenol are based on its downstream application. The phenol content is generally over 99%, the water content below 0.1%. For ε-caprolactam and bisphenol A production, only a low level of carbonyl compounds is in fact tolerable. The nature of by-products accordingly depends on the respective synthesis route. Phenol produced from cumene contains acetophenone and α-methylstyrene as co-products. Phenol manufactured by the *Raschig* method contains small amounts of chlorophenol; tar phenols contain minor proportions of nitrogen and sulfur components.

Table 5.5 summarizes the production figures for the major phenol producers; the largest manufacturers are *Allied-Signal, Shell* (USA) and *Phenolchemie* (West Germany).

Table 5.5: Production of phenol (1985)

	(1,000 t)
USA	1,250
Brazil	100
France	125
West Germany	410
Italy	285
Netherlands/Belgium	100
Great Britain	125
Spain	80
Soviet Union	520
Romania	100
Japan	260
Others	145
Total production	3,500

The most important applications for phenol are phenolic resins and the production of caprolactam and bisphenol A. Smaller amounts of phenol are used in the production of adipic acid, nonylphenol, acetylsalicylic acid, dodecylphenol, chlorophenols and xylenols.

Figure 5.18 shows the major uses for phenol in Western Europe, Japan and the USA.

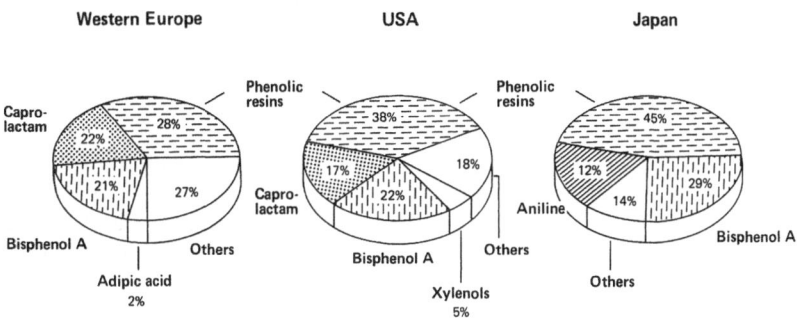

Figure 5.18: Main uses for phenol (1985)

5.3.4 Phenol derivatives

Phenol displays higher reactivity than benzene, due to the presence of its OH group, and this accounts for many of its uses.

This high reactivity is especially essential for the production of phenol/formaldehyde condensation products (Bakelite resins); phenol/formaldehyde resins were the first synthetic plastics, with production patented by Leo H. Baekeland in 1907. In addition to phenol, cresols, xylenols and long-chain alkylphenols are used to obtain special quality characteristics. Phenolic resins are used widely for numerous applications. Phenolic resin production in Western Europe in 1985 was around 500,000 t, in the USA 1,150,000 t and in Japan 325,000 t.

Phenol/formaldehyde resin

In terms of quantity, other polymer building blocks, such as bisphenol A and caprolactam are also among the most important phenol products.

5.3.4.1 Bisphenol A

Bisphenol A was first obtained by Alexander P. Dianin in 1891, by acid catalysed condensation of phenol and acetone.

In some European countries, the compound was named Dian, after its Russian discoverer. As a cost-effective 'bifunctional' phenol, bisphenol A is particularly suited for the production of resins, as a result of its high reactivity and bifunctionality. Bisphenol A is complemented by other bisphenols such as bisphenol F and bisphenol S in the production of resins.

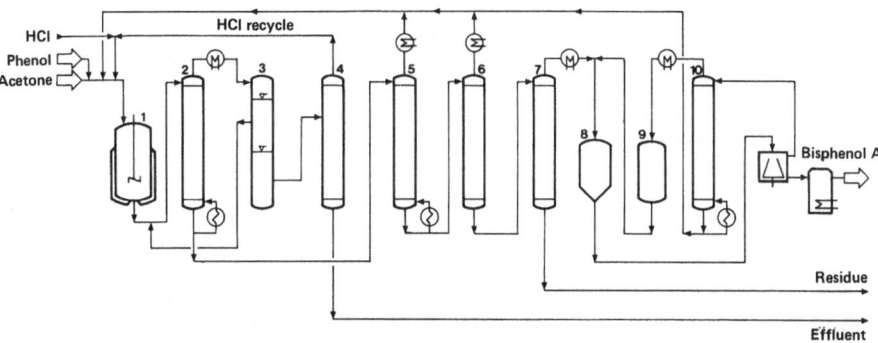

Figure 5.19 shows a flow diagram for the production of bisphenol A.

1 Reactor; **2** Hydrochloric acid column; **3** Hydrochloric acid separator; **4** HCl recovery; **5** Phenol column; **6** Isomers column; **7** Bisphenol A column; **8** Bisphenol A crystallizer; **9** Solvent tank; **10** Solvent recovery

Figure 5.19: Flow diagram for the production of bisphenol A

The acid-catalysed reaction of acetone with phenol is generally carried out at temperatures of from 50 to 90 °C with hydrogen chloride or on sulfonated cross-linked polystyrene arranged as a fixed bed, with a 15-fold molar excess of phenol. When the hydrogen chloride has been separated by distillation or neutralization, the bisphenol A crystallizes out as an adduct with phenol. The adduct is split thermally by distillation. Subsequent refining of bisphenol A is carried out by recrystallization from aromatics or a heptane-aromatics mixture; the yield is between 80 and 95%. High purity bisphenol A can be recovered by melt crystallization in a mixture with phenol. Figure 5.20 shows the phase diagram for the crystallization of bisphenol A from phenol. Crystallization gives a 1:1 adduct, which can be refined by distillation to a pure product containing up to 99.9% bisphenol A.

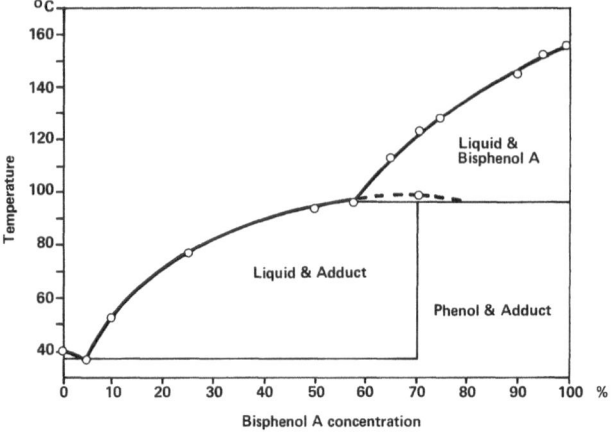

Figure 5.20: Phase diagram for the crystallization of bisphenol A

A satisfactory yield can also be obtained in the production of bisphenol A by the reaction of phenol with methylacetylene on ion-exchange resins. However, this method has not yet been applied, because of the high price of methylacetylene. Other methods which have no industrial use at present include the reaction of p-isopropenylphenol with phenol in the presence of Friedel-Crafts catalysts which, at low temperature, gives almost quantitative yields of bisphenol A.

By using alkali, (e.g. NaOH), the reaction can be reversed at 190 to 240 °C under reduced pressure to produce the educts.

Bisphenol A is mainly used to manufacture epoxy resins by the reaction with epichlorohydrin in the presence of caustic soda at 40 to 60 °C. This yields resins with a molecular weight between 450 and 4000. The molecular weight of the resins increases as the epichlorohydrin/bisphenol A ratio falls.

[structure: —[—O—C₆H₄—C(CH₃)₂—C₆H₄—O—CH₂—CH(OH)—CH₂—]ₙ—]

A further rapidly growing application for bisphenol A is the production of polycarbonates by reaction with phosgene. Polycarbonates are widely used as thermoplastic construction materials.

n HO—C₆H₄—C(CH₃)₂—C₆H₄—OH + n COCl₂ $\xrightarrow[-2n\ NaCl,\ -2n\ H_2O]{+2n\ NaOH}$ —[—O—C₆H₄—C(CH₃)₂—C₆H₄—O—C(=O)—]ₙ—

Bisphenol A is also used in the manufacture of the high-temperature resistant polysulfone plastic, Udel (*Union Carbide*), which is produced by the reaction of the di-potassium salt of bisphenol A with 4,4'-dichlorodiphenyl sulfone.

n KO—C₆H₄—C(CH₃)₂—C₆H₄—OK + n Cl—C₆H₄—S(=O)₂—C₆H₄—Cl $\xrightarrow{-2n\ KCl}$

—[—O—C₆H₄—C(CH₃)₂—C₆H₄—O—C₆H₄—S(=O)₂—C₆H₄—]ₙ—

Udel

The consumption of bisphenol A in 1985 in the USA was 350,000 t, in Japan 100,000 t and in Western Europe 280,000 t.

Care should be taken when dealing with bisphenol A to avoid dust explosions.

5.3.4.2 Cyclohexanol and Cyclohexanone

Cyclohexanol and cyclohexanone are produced from phenol as intermediates for synthetic fibers (Nylon 66, Nylon 6) obtained via adipic acid and caprolactam respectively.

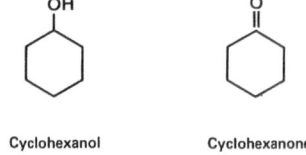

Cyclohexanol   Cyclohexanone

However, the production of caprolactam and adipic acid is predominantly based on cyclohexane (see Chapter 5.4). Cyclohexanol can be produced by catalytic hydrogenation of phenol. The hydrogenation of phenol was first described by Paul Sabatier and Jean Baptiste Senderens in 1904. Figure 5.21 shows a flow diagram for the hydrogenation of phenol.

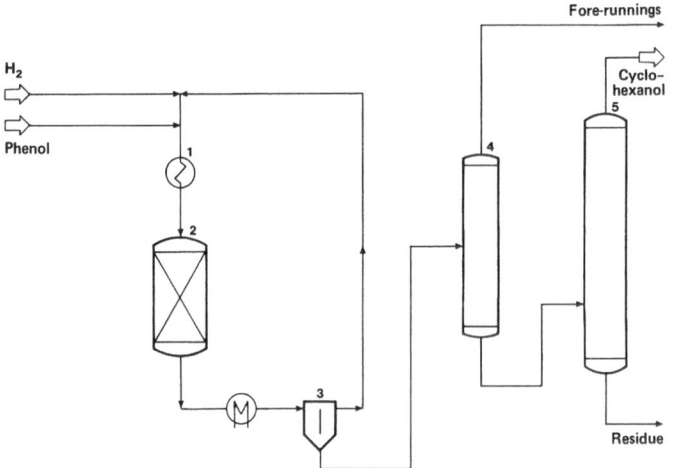

1 Phenol vaporizer; **2** Reactor; **3** Cyclohexanol separator; **4** Fore-runnings column; **5** Cyclohexanol column

Figure 5.21: Flow diagram for the hydrogenation of phenol

Phenol is vaporized with recycled and make-up hydrogen and hydrogenated by excess hydrogen at temperatures of 120 to 200 °C and 20 bar on silica or aluminum oxide catalysts, which are modified with nickel. The cyclohexanol is separated by condensation. The yield of cyclohexanol is almost quantitative.

Cyclohexanone on the other hand is produced on a larger scale from phenol by catalytic hydrogenation. Phenol is fed in the gas phase with hydrogen at 140 to 170 °C, through a catalyst bed at atmospheric pressure. The catalyst generally contains palladium, in concentrations of 0.2 to 0.5 wt% on a zeolite carrier. The yield is over 95% at quantitative conversion.

Figure 5.22 shows the flow diagram for catalytic hydrogenation of phenol to cyclohexanone.

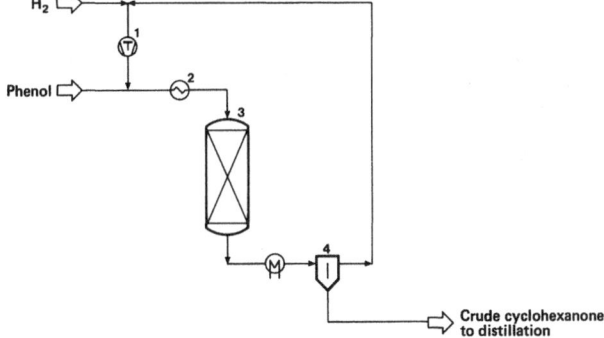

1 Hydrogen circulation pump; **2** Phenol vaporizer; **3** Reactor; **4** Cyclohexanone separator

Figure 5.22: Flow diagram for the production of cyclohexanone from phenol

As a possible alternative, hydrogenation can be carried out in the liquid phase at 175 °C and a pressure of 13 bar on Pd/activated carbon catalysts (*Allied-Signal*).

Over 200,000 t of phenol are used annually in the USA (*Allied-Signal, Monsanto*) and Western Europe (*Montedison, DSM*) for the production of caprolactam.

Caprolactam

5.3.4.3 Alkylphenols

The most important commercial alkylphenols are the methyl derivatives (the cresols and xylenols), butylated phenols, predominantly tertiary butylphenols, and phenols with long alkyl chains, for example nonylphenol and dodecylphenol. The latter are commonly used in the production of surfactants.

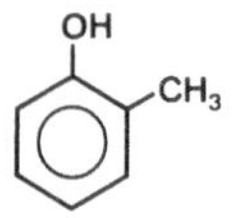

o-Cresol

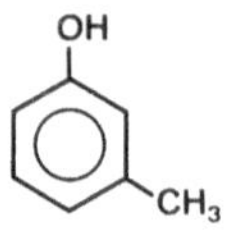

m-Cresol

p-Cresol

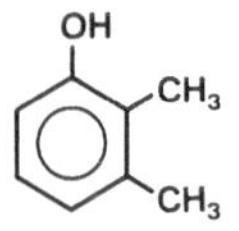

2,3-Xylenol

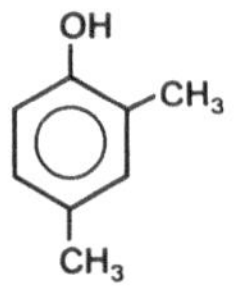

2,4-Xylenol

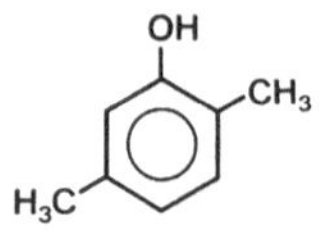

2,5-Xylenol

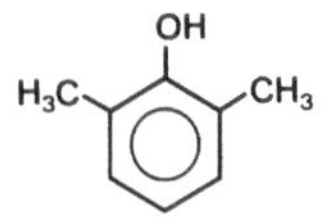

2,6-Xylenol

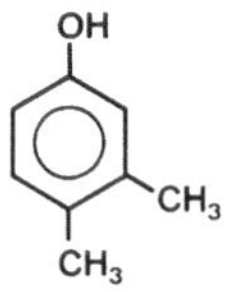

3,4-Xylenol

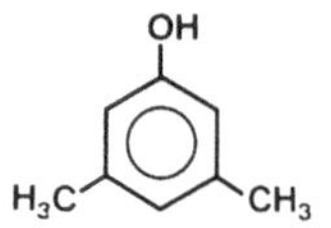

3,5-Xylenol

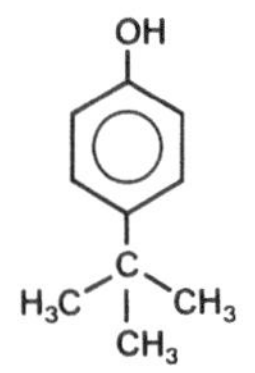

p-tert-Butylphenol

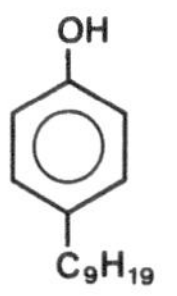

p-Nonylphenol

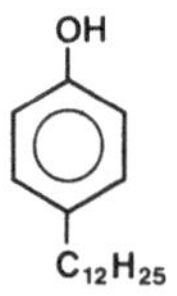

p-Dodecylphenol

As a result of the high reactivity of phenol, alkylation takes place with olefins and alcohols under relatively mild conditions. Methylphenols, i.e. o-, m- and p-cresol, can be produced as a mixture by the alkylation of phenol with methanol. Since reactivity is increased by the introduction of the methyl group, alkylation of cresols to produce xylenols occurs under even milder reaction conditions. The long-chain alkylphenols are also produced by alkylation of phenol.

5.3.4.3.1 Cresols

The mono-methyl derivatives of phenol, the cresols, were discovered in coal tar in 1854 by Alexander Wilhelm Williamson. Coal tar then became the most important source of cresols for a century. Until the mid-1960's, the 'natural' sources for cresols were largely adequate.

The proportion of cresols in the 'tar acids' of coal tar is between 40 and 50%; cresols are recovered jointly with phenol from the carbolic oil fraction of coal tar. A second source of cresols are refinery waste streams, arising during processing of heavy naphtha fractions from crackers, which contain about 60% cresol.

A third 'natural' source of alkylated phenols is coal gasification with tar by-products (*SASOL*), complemented by the future possibility of cresols in the gasoline and middle-oil cuts from coal liquefaction.

In the commercial synthesis of cresols, the most common processes are alkaline hydrolysis of chlorotoluene, the cleavage of cymene hydroperoxide and the alkylation of phenol in the gas phase with methanol. Alkali fusion of toluenesulfonic acids has largely lost its earlier significance, because of the inorganic salts which are co-produced.

For the manufacture of cresols by alkaline hydrolysis of chlorotoluene, toluene firstly reacts with chlorine, e.g. at 30 °C using $FeCl_3/S_2Cl_2$ as a catalyst, to yield a chlorotoluene mixture (o/p-ratio 1:1).

The isomer mixture is hydrolysed in a long tubular reactor (nickel-steel alloy) under pressure of 280 to 300 bar, at 390 °C, yielding o-, m- and p-cresol in the ratio 1:2:1, respectively. The higher proportion of m-cresol can be explained by an aryne mechanism.

When the lower-boiling o-cresol (boiling point 191.0 °C) has been distilled off, a m-/p-cresol mixture remains, with 70% m-content, which is difficult to separate by either crystallization or distillation (202.0 °C, and 201.9 °C resp.).

The phase diagram (Figure 5.23) for the separation of m- and p-cresol shows two eutectics, so that the separation of the cresol isomers is only possible with a p-cresol concentration of over 60%, or a m-cresol concentration of over 88%; the application of pressure can facilitate the separation of the cresol isomers.

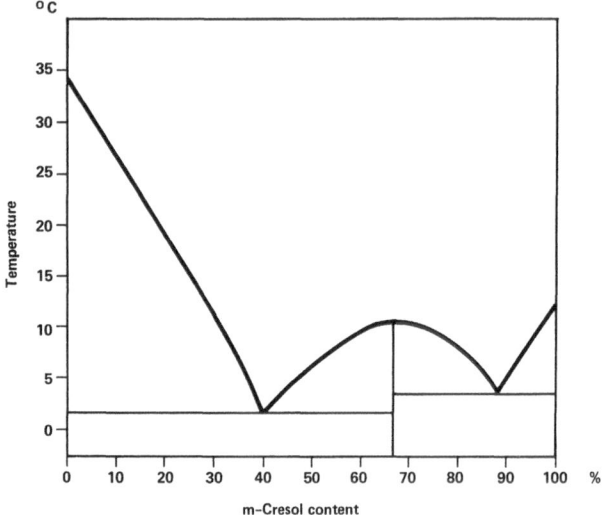

Figure 5.23: Phase diagram for the separation of m-/p-cresol by crystallization

If the separation of the isomers is carried out at the chlorotoluene stage (distillation/crystallization), an o-/m-cresol mixture arises during hydrolysis of o-chlorotoluene in the ratio 1:1, from which pure m-cresol can be produced, whereas the hydrolysis of p-chlorotoluene gives a m-/p-cresol mixture.

In Japan in particular, m- and p-cresol are synthesized by oxidation of cymene (*Sumitomo Chemical* and *Mitsui Petrochemical*). Friedel-Crafts propylation of toluene at 60 to 80 °C catalysed by $AlCl_3$ yields an o-, m-, p-cymene mixture.

The optimum isomer distribution requires a low o-cymene content, since o-cymene is intricate to oxidize and inhibits the oxidation of the other cymenes. In practice, a mixture of 3% o-cymene, 64% m-cymene and 33% p-cymene is used. The oxidation of the cymene mixture is carried out to a peroxide content of around 20%, whereas the degree of oxidation of cumene in phenol synthesis is around 30%. In addition to cresols, the acid-cleaved reaction mixture contains acetone and unconverted cymene together with a wide range of co-products, such as isopropylbenzaldehyde, isopropylbenzyl alcohol, methyl acetophenone and isopropyltolyl alcohol. The m-/p-cresol mixture is over 99.5% pure; the m-/p-ratio is 1.5:1.

Cymene can also be recovered from renewable raw materials such as turpentine oil; this involves the dehydrogenation of monocyclic terpenes (menthane).

Production and uses of benzene derivatives 167

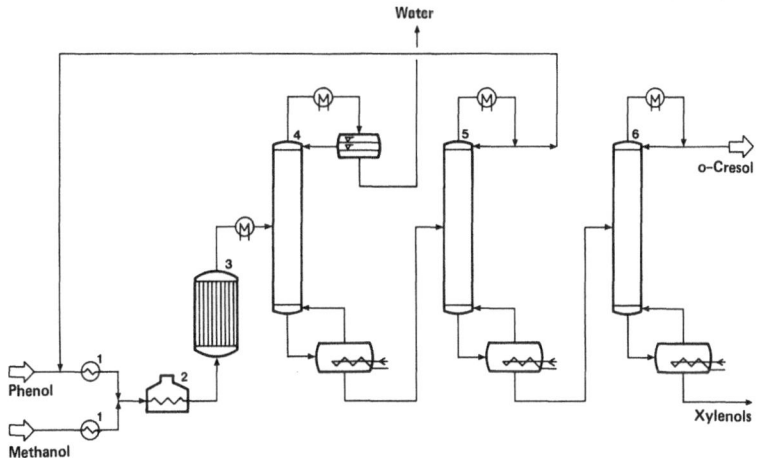

p-Cresol production based on this raw material was carried out on an industrial scale in the USA (*Hercules*) up to the early 1970's.

Methylation of phenol is commonly used to produce o-cresol and 2,6-xylenol. It is carried out in the gas phase at slightly increased pressure and temperatures of from 300 to 400 °C with Al_2O_3 catalysts in multitubular reactors. After removing the water, subsequent distillation separates the dried mixture into the products anisole/phenol, 99%-pure o-cresol and 2,6-xylenol. Phenol and anisole are recycled. Higher alkylated phenols can be recovered from the distillation residue.

Figure 5.24 shows a typical process for the production of o-cresol.

1 Vaporizer; 2 Heater; 3 Multitubular reactor; 4 Dewatering column; 5 Phenol column; 6 o-Cresol column

Figure 5.24: Production of o-cresol and 2,6-xylenol by methylation of phenol

Cresols can also be produced by liquid-phase alkylation. Phenol can be converted into cresols and xylenols with methanol, using aluminum methylate as catalyst, at 350 to 400 °C under pressure. The main product of this reaction is o-cresol. In a process used by *Union Rheinische Braunkohlen Kraftstoff*, the operation is carried out at temperatures from 200 to 240 °C and pressures at around 25 bar, with dilute zinc halide/hydrogen halide solution as catalyst.

In common with the *Raschig* process for producing phenol, cresols can also be synthesized by the oxychlorination of toluene.

An interesting process for the manufacture of p-cresol, which shows that the build-up of aromatics from low molecular components should always be considered as an alternative, is the Diels-Alder reaction of isoprene and vinyl acetate. This process, however, has not been exploited on an industrial scale.

o-Cresol is mainly used in the production of 4-chloro-o-cresol, from which, with chloroacetic acid or 2-chloropropionic acid the hormone-type herbicides such as 4-chloro-2-methylphenoxyacetic acid (MCPA) and 2-(4-chloro-2-methylphenoxy)-propionic acid (MCPP) are produced.

The intermediate product, 4-chloro-2-methylphenol, is obtained by chlorination of o-cresol, e.g. with sulfuryl chloride.

The production of 4-chloro-2-methylphenoxycarboxylic acids in Western Europe in 1985 was around 25,000t; these cresol derivatives thus belong to the most important plant protection agents.

The alkylation of o-cresol with propylene yields carvacrol (2-methyl-5-isopropylphenol), which is used as an antiseptic.

m-Cresol is mainly used for the production of thymol, which is obtained by isopropylation at 360 °C and 50 bar. Hydrogenation of thymol leads to menthol, a component of fragrances with a peppermint odor.

Production and uses of benzene derivatives 169

Carvacrol   Thymol   Menthol

In addition, m-cresol is used to produce plant protection agents such as fenitrothion. The manufacturing process involves nitrosation of m-cresol with nitrous acid esters in isopropanol, and subsequent oxidation of the nitroso compounds with nitric acid to yield 4-nitro-m-cresol (or its sodium salt), which, when acted upon by O,O-dimethyl thiophosphoric acid chloride, gives fenitrothion.

Fenitrothion

Another important derivative of m-cresol used in the manufacture of plant protection agents is m-phenoxytoluene, which can be produced from m-cresol and chloro- or bromobenzene at temperatures of 200 °C, with copper catalysts. m-Phenoxytoluene is converted into m-phenoxybenzoic acid methyl ester by oxidation with a cobalt acetate/KBr catalyst and subsequent esterification; m-phenoxybenzoic acid methyl ester serves as an intermediate in the production of m-phenoxybenzaldehyde, which is used as the raw material in the production of the synthetic pyrethroid insecticide, fenvalerate (see Chapter 6.3.2). The cyanohydrin is formed in-situ, then made to react with 2-isopropyl-(4-chlorophenyl) acetic acid chloride to yield fenvalerate, which was developed by *Sumitomo Chemical* in 1972. Pyrethroid insecticides are distinguished by their low toxicity and high activity.

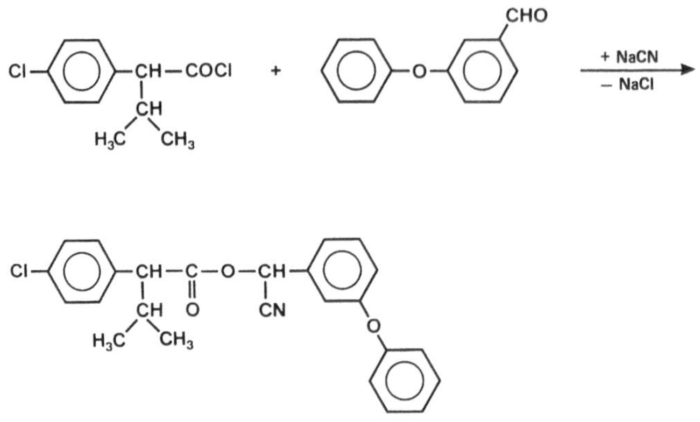

[Reaction scheme: 4-chloro-α-isopropylphenylacetyl chloride + m-phenoxybenzaldehyde, with + NaCN / − NaCl, yielding Fenvalerate]

Fenvalerate

Other important pyrethroid insecticides based on m-phenoxybenzaldehyde are deltamethrin (*Roussel-Uclaf*) and the chlorine analogue cypermethrin.

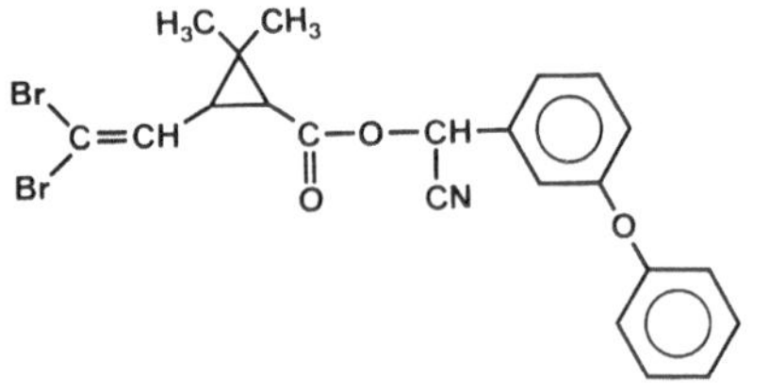

Deltamethrin

p-Cresol is principally used in the production of 2,6-di-tert-butyl-p-cresol (BHT), which has a wide range of uses as an antioxidant and preservative for plastics and rubber, and in food production.

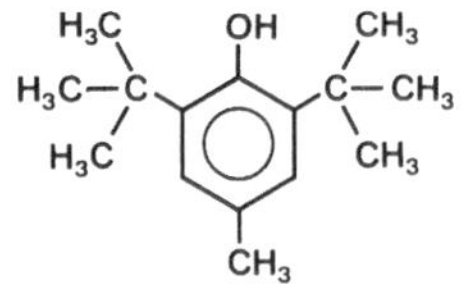

BHT

BHT can be produced by direct dialkylation of p-cresol with isobutene with sulfuric acid catalysis at 50 to 80 °C, or by monobutylation of m-/p-cresol mixtures. In the latter method, it is necessary to separate the monobutylated m-/p-mixture and further butylate the 2-tert-butyl-p-cresol to 2,6-di-tert-butyl-p-cresol.

BHT consumption in Western Europe and the USA in 1985 was around 9,000 t. Phosphate esters, which are used as additives for lubricants, are produced from m-/p-cresol mixtures.

Production of pure cresols in the Western world in 1985 amounted to ca. 70,000 t o-cresol and 20,000 t p-cresol; in addition, around 50,000 t of mixed cresols and 80,000 t m-/p-cresol mixtures were produced.

5.3.4.3.2 Xylenols

Like the cresols, the xylenols are recovered from coal tar and petroleum cracker waste stream although the largest quantities are produced synthetically. (Isomerization with Friedel-Crafts catalysts allows the isomer ratio to be changed). The most important of the six xylenol isomers in terms of quantity, is 2,6-dimethylphenol, which is produced by methylation of phenol with excess methanol. 2,6-Xylenol, which is produced worldwide in quantities of around 100,000 tpa (of which 75,000 t are manufactured in the USA), is predominantly used in the production of polydimethylphenylene oxide (PPO) resins. The dehydrogenation of 2,6-dimethylphenol to polydimethylphenylene oxide occurs at ambient temperature by passage of oxygen with a CuCl/pyridine catalyst; the largest producer is *General Electric*.

Polydimethylphenylene oxide

2,6-Xylenol reacts with ammonia to yield 2,6-xylidine, a base material for the fungicide metalaxyl (*Ciba-Geigy*).

Metalaxyl

3,5-Dimethylphenol is not directly accessible by selective alkylation of phenol with methanol; it is produced by the aromatization of isophorone at 540 to 650 °C on $Cr_2O_3/K_2O/Al_2O_3$ catalysts, or chrome-nickel steel.

Isophorone

Figure 5.25 shows the process operated by *Rütgerswerke* at 520 to 540 °C and pressures from 10 to 15 bar. The reaction product is refined in a way similar to the recovery of phenols from coal tar.

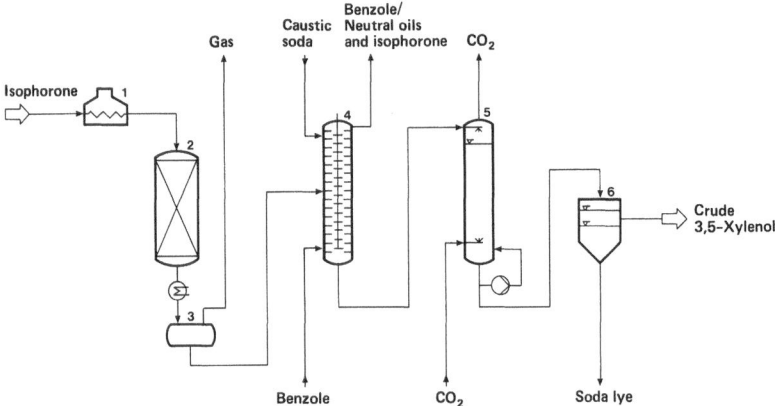

1 Vaporizer/Heater; 2 Reactor; 3 Gas separator; 4 RDC-extractor; 5 Saturation column; 6 Phase-separation vessel

Figure 5.25: Flow diagram for the aromatization of isophorone to 3,5-dimethylphenol

An important product from 3,5-dimethylphenol is the insecticide methiocarb (*Bayer*), which is manufactured by the reaction of 3,5-dimethylphenol with methylsulfenyl chloride to yield 4-methylmercapto-3,5-xylenol followed by reaction with methyl isocyanate.

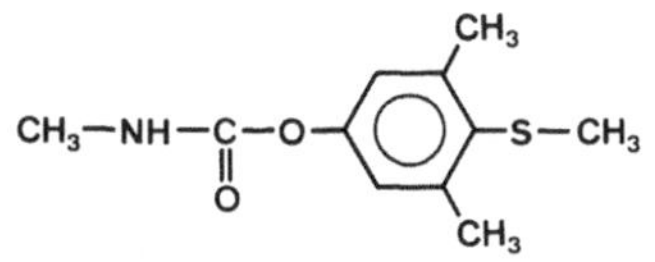

Methiocarb

3,5-Dimethylphenol is also used as a building block in the synthesis of vitamin E (α-tocopherol) and, after chlorination to 4-chloro- and 2,4-dichloro-3,5-dimethylphenol, as a disinfectant and commercial preservative; in addition, ammonolysis yields 3,5-dimethylaniline, an intermediate for pigments.

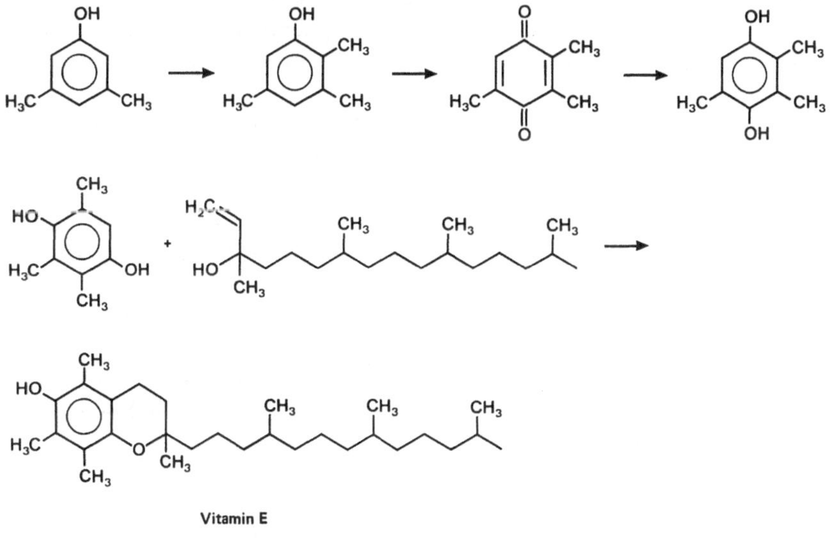

Vitamin E

Vitamin E, which has a worldwide production of around 7,000 tpa, is also produced from vegetable oils (deodorizing process) to the extent of about 10% of the total, and from 2,3,5-trimethylaniline (see Chapter 8.1).

A route starting from purely aliphatic chemicals includes condensation of crotonaldehyde with diethyl ketone for the production of the 2,3,5-trimethyl hydroquinone.

$$CH_3-CH=CH-CHO + C_2H_5-\underset{\underset{O}{\|}}{C}-C_2H_5 \xrightarrow{-H_2O}$$

2,5,6-Trimethyl-2-cyclohexen-1-one → 2,3,5-Trimethylhydroquinone

Xylenol mixtures are used as solvents and disinfectants.

5.3.4.3.3 Higher alkyl phenols

Among the tertiary-butyl phenols, the o- and p-derivatives and 2,6-di-tert-butylphenol are commerically important. They are used in the production of antioxidants. o-tert-Butylphenol and 2,6-di-tert-butylphenol are produced by alkylation of phenol with isobutene at a reaction temperature of 100 °C in the presence of aluminum phenolate as catalyst.

Suitable catalysts for the butylation of phenol with isobutene to p-tert-butylphenol are sulfuric acid, phosphoric acid, boron trifluoride or macroporous, pelleted, strongly acidic sulfonated styrene/divinylbenzene resins.

Among the long-chain alkylphenols, p-tert-octylphenol, nonylphenol and dodecylphenol have special commercial significance. Alkylation can take place with straight-chain or branched olefins.

To produce p-tert-octylphenol, phenol and diisobutylene (a mixture of 2,4,4-trimethyl-1-pentene and 2,4,4-trimethyl-2-pentene) are fed through a bed of ion-exchange resin in the ratio 1.5:1; the resin is maintained at a temperature of 100 to 105 °C by internal cooling tubes. The alkylation is restricted to 95% diisobutylene conversion, to avoid the formation of undesirable by-products. The reaction mixture, which is separated into its constituents by vacuum distillation, consists of 93 to 96% p-tert-octylphenol; the concentration of o-tert-octylphenol is between 2 and 3%.

Nonylphenol is produced by the reaction of phenol with propylene trimer at a temperature of 70 to 125 °C. An ion-exchange resin is used as catalyst. The reaction is normally carried out in two stages to avoid overheating.

Production of nonylphenol in 1985 was 80,000 t in the USA, and around 55,000 t in Western Europe.

The reaction of nonylphenol with ethylene oxide yields non-ionic surfactants, which are used as raw materials for detergents and in pesticide formulations.

Dodecylphenol, in the form of its calcium or magnesium sulfonate, is used as an additive for lubricants. It is produced by the reaction of phenol with propylene tetramer in the molar ratio 3:1 at a temperature of 100 °C on ion-exchange resins. West European production in 1985 was around 20,000 t.

Higher alkylated phenols are obtained by the interaction of phenol and C_{16}-C_{18} olefins from wax cracking. The corresponding alkylated salicylic acids (formed by Kolbe-Schmitt carboxylation) are lubricants for diesel engines.

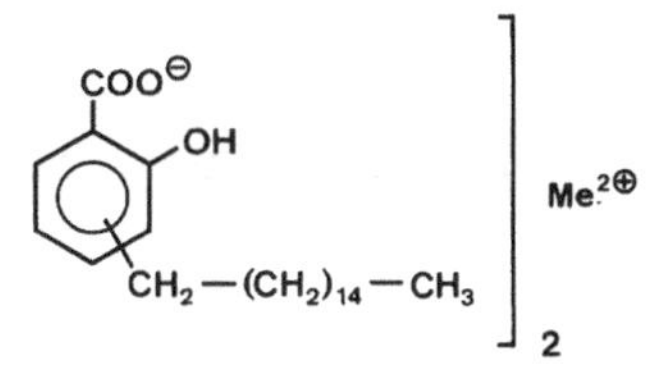

n-Hexadecylsalicylate (structure shown with COO⁻, OH, CH₂–(CH₂)₁₄–CH₃ substituents, Me²⁺ counterion, subscript 2)

The most important application for long-chain alkylphenols, following hydroxyethylation with ethylene oxide, is the production of raw materials for detergents, wetting agents and emulsifiers. Use of these is declining, however, because of their poor biodegradability. Nevertheless, as emulsifiers, they can only be replaced with difficulty by aliphatic fatty alcohol ethoxylates.

5.3.4.4 Salicylic acid

Salicylic acid was first obtained in 1838 by Raphael Piria, by the reaction between salicylaldehyde and potassium hydroxide.

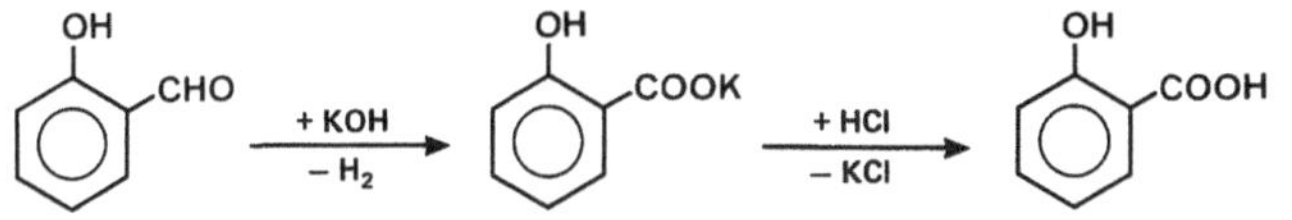

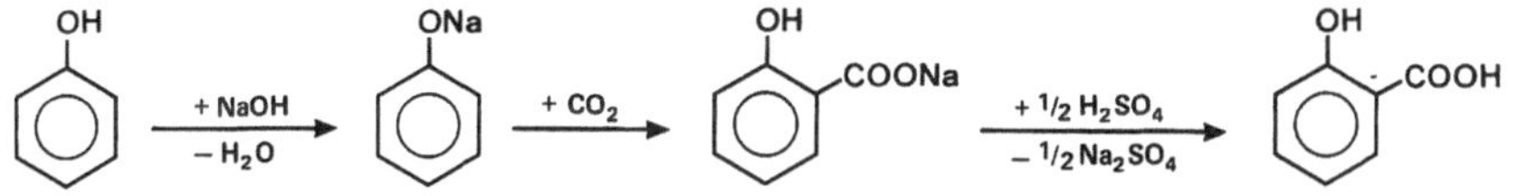

[Reaction scheme: salicylaldehyde (OH, CHO) + KOH, −H₂ → potassium salicylate (OH, COOK) + HCl, −KCl → salicylic acid (OH, COOH)]

[Reaction scheme: phenol (OH) + NaOH, −H₂O → sodium phenolate (ONa) + CO₂ → sodium salicylate (OH, COONa) + ½ H₂SO₄, −½ Na₂SO₄ → salicylic acid (OH, COOH)]

The presently dominant process for the production of salicylic acid is based on the Kolbe-Schmitt synthesis. Hermann Kolbe and E. Lautemann described the reaction of carbon dioxide with sodium phenolate to produce sodium salicylate and its subsequent reaction with sulfuric acid to yield salicylic acid, in 1860. However, as a result of the formation of di-sodium salicylate in this process, only half the phenol was converted into salicylic acid. Rudolf Schmitt modified the synthesis in 1884, by first bringing into contact carbon dioxide with cold, dry sodium phenolate, then heating the mixture under pressure to 120 to 140 °C.

In the process commonly used today, carbon dioxide is fed at a pressure of 6 to 7 bar and at a temperature of 100 °C over dry sodium phenolate, which is produced from phenol and 50% aqueous sodium hydroxide solution followed by vacuum evaporation of the water. The reaction mixture is then carboxylated at 150 to 170 °C until no more CO_2 uptake occurs. The crude salicylic acid is recovered after adding water and sulfuric acid. Refining can be carried out with activated carbon containing zinc dust to remove colored impurities, or by crystallization of the hexahydrate. Salicylic acid is manufactured from the purified sodium salicylate by acidification with sulfuric acid and subsequent vacuum sublimation.

The production of salicylic acid in the United States in 1985 was around 15.000t. West European production is estimated at around 20.000t; the largest producers are *Monsanto* and *Rhône Poulenc*.

Over one-half of the salicylic acid is used to manufacture acetylsalicylic acid, quantitatively one of the most important analgesic pharmaceuticals. Acetylsalicylic acid was first produced in 1853 from acetyl chloride and sodium salicylate. Its introduction as a medicament occurred in Germany in 1899 (Aspirin) and in the USA in 1900. Acetylsalicylic acid is now produced by reaction between salicylic acid and acetic anhydride in stainless steel or enamelled reactors at a temperature below 98 °C (duration of reaction: 2 to 3 hours). The reaction mixture is refined by crystallization at 0 °C.

Other important derivatives of salicylic acid are the amyl and methyl esters, which are used as fragrances.

5.3.4.5 Chlorinated phenols

Chlorinated phenols are commonly used as intermediates for the manufacture of agricultural chemicals and wood preservatives. o-Chlorophenol is obtained as the main product from chlorinating of phenol with NaOCl/HCl; it is used to manufacture pentachlorophenol and the insecticide profenofos.

Profenofos

p-Chlorophenol can be produced with high selectivity by reaction of phenol with sulfuryl chloride (SO_2Cl_2) in the presence of $FeCl_3$ or $AlCl_3$.

An important product of p-chlorophenol is the fungicide triadimefon (*Bayer*), which is obtained from α-bromopinacolone and sodium p-chlorophenoxide with subsequent bromination and final treatment with 1,2,4-triazole.

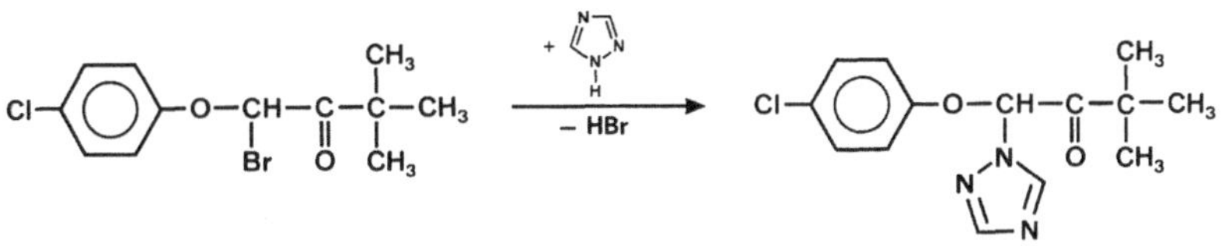

[Reaction scheme showing conversion to Triadimefon with loss of HBr]

Triadimefon

Commercially the most important polychlorinated phenols are 2,4-dichlorophenol, 2,4,5-trichlorophenol and pentachlorophenol.

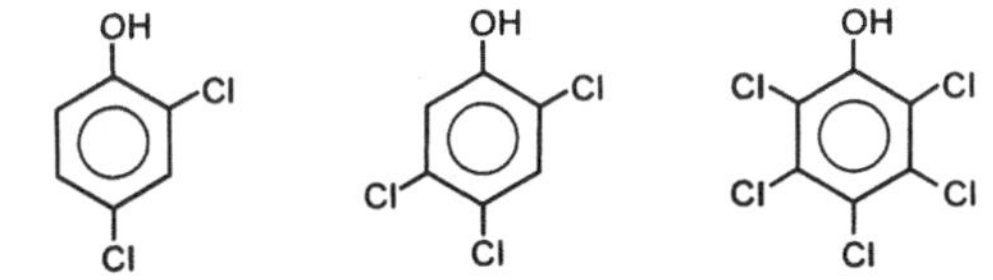

[Structures of 2,4-Dichlorophenol, 2,4,5-Trichlorophenol, and Pentachlorophenol]

2,4-Dichlorophenol is produced by chlorination of phenol in a polar solvent such as water or acetic acid at temperatures from 70 to 80 °C to minimize further chlorination; (when using non-polar solvents with an excess of chlorine, 2,4,6-trichlorophenol is the main reaction product). The crude product is purified by fractional distillation. West European production of 2,4-dichlorophenol is around 20,000 tpa. 2,4-Dichlorophenol is used mainly to produce 2,4-dichlorophenoxyacetic acid (2,4-D), first used as a herbicide in 1942 and obtained by the reaction of 2,4-dichlorophenol and monochloroacetic acid at 80 to 100 °C and a pH value of 7 to 11.

The corresponding propionic acid (2,4-DP) is also manufactured in similar volume. Total West European production in 1985 was around 20,000 t; consumption in the USA in 1985 was of a similar order of magnitude.

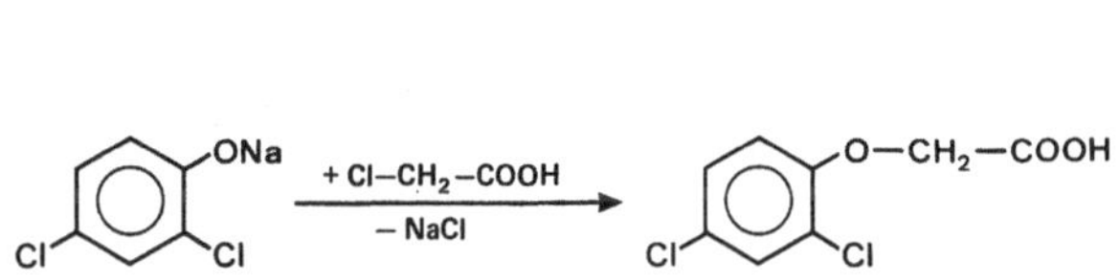

[Reaction scheme: sodium 2,4-dichlorophenolate + Cl–CH₂–COOH → 2,4-D, with loss of NaCl]

2,4-D

Figure 5.26 shows the flow diagram for the production of 2,4-dichlorophenoxyacetic acid.

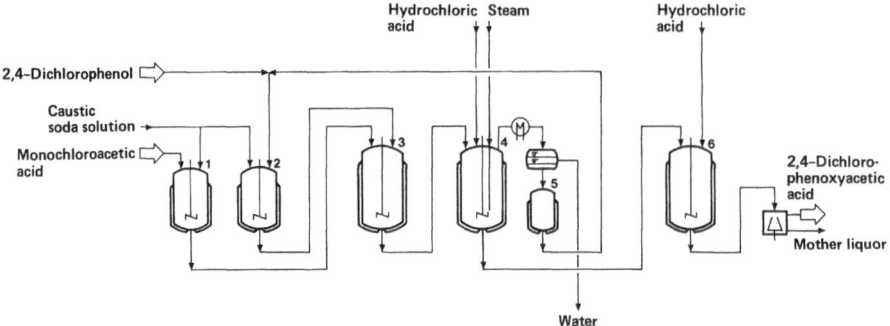

1 Neutralization vessel; **2** Dissolution vessel; **3** Reactor; **4** Steam distillation; **5** 2,4-Dichlorophenol storage tank; **6** Precipitation vessel

Figure 5.26: Flow diagram for the production of 2,4-D

2,4-D and 2,4-DP, together with the cresol derivatives MCPP and MCPA, described in Chapter 5.3.4.3.1, belong to the most important group of aromatic plant protection agents in terms of quantity, and are used principally for weed control. An interesting new development is the use of highly active optically-pure D-forms of the phenoxypropionic acid derivatives 2,4-DP and MCPP, which enable the ecological side-effects to be noticeably reduced; this process uses biotechnologically produced D-lactic acid (*R*-acid) as a building block.

D–Lactic acid

Another important plant protection agent based on 2,4-dichlorophenol is diclofop-methyl, which is produced via the intermediate nitrofen (from 2,4-dichlorophenol and p-nitrochlorobenzene) by reduction of the nitro group, diazotization, and hydrolysis of the diazonium salt to yield the phenol with subsequent reaction with 2-chloropropionic acid methyl ester.

Diclofop–methyl

Production of this plant protection agent, developed by *Hoechst,* was around 5,000 t in 1985.

2,4,5-Trichlorophenol is produced via the chlorination of benzene with 4 mols of chlorine to yield the intermediate 1,2,4,5-tetrachlorobenzene. Tetrachlorobenzene is hydrolyzed with sodium hydroxide in ethylene glycol or methanol at maximum temperatures of 140 °C. During this stage of the process, care must be taken to control the temperature closely, to avoid the formation of 2,3,7,8-tetrachloro-dibenzo-p-dioxin, one of the most toxic of the 75 possible chlorodioxins. 2,4,5-Trichlorophenol is used in production of the herbicide 2,4,5-trichlorophenoxy-acetic acid (2,4,5-T), which is obtained by reaction of sodium chloroacetate with 2,4,5-trichlorophenol.

2,3,7,8-Tetrachloro–dibenzo–p–dioxin

Pentachlorophenol is manufactured, typically, by the chlorination of molten phenol at temperatures from 100 to 180 °C in the presence of a Friedel-Crafts catalyst such as ferric chloride or aluminum chloride. The reaction time is up to 15 hours. The crude pentachlorophenol is then refined by distillation; for a long time it was one of the most important wood preservatives, also in the form of its sodium salt.

An alternative method for the production of sodium pentachlorophenoxide is the hydrolysis of hexachlorobenzene in alkaline solution at 240 °C under pressure. This process is declining in importance, however. World production of pentachlorophenol in 1985 was around 35,000 t.

The application of highly-chlorinated phenols is in rapid decline; production is banned in a number of countries, due to their poor biodegradability and the possibility of producing dioxins and dibenzofurans in their manufacture.

5.3.4.6 Nitrophenols

Nitrophenols are less important as intermediates than chlorophenols. Nitrophenols can be obtained by nitration of phenol or, preferably, by nucleophilic substitution of the corresponding chloronitrobenzenes (see Chapter 5.8.1) with alkali solution. Nitration of phenol yields an o-/p-mixture consisting of around 55% o-nitrophenol and 45% p-nitrophenol; separation of the isomers is possible by steam distillation.

o-Nitrophenol is usually reduced by using palladium/activated carbon to o-aminophenol, which is used as the raw material in the production of dyes and plant protection agents, such as the insecticide phosalone. In this case, o-aminophenol is converted to benzoxazol-2-one with urea. Chlorination and chloromethylation gives 6-chloro-3-chloromethylbenzoxazol-2-one, which yields phosalone (*Rhône Poulenc*) by further reaction with O,O-diethyl dithiophosphoric acid (or its salt).

Phosalone

p-Nitrophenol is principally employed as an intermediate in the production of p-aminophenol, which is used in the production of p-acetylaminophenol (paracetamol), a mild analgesic, as a photographic developer and as a dyestuffs component. However, the overwhelming proportion of p-aminophenol is not based on p-nitrophenol, but is produced from nitrobenzene by catalytic reduction (Pt-catalyst) in a sulfuric acid medium via the intermediate phenylhydroxylamine. This route is used especially by the largest paracetamol producer, *Mallinckrodt* (USA). An additional route to paracetamol the world market of which is some 27,000 tpa starts with hydrolysis of p-nitrochlorobenzene.

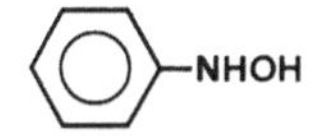

Phenylhydroxylamine

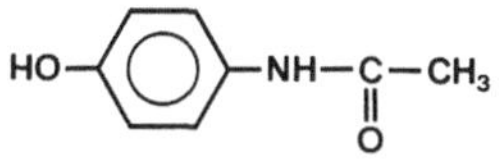

Paracetamol

Methyl parathion, a broad-spectrum insecticide, is produced by the reaction of sodium p-nitrophenolate and O,O-dimethyl thiophosphoryl chloride.

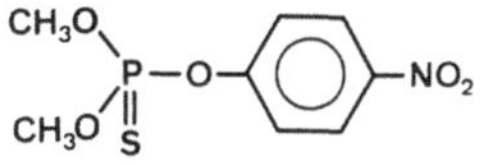

Methyl parathion

The diethyl derivative also serves as an insecticide; total world production of these phosphoric acid esters is estimated to be around 20,000 tpa.

Nitration of phenol in sulfuric acid produces 2,4,6-trinitrophenol (picric acid). Picric acid is used predominantly as an intermediate in the production of picramic acid, a raw material for dyestuffs; picric acid has lost its earlier importance as an explosive.

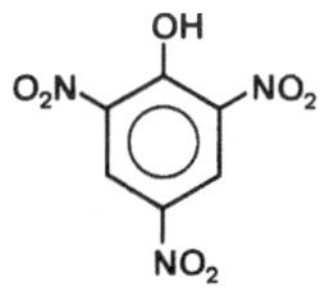

Picric acid

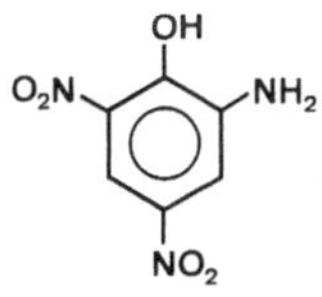

Picramic acid

5.3.4.7 Other phenol derivatives

The Kolbe-Schmitt reaction of potassium phenoxide with CO_2 at 200 °C gives an 80% yield of p-hydroxybenzoic acid, which is used in the form of the propyl and butyl esters as a preservative in cosmetics and pharmaceuticals.

p-Hydroxybenzoic acid propyl ester

(structure: benzene ring with COOC₃H₇ para to OH)

p-Hydroxybenzoic acid is used to a small extent for the production of p-hydroxybenzonitrile (via ammonolysis) an intermediate in the production of the herbicides bromoxynil and ioxynil.

Bromoxynil: 4-hydroxy-3,5-dibromobenzonitrile

Ioxynil: 4-hydroxy-3,5-diiodobenzonitrile

p-Hydroxybenzonitrile can also be obtained through p-hydroxybenzaldehyde, which can be produced by reacting glyoxylic acid and phenol.

p-Hydroxybenzaldehyde is converted with hydroxylamine to the oxime, which yields the nitrile by dehydration.

$$\text{PhOH} + \text{OHC-COOH} \xrightarrow[-\text{CO}]{-\text{H}_2\text{O}} \text{p-HO-C}_6\text{H}_4\text{-CHO} \xrightarrow[-\text{H}_2\text{O}]{+\text{NH}_2\text{-OH}} \text{p-HO-C}_6\text{H}_4\text{-CH=NOH} \xrightarrow{-\text{H}_2\text{O}} \text{p-HO-C}_6\text{H}_4\text{-CN}$$

Sodium phenoxide reacts with chloroacetic acid to yield phenoxyacetic acid, used in the production of penicillin V.

Phenoxyacetic acid: C₆H₅−O−CH₂−COOH

Penicillin V: phenoxyacetyl-aminopenicillanic acid

The reaction of phenol with chloroform and alkali (dichlorocarbene) at 65 to 70 °C gives salicylaldehyde (Reimer-Tiemann synthesis), from which coumarin is produced by reaction with acetic anhydride and sodium acetate at 135 to 155 °C (Perkin reaction). Coumarin is used as a perfume and fragrance.

5.3.4.8 Polyhydric phenols

From the commercial point of view, polyhydric phenols such as di- and trihydroxybenzene compounds are less important than phenol. In addition to synthesis, they can also be recovered from low-temperature tars, especially from the pyrolysis of brown coal.

Catechol (o-dihydroxybenzene) was first discovered by Hugo Reinsch in 1839, by distilling catechin, a tannin-like cotton colorant. Since it was produced by destructive distillation, it was originally called pyrocatechol.

The methyl ether of catechol, guaiacol, is contained in beechwood tar; catechol can be produced from it by distillation with hydrobromic acid.

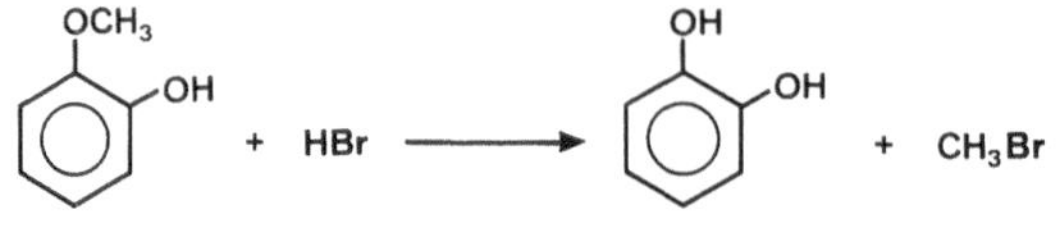

$$\text{Guaiacol} + HBr \longrightarrow \text{catechol} + CH_3Br$$

Guaiacol

Outdated methods for producing catechol are based on o-sulfonated or o-halogenated phenols. These phenol derivatives are hydrolyzed under pressure with alkali and the reaction product is extracted with a solvent such as ether. The crude catechol is then refined by distillation.

By present routes, catechol arises as a co-product in the synthesis of hydroquinone from phenol by oxidation with peracids. Catechol is used as an antioxidant, principally for turpentine oil and perfumes. For the latter application, especially in the USA, 4-tert-butylcatechol, obtained by butylation, is used.

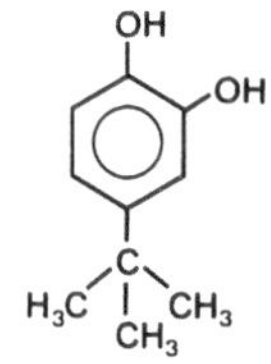

4-tert-Butylcatechol

Catechol is also used as a starting material for the production of the perfumery and flavoring agent, piperonal (3,4-methylenedioxybenzaldehyde).

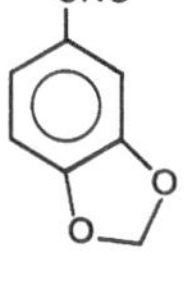

Piperonal

An important plant protection agent based on catechol is carbofuran, manufactured by *Bayer* and *FMC*. It is obtained by the reaction of catechol with methallyl chloride, followed by Claisen rearrangement at around 200 °C and ring closure to the respective benzofuran, with subsequent reaction of the hydroxyl group with methyl isocyanate in the presence of triethanolamine. (An alternative method of synthesis is based on o-nitrophenol).

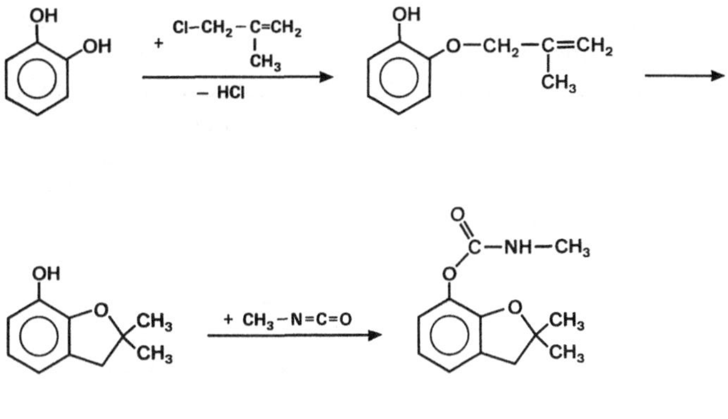

[Reaction scheme: catechol + Cl–CH₂–C(CH₃)=CH₂ → 2-(2-methylallyloxy)phenol, –HCl]

[Reaction scheme: cyclization to 2,2-dimethyl-2,3-dihydrobenzofuran-7-ol + CH₃–N=C=O → Carbofuran]

Carbofuran

Another catechol-based insecticide is propoxur (*Bayer*), which is obtained from 1-hydroxy-2-isopropoxybenzene and methyl isocyanate.

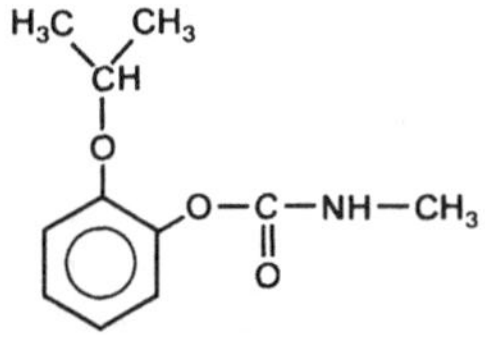

Propoxur

Catechol is used as a raw material in the production of vanillin and veratrole, which is commonly used in the synthesis of papaverine (see Chapter 14.5.4).

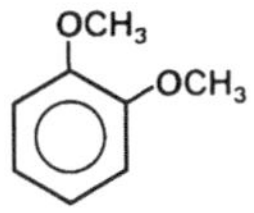

Veratrole

Production of catechol in Western Europe in 1985 was around 7,000 t.

Resorcinol (m-dihydroxybenzene) is of greater commercial importance than catechol. It was first produced by Wilhelm Körner in 1868 by alkali fusion of iodophenol. Current industrial processes for resorcinol production are based either on benzene-m-disulfonic acid, which is converted to the dihydroxy compound by alkali fusion (*Koppers, Hoechst*), or on m-diisopropylbenzene, which is oxidized to resorcinol analogously to Hock's phenol synthesis.

In the *Hoechst* resorcinol process, liquid SO_3, benzene and Na_2SO_4 (to prevent build-up of sulfones) react together at 150 °C to form benzene-1,3-disulfonic acid; the disulfonic acid is neutralized in a fluidized bed dryer with concentrated caustic lye. Subsequent fusion with sodium hydroxide in electrically-heated cast-iron reactors at 350 °C yields the resorcinol disodium salt which is dissolved in water. After acidification with sulfuric acid, resorcinol is extracted from the aqueous phase with a suitable solvent such as diisopropyl ether and refined by vacuum distillation.

Figure 5.27 shows a process diagram of the *Hoechst* process.

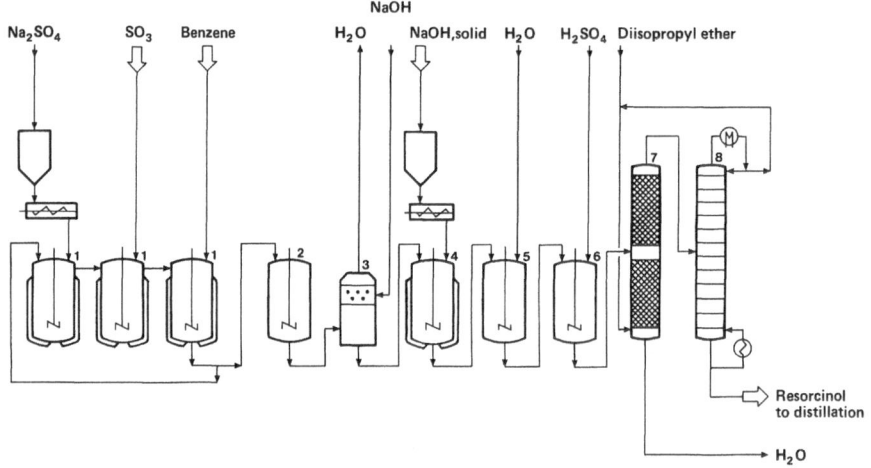

1 Cascade reactor; 2 Dosing vessel; 3 Fluidized bed dryer; 4 Fusion reactor; 5 Dissolution vessel; 6 Precipitation vessel; 7 Extraction column; 8 Distillation column

Figure 5.27: Resorcinol production by alkali fusion of benzene-1,3-disulfonic acid

Because of the large portions of inorganic reagents, attempts have been made to replace this process with a purely organic one. The oxidative cleavage of m-diisopropylbenzene into acetone and resorcinol has not yet been widely used, however, since the number of possible by-products is fundamentally higher than with cumene oxidation for the production of phenol, because of the presence of two isopropyl groups. Furthermore, the reaction rate is significantly lower, so that only a low productivity per unit volume can be achieved. The instability of the diperoxide, moreover, means that special consideration must be given to safety precautions. The process is used exclusively in Japan by *Mitsui Petrochemical* and *Sumitomo Chemical*.

World production of resorcinol in 1985 was around 25,000 t. Resorcinol resins, obtained from resorcinol and formaldehyde, are used as special adhesives in the timber industry and as tackifiers, usually in combination with vinylpyridine latex for steel cord tires.

Reaction of resorcinol with benzotrichloride or benzoyl chloride yields 2,4-dihydroxybenzophenone, which, with octyl chloride, forms 2-hydroxy-4-octyloxybenzophenone, one of the most important UV absorbers for the stabilization of polyolefin films.

2-Hydroxy-4-octyloxybenzophenone

Sumitomo Chemical has developed a new process for the production of m-aminophenol from resorcinol. Resorcinol is made to react with ammonia to yield m-aminophenol which may then be transformed, e.g. with CO_2 at 5 to 10 bar in an alkaline medium into 4-aminosalicylic acid, an anti-tuberculosis drug.

m-Aminophenol   4-Aminosalicylic acid

To date, m-aminophenol has mainly been produced by alkali fusion of 3-aminobenzenesulfonic acid (metanilic acid).

Hydroquinone (p-dihydroxybenzene) was first produced by Pierre Joseph Pelletier and Jean Caventou in 1820, by dry distillation of quinic acid. The structure was elucidated by Friedrich Wöhler in 1844.

Quinic Acid

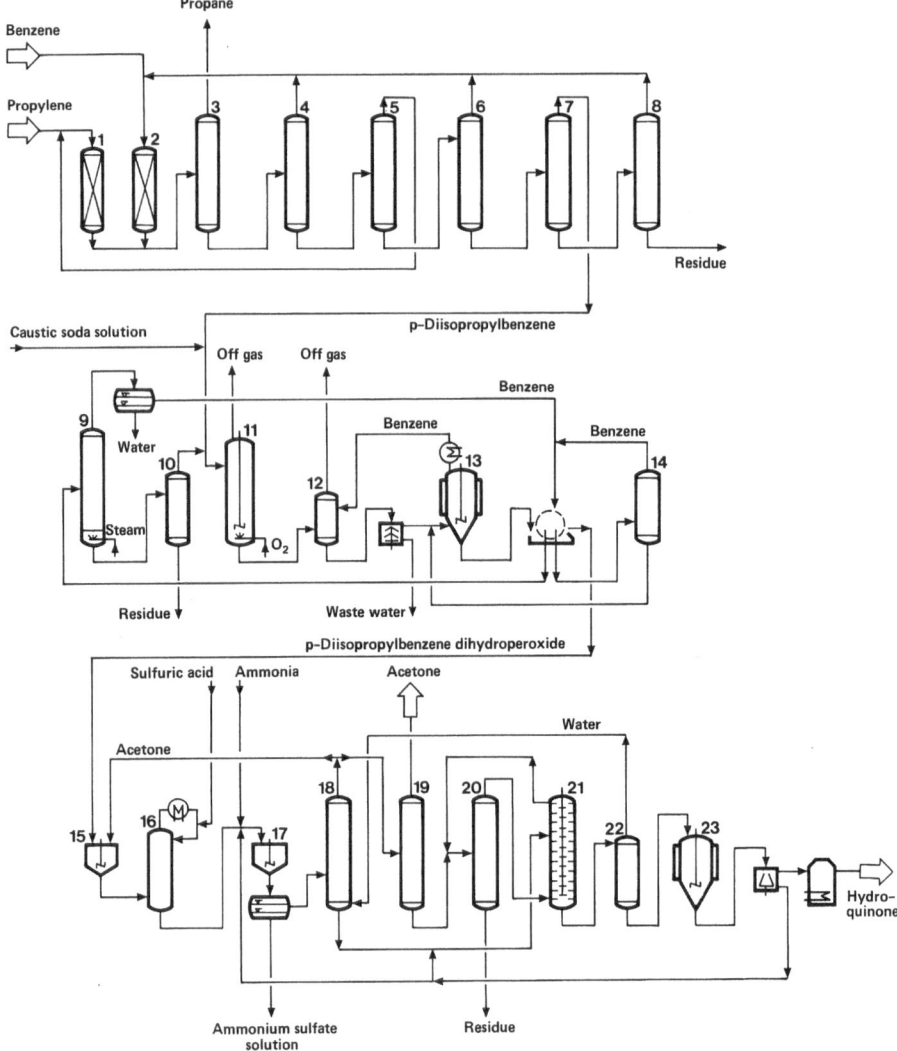

1 Alkylation reactor; 2 Transalkylation reactor; 3 Depropanizer; 4 Benzene column; 5 Cumene column; 6 o-/m-Diisopropylbenzene column; 7 p-Diisopropylbenzene column; 8 Triisopropylbenzene column; 9 Benzene recovery; 10 Monohydroperoxide stripping column; 11 Oxidation reactor; 12 Depressurization vessel; 13 Crystallizer; 14 Benzene recovery; 15 Dissolving vessel; 16 Conversion reactor; 17 Neutralization vessel; 18 Anti-solvent column; 19 Acetone column; 20 Benzene recovery; 21 Extractor; 22 Stripping column; 23 Crystallizer

Figure 5.28: Production of hydroquinone by oxidative cleavage of p-diisopropylbenzene

The classical method for preparing hydroquinone is based on the oxidation of aniline with manganese dioxide or sodium dichromate in sulfuric acid. The quinone which is obtained as an intermediate is reduced to hydroquinone with iron filings in dilute hydrochloric acid. This process, which is still used e.g. in India and China, is characterized by the simultaneous production of large amounts of manganese or chromium and iron salts, together with ammonium sulfate. For this reason, oxidative cleavage of p-diisopropylbenzene similarly to phenol synthesis was developed for the production of hydroquinone.

This process, illustrated by the flow diagram in Figure 5.28, has been developed to commercial viability by, among others, *Goodyear* in the United States, and *Mitsubishi Petrochemical* in Japan.

In a *Rhône Poulenc* process, hydroquinone and catechol are produced by the reaction of phenol and performic acid or other peracids in the presence of phosphoric acid. The yield of the co-product, catechol, can only be controlled to a limited extent in this route.

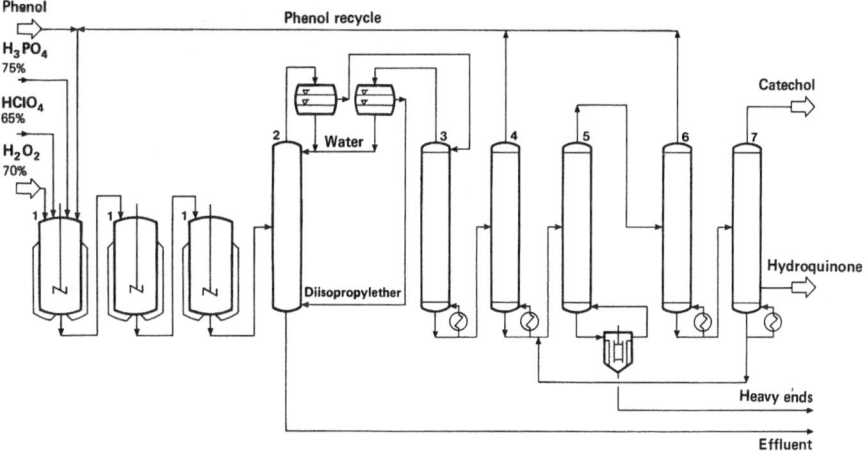

Figure 5.29 shows the flow diagram for the *Rhône Poulenc* process.

1 Cascade reactor; **2** Extractor; **3** Diisopropyl ether/water separation; **4** Phenol recovery column; **5** Heavy ends column; **6** Upgrading column; **7** Catechol/hydroquinone column

Figure 5.29: Process for the production of hydroquinone and catechol by oxidation of phenol

Electrochemical oxidation of benzene, which is very selective, can be used to produce high-purity hydroquinone. However, this process has had no commercial application to date.

World production of hydroquinone is around 25,000 tpa. Hydroquinone is predominantly used in photography as a developer and is employed to produce antioxidants by reaction with amines (see Chapter 6.1.1). It is also a polymerization inhibitor and a dyestuffs intermediate.

Hydroquinone can be used as a raw material for the production of quinacridone pigments (see Chapter 7.3.2) after carboxylation (Kolbe-Schmitt) to 2,5-dihydroxyterephthalic acid and further reaction with arylamines (e.g. aniline).

Reaction of the dipotassium salt of hydroquinone with 4,4'-difluorodiphenyl ketone yields the high-temperature resistant plastic polyetheretherketone (PEEK).

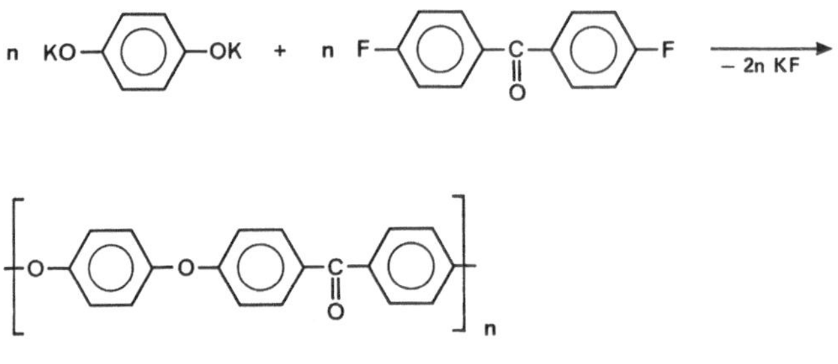

PEEK

Sumitomo Chemical has developed a new process for the manufacture of phloroglucinol which is used in relatively small amounts as a pharmaceutical intermediate. It is based on the oxidative cleavage of 1,3,5-triisopropylbenzene, in a method similar to the Hock process for the production of phenol.

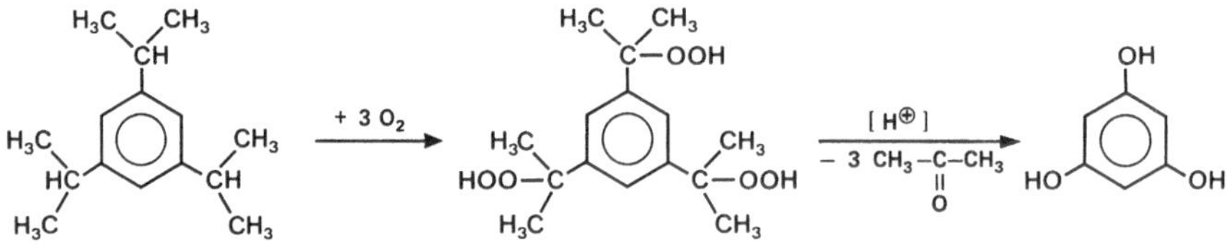

Phloroglucinol

5.4 Benzene hydrogenation – cyclohexane

Cyclohexane is an important intermediate in the production of Nylon 6 via caprolactam, or Nylon 66 via adipic acid. Demand for cyclohexane rose considerably with the production of nylon fibers, which began in 1937 with *Du Pont*, followed by *IG Farbenindustrie*. It is obtained either by refining appropriate petroleum fractions, or by benzene hydrogenation.

In comparison with benzene hydrogenation, recovery of cyclohexane by distillation of petroleum fractions is only competitive in a few individual cases (e.g. *Phillips Petroleum*/Borger, Tx), since separation of co-boiling compounds such as methylcyclopentane imposes high energy requirements. Cyclohexane produced by this method is generally only 85% pure, but higher degrees of purity can be obtained by extractive distillation.

Cyclohexane is produced predominantly by hydrogenation of benzene. For thermodynamic reasons, the temperature should not exceed 220 °C, since the benzene content then becomes too large, because of the position of the equilibrium.

$$\text{C}_6\text{H}_6 + 3\,\text{H}_2 \longrightarrow \text{C}_6\text{H}_{12} \qquad \Delta H_{298} = -206 \text{ kJ/Mol}$$

Whereas older hydrogenation processes were operated with sulfur-resistant catalysts because of the use of sulfur-containing benzene, the new methods, which use nickel and noble metal catalysts, require a very pure benzene with a sulfur

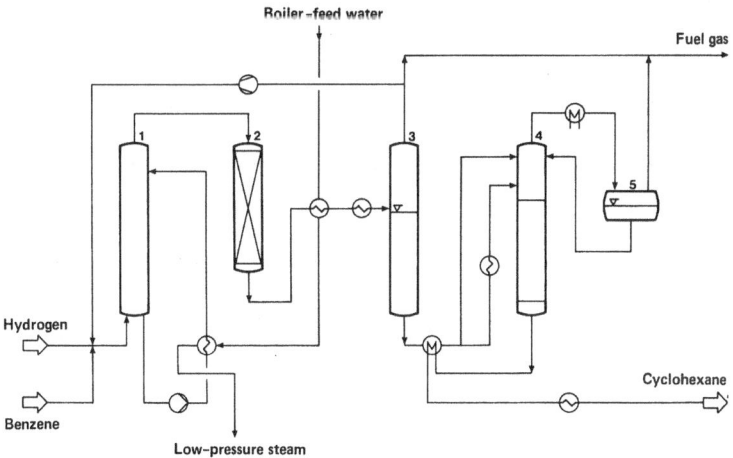

1 Main reactor; **2** Finishing reactor; **3** Separator; **4** Stabilizer; **5** Separator

Figure 5.30: Process for the production of cyclohexane by liquid phase hydrogenation of benzene (suspended catalyst)

content below 1 ppm. Hydrogenation occurs in the gas phase with a fixed-bed catalyst, or in the liquid phase with a suspended catalyst. Figure 5.30 shows the flow diagram of the widely used liquid-phase process developed by *Institut Francais du Pétrole (IFP)*.

In this process, benzene and hydrogen are fed into the reactor without preheating. The catalyst (Raney nickel) is suspended in liquid cyclohexane. Heat is removed by circulation through an external heat exchanger. The cyclohexane produced is transferred in vapor form from the reactor with a slight excess of hydrogen into a post-hydrogenation reactor. The main reactor operates by the bubble-column principle, the finishing reactor as a fixed-bed reactor. To avoid the isomerization of cyclohexane to methylcyclopentane it is necessary to observe an upper temperature limit of around 230 °C; the yield is 99.8%.

An example of a liquid-phase process with a fixed-bed reactor (Pt-catalyst) for benzene hydrogenation is the Hydra process developed by *UOP*, shown by the flow diagram in Figure 5.31.

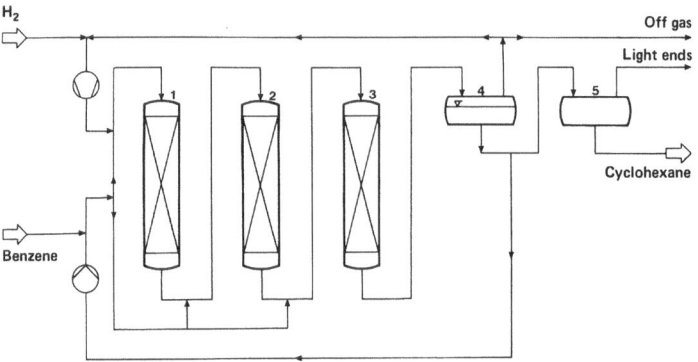

1-3 Hydrogenation reactors; 4 Gas separator; 5 Depressurizing vessel

Figure 5.31: Process for the production of cyclohexane by the liquid-phase hydrogenation of benzene (fixed bed catalyst)

The reaction temperature varies between 200 and 350 °C, with a pressure in the reactor of 30 bar; the hydrogen/benzene ratio is 3:1.

The Hydra process was originally based on platinum catalysts, modified with a lithium salt; more recently, nickel catalysts have been used. Maintaining a low residence time ensures that, despite the high reaction temperature, the cyclohexane/methylcyclopentane equilibrium cannot be achieved.

Gas-phase hydrogenation of benzene, used in isolated cases (e.g. Bexane process/*DSM*), is carried out at temperatures up to 370 °C and a pressure of 30 bar on Pt-catalysts.

The purity of the benzene used is of particular importance for the quality of the cyclohexane produced. The concentration of pure cyclohexane is usually 99.5%; the benzene content is below 300 ppm, while the level of aliphatics is limited to 5,000 ppm.

Table 5.6 summarizes the production data for the major cyclohexane manufacturing countries. The largest producers are *Texaco* (USA), *Phillips Petroleum* (Puerto Rico), *ICI* (UK) and *Exxon* (Netherlands).

Table 5.6: Production of cyclohexane (1985)

	(1,000 t)
USA	790
Canada	90
Brazil	50
West Germany	160
Italy	45
Belgium/Netherlands	265
Great Britain	240
Japan	530
Others	530
Total production	2,700

The most important products of cyclohexane are the polyamide building blocks, adipic acid and caprolactam, which are obtained by oxidation of cyclohexanol or by the formation of the oxime from cyclohexanone and subsequent Beckmann rearrangement.

$$HOOC-(CH_2)_4-COOH$$

Adipic acid

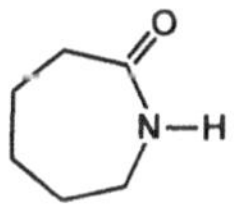

ε-Caprolactam

5.5 Nitrobenzene and aniline

Nitrobenzene and aniline, the latter originally recovered from coal-tar bases, were among the most important aromatics during the developmental phase of industrial dyestuffs chemistry in the middle of the 19th century.

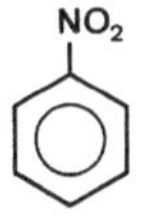

Nitrobenzene

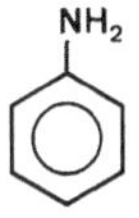

Aniline

5.5.1 Nitrobenzene

Nitrobenzene was first obtained by Eilhard Mitscherlich in 1834, by nitration of benzene. In England in 1847, its production from coal-tar benzene was patented and the technology developed by Charles B. Mansfield; manufacture began in France in 1848.

Nitrobenzene is used mainly as an intermediate in the production of aniline. It also forms the basis for the production of 3-chloronitrobenzene, m-nitrobenzene-sulfonic acid and p-aminophenol; its earlier importance in the production of the dyestuff intermediate benzidine is insignificant today, due to the carcinogenicity of the latter. The industrial production of nitrobenzene is achieved by isothermal nitration of benzene. Mixed nitration acid, composed of 40% HNO_3, 40% H_2SO_4 and 20% water, is used as the nitrating medium. Whereas the process was originally carried out batchwise, continuous operation predominates today in plants with a capacity up to 100,000 tpa. Stainless steel equipment is used, which is resistant to corrosion due to a passivation effect.

Figure 5.32 shows the flow diagram for the continuous nitration process developed by *Bofors Nobel Chemie*.

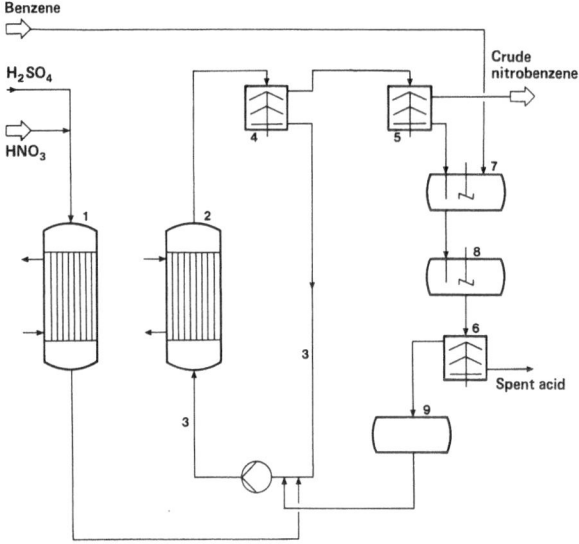

1 and 2 Cooler; 3 Pump circulation reactor; 4-6 Centrifugal separators; 7 and 8 Stirred vessels; 9 Benzene storage tank

Figure 5.32: Flow diagram for the continuous nitration of benzene

Since nitration proceeds heterogeneously between the aqueous and organic layer, the benzene and mixed acid are mixed intensively. Nitration then takes place predominantly in the acid phase. The reaction, which is carried out at 60 °C, operates with a HNO_3/benzene molar ratio of 0.94 to 0.98. The yield is around 98% based on HNO_3. Oxygen compounds of nitrobenzene with phenolic structures arise as by-products; they are rendered soluble by treatment of the reaction product with dilute alkali. Safety measures are important in commercial nitration, to avoid explosions. Since nitration is exothermic and a breakdown of the nitroproducts can occur at higher temperatures, exact temperature control, thorough mixing and proper heat removal are of fundamental importance. Nitration reactors are therefore frequently operated with an atmosphere of nitrogen.

Table 5.7 summarizes production capacities for the major nitrobenzene-producing countries. The largest manufacturers are *Bayer* (West Germany), *Du Pont, First Chemical* and *ICI/Rubicon Chemicals* (USA).

Table 5.7: Production capacities for nitrobenzene (1985)

	(1,000 t)
USA	650
Brazil	15
Belgium	200
West Germany	240
Portugal	70
Great Britain	160
Japan	90
India	20
Other countries	255
Total capacity, Western world	1,700

In addition to the manufacture of aniline (see Chapter 5.5.2), nitrobenzene is used, to a lesser extent, to produce m-nitrochlorobenzene and m-nitrobenzenesulfonic acid, and for the production of p-aminophenol (see Chapter 5.3.4.6)

m-Nitrochlorobenzene   m-Nitrobenzene-sulfonic acid

m-Nitrochlorobenzene can only be obtained in low yield by nitration of chlorobenzene. It is therefore preferable to produce it by chlorination of nitrobenzene in the presence of Lewis acids, such as iron-III-chloride; the crude reaction product is refined by distillation. m-Nitrochlorobenzene is used predominantly for the production of m-chloroaniline, a raw material used in the manufacture of plant protection agents and pharmaceuticals, such as 4,7-dichloroquinoline (see Chapter 14.5.3).

m-Nitrobenzenesulfonic acid is obtained with high selectivity by the sulfonation of nitrobenzene. It is mainly used to produce m-aminophenol by alkali fusion.

m-Aminophenol (see Chapter 5.3.4.8) finds application in the production of plant protection agents, pharmaceuticals and dyestuffs, as well as for the production of 3,4'-diaminodiphenyl ether, which is used for the production of a high-value aramid fiber (Technora, *Teijin*).

Technora

5.5.2 Aniline

Clearly the most important product from nitrobenzene is aniline; it was first obtained by Otto Unverdorben in 1826 by distilling indigo and was named 'krystallin', since it easily forms crystalline salts with sulfuric acid. The method introduced by Nikolai N. Zinin in 1841 of reducing nitrobenzene, later modified by Pierre I. A. Béchamp employing with iron and hot hydrochloric acid is still used on a large scale today, since the resultant iron oxides can be used as pigments.

The reduction, carried out as a batch process, gives a crude aniline which can be refined by distillation; the yield of aniline is around 99%.

$$4\,C_6H_5NO_2 + 9\,Fe + 4\,H_2O \xrightarrow{[HCl]} 4\,C_6H_5NH_2 + 3\,Fe_3O_4$$

A more recent method for producing aniline is continuous gas-phase hydrogenation of nitrobenzene in fluidized or fixed bed reactions.

Figure 5.33 shows the fluidized bed process developed by *American Cyanamid*, which operates at a reaction temperature of 270 °C and a pressure of 1.8 bar. Silica gel modified with copper is used as catalyst. The yield from this process is also around 99%; the *BASF* process operates in a similar way.

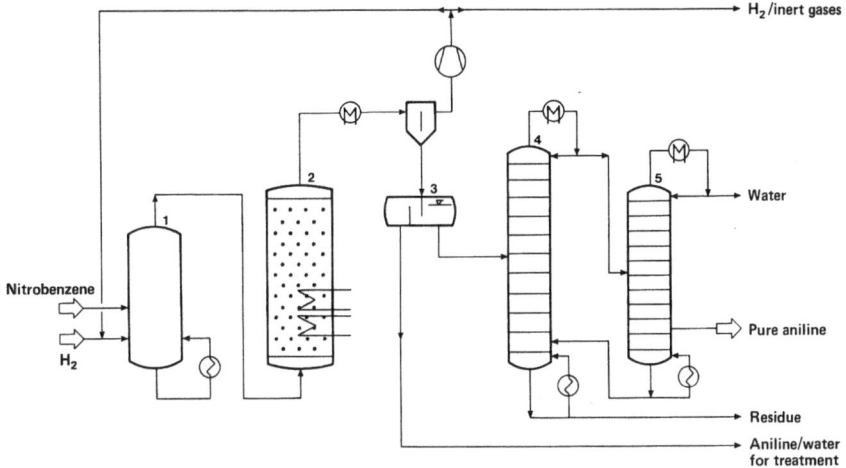

1 Saturator; **2** Fluidized-bed reactor; **3** Separator; **4** Crude aniline fractionation; **5** Dewatering column

Figure 5.33: Production of aniline by gas-phase hydrogenation in the fluidized-bed process

Typical of gas-phase hydrogenation in a fixed bed is the process developed by *Lonza/Alusuisse* (Figure 5.34). Here, nitrobenzene is sprayed into hydrogen at 2 to 15 bar in the ratio 1.5 to 1.6. The reaction mixture then enters the gas cycle, which is heated to 150 to 300 °C. The catalyst is copper on pumice. The *Bayer* process, which operates with a fixed-bed catalyst, uses NiS catalysts; the reaction takes place at 300 to 470 °C.

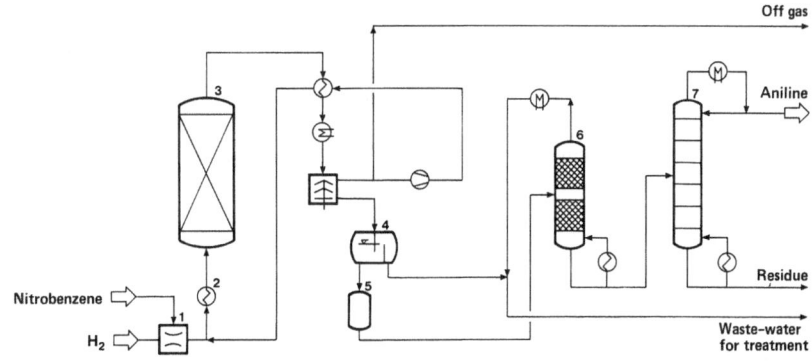

1 Sprayer; **2** Vaporizer; **3** Fixed bed reactor; **4** Separator; **5** Crude product storage tank; **6** Dewatering column; **7** Distillation

Figure 5.34: Production of aniline by gas-phase hydrogenation in the fixed bed process

With the availability of relatively inexpensive phenol from the Hock synthesis by cumene oxidation, it has also become economical to produce aniline by gas-phase ammonolysis (400 °C, 200 bar) of phenol.

$$C_6H_5OH + NH_3 \longrightarrow C_6H_5NH_2 + H_2O \quad \Delta H = -10 \text{ kJ/Mol}$$

The process developed by *Halcon*, using a fixed-bed catalyst, is operated by *Mitsui Petrochemical*/Japan and *Aristech*/USA. Silica/alumina or silicon boroxide with added tungsten and vanadium are used as catalysts. Diphenyl- and triphenylamine, as well as carbazole, occur as by-products of the reaction. The particular advantage of the process lies in the low investment costs in comparison with processes based on nitrobenzene.

Figure 5.35 shows the flow diagram of the *Halcon* fixed bed process.

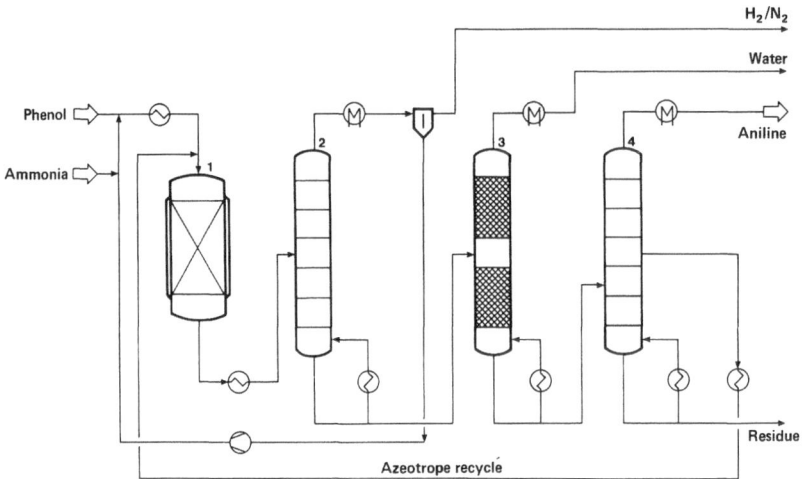

1 Fixed bed reactor; 2 Column for NH_3 recovery; 3 Drying column; 4 Aniline fractionation column

Figure 5.35: Process for the continuous production of aniline by gas-phase ammonolysis of phenol

The ammonolysis of chlorobenzene previously operated by *Dow* in the USA, analogous to the synthesis of phenol from chlorobenzene, has had no commercial importance for some years. Table 5.8 summarizes the production capacities of the major aniline producer countries; it is striking that aniline production is limited to a small number of countries.

Table 5.8: Production capacities for aniline (1985)

	(1,000 t)
USA	570
Belgium	110
West Germany	200
Great Britain	115
Portugal	50
Japan	175
Switzerland	15
Other countries	165
Total production	1,400

Consumption of aniline in the USA in 1985 was around 350,000 t.

5.5.3 Aniline derivatives

The most important uses for aniline today are the production of 4,4'-diphenylmethane diisocyanate (MDI) and chemicals used in rubber processing, especially as vulcanization accelerators or antioxidants. Additional applications, which were dominant in former times, e.g. dyestuffs, plant protection agents and raw materials for fibers, complete the range of aniline applications.

The most important commercial intermediates from aniline are the mixture of 4,4'- and 2,4'-diaminodiphenylmethane, obtained by reaction with formaldehyde, and the higher polyamines as well as 2-mercaptobenzothiazole and cyclohexylamine, used as rubber additives. N,N-Dialkylanilines are commonly used for the production of dyestuffs, while phenylhydrazine is used as an intermediate in the manufacture of plant protection agents, pharmaceuticals and dyestuffs. Other important aniline products are sulfanilic acid and acetanilide.

5.5.3.1 4,4'-Diphenylmethane diisocyanate (MDI)

4,4'-Diphenylmethane diisocyanate is the second-most important aromatic isocyanate, next to toluene diisocyanate (see Chapter 6.1.2).

The reaction of formaldehyde with aniline in the presence of acids (e.g. HCl) also includes an attack of the formaldehyde on the amino nitrogen, which can then further react with aniline to give p-aminobenzylaniline and higher-molecular secondary amines. In the second reaction stage, these secondary amines are 'isomerized' to primary amines, with a reaction time of around 4 hours.

H₂N—⟨○⟩—CH₂—⟨○⟩—NH₂   ⟨○⟩—NH—CH₂—⟨○⟩—NH₂

4,4'-Diaminodiphenylmethane   p-Aminobenzylaniline

200 Production and uses of benzene derivatives

$$\langle\bigcirc\rangle-NH{-}[CH_2{-}\langle\bigcirc\rangle{-}NH]_n{-}CH_2{-}\langle\bigcirc\rangle-NH_2$$

Polyamine mixture

In the acid-catalyzed process, condensation in the first stage is carried out at 50 to 70 °C, while in the second stage a temperature of around 105 °C and a pressure of 3 bar are used. Subsequent neutralization is carried out at 90 to 100 °C, to keep the polyamines in the liquid state. The crude amine mixture is refined by distillation, with around 60% 4,4'-diamine, 5% 2,4'-diamine and 35% polyamines being recovered.

Increasing the temperature and raising the aniline/formaldehyde ratio favors the formation of the 2,4'-isomer, at the expense of the 4,4'-isomer.

Figure 5.36 shows a flow diagram for the production of diaminodiphenylmethane.

Following the distillation of the amines to recover 4,4'-diaminodiphenylmethane and the polyamine fraction, a reaction with phosgene is performed at temperatures up to 140 °C in solvents such as dichlorobenzene.

$$H_2N-\langle\bigcirc\rangle-CH_2-\langle\bigcirc\rangle-NH_2 + 2\ COCl_2 \longrightarrow$$

$$O{=}C{=}N-\langle\bigcirc\rangle-CH_2-\langle\bigcirc\rangle-N{=}C{=}O + 4\ HCl$$

The MDI is concentrated by removing the solvent in thin-film evaporators; it is used with, e.g., polyhydric alcohols to yield the widely-used polyurethane plastics.

Consumption of MDI in the USA in 1985 was around 260,000 t; *Dow* and *Mobay/Bayer* operate the largest capacities.

5.5.3.2 Cyclohexylamine and benzothiazole derivatives

The second most important application for aniline is the manufacture of chemicals for the rubber industry in the form of cyclohexylamine and benzothiazole derivatives.

Cyclohexylamine is generally produced by catalytic liquid-phase hydrogenation of aniline under high pressure (60 bar) at 230 °C on cobalt catalysts. Alternative methods of synthesis are the ammonolysis of cyclohexanol at 200 bar and 220 °C on a calcium silicate catalyst, the catalytic ammonolysis of cyclohexyl chloride and the reduction of nitrocyclohexane.

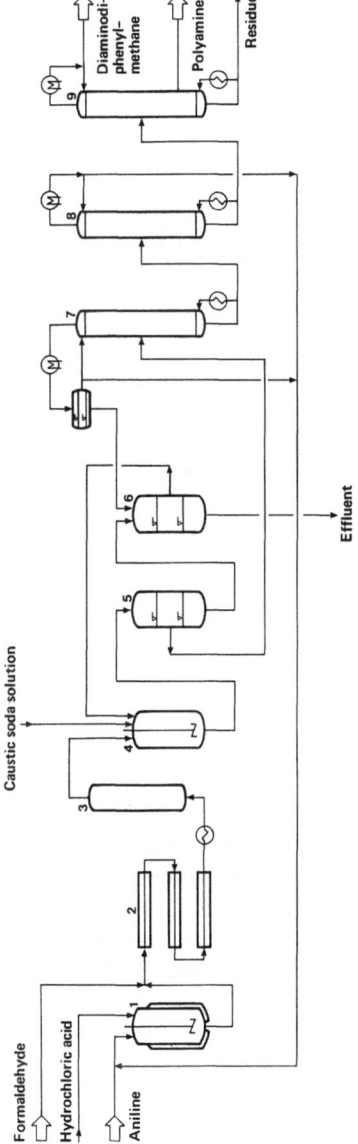

Figure 5.36: Flow diagram for the production of diaminodiphenylmethane

1 Reactor; **2** Tube reactor; **3** Finishing reactor; **4** Neutralization vessel; **5** and **6** Phase separators; **7** Dewatering column; **8** Aniline column; **9** Amine column

In Western Europe, hydrogenation of aniline predominates; this method is also employed by the largest US producer, *Air Products*. Production of cyclohexylamine in Western Europe in 1985 was around 10,000t; USA production in 1985 amounted to 4,000t.

One of the main applications for cyclohexylamine is in the manufacture of the vulcanization accelerator, cyclohexylbenzothiazolesulfenamide (CBS) from cyclohexylamine and 2-mercaptobenzothiazole (MBT). The dicyclohexylamine derivative is also used as a vulcanization accelerator (DCBS).

MBT + cyclohexylamine → CBS

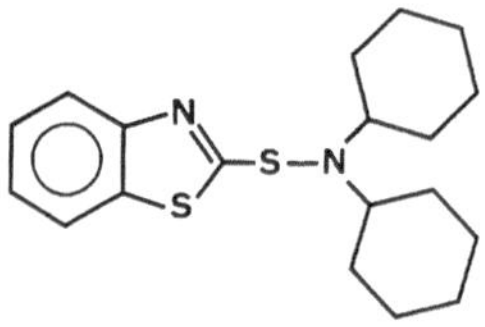

DCBS

2-Mercaptobenzothiazole is obtained from aniline, carbon disulfide and sulfur in a continuous reaction at 285 °C and a pressure of 150 bar. The largest producers in Western Europe are *Bayer* and *Monsanto*.

Other important vulcanization accelerators based on 2-mercaptobenzothiazole are dibenzothiazyl disulfide (MBTS), obtained by the oxidation from MBT (e.g. with chlorine), and 2-morpholinobenzothiazylsulfenamide (MBS), (which more accurately should be referred to as N-(oxydiethylene)-2-benzothiazylsulfenamide).

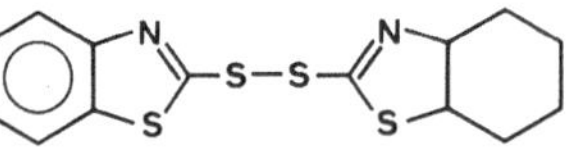

MBTS 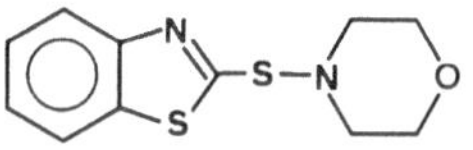   MBS

Mercaptobenzothiazole is also used in the production of the herbicide methabenzthiazuron (*Bayer*), via the intermediate 2-(methylamino)-benzothiazole.

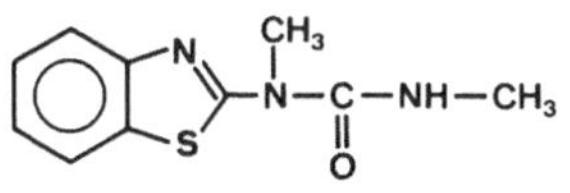

Methabenzthiazuron

Another application of cyclohexylamine is in the manufacture of sodium and calcium cyclohexylsulfamates, used as sweeteners. Cyclohexylamine is converted with SO_3 in the presence of an excess of a salt-forming tertiary amine. Subsequent reaction of the amine salt with sodium hydroxide yields sodium cyclohexylsulfamate (sodium cyclamate). The calcium salt is produced analogously using calcium hydroxide.

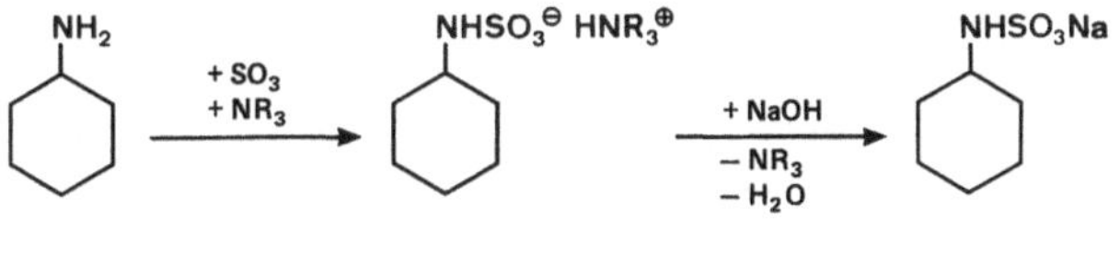

Sodium cyclamate

5.5.3.3 Secondary and tertiary aniline bases

The N-alkyl derivatives of aniline, such as N-methylaniline and N,N-dimethylaniline, are produced by the reaction of aniline at around 200 °C with an excess of the appropriate alcohol, under pressure (30 to 50 bar) in the presence of an acid catalyst (H_2SO_4, H_3PO_4); they are commonly used in the production of dyestuffs.

Production of N,N-dimethylaniline in Western Europe in 1985 was around 9,000 t. Among the most important compounds based on N,N-dimethylaniline is Michler's ketone, which is used to manufacture triphenylmethane dyes such as Basic Violet 3 (Crystal Violet) or Basic Green 4 (Malachite Green), which is obtainable by reaction with benzaldehyde in a sulfuric acid medium, followed by oxidation.

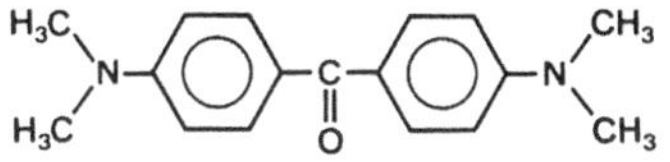

Michler's ketone

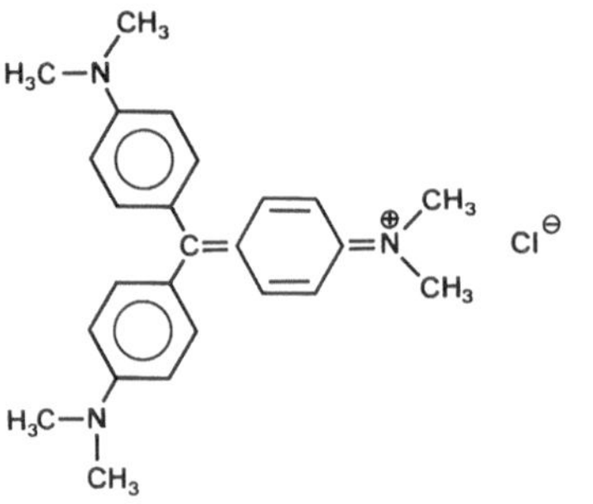

Basic Violet 3

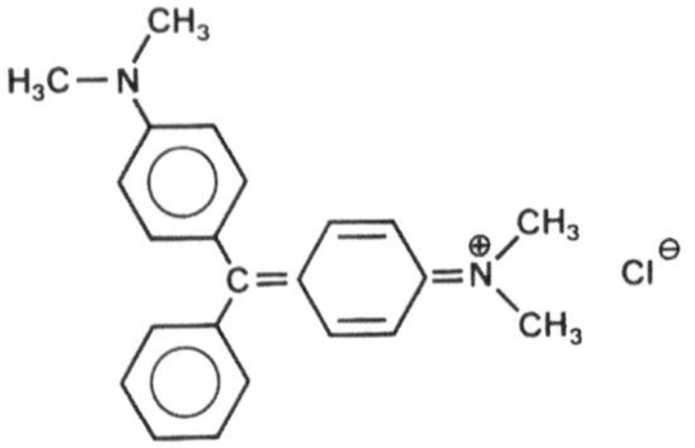

Basic Green 4

Diphenylamine is produced by the condensation of aniline in the presence of small amounts of mineral acid, with the release of ammonia. The crude diphenylamine/aniline mixture is refined by fractional distillation. Diphenylamine can also be produced by the reaction of chlorobenzene and aniline under pressure; Cu_2O/KCl is used as a catalyst.

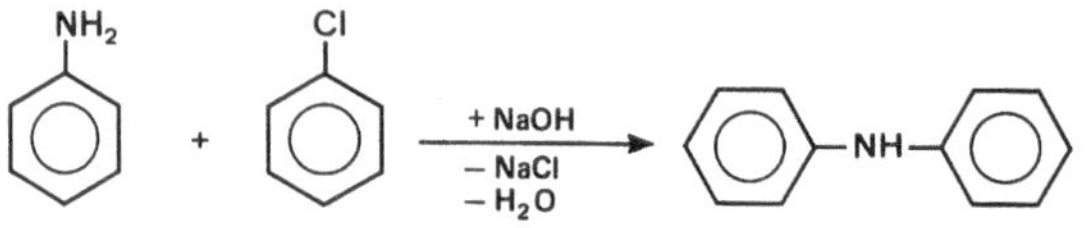

Diphenylamine is used as an intermediate in the production of dyes and in the form of N-nitrosodiphenylamine as a vulcanization retarder. In some countries, e.g. in West Germany, the application of this nitroso compound is banned, because of its carcinogenicity, particularly in combination with secondary amines.

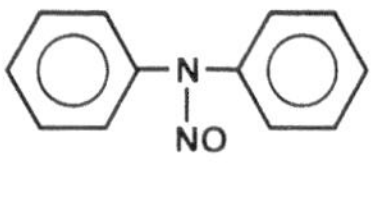

N-Nitrosodiphenylamine

Phenothiazine is produced in 90% yield by the reaction of diphenylamine and sulfur, using iodine as a catalyst, at 180 °C. It is used as a raw material in the production of antioxidants and psychotropic drugs, such as promazine.

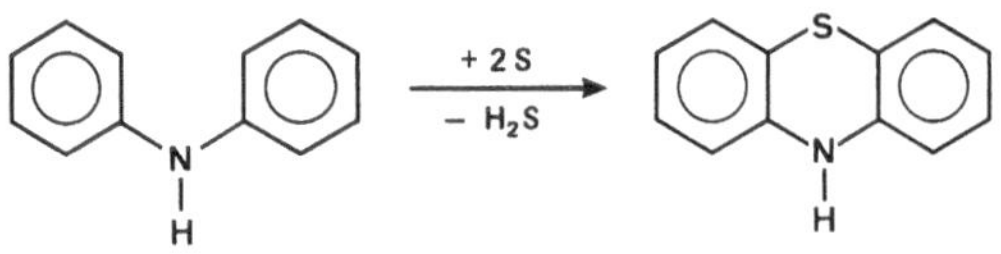

Phenothiazine

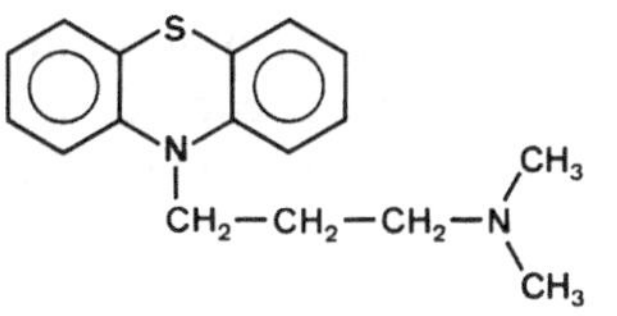

Promazine

5.5.3.4 Other aniline derivatives

Acetanilide is obtained from aniline and acetic anhydride. It is mainly used in the production of sulfonamide chemotherapeutics; those are obtained by the reaction of acetanilide with chlorosulfonic acid, followed by reaction with a primary amine and subsequent deacylation.

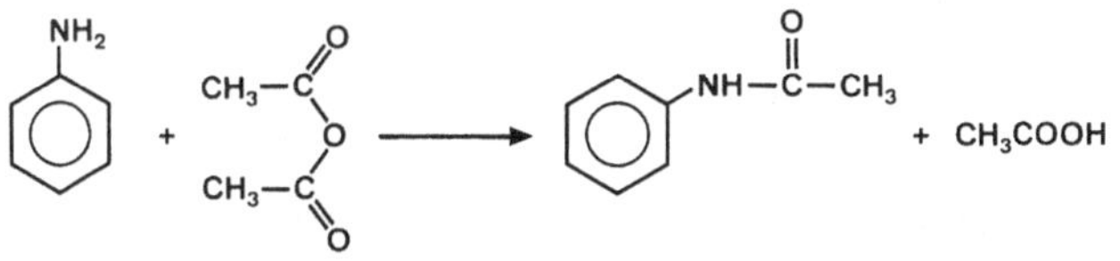

Acetanilide

In human medicine, sulfamethoxazole (in combination with trimethoprim) is commonly used as an important antimetabolite for the treatment of bacterial infections, while other sulfonamides are mainly restricted to veterinary medicine, because of possible side effects.

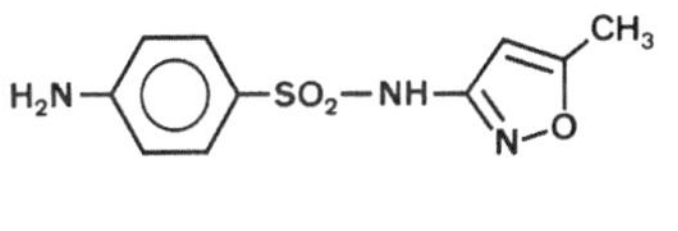

Sulfamethoxazole 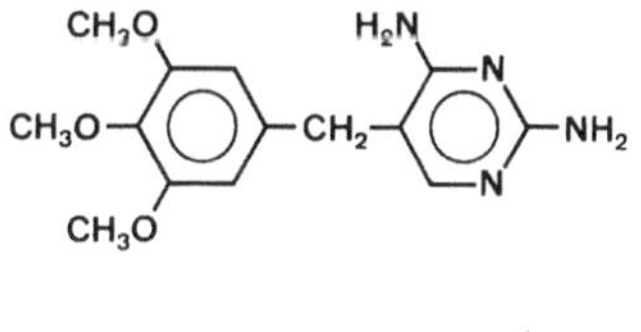   Trimethoprim

Production of 4,4'-diaminobiphenyl (benzidine) and its derivatives is normally carried out by reduction of the corresponding nitrobenzenes with zinc powder in high-boiling solvents, such as o-dichlorobenzene, to yield the respective hydrazobenzene, which is converted into the benzidine form in acid solution. Since benzidine is carcinogenic, special care should be taken in its production; manufacture is banned in many countries, because of its toxicity.

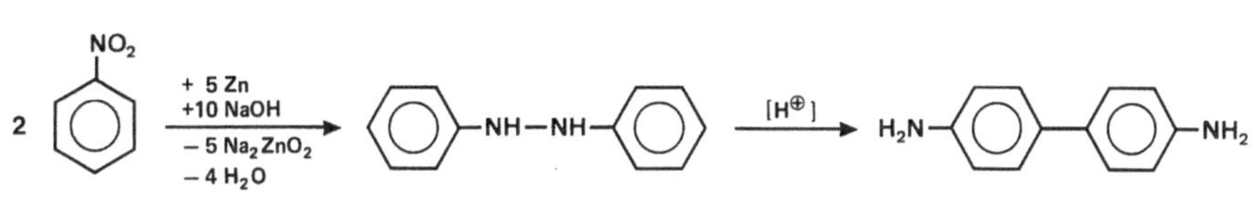

Benzidine

Benzidine derivatives, especially 3,3'-dichloro- and 3,3'-dimethylbenzidine, are important intermediates used in the production of azo dyes, o-nitrochlorobenzene and o-nitrotoluene being the respective starting materials for these two examples.

Selective introduction of alkyl groups to the benzene nucleus of aniline, as, for example, in the production of 2,6-diethylaniline, is not economically viable by normal electrophilic substitution of aniline since the amino nitrogen is also attacked electrophilically. 2,6-Diethylaniline is therefore produced with organometallic catalysts, such as diethylaluminum chloride. The reaction with ethylene under a pressure of 70 bar at a temperature of around 150 °C produces 2,6-diethylaniline in a yield of over 85%. 2,6-Diethylaniline is used mainly for the production of the herbicide alachlor, which is consumed in the USA in quantities of around 42,000 tpa; the largest producer is *Monsanto*. 2,6-Diethylaniline is made to react with formaldehyde to yield 2,6-diethylmethyleneaniline, which is then converted into alachlor by treatment with chloroacetyl chloride and methanol.

Alachlor

Phenylhydrazine is produced by diazotization of aniline and subsequent reduction with sodium hydrogen sulfite.

Phenylhydrazine

West European production was around 8,000 t in 1985. The main use of phenylhydrazine is in the production of 1-phenyl-3-methyl-pyrazolin-(5)-one (PMP), which is obtained by reaction with acetoacetic acid methyl ester or acetoacetamide.

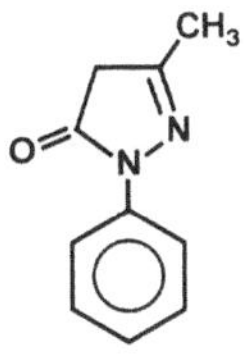

1-Phenyl-3-methylpyrazol-(5)-one

N-Alkylation of PMP with dimethyl sulfate yields 1-phenyl-2,3-dimethylpyrazolin-(5)-one, a pain-relieving medicament (antipyrine).

PMP is also used as a raw material in the manufacture of azo pigments such as the important Pigment Orange 13.

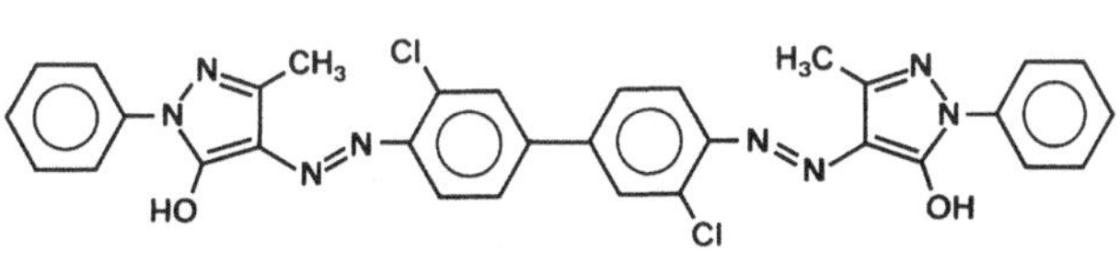

Pigment Orange 13

Another important derivative of phenylhydrazine is the plant protection agent pyrazon, produced by *BASF* from phenylhydrazine and mucochloric anhydride, with subsequent replacement of one chlorine by ammonia.

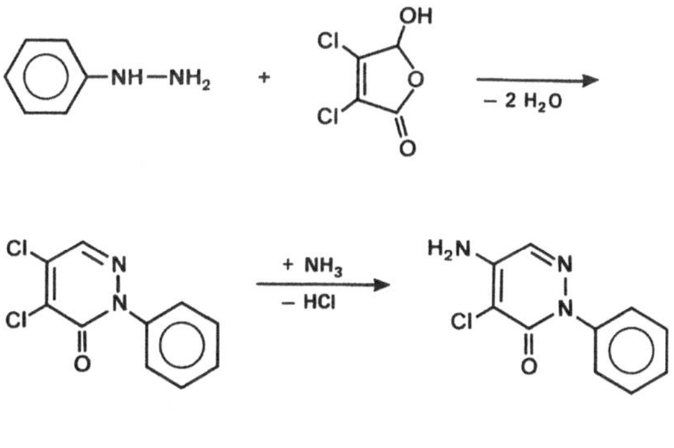

Pyrazon

Although the first commercially important synthesis of indigo was based on phthalic anhydride, aniline is today the starting point for the large-scale synthesis of indigo. In this process, aniline is condensed with formaldehyde and converted into N-phenylglycine with cyanide. The treatment of N-phenylglycine with sodium amide yields indoxyl, which in turn gives indigo on oxidation.

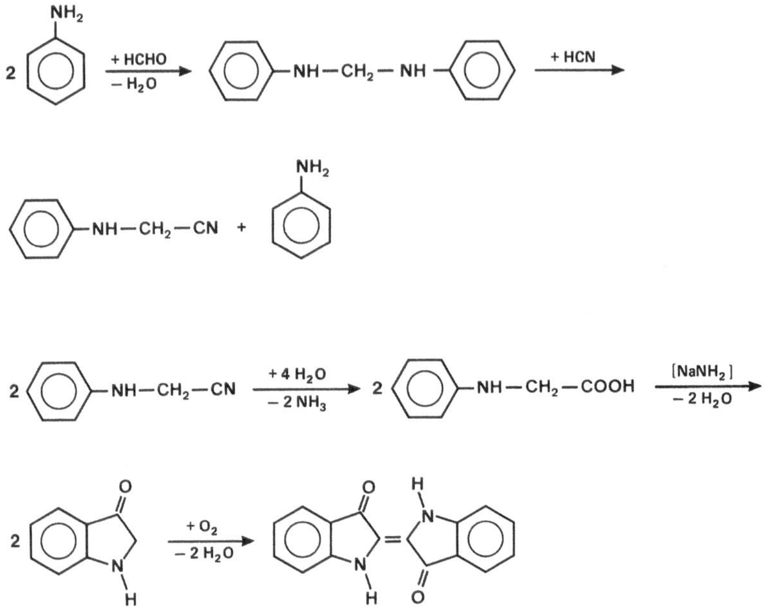

Production and uses of benzene derivatives

[Reaction scheme: 2 aniline + HCHO → PhNH–CH₂–NH–Ph (−H₂O), then + HCN]

[Reaction scheme: Ph–NH–CH₂–CN + aniline]

[Reaction scheme: 2 Ph–NH–CH₂–CN + 4 H₂O → 2 Ph–NH–CH₂–COOH (−2 NH₃), then [NaNH₂], −2 H₂O]

[Reaction scheme: 2 Indoxyl + O₂ → Indigo (−2 H₂O)]

Indoxyl   Indigo

However, the former position of indigo as the 'King of dyestuffs' is no longer valid today.

Sulfanilic acid is produced by heating aniline hydrogen sulfate at 170 to 180 °C. It is used in the manufacture of optical brighteners and azo dyes such as tartrazine (Acid Yellow 23) and Acid Orange 7 (see Chapter 9.3.3.2).

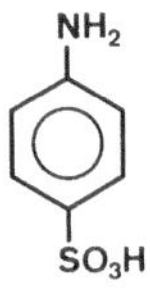

Sulfanilic acid 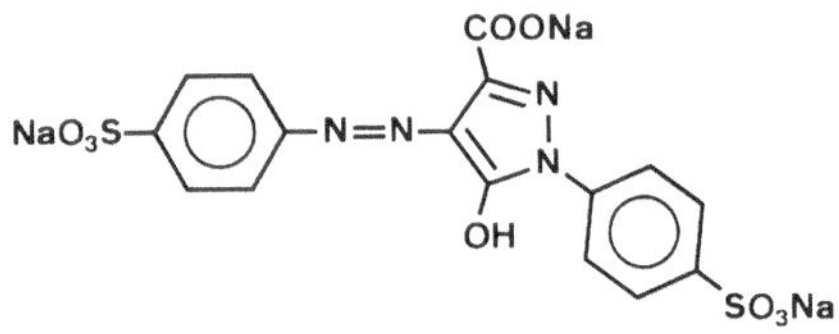   Tartrazine

The condensation of aniline with cyanogen chloride yields N,N'-diphenylguanidine (DPG), which is used as a vulcanization accelerator.

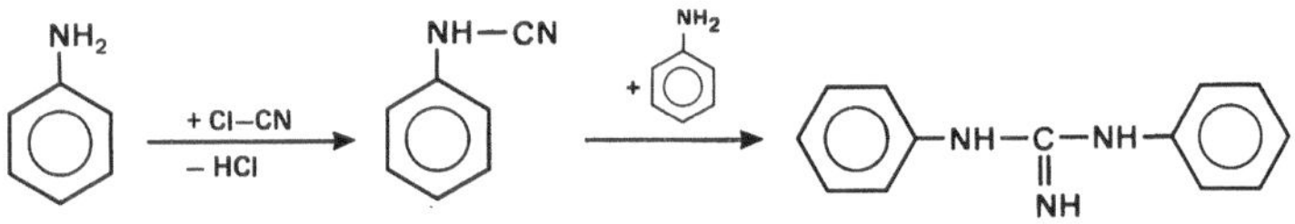

[Reaction scheme: aniline + Cl–CN → Ph–NH–CN (−HCl) + aniline → Ph–NH–C(=NH)–NH–Ph]

N,N'-Diphenylguanidine

In spite of the current dominance of applications for aniline in the production of polyurethanes and chemicals for rubber processing, aniline has remained an important raw material for the dyestuffs industry. It is used today for production of the dyes Solvent Red 19 and Solvent Red 23, which are used in West Germany and France as markers for heating oil, to distinguish it from diesel oil; methylnaphthalene is used as a solvent for the dyes.

Solvent Red 19 is obtained by the diazotization of aniline and conversion into p-aminoazobenzene followed by diazotization and reaction with N-ethyl-2-aminonaphthalene, while Solvent Red 23 is obtained by the reaction of diazotized p-aminoazobenzene with β-naphthol.

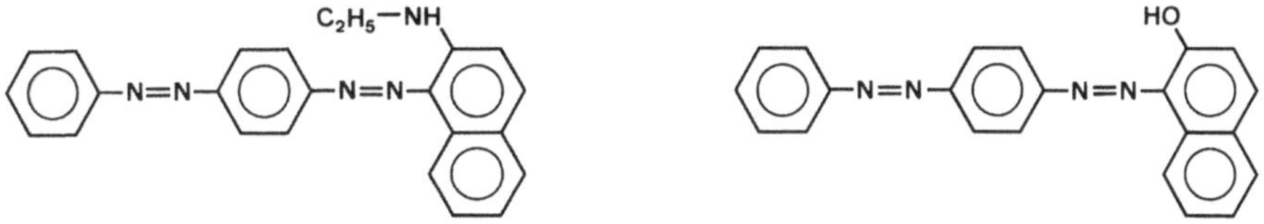

Solvent Red 19

Solvent Red 23

Diazotized aniline is also used as a starting material for the production of fluorobenzene, which is produced only in small quantities by thermal decomposition of the diazonium fluoride.

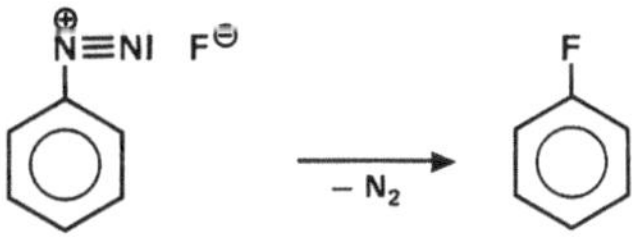

The reaction of diketene in dilute acetic acid with aniline yields acetoacetanilide from which azo pigments such as Pigment Yellow 1 are derived. An important azo dye based on this raw material is Acid Yellow 151.(Likewise, homologs of aniline (e.g. xylidines) are used for the production of acetoacetanilide pigments such as Pigment Yellow 13.)

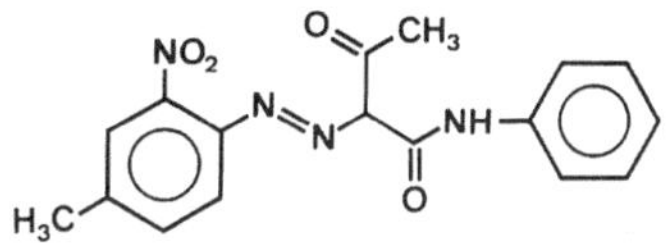

Pigment Yellow 1

Acid Yellow 151

Pigment Yellow 13

5.6 Alkylbenzenes and alkylbenzene sulfonates

Monoalkylbenzenes with a long alkyl chain ($\geq C_6$) in the form of alkylbenzene sulfonates are the most important raw materials for the production of ionic detergents.

The development of organic surfactants and dispersing agents was based in the first half of the last century on renewable raw materials. Friedlieb Ferdinand Runge first produced sulfated olive oils in 1834, followed in 1875 by the production of Turkey red oil by sulfating castor oil. The development of sulfonates based on alkylaromatics goes back to a discovery by Fritz Günther (*BASF*) in 1917. While attempting to produce diisopropyl ether, Günther used naphthalenesulfonic acid as a condensation agent, instead of sulfuric acid. On examining the reaction product, he found that the isopropyl alcohol had reacted with the naphthalenesulfonate salt and the product displayed surprising wetting and emulsifying power in dilute solution; the surfactant became well known as 'Nekal'.

Production of alkylbenzene derivatives in the 1940's was linked to the Fischer-Tropsch synthesis. In 1941, *IG Farbenindustrie* began to produce chloroparaffins by chlorination of a Fischer-Tropsch alkane mixture rich in n-paraffins with an average chain length of C_{14}; the chloroparaffin was made to react with benzene under Friedel-Crafts conditions and subsequently sulfonated to yield an alkylbenzene sulfonate. The product was given the trade name 'Igepal NA'.

Modern benzene-based surfactants have alkyl groups with 10 to 14 carbon atoms, since with fewer than 6 carbon atoms in the alkyl group the alkylbenzene sulfonates are not sufficiently surface active, and alkylbenzene sulfonates with 15 and more carbon atoms in the alkyl group are difficult to dissolve in water, although they are soluble in organic media.

There are two types of alkylbenzene sulfonates, namely alkylbenzenes with linear alkyl groups (LAS) and alkylbenzenes with branched alkyl groups (ABS or TPS).

$CH_3-(CH_2)_4-CH-(CH_2)_5-CH_3$ — benzene ring — SO_3H

LAS

$CH_3-CH(CH_3)-CH_2-C(CH_3)_2-CH_2-CH(CH_3)-CH(CH_3)-CH_3$ — benzene ring — SO_3H

TPS

The branched C_{12} alkylbenzenes, produced predominantly between 1950 and 1970, were obtained from propylene tetramer and benzene at temperatures from 20 to 50 °C, with a reaction time of a few minutes, using Friedel-Crafts catalysts (HF or $AlCl_3$). To avoid polyalkylation as far as possible, the operation was carried out with a 5 to 10-fold excess of benzene. The yield of dodecylbenzene was 70 to 80%.

Since linear alkylbenzene derivatives are more readily biodegradable than branched ones, the use of branch-chain alkylbenzene sulfonates has sharply declined with the introduction of respective regulations since the mid-1960's (e.g. the West German 'Detergents Law' of 1962).

Linear alkylbenzene sulfonates are produced from the reaction of benzene and secondary monochloroparaffins; the latter are obtained by chlorination of n-paraffins. A requirement of this process was the availability of pure n-paraffins, which was made possible by the introduction of molecular sieves. The n-paraffins can be separated with molecular sieves from kerosene or gas oil, since they have a smaller diameter (ca. 4.9 Å) than the branched paraffins.

Besides alkylation with chloroparaffins, reaction with olefins is of particular importance today. Linear alkylbenzenes are mainly produced by the hydrogen fluoride process with dehydrochlorination, as shown in the flow diagram in Figure 5.37.

The chlorination of paraffins first produces a so-called 'chloro-oil', which contains around 30% alkyl chlorides and 70% paraffins. The 'chloro-oil' is dehydrochlorinated in a dehydrochlorination column, with a bottom temperature of around 300 °C. The resulting olefin/paraffin mixture is mixed with a large molar excess of benzene and fed into the reactor, which is fitted with a powerful stirrer and cooling pipes. Here, the benzene is alkylated in the presence of hydrogen fluoride at a temperature below 50 °C. The reaction product is then separated into two layers in a separation vessel: the upper layer, the crude alkylate, is split by distillation into benzene, an inter-cut, paraffin, alkylbenzene and a higher-boiling tail product. The hydrogen fluoride, which is present in the lower layer, is recirculated.

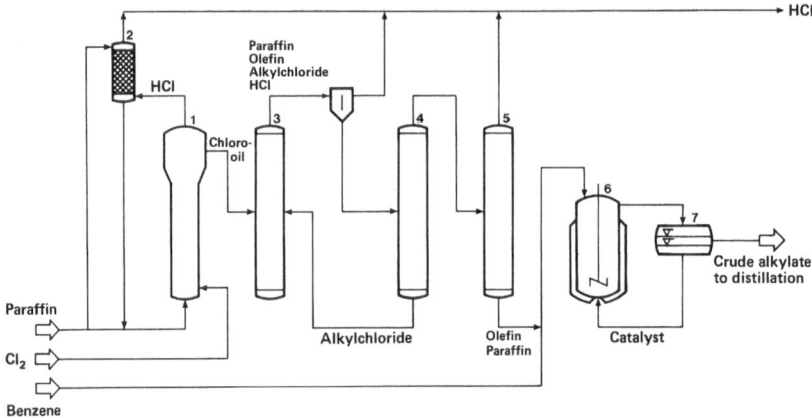

1 Chlorination tower; 2 Hydrogen chloride scrubber; 3 Dehydrochlorination column; 4 Alkyl chloride separation column; 5 Degassing column; 6 Alkylation reactor; 7 Separator

Figure 5.37: Production of linear alkylbenzenes by the hydrogen fluoride method

Alkylation with aluminum chloride is in competition with the hydrogen fluoride process. When using aluminum chloride as the catalyst with linear alkyl chlorides or olefins, an isomer mixture is produced in which the 2-phenylalkanes predominate and 3-, 4- and 5-phenylalkanes are present in smaller amounts. In the hydrogen fluoride process, on the other hand, the 3-, 4- and 5-phenylalkanes are the main isomers.

Long-chain alkylbenzenes are converted into the corresponding alkylbenzene sulfonates by sulfonation with sulfuric acid or sulfur trioxide. Sulfonation with

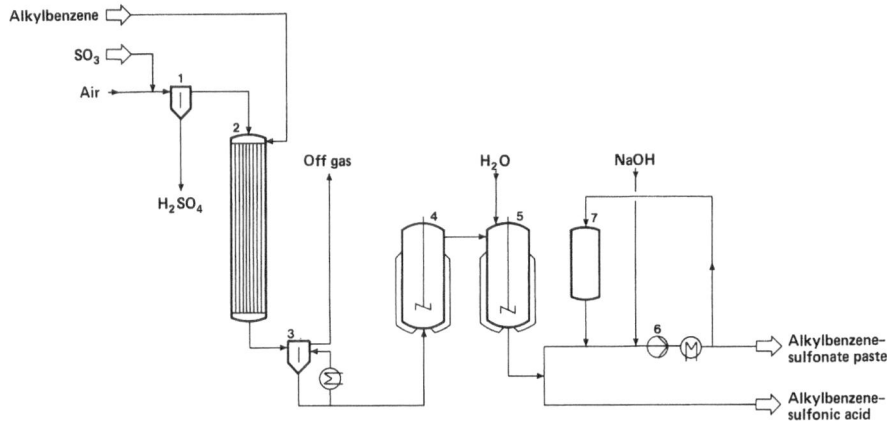

1 Mist separator; 2 Falling-film sulfonation reactor; 3 Gas separator; 4 Finishing reactor; 5 Hydrolyser; 6 Blender; 7 Storage tank

Figure 5.38: Continuous production of alkylbenzene sulfonates

sulfur trioxide has advantages over the sulfuric acid process in terms of faster reaction; additionally the production of spent sulfuric acid is avoided.

Figure 5.38 shows the flow diagram for the sulfonation of alkylbenzene with sulfur trioxide.

Sulfur trioxide is fed in in a slight excess. Residence time for liquid and gas in the reactor is several minutes and the reaction temperature is 40 to 50 °C.

The sulfonic acids are neutralized after recovery; they are used predominantly as surfactants in the form of their sodium salts. Since sulfonic acids form highly viscous gels with 10 to 50% water, intensive mixing is required during the neutralisation. The sulfonic acids, which are initally colored brown, lighten noticeably with neutralization. Further lightening of the pale-brown sulfonates can be achieved by adding 1 to 3% hypochlorite (100 to 150 g of active Cl_2/l).

Table 5.9 summarizes production capacities for the major alkylbenzene producer countries. The largest manufacturers are *Vista, Monsanto* (USA), *EniChem* (Italy) and *Petroquimica Española* (Spain).

Table 5.9: Production capacities for alkylbenzenes (1985)

	(1,000 t)
USA	350
Mexico	140
France	150
West Germany	140
Italy	250
Great Britain	100
Spain	165
Yugoslavia	65
Japan	185
China (peopl. rep.)	50
Other countries	405
Total capacity	2,000

5.7 Maleic anhydride

Maleic anhydride (MA) is an important raw material in the production of alkyd and polyester resins. It was first obtained by Nikolas Louis Vauquelin in 1817, by heating maleic acid to over 140 °C. In 1905, Richard Kempf obtained maleic acid by the oxidation of benzoquinone. The first patents covering the production of maleic anhydride from benzene originate from John M. Weiss and Charles R. Downs in 1918. The oxidation of benzene remains a feasible route to maleic anhydride even today, although since around 1975, n-butane and n-butylene have increasingly replaced benzene as raw materials. n-Butane and n-butylene are available as co-products in steam cracking of naphtha and from natural gas condensates.

Production and uses of benzene derivatives

$$\text{C}_6\text{H}_6 + 4\tfrac{1}{2}\,\text{O}_2 \longrightarrow \text{C}_4\text{H}_2\text{O}_3 + 2\,\text{CO}_2 + 2\,\text{H}_2\text{O}$$

The method for producing maleic anhydride by the oxidation of benzene is largely analogous to processes for producing phthalic anhydride from o-xylene (see Chapter 7.1.1 and Figure 5.39).

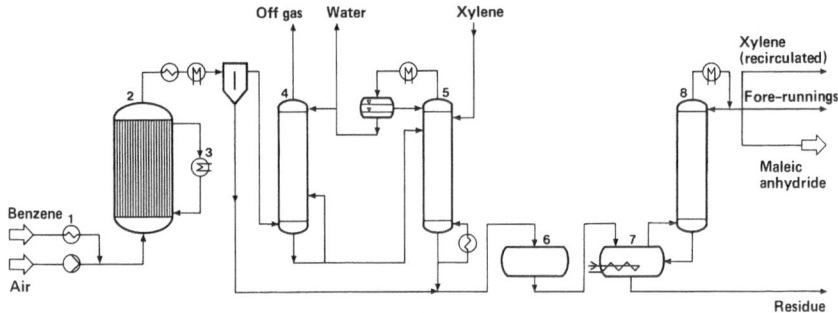

1 Benzene vaporizer; **2** Multitubular reactor; **3** Salt bath cooler; **4** Gas scrubber; **5** Dehydration column; **6** Storage tank; **7** Distillation boiler; **8** Distillation column

Figure 5.39: Production of maleic anhydride by the oxidation of benzene

Benzene vapor is mixed with air (benzene concentration: 1 to 1.4 mol%) and fed over a fixed-bed catalyst contained in a multitubular reactor. The tubes, which are arranged vertically, have a diameter of 20 to 50 mm; one reactor contains up to 20,000 tubes. The reaction temperature is kept at 350 to 400 °C by cooling with a salt bath; the pressure is 1 to 2 bar.

The reaction products leaving the reactor are first quenched to below 200 °C, then cooled further in a condenser to around the dew point (55 to 65 °C). The uncondensed maleic anhydride is recovered in an attached absorber. The off gas, carrying unconverted benzene, can be processed in a benzene-adsorption plant. The crude maleic anhydride is refined by distillation with o-xylene as an entrainer. Care should be taken in operating the dehydrating column that the temperature does not exceed 130 °C, in order to restrict the undesirable isomerization of maleic acid to fumaric acid.

In addition to condensation, the oxidation product may also be collected in solid form, by cooling to around 40 °C. In this process, the crude maleic anhydride solidifies on the surface of the cooler in the form of needle-shaped crystals. At least two such collectors (switch condensers) are needed. The full collector is heated to 60 to 80 °C and the molten MA drawn off into a collecting vessel for further refining by distillation.

Vanadium pentoxide is generally used as a catalyst, modified with molybdenum or tungsten oxide to improve the yield. The service life of the catalyst is 1 to 3 years. The yield of maleic anhydride is around 70 mol% or 90 wt%. By-products include small amounts of phenols, aldehydes and carboxylic acids; moreover, about 20% of the benzene is converted into CO_2.

Since in the oxidation of benzene to maleic anhydride, two carbon atoms are lost as CO_2, attempts were made early on to produce maleic anhydride from C_4 hydrocarbons. With the development of the petrochemical industry, large quantities of C_4-cuts became available, from which butene and, by hydrogenation, butane, could be recovered. The use of benzene is in decline, particularly in the USA, since benzene is now produced there, among other routes, by dealkylation of toluene, whereas C_4-components are readily available from cat-cracker and ethylene plants.

$$CH_3-CH_2-CH_2-CH_3 \;+\; 3\tfrac{1}{2}\,O_2 \;\longrightarrow\; \text{(maleic anhydride)} \;+\; 4\,H_2O$$

$$CH_3-CH=CH-CH_3 \;+\; 3\,O_2 \;\longrightarrow\; \text{(maleic anhydride)} \;+\; 3\,H_2O$$

Figure 5.40 shows the flow diagram for the oxidation of butane.

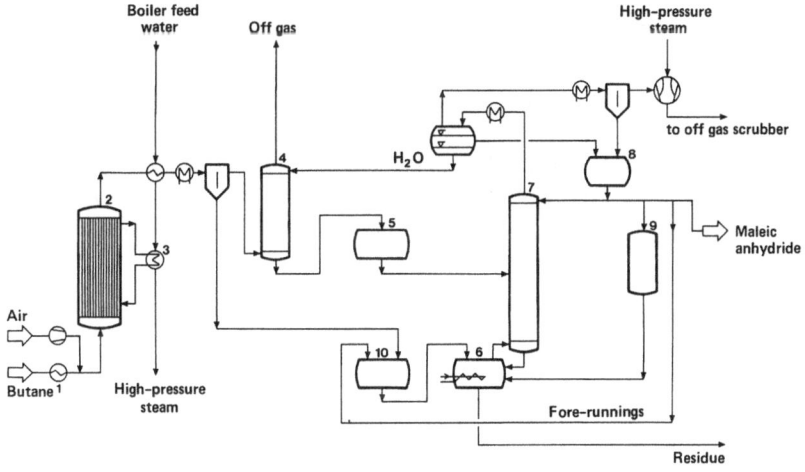

1 Butane vaporizer; **2** Multitubular reactor; **3** Salt bath cooler; **4** Gas scrubber; **5** Maleic acid solution vessel; **6** Batch distillation boiler; **7** Distillation column/dehydration column; **8** Distillate receiver; **9** Entrainer vessel; **10** Crude MA container

Figure 5.40: Production of maleic anhydride by the catalytic fixed-bed oxidation of butane

The process resembles the oxidation of benzene, but the concentration of maleic anhydride in the reaction product is lower. The reaction rate is also lower than for oxidation of benzene, so that a larger reactor (by a factor of 1.2) and bigger compressor with correspondingly higher investment are needed. Some MA plants have recently been revamped for both feedstocks (dual-feed). A disadvantage of butane oxidation is the low yield, which is around 50 to 55 mol %.

A fluidized bed process has recently been developed for producing maleic anhydride from butane and is distinguished by better heat dissipation, lower maintenance costs and reduced investment; it can also be applied to butene. Figure 5.41 shows the diagram for butane oxidation.

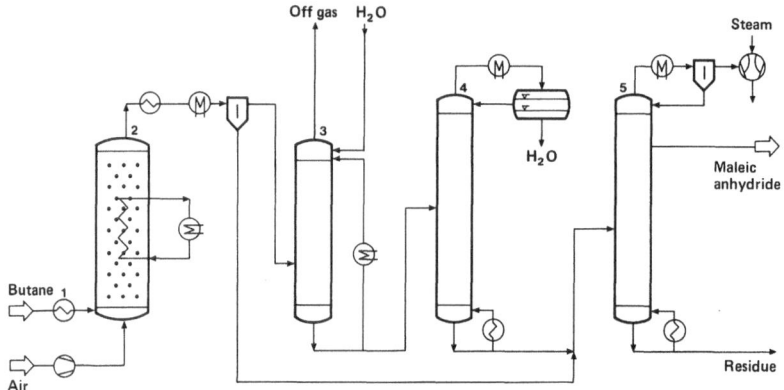

1 Butane vaporizer; 2 Fluidized bed reactor; 3 Gas scrubber; 4 Dehydration column; 5 Distillation column

Figure 5.41: Production of maleic anhydride by the catalytic oxidation of butane in a fluidized bed

In Japan, *Mitsubishi Chemical* operates a fluidized-bed plant (18,000 tpa) based on butene.

Table 5.10 summarizes production figures for the major maleic anhydride producer countries. The largest capacities are operated by *Alusuisse Italia* (Italy), *Nihon Iyoryu Kogyo* (Japan), *CdF-Chimie* (France) and *Monsanto* (USA).

Table 5.10: Production of maleic anhydride (1985)

	(1,000 t)
USA	170
France	20
West Germany	45
Italy	40
Great Britain	15
Japan	70
USSR	80
Other countries	60
Total production	500

More than one-half of maleic anhydride production is used to make unsaturated polyester resins by the reaction with glycols. Unsaturated polyester resins are mainly used as reinforcement plastics for structural components in the building industry, for tanks and for electrical articles.

Maleic anhydride is also used as a raw material for the production of lubricant additives and to make fumaric acid, a raw material for polyester and mixed polymers.

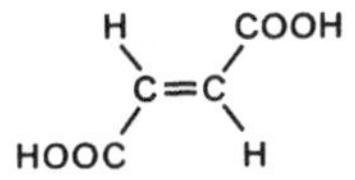

$$\underset{HOOC}{\overset{H}{\diagdown}}C=C\underset{H}{\overset{COOH}{\diagup}}$$

Fumaric acid

Derivatives of maleic anhydride such as captan (*Chevron*), are also used as plant protection agents. Captan is obtained from maleic anhydride by Diels-Alder reaction with butadiene to give cis-1,2,3,6-tetrahydrophthalic anhydride, followed by ammonolysis and treatment of the cis-1,2,3,6-tetrahydrophthalimide with trichloromethanesulfenyl chloride.

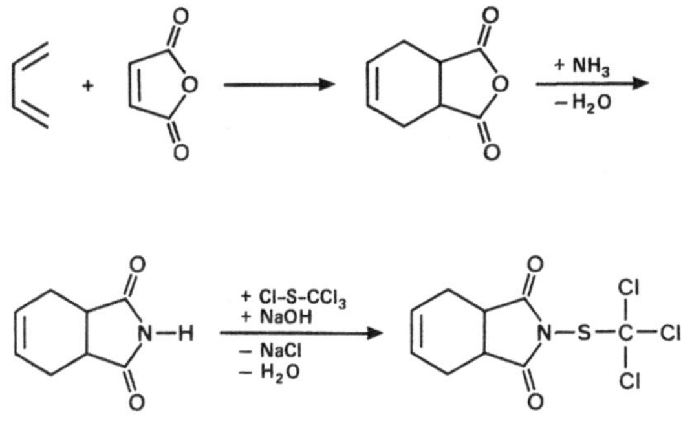

Captan

Maleic anhydride is also used, in competition with acetylene, as a starting material for the production of 1,4-butanediol. Partial hydrogenation of maleic anhydride in the liquid phase (250 °C, 140 bar) on a nickel catalyst yields γ-butyrolactone, a starting material for Vitamin B_1.

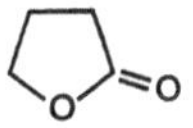

δ-Butyrolactone

5.8 Chlorobenzenes

5.8.1 Chlorobenzene

Chlorobenzene was one of the first basic organic chemicals produced on a large scale; it was already manufactured in 1909 by *United Alkali Co.,* Wipnes/USA. The importance of chlorobenzene rose dramatically during World War I, since it was needed as an intermediate in the production of phenol for the manufacture of picric acid. Other uses, now principally of historic importance, are the production of DDT (1,1,1-trichloro-2,2-di(4-chlorophenyl)ethane) from chlorobenzene and chloral in the presence of sulfuric acid, first synthesized in 1874 by Othmar Zeidler, and the ammonolysis to produce aniline.

The reaction of benzene and chlorine is carried out with Friedel-Crafts catalysts such as $FeCl_3$ or HCl, under very mild conditions. Low molar ratios of chlorine to benzene favor monosubstitution. The production of monochlorobenzene is therefore performed with a chlorine/benzene molar ratio of 0.6, at temperatures of 30 to 80 °C.

Figure 5.42 shows the formation of polychlorinated benzenes in relation to the molar ratio with batch chlorination, chlorination in continuous stirred tank reactors and in a two-cascade reactor.

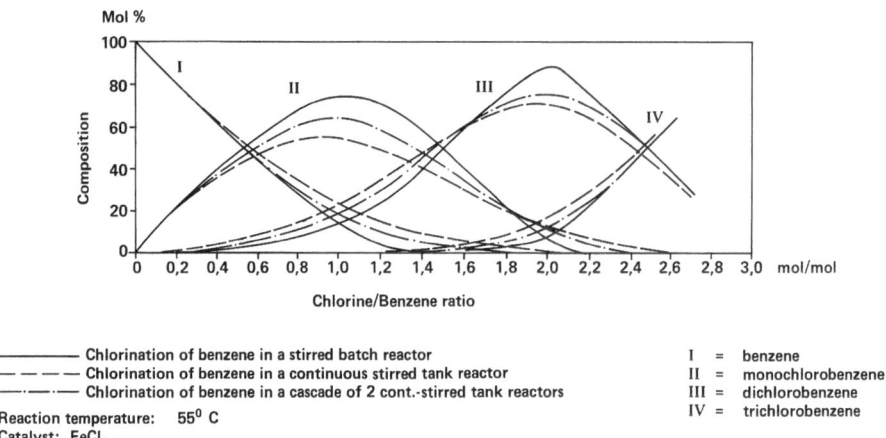

———— Chlorination of benzene in a stirred batch reactor
– – – – Chlorination of benzene in a continuous stirred tank reactor
–·–·– Chlorination of benzene in a cascade of 2 cont.-stirred tank reactors

Reaction temperature: 55° C
Catalyst: $FeCl_3$

I = benzene
II = monochlorobenzene
III = dichlorobenzene
IV = trichlorobenzene

Figure 5.42: Chlorination of benzene in relation to molar ratio and type of reactor

For the industrial application of the reaction it is necessary to use dry reactants, since water deactivates the catalyst and causes corrosion damage as a result of the formation of hydrochloric acid.

The flow diagram for the continuous production of chlorobenzene is shown in Figure 5.43.

Benzene, which is pre-dried to a residual water content of 35 ppm, is brought into contact with dry HCl gas, produced in the chlorination process, in a spray

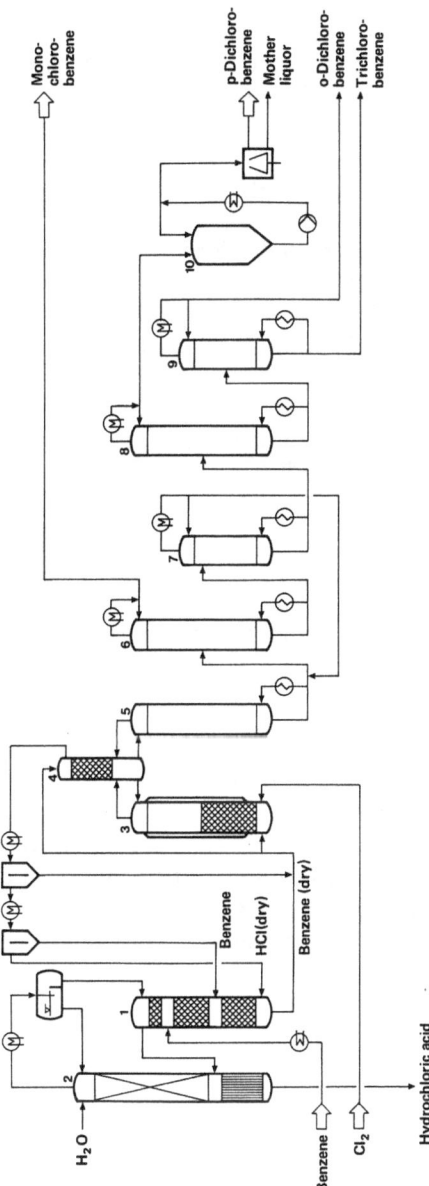

Figure 5.43: Continuous production of chlorobenzene

1 Benzene drying (Hausmann column); 2 Adiabatic HCl absorption; 3 Reactor; 4 and 5 Benzene and HCl stripping columns; 6 Chlorobenzene main column; 7 Chlorobenzene side column; 8 p-Dichlorobenzene column; 9 o-Dichlorobenzene column; 10 Crystallizer

column, in order to obtain virtually water-free benzene which is then fed into the chlorination reactor.

With a chlorine/benzene molar ratio of 0.6, the crude product consists of 30% benzene, 60% monochlorobenzene, 3% o-dichlorobenzene and 7% p-dichlorobenzene. The reaction can be controlled by adjusting the reaction conditions, so that up to 95% monochlorobenzene is obtained.

Monochlorobenzene is recovered from the reaction product by atmospheric distillation in the chlorobenzene column (around 30 trays). The residue is separated in the chlorobenzene side-column (around 12 trays/ 200 mbar) into an overhead fraction rich in monochlorobenzene together with a residue rich in dichlorobenzenes. The dichlorobenzenes are further refined in the p-dichlorobenzene column (150 mbar); p-dichlorobenzene is recovered by crystallization of the overhead product, while the bottoms are redistilled in the o-dichlorobenzene column.

To avoid corrosion, cast iron, enamel or nickel-clad reactors, and plastic HCl pipelines (e.g. from phenol-formaldehyde resins or polyvinylidene fluoride) or glass or enamel pipes are employed; iron Raschig rings are used as the catalyst, placed as a layer on the reactor base.

In addition to Friedel-Crafts chlorination of benzene, chlorobenzene can also be produced by oxychlorination, using hydrogen chloride; in the process developed by *Gulf*, phenol is obtained as the end product.

$$\text{C}_6\text{H}_6 + \text{HCl} + \tfrac{1}{2}\text{O}_2 \xrightarrow[-\text{H}_2\text{O}]{[\text{HNO}_3]} \text{C}_6\text{H}_5\text{Cl} \xrightarrow[-\text{HCl}]{+\text{H}_2\text{O}} \text{C}_6\text{H}_5\text{OH}$$

Figure 5.44 shows the flow diagram for this HNO_3-catalyzed liquid-phase process.

Unlike the *Raschig* process (see Chapter 5.3.2), in which chlorobenzene is obtained by the oxychlorination of benzene at 240 °C with benzene conversion of 10 to 15%, the *Gulf* process operates with satisfactory selectivity and high conversions, even at relatively low temperatures (60 to 150 °C).

Table 5.11 summarizes the capacities of major chlorobenzene producer countries.

Table 5.11: Chlorobenzene production capacities (1985)

	(1,000 t)
USA	170
West Germany	180
Italy	20
France	20
Japan	55
Other countries	15
Total capacity, Western world	460

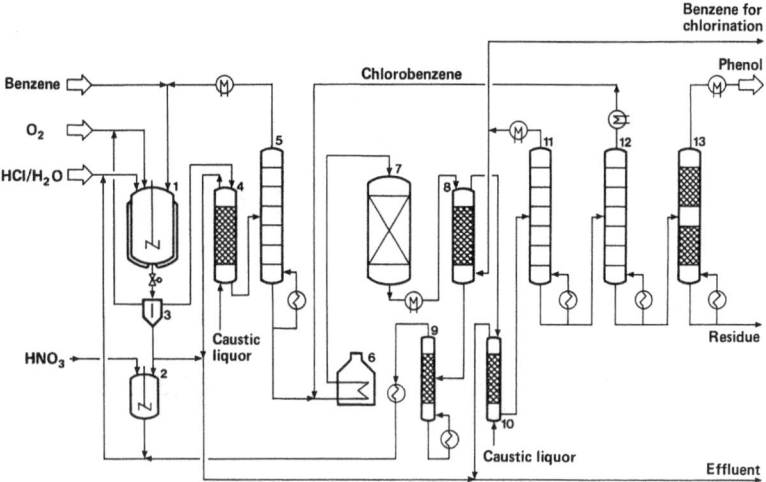

1 Oxychlorination reactor; 2 Catalyst vessel; 3 Separator; 4 Caustic scrubber; 5 Benzene fractionation column; 6 Heater; 7 Hydrolysis reactor; 8 Extraction column; 9 HCl stripping column; 10 Caustic scrubber; 11 Benzene column; 12 Chlorobenzene column; 13 Phenol column

Figure 5.44: Flow diagram for the production of phenol by oxychlorination

Consumption of monochlorobenzene in the USA in 1985 was 110,000 t.

The industrial importance of monochlorobenzene is in sharp decline, since, from the mid 1970's, it has largely been replaced as a starting material for the production of phenol, and consumption of DDT has fallen dramatically because of its poor biodegradability and the ban on its use in many countries (e.g. in the USA in 1973). Chlorobenzene has also lost its importance for aniline production in recent years.

The insecticidal effect of DDT was discovered by Paul Müller (*Ciba Geigy*) in 1939.

$$2\ C_6H_5Cl + CCl_3-CHO \xrightarrow{-H_2O} Cl-C_6H_4-CH(CCl_3)-C_6H_4-Cl$$

DDT

The discovery of a synthetic insecticide was extremely important in the early 1940's, especially in Great Britain and the USA, since the derris extract rotenone, which had been imported from Malaysia for many years, was no longer available during World War II.

Rotenone

DDT was one of the most important insecticides, applied with great success, particulary in the struggle against malaria.

5.8.1.1 Nitrochlorobenzenes

Chlorobenzenes remain important intermediates for the production of nitrated chlorine compounds, particularly o- and p-nitrochlorobenzene.

o-Nitrochlorobenzene   p-Nitrochlorobenzene

The nitration of chlorobenzene with a mixed acid reagent comprising around 35% nitric acid, 53% sulfuric acid and 12% water, at temperatures of from 40 to 80 °C and a HNO_3/chlorobenzene molar ratio of around 1 gives a 98% yield of an isomeric mixture consisting of around 33% o-chloronitrobenzene, 66% p-chloronitrobenzene and 1% m-chloronitrobenzene.

The isomers are separated by a combination of distillation and crystallization, the lower melting point of o-chloronitrobenzene (33 °C) facilitates separation by crystallization from p-chloronitrobenzene, which melts at 83.5 °C.

Production of o- and p-chloronitrobenzene in West Germany is around 60,000 tpa; in the USA some 40,000 tpa are produced; they are very versatile intermediates (Figure 5.45).

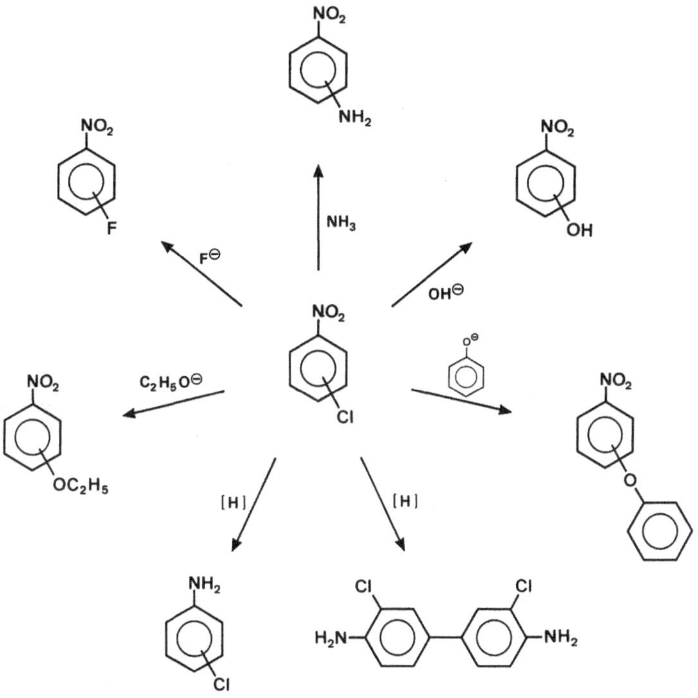

Figure 5.45: Intermediates derived from o-/p-chloronitrobenzene

One of the uses of o-chloronitrobenzene is in the production of o-chloroaniline, which is obtained by catalytic reduction with sulfided palladium/active carbon catalysts. These catalysts enable dechlorination to be suppressed.

3,3′,4,4′-Tetraaminobiphenyl (3,3′-diaminobenzidine), produced from o-chloronitrobenzene, is used in the manufacture of high-temperature resistant benzimidazole polymers (PBI), which are produced by reaction with diphenyl isophthalate (*Hoechst-Celanese*).

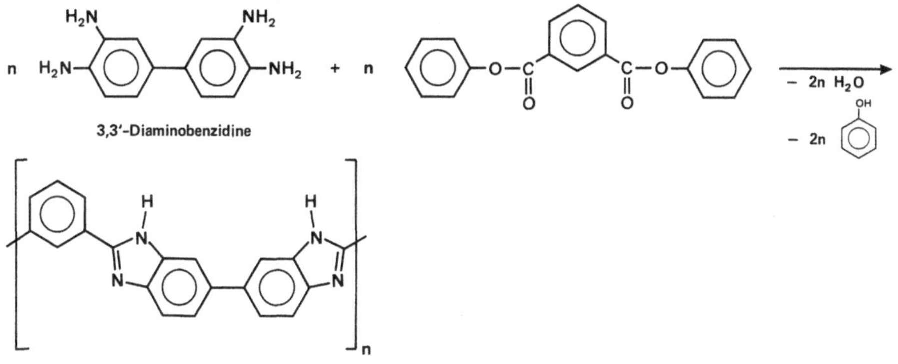

Poly-1,3-phenylene-2,2′-(5,5′-bis-benzimidazole)

Nucleophilic substitution of o-nitrochlorobenzene with ammonia yields o-nitroaniline, which is used to make dyestuffs and plant protection agents as well as UV-stablizers.

Figure 5.46 shows the flow diagram for ammonolysis of chloronitrobenzene to nitroaniline.

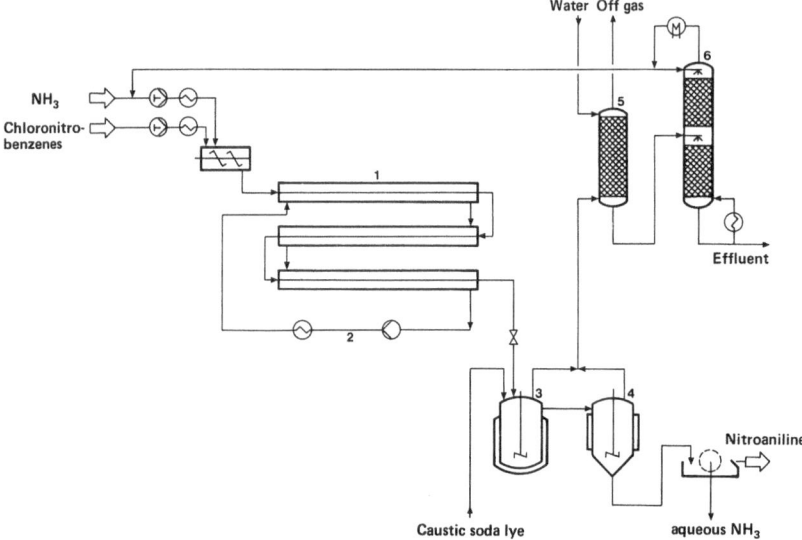

1 Tube reactor; 2 Heating/cooling circulation; 3 Flash vessel; 4 Crystallizer;
5 Waste gas scrubber; 6 NH₃ distillation column

Figure 5.46: Flow diagram for ammonolysis of chloronitrobenzene

In the continuous process, chloronitrobenzene, liquid ammonia and aqueous ammonia solution interact in a long tube reactor under high pressure (200 bar) at 200 °C. The ammoniacal effluent, which occurs after the depressurization, is stripped of ammonia by pressure distillation; this produces a concentrated aqueous NH_3 solution which can be re-used.

o-Nitroaniline is converted into o-phenylenediamine by catalytic reduction on a palladium/activated carbon catalyst; it is used in the production of plant protection agents such as the fungicide carbendazim and its derivative, benomyl, and for the manufacture of the fungicide thiophanate-methyl (*Nippon Soda*).

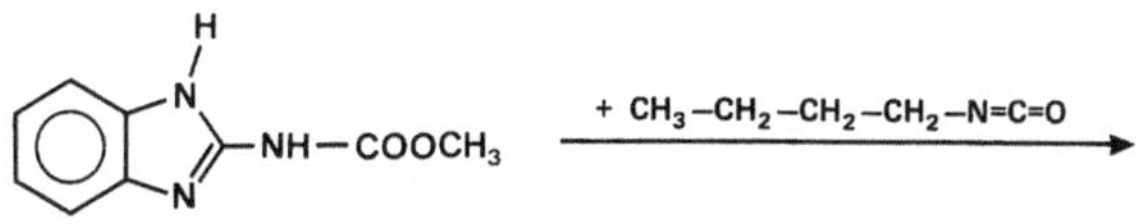

[Carbendazim structure] + CH₃–CH₂–CH₂–CH₂–N=C=O →

Carbendazim

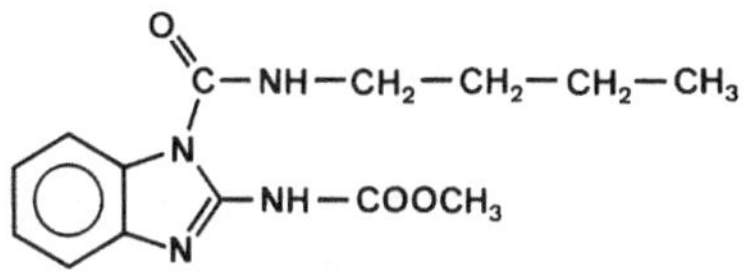

Benomyl

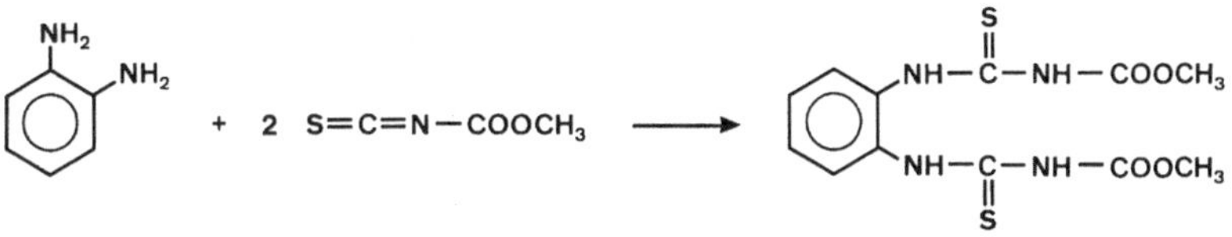

[o-phenylenediamine] + 2 S=C=N–COOCH₃ → Thiophanate-methyl

Additionally, o-phenylenediamine serves as a starting material for benzimidazolone pigments, such as Pigment Orange 36, which is obtained via the versatile coupling component 5-acetoacetylaminobenzimidazolone.

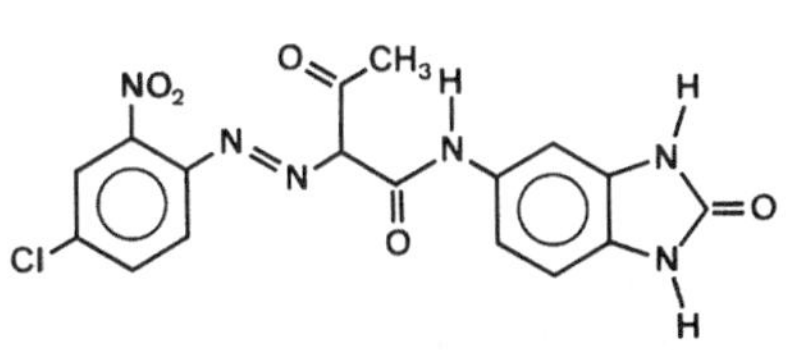

Pigment Orange 36

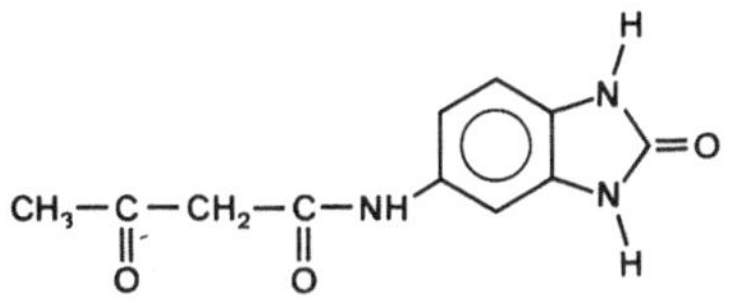

$CH_3-C(=O)-CH_2-C(=O)-NH-$[benzimidazolone]

5-Acetoacetylaminobenzimidazolone

Another widely-used coupling component for the production of red pigments is 5-(2′-hydroxy-3′-naphthoyl)-aminobenzimidazolone, which is employed, for example, in the production of Pigment Red 171.

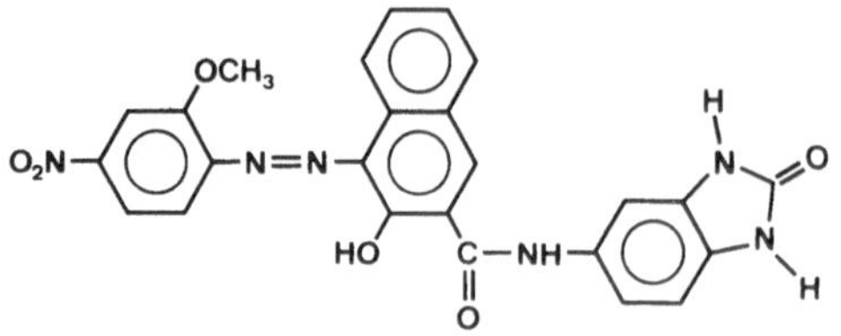

Pigment Red 171

p-Chloronitrobenzene is widely used in the synthesis of p-nitrophenol, p-chloroaniline, p-nitroaniline and p-nitrodiphenyl ethers (e.g. nitrofen).

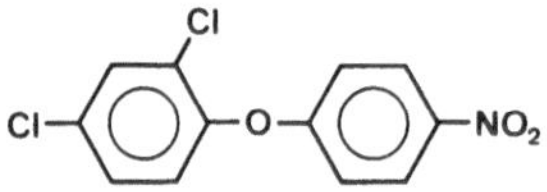

Nitrofen

Nitrofen was introduced as a weedkiller in the early 60-ies by *Rohm & Haas*.

A traditional product is Pigment Red 1 (Para Red), which is obtained from diazotized p-nitroaniline and 2-naphthol.

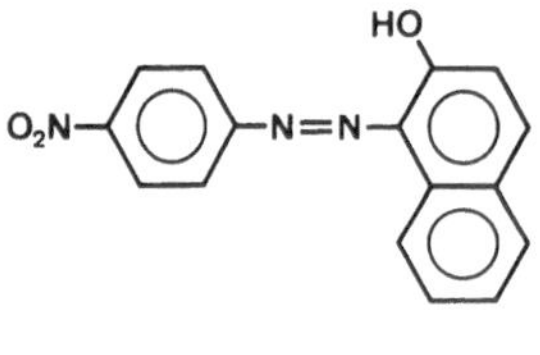

Pigment Red 1

Condensation of aniline with p-nitrochlorobenzene produces 4-nitrodiphenylamine (PNDPA), an important intermediate in the production of antioxidants. Reductive N-alkylation of PNDPA gives anti-ageing materials for rubber, the most important of which are IPPD and 6PPD.

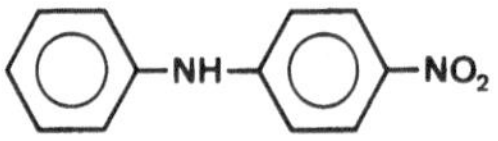

4-Nitrodiphenylamine 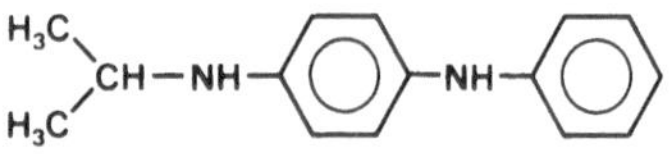   *IPPD*

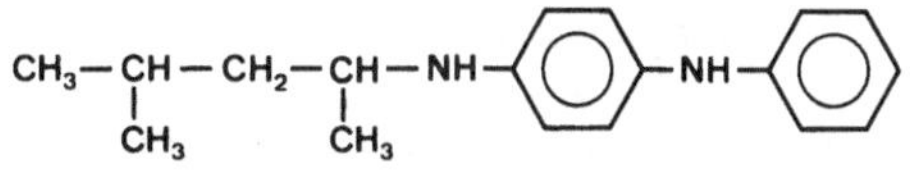

$$CH_3-CH-CH_2-CH-NH-\underset{}{\bigcirc}-NH-\underset{}{\bigcirc}$$
$$\overset{|}{CH_3}\overset{|}{CH_3}$$

6PPD

Production of these PNDPA derivatives, which are made mainly by *Monsanto* and *Bayer,* is around 20,000 tpa in Western Europe.

Reduction of p-nitroaniline yields p-phenylenediamine.

p-Phenylenediamine

p-Phenylenediamine is an intermediate used in the manufacture of dyestuffs such as Disperse Yellow 3, obtained from p-cresol and p-phenylenediamine.

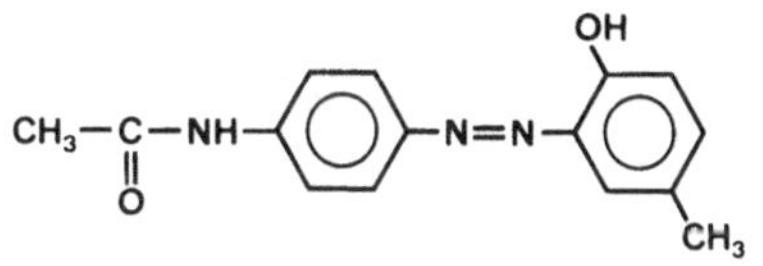

Disperse Yellow 3

Another dyestuff based on p-phenylenediamine is Safranine B Extra (C.I.50200), an azine dye which is obtained by reaction with aniline, via an intermediate indamine stage; the earlier importance of safranine dyes however has diminished.

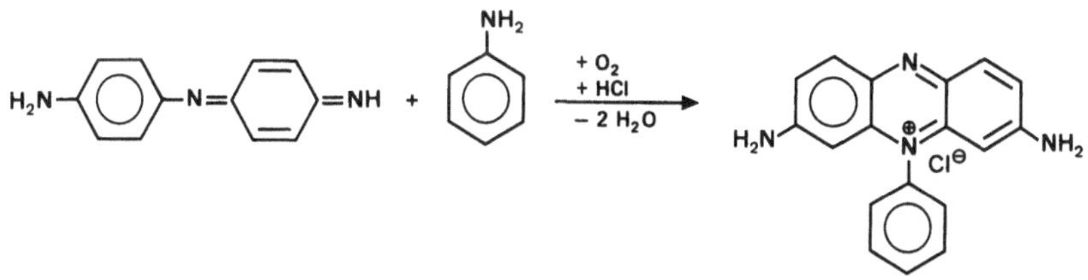

Safranine B Extra

Polycondensation of p-phenylenediamine with terephthaloyl dichloride produces aromatic polyamides, which display liquid-crystal characteristics. The spinning of these polymers from hexamethylphosphoramide or sulfuric acid solution yields high-value aramid fibers (Kevlar (*Du Pont*), Twaron (*Akzo*)), which are distinguished by their high temperature resistance (see Chapter 7.3.1).

A possible route to the tranquilizer diazepam (*Hoffmann-la Roche*) starts from p-chloroaniline.

5.8.2 Dichlorobenzenes

Among the dichlorinated benzenes, o-dichlorobenzene and p-dichlorobenzene are of commercial importance. The ratio of p- to o-dichlorobenzene formed during the chlorination of benzene depends on the catalyst and reaction conditions and varies between 1 and 5. As a result of the directive effect of the first chlorine atom in the chlorobenzene molecule, the proportion of m-dichlorobenzene is very low. (m-Dichlorobenzene can be obtained by denitrochlorination of m-dinitrobenzene or by isomerization in the presence of HCl and aluminum chloride at 120 °C from o- or p-dichlorobenzene). The dichlorobenzene isomers are separated by a combination of distillation and crystallization. Since the boiling points of o- and p-dichlorobenzene lie closely together (179.0/173.7 °C), separation by distillation alone is difficult; crystallization, however, is facilitated by their different melting points (-17.6/53.0 °C), and can be carried out very efficiently (see Chapter 5.8.1). Figure 5.47 shows the phase diagram for o-/p-dichlorobenzene with the eutectic point at around 12% p-dichlorobenzene.

Figure 5.47: Phase diagram for o-/p-dichlorobenzene

Production of o-dichlorobenzene is around 80,000 tpa. The largest producers are the United States, Japan and West Germany.

o-Dichlorobenzene is used as an inert solvent, e.g. in the manufacture of toluene diisocyanates. It is also employed as the starting material for the production of 3,4-dichloroaniline, an intermediate for the synthesis of dyes and plant protection agents (herbicides) such as linuron, which is produced in quantities of 4,000 tpa by the reaction of 3,4-dichlorophenyl isocyanate and O,N-dimethylhydroxylamine.

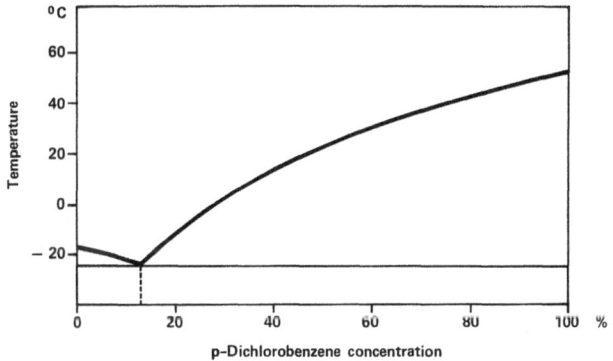

Linuron

Another herbicide based on 3,4-dichloroaniline, produced in similar quantities, is diuron (*Du Pont*), which can be obtained by the reaction of 3,4-dichloroaniline, urea and dimethylamine accompanied by the liberation of ammonia.

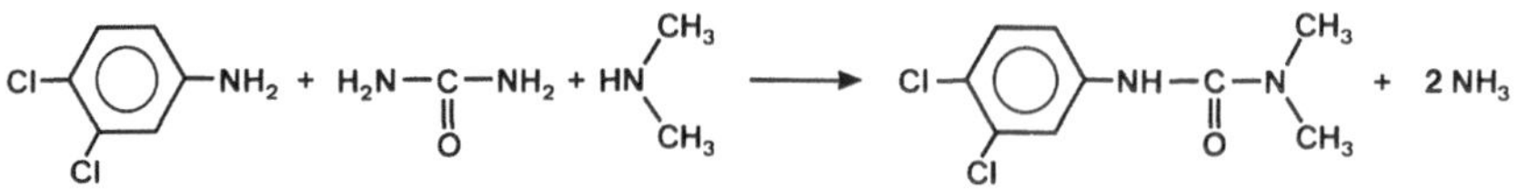

Diuron

3,4-Dichloroaniline and propionic acid react together in the presence of thionyl chloride to yield propanil, a herbicide introduced by *Rohm & Haas*.

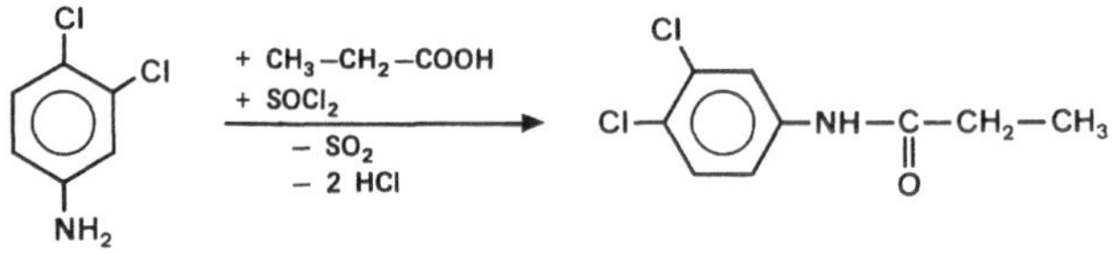

Propanil

The reaction of 3,4-dichloronitrobenzene with KF in aprotic solvents such as sulfolane gives 3-chloro-4-fluoronitrobenzene, which is converted into the *Shell* herbicide flamprop-methyl by reduction, reaction with chloropropionic acid, esterification and subsequent acylation with benzoyl chloride.

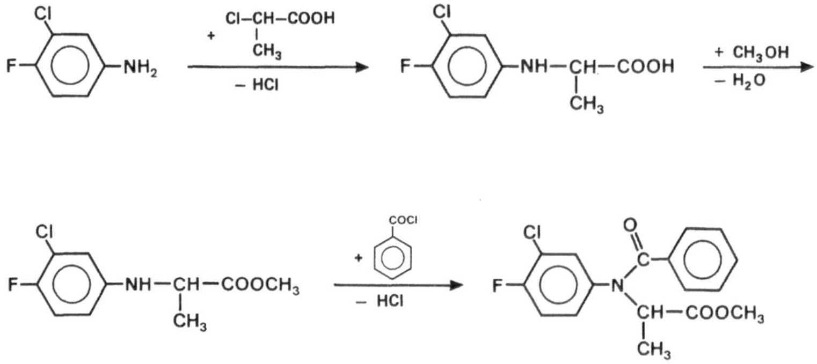

Flamprop-methyl

p-Dichlorobenzene is produced worldwide in a quantity of around 75,000 tpa; consumption in the USA in 1985 was 25,000 t. Apart from its use as a disinfectant and deodorant, it has been used for several years to produce polyphenylene sulfide (PPS), which is manufactured by the reaction of p-dichlorobenzene with sodium sulfide.

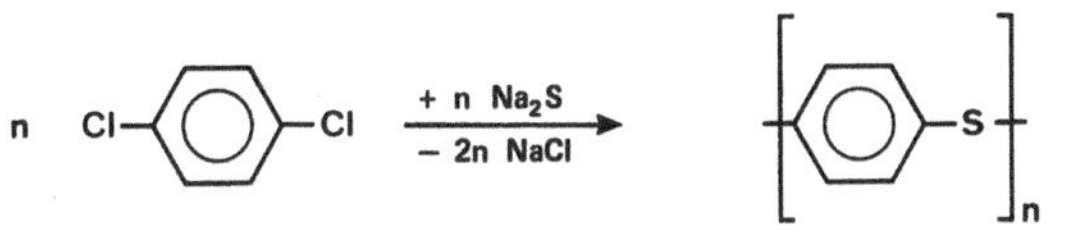

Polyphenylene sulfide

Polyphenylene sulfide is used in the manufacture of thermoplastic resins, which are marked by high long-term temperature stability (260 °C) and good chemical resistance (*Phillips Petroleum*, USA; *Kureha*, Japan).

The nitration of p-dichlorobenzene at 30 to 65 °C gives high yields of 1,4-dichloro-2-nitrobenzene, a possible starting material in the manufacture of Pigment Red 88.

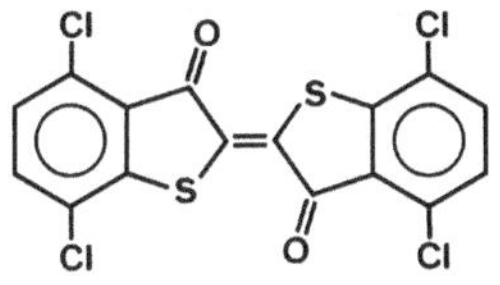

Pigment Red 88

The thioindigo derivative Pigment Red 88 can be obtained by initial reaction of p-dichlorobenzene with chlorosulfonic acid followed by reduction with zinc, S-alkylation with chloroacetic acid, and ring closure with AlCl$_3$ to yield the intermediate 2,5-dichlorothioindoxyl. Oxidation with atmospheric oxygen then yields Pigment Red 88.

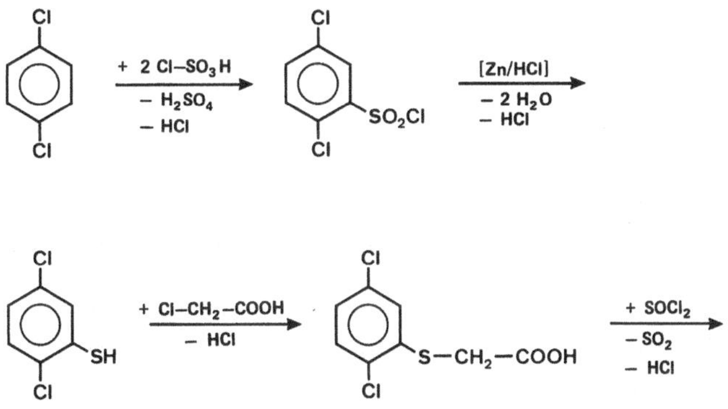

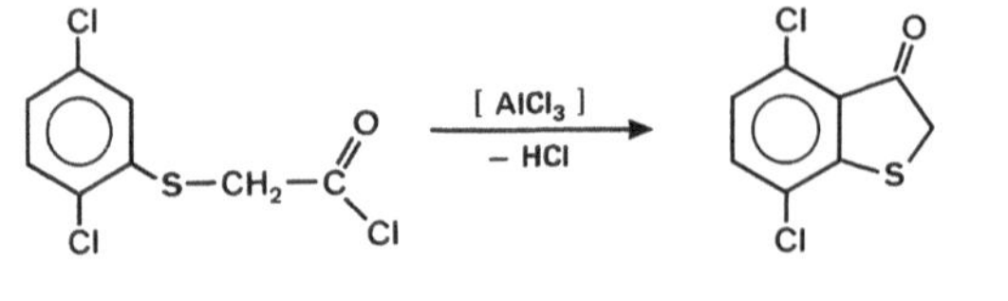

2,5-Dichlorothioindoxyl

m-Dichlorobenzene is used as the base for production of the plant protection agent iprodione; the bromination of m-dichlorobenzene at 10 to 40 °C in the presence of $AlCl_3$ to 1-bromo-2,4-dichlorobenzene, followed by isomerization leads to 1-bromo-3,5-dichlorobenzene. 3,5-Dichloroaniline is produced by reaction with ammonia, and is converted into the isocyanate and finally with glycine to the respective hydantoin. Further reaction with isopropyl isocyanate leads to iprodione (*Rhône Poulenc*).

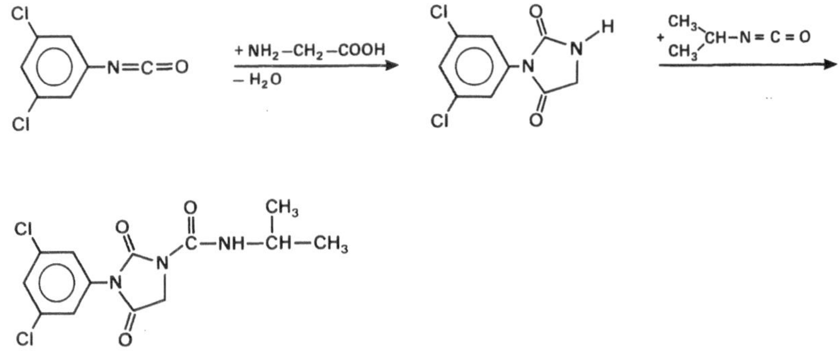

Iprodione

5.8.3 Hexachlorocyclohexane

A chlorinated benzene derivative, which is only of historical importance now in the major industrialized countries, is hexachlorocyclohexane, of which the γ-isomer, lindane, was for some time widely used as a plant protection agent. Hexachlorocyclohexane (HCH) was first produced by Michael Faraday in 1825 by adding chlorine to benzene under the action of sunlight. In 1940, *ICI* discovered the insecticidal effect of hexachlorocyclohexane.

Figure 5.48 shows the eight isomeric structures of hexachlorocyclohexane.

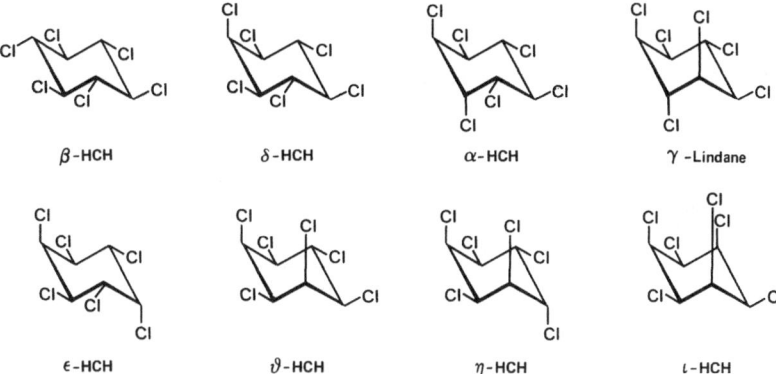

Figure 5.48: Structures of hexachlorocyclohexane isomers

Technical hexachlorocyclohexane contains around 65% of the α-isomer, 7% of the β-isomer, 14% of the γ-isomer, 4% of the ε-isomer and 10% of the remaining isomers.

Hexachlorocyclohexane is produced industrially by the reaction of an excess of benzene with chlorine at 15 to 25 °C in a glass reactor under the influence of UV light at atmospheric pressure. Care must be taken to exclude oxygen and catalysts, such as iron, which favor substitution. When 5 to 8% of the benzene has been chlorinated, the ß-isomer begins to precipitate. After benzene and unreacted chlorine have been evaporated at temperatures of from 85 to 88 °C, a hexachlorocyclohexane mixture with a γ-isomer content of 12 to 14% remains. The γ-isomer is recovered by fractional crystallization. The other isomers are converted into tri- and tetrachlorobenzenes by thermal or catalytic dechlorination; tri- and tetra chlorobenzenes are used as the precursors for 2,4,5-trichlorophenol (see Chapter 5.3.4.5).

Because of the poor biodegradability and the tendency to concentrate in the food cycle, the importance of hexachlorocyclohexane as an insecticide has declined sharply. World wide production is currently estimated at only 10,000 tpa.

5.9 Process review

Table 5.12 shows the major processes for producing benzene derivatives in summary form.

Table 5.12: Summary of major processes for the production of benzene derivatives

Process	Target product	Process conditions			Reaction components	Other characteristics
		Pressure (bar)	Temp. (°C)	Catalyst		
1. Alkylation:						
Benzene alkylation (Union Carbide/Badger)	Ethylbenzene	2–4	125–140	$AlCl_3$	Benzene/ ethylene	Liquid-phase reaction; corrosive medium
Benzene alkylation (Mobil/Badger)	Ethylbenzene	20	420–430	Zeolite	Benzene/ ethylene	Gas-phase reaction; low corrosion
Benzene alkylation	Cumene	3–10	250–350	Phosphoric acid/silicate	Benzene/ propylene	Gas-phase reaction; liquid-phase reaction also possible with $AlCl_3$ at 50–70°C
Phenol alkylation	2,6-xylenol	1–2	300–400	Al_2O_3	Phenol/ methanol	Gas-phase reaction; can also be used to produce cresols
2. Dehydrogenation:						
BASF-process	Styrene	atmos.	580–590	Fe_2O_3	Ethylbenzene	Isothermal process
DOW-process	Styrene	atmos.	570–640	Fe_2O_3	Ethylbenzene	Adiabatic process
3. Oxidation:						
Hock-process	Phenol	6	90–100	–	Cumene/ O_2 (air)	Most important phenol synthesis
DOW-process	Phenol	5–10	150–170	Cu-salts	Toluene/ O_2 (air)	Only isolated usage
Benzene oxidation	Maleic anhydride	1–2	350–400	V_2O_5	Benzene/ O_2 (air)	Fixed-bed process
Butane oxidation	Maleic anhydride	1–2	350–400	V_2O_5	Butane/ O_2 (air)	Fixed-bed process; increasing importance vs. benzene oxidation
4. Condensation:						
Hooker-process	Bisphenol A	atmos.	50–90	HCl	Phenol/ acetone	Both reaction components can be produced by Hock synthesis
5. Hydrogenation:						
Phenol hydrogenation	Cyclohexanone	atmos.	140–170	Pd	Phenol/ H_2	Gas-phase reaction
Phenol hydrogenation	Cyclohexanol	10–20	120–200	Ni/ SiO_2/ Al_2O_3	Phenol/ H_2	Gas-phase reaction

Table 5.12 (continued)

Process	Target product	Process conditions			Reaction components	Other characteristics
		Pressure (bar)	Temp. (°C)	Catalyst		
Benzene hydrogenation (*IFP*)	Cyclohexane	20–50	150–200	Ni	Benzene/ H_2	Liquid-phase reaction
Nitrobenzene hydrogenation	Aniline	1.8	270	Cu	Nitrobenzene/H_2	Fluidized-bed process
6. Alkali fusion:						
Hoechst-process	Resorcinol	atmos.	350	–	Benzene-1,3-disulfonic acid/NaOH	Process analogous to earlier phenol synthesis from benzenesulfonic acid
7. Nitration:						
Bofors-Nobel-process	Nitrobenzene	atmos.	60	–	Benzene/ HNO_3/ H_2SO_4	Nitration can also be carried out batch-wise
8. Sulfonation:						
Alkylbenzene sulfonation	Surfactants	atmos.	40–50	–	Alkylbenzenes/SO_3	Falling film sulfonation
9. Chlorination:						
Benzene chlorination	Chlorobenzene	atmos.	30–40	$AlCl_3$/ $FeCl_3$	Benzene/Cl_2	Continuous or batch process; degree of chlorination dependent on process conditions

6 Production and uses of toluene derivatives

Basically there are three classes of reactions of toluene, namely electrophilic substitution, side-chain reactions and cleavage of the methyl-phenyl bond. From an industrial point of view the most important electrophilic substitutions are nitration and chlorination; among the side-chain reactions oxidation and chlorination are of large industrial significance. Cleavage of the bond between the methyl group and the aromatic ring of toluene occurs during dealkylation to produce benzene, with biphenyl as a by-product (see Chapters 4.4.1 and 10.1).

The predominant uses for toluene are, depending on location, the production of benzene by dealkylation (predominantly in the USA), the application as an aromatic solvent, the production of toluene diisocyanate through nitrotoluene intermediates and the manufacture of the oxidation products benzoic acid, benzaldehyde and benzyl alcohol.

The production of toluene-derived phenol and ε-caprolactam is carried out only in rare cases.

Versatile aromatic intermediates, especially in the manufacture of plant protection agents and dyestuffs, are the chlorinated toluene derivatives o- and p-chlorotoluene, benzyl chloride, and benzotrichloride, together with toluene nitro-derivatives. Sulfonic acids of toluene have extensive applications as surfactants.

Figure 6.1 shows the most important applications for toluene in the USA, Western Europe and Japan.

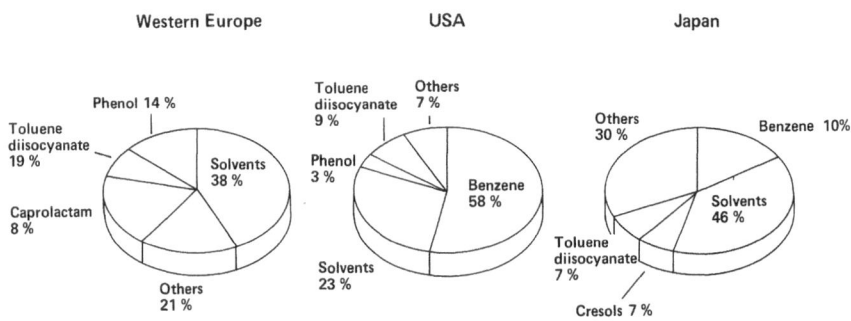

Figure 6.1: Major uses of toluene in Western Europe, the USA and Japan (1985)

6.1 Nitro-derivatives of toluene

6.1.1 Mononitrotoluene and its products

The nitration of toluene leads to mono-, di- and trinitrotoluene derivatives, depending on the reaction conditions.

Technical nitration of toluene is carried out in a manner similar to that of benzene nitration; however, since the reactivity is higher because of the presence of the methyl group, lower temperatures suffice.

The reaction temperature should not exceed 60 °C, since above this limit both the methyl group and the aromatic nucleus are attacked oxidatively by the nitration mixture; this leads to an increased formation of by-products, including nitrophenols, nitrocresols, phenylnitromethane and tetranitromethane.

Nitrophenols · Nitrocresols · Phenylnitromethane · Tetranitromethane

The mononitration of toluene is carried out at temperatures between 30 and 45 °C and with relatively low $NO_2^{\oplus}$ concentrations; a nitric acid/sulfuric acid mixture (20/60) produces 57 to 60% 2-nitrotoluene, 3 to 4% 3-nitrotoluene and 37 to 40% 4-nitrotoluene. The yield of nitrotoluenes is 97 to 98%, with approx. 0.1% dinitrotoluene and small amounts of nitrocresols arising as by-products. (The nitration can be carried out batchwise or continuously.)

The three nitrotoluene isomers are separated by sequential distillation and crystallization. 2-Nitrotoluene can be recovered as a highly concentrated overhead fraction because of its lower boiling point (221.7 °C). The higher-boiling 3- and 4-nitrotoluenes (232.6 °C and 238.4 °C respectively) are separated industrially by crystallization from the melt, which, with a difference in crystallizing points of 35.3 °C (16.1 °C and 51.4 °C respectively), leads to extremely pure isomers.

2-Nitrotoluene · 3-Nitrotoluene · 4-Nitrotoluene

The capacity for mononitrotoluene production in the Western world is around 200,000 tpa; *Bayer* operates by far the largest capacities.

Production and uses of toluene derivatives

The main application of 2-nitrotoluene is its further nitration to 2,4/2,6-dinitrotoluene mixtures, which are used in the production of toluene diisocyanate. Other important secondary reactions of 2-nitrotoluene are further nitration to trinitrotoluene, reduction to 2-toluidine and the production of 2-tolidine by benzidine rearrangement.

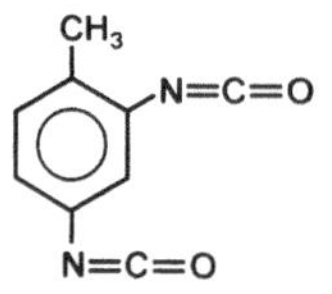

2,4-Toluene diisocyanate

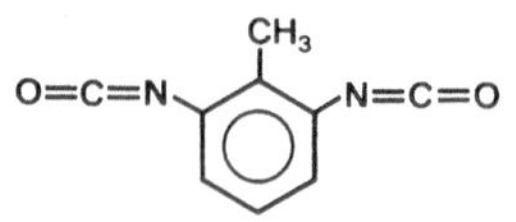

2,6-Toluene diisocyanate

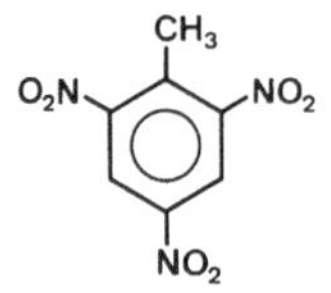

2,4,6-Trinitrotoluene

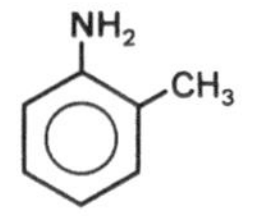

2-Toluidine

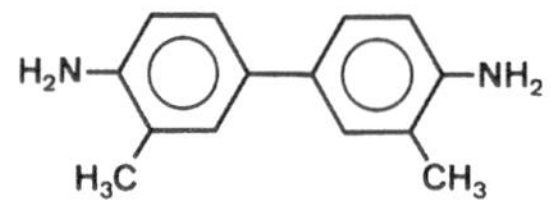

2-Tolidine

2-Nitrotoluene used to be an undesirable by-product from toluene nitration, but since the introduction, in 1974, of the herbicide metolachlor by *Ciba Geigy*, it is in considerably greater demand.

The main product from 2-nitrotoluene is 2-toluidine, which can be manufactured by batchwise or continuous reduction. In the batch process moderate hydrogen pressures of 20–50 bar are employed, while higher pressures, in excess of 100 bar are used in continuous hydrogenation. The catalysts are commonly nickel or noble metals such as palladium/active carbon.

In the production of metolachlor, 2-toluidine is ethylated with organometallic aluminum reagents, the resultant 2-ethyl-6-methylaniline is made to react with

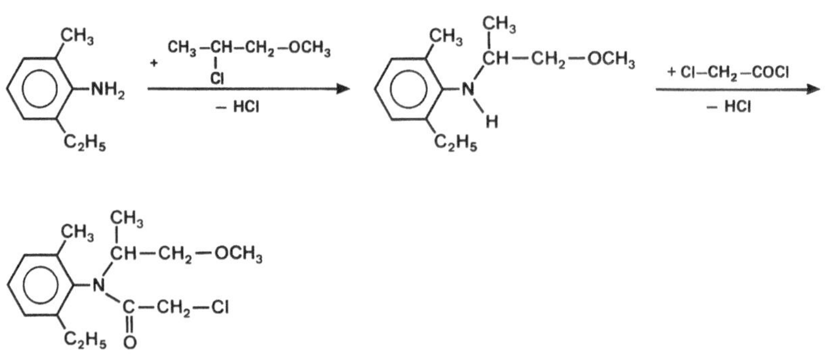

Metolachlor

2-chloro-1-methoxypropane and the reaction product then condensed with chloroacetyl chloride. Production of metolachlor in 1985 totalled 17,000 t.

Another important product from 2-toluidine is a mixture of arylated p-phenylenediamines, which is widely used as an antioxidant material in rubber processing; it is manufactured by the reaction of hydroquinone, aniline and 2-toluidine and consists of N,N'-ditolyl- and N,N'-diphenyl-p-phenylenediamine (DTPD, DPPD) and N-o-tolyl-N'-phenyl-p-phenylenediamine *(Goodyear)*.

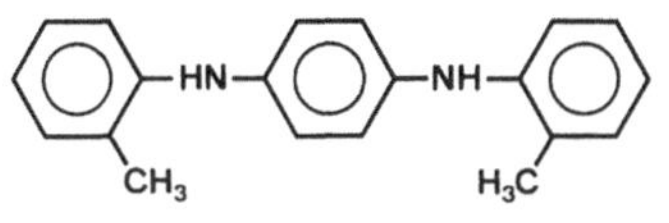

DTPD

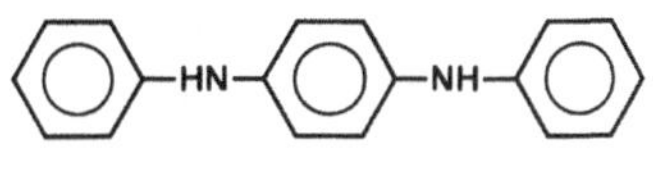

DPPD

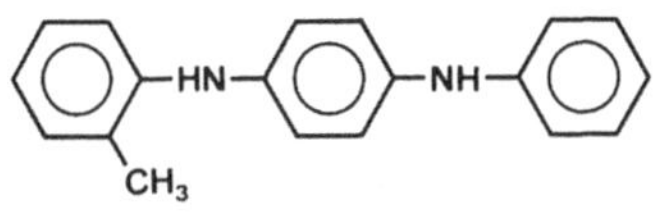

N-o-Tolyl-N'-phenyl-p-phenylenediamine

A reaction between o-toluidine and diketene yields N-acetoacetyl-2-toluidide, a base material for the production of pigments such as Pigment Yellow 14.

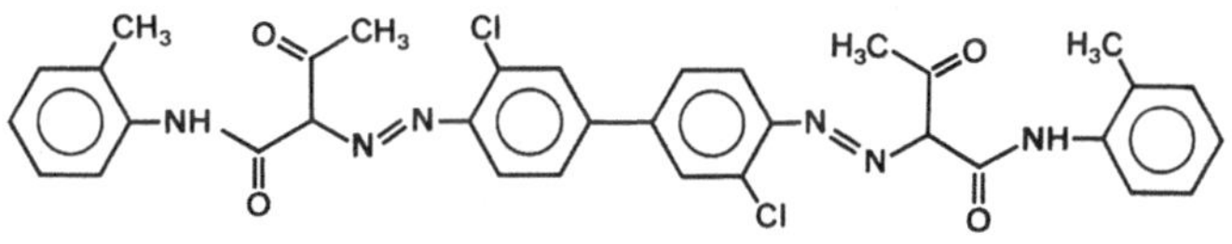

Pigment Yellow 14

o-Tolidine is another intermediate based on o-toluidine and is used in the production of dyes and pigments. o-Tolidine is manufactured from 2-nitrotoluene by reduction via the corresponding hydrazo compound, followed by benzidine rearrangement.

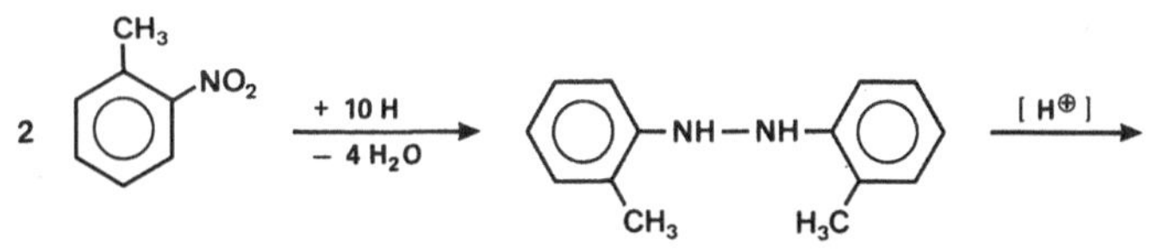

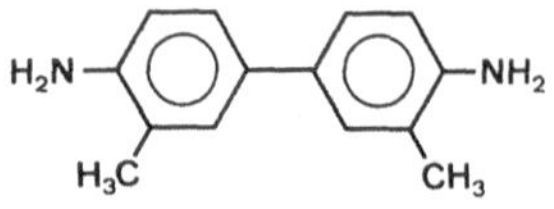

o-Tolidine

An important product from o-tolidine is Acid Red 114.

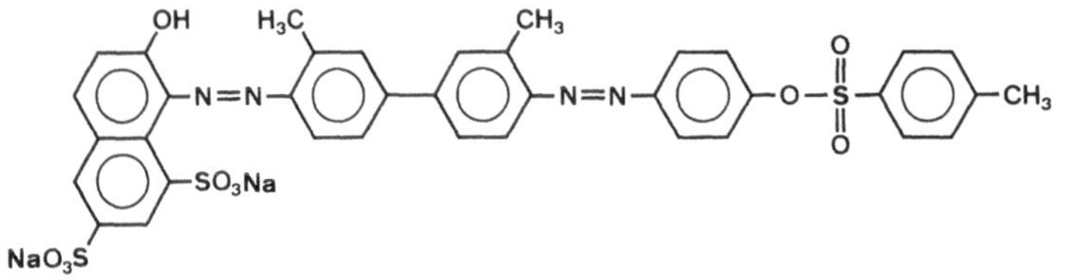

Acid Red 114

The catalytic reduction of 3-nitrotoluene yields 3-toluidine, an intermediate for the manufacture of azo dyes such as Disperse Red 65.

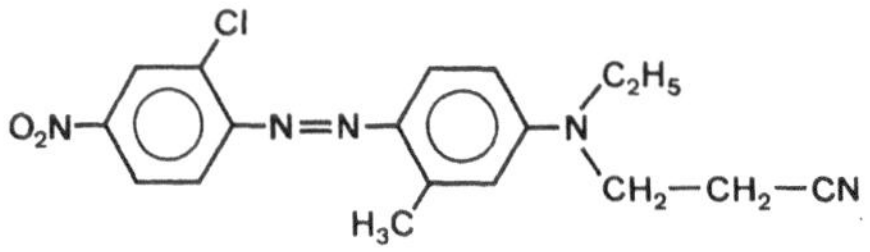

Disperse Red 65

After conversion into the corresponding isocyanate, 3-toluidine is also used in the production of the herbicide phenmedipham, developed by *Schering*, and

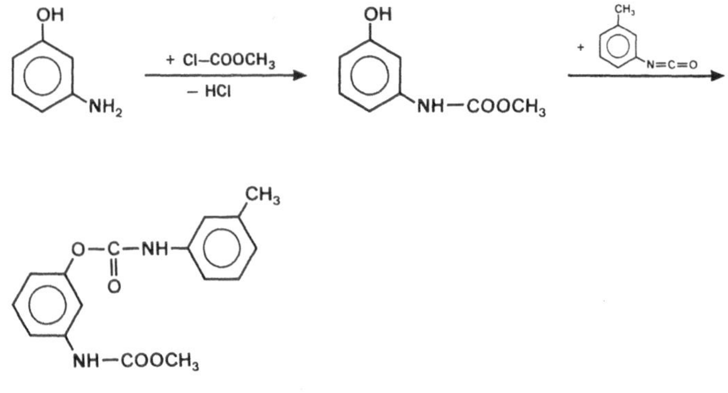

Phenmedipham

obtained by the reaction of 3-methoxycarbonylaminophenol, a product of 3-aminophenol (see Chapter 5.3.4.8) with 3-tolyl isocyanate.

4-Nitrotoluene is used mainly in the manufacture of pure 2,4-toluene diisocyanate. The second most important product is 4-nitrotoluene-2-sulfonic acid, obtained by sulfonation, which is converted into 4,4'-dinitrostilbene-2,2'-disulfonic acid (DNSDSA) by oxidation with sodium hypochlorite. The reduction of DNSDSA yields the corresponding amino-compound (DASDSA), which is used in the manufacture of optical brighteners and dyes.

4,4'-Diaminostilbene-2,2'-disulfonic acid (DASDSA)

For the production of the latter classes of compounds, DASDSA reacts with 2 mols of 2,4,6-trichloro-1,3,5-triazine, and the chlorine group of the triazine moieties is substituted with amines such as aniline or morpholine.

N,N'-Di [2-(4,6-dichlorotriazino)]-4,4'-diaminostilbene-2,2'-disulfonic acid

In addition, 4-nitrotoluene-2-sulfonic acid is used as a raw material in the production of dyes.

Around 4,000 t of 4-toluidine were produced in Western Europe in 1985 by the catalytic reduction of 4-nitrotoluene. It is used as an intermediate in the manufacture of organic pigments, for example those based on 3-nitro-4-aminotoluene, which is obtained by the acylation of 4-toluidine followed by nitration and removal of the protecting acyl group. Examples of important pigments based on this raw material are Pigment Yellow 1 (see Chapter 4.5.3.4) and Pigment Red 3.

Pigment Red 3

The sulfonation of 4-toluidine yields 4-aminotoluene-3-sulfonic acid (4B acid). This acid is used in the production of one of the most important red pigments, Pigment Red 57:1.

Pigment Red 57 : 1

The chlorination of 4-nitrotoluene and subsequent catalytic reduction on sulfided Pd/active carbon catalysts yields 3-chloro-4-methylaniline, from which the herbicide chlortoluron is made by initial conversion into the isocyanate and subsequent reaction with dimethylamine.

Chlortoluron

6.1.2 Dinitrotoluenes and their derivatives

In terms of volume, the dinitration of toluene is of greater importance than mononitration. Reaction conditions for the dinitration are more severe than those for mononitration, with temperatures between 65 and 70 °C. A nitrating mixture with 65% sulfuric acid, 25% nitric acid and 10% water is used and nitration may be carried out as either a continuous or batch process.

When working up the products from dinitration, the reaction mixture is settled and separated into a dinitrotoluene-rich stream and a spent-acid stream. The dinitrotoluene stream is refined by alkali washing and crystallization. The isomer composition is 20% 2,6-dinitrotoluene, 76% 2,4-dinitrotoluene, 0.6% 3,5-dinitrotoluene and small quantities of 2,5-dinitrotoluene (80/20 DNT). The yield from dinitration is around 96 to 98%; the commercial product has a crystallizing point of 55 to 58 °C. The decomposition of dinitrotoluenes at higher temperatures is highly exothermic, and hence they should not be stored at a temperature exceeding 75 °C.

Dinitrotoluenes are used predominantly for the manufacture of toluene diisocyanate (TDI), which is produced from the corresponding diaminotoluenes.

Reduction of dinitrotoluenes is carried out industrially with Raney-nickel or palladium/carbon catalysts, under a hydrogen pressure of 70 bar and at a temperature of up to 150 °C, in a cascade or loop reactor. The concentration of dinitrotoluene in the reaction mixture is kept very low to achieve yields of over 99%. The reaction mixture is divided into a recycle stream and a residue, from which the catalyst is separated in a filter or cyclone. Distillation of the reaction product then yields the components methanol, water, diamines and a residue. Distillation of the water-free amines is carried out under vacuum (approx. 50 mbar), with 2,3- and 3,4-toluenediamine being separated in a 'fronts' column. A tar-like residue is removed as a bottom product in the main column, while a mixture containing 2,4- and 2,6-toluenediamine is collected as an overhead fraction.

It is particularly important to remove 2,3- and 3,4-toluenediamine by distillation since, in the subsequent reaction with phosgene, these compounds would produce benzimidazolones which, having active amide hydrogen atoms, can react with the principal product, toluene diisocyanate, and reduce the yield.

4-Methylbenzimidazolone   5-Methylbenzimidazolone

Figure 6.2 shows the flow diagram for the reduction of dinitrotoluene in methanol solution.

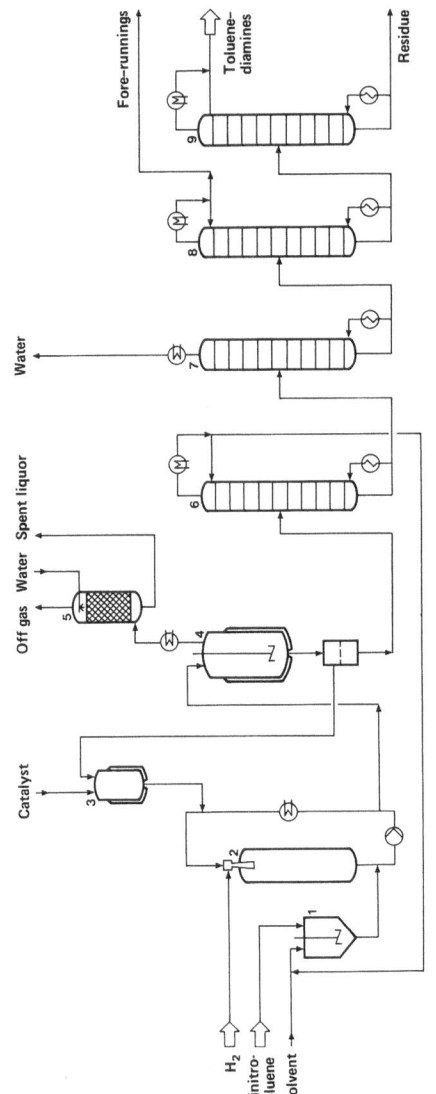

Figure 6.2: Flow diagram for the reduction of dinitrotoluene to toluenediamine

1 Dissolving vessel; **2** Loop reactor; **3** Catalyst vessel; **4** Pressure-let-down vessel; **5** Scrubber; **6** Solvent recovery; **7** Dewatering column; **8** and **9** Fractionating columns

The reaction of toluenediamine with phosgene can be carried out in solution or in the gas-phase. Less active aromatics with high boiling points, such as o-dichlorobenzene, are used as solvents.

Figure 6.3 depicts the production of toluene diisocyanate at atmospheric pressure. In this process, a 10 to 20% solution of toluenediamine in o-dichlorobenzene reacts with a 25 to 50% phosgene solution. The temperature of the exothermic reaction rises from approx. 5 °C (carbamoyl chloride formation) to approx. 170 °C (isocyanate formation).

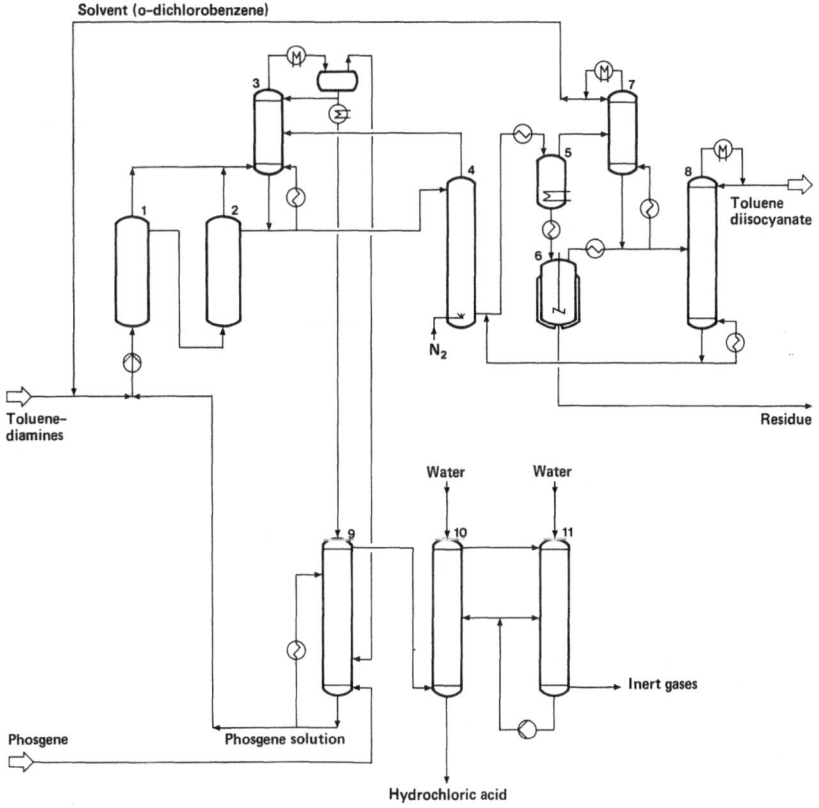

1 and 2 Reactors; 3 Washing column; 4 Stripping column; 5 Pre-evaporator; 6 Final evaporator; 7 Solvent column; 8 Isocyanate column; 9 Phosgene extraction column; 10 HCl-absorber; 11 Phosgene decomposer

Figure 6.3: Flow diagram for the production of toluene diisocyanate

In process engineering terms, the production of toluene diisocyanate is a very complex process, requiring extensive specialist know-how, particularly, because of the low permitted limits for the concentration of toluene diisocyanate at the workplace, the necessary safety procedures in handling phosgene, corrosion problems and precipitation of salts.

Toluene diisocyanate is used in the manufacture of polyurethanes, which are produced by reaction with polyols and polyesters.

Polyurethane   R : Polyester segment

Table 6.1 summarizes capacities of the most important toluene diisocyanate producer countries; *Bayer* operates the largest capacities.

Table 6.1: Capacities of the most important toluene diisocyanate producers (1985)

	(1,000 t)
USA	300
Brazil	30
Belgium	30
France	120
West Germany	150
Italy	60
Japan	80
Others	100
Total capacity in Western world	870

TDI production in the USA in 1985 was 280,000 t, in Western Europe 290,000 t and in Japan 80,000 t.

6.1.3 Trinitrotoluene compounds

In comparison with the production of dinitrotoluene compounds, the production of trinitrotoluene is of relatively little importance today. Either toluene or surplus 2-nitrotoluene can be used as raw materials.

The manufacture of 2,4,6-trinitrotoluene (TNT), a traditionally important explosive, is mainly continuous, with units of 4 to 6 nitration reactors and separators for the separation of sparingly soluble nitrotoluenes in the nitration mixture. The reactants toluene and nitrating acid (16.5% HNO_3, 83.5% H_2SO_4) are fed countercurrently into the cascade reactor. In the first reactor, mononitrotoluene is the predominant product, while in subsequent reactors further nitration takes place. The reaction proceeds with an increase in temperature from 50 to 100 °C.

The crude trinitrotoluene is worked up by washing with water and soda solution or with sodium sulfite solution to remove the less stable asymmetric isomers which lower the crystallizing point of 2,4,6-TNT. The electron-withdrawing effect of the nitro groups of the asymmetric isomers facilitates nucleophilic substitution by $OH^\ominus$ or $HSO_3^\ominus$ groups, even at ambient temperature, so that the reaction products are water-soluble. The crystallizing point of the purified TNT should be at least 80.6 °C. When mixed with 20% aluminum, TNT is marketed under the trade name 'Tritonal', and when mixed with 50% ammonium nitrate, it is sold under the trade name 'Amatol'.

6.2 Benzoic acid

Whereas the nitro derivatives of benzene are produced by electrophilic aromatic substitution, further important derivatives of toluene are predominantly obtained through reactions of the methyl group; they include the production of oxidation products such as benzoic acid and the side-chain chlorinated toluene compounds.

In common with many other aromatic compounds, benzoic acid was first discovered in a renewable raw material, namely gum benzoin, by Blaise de Vigenère as early as in the 16th century. Carl Wilhelm Scheele further studied this raw material in 1755, and it remained the main source for medicinal benzoic acid until the mid-nineteenth century. The first technical synthesis of benzoic acid was based on naphthalene, via the intermediate phthalic anhydride; this synthesis was introduced in 1863. In 1877, August Wilhelm von Hofmann reported the synthesis of benzoic acid from hippuric acid, which is present in the urine of herbivores.

The production of benzoic acid from naphthalene, practised up to the end of World War II, is possible in one stage, without the isolation of the intermediate phthalic acid. Oxidation of naphthalene is carried out at 340 °C with zinc oxide. This decarboxylates the phthalic acid in the gas-phase; however, conversion in this reaction is not complete, and the benzoic acid must be separated from the phthalic acid. Separation is possible by dissolving the phthalic acid in water.

Decarboxylation of phthalic acid in the liquid-phase also leads to the production of benzoic acid. In this process, developed by *Monsanto,* liquid phthalic anhydride is converted over nickel oxide or copper oxide with the introduction of steam at 220 °C.

Nowadays, benzoic acid is obtained on a commercial scale virtually exclusively by the oxidation of toluene with air.

The flow diagram for liquid-phase oxidation is shown in Chapter 5.3.2, within the description of the production of phenol.

Other processes for the manufacture of benzoic acid, such as hydrolysis of benzonitrile, which is present in coal tar, the oxidation of biphenyl and the hydrolysis of benzotrichloride are of no industrial significance today.

Benzonitrile Biphenyl Benzotrichloride

The main field of application for benzoic acid is the production of phenol; however, the significance of this phenol route has declined in recent years. Additionally, benzoic acid is used in the production of benzoyl chloride and sodium benzoate. In Italy, benzoic acid is used as the raw material for the production of ε-caprolactam, in a process developed by *Snia Viscosa*. This involves hydrogenating benzoic acid at 170 °C and 15 bar over a palladium catalyst, purifying the cyclohexane carboxylic acid by distillation followed by its reaction with nitrosylsulfuric acid to yield ε-caprolactam.

Benzoic acid production in Western Europe in 1985 was approx. 25,000 t, and in the USA, including feedstock for the production of phenol, approx. 80,000 t.

Sodium benzoate, which is obtained by the neutralizing of benzoic acid with NaOH, is used as a preservative in foodstuffs such as sauces, syrups and fruit juices. It also acts as a corrosion inhibitor in glycol-based anti-freeze, for which some 3,500 t were consumed in Western Europe in 1985.

Benzoic acid esters, such as 1,3-propyleneglycol dibenzoate, are important plasticizers for polyurethanes.

Benzyl benzoate is used in the perfume industry and, because of its effect in dilating blood vessels and reducing convulsions, it is also a component of preparations for the treatment of asthma.

The reaction of benzoic acid with benzotrichloride produces benzoyl chloride, which is used predominantly in the production of dibenzoyl peroxide.

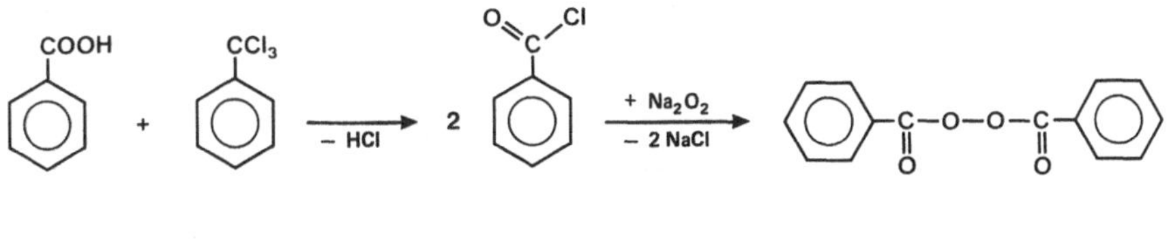

Dibenzoyl peroxide

West European production of benzoyl chloride in 1985 was around 15,000 t. In addition to the dibenzoyl peroxide obtained through reaction with Na_2O_2 and used as a radical initiator for the production of plastics (PVC, polyethylene, polystyrene), benzoyl chloride is used mainly in the production of plant protection agents such as the *Bayer* herbicide metamitron.

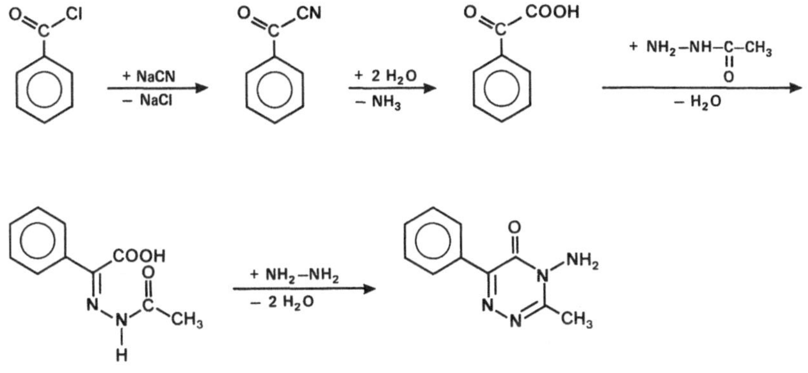

Metamitron

Benzophenone is obtained by the reaction of benzoyl chloride with benzene under Friedel-Crafts reaction conditions. Benzophenone is used as a perfume component and as an additive in the production of printing inks.

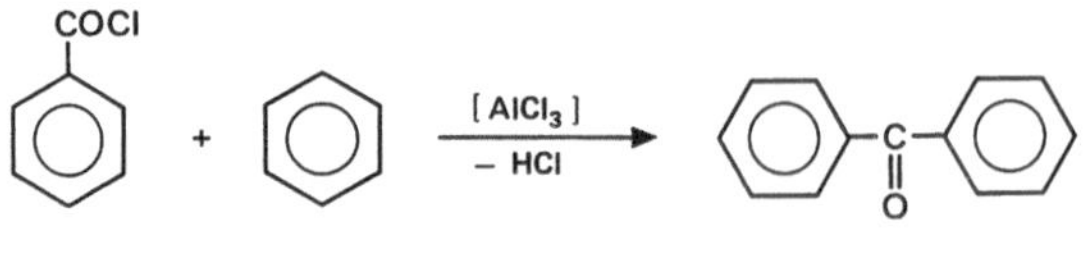

250 Production and uses of toluene derivatives

$$\text{PhCOCl} + \text{PhH} \xrightarrow[-\text{HCl}]{[\text{AlCl}_3]} \text{Ph-CO-Ph}$$

Benzophenone

The sulfonation of benzoic acid and alkali fusion produces 3-hydroxybenzoic acid. 3-Hydroxybenzoic acid is an intermediate in the manufacture of the herbicide acifluorfen. (3-Hydroxybenzoic acid can also be produced from m-cresol by oxidation.)

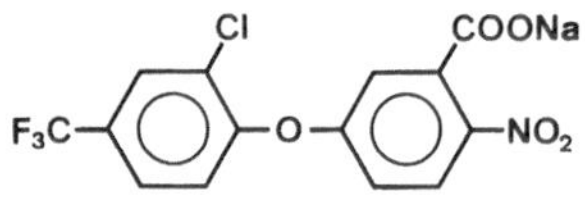

Acifluorfen: F_3C–(2-Cl-phenyl)–O–(2-COONa, 4-NO$_2$-phenyl)

Acifluorfen

6.3 Chlorine derivatives of toluene

The chlorination of toluene can be carried out in the side chain and in the aromatic nucleus; both product groups are of commercial significance, although side-chain-chlorinated toluenes are predominant in terms of quantity.

The two types of reaction follow different reaction mechanisms. Side-chain chlorination occurs by a radical chain reaction mechanism, nuclear chlorination by electrophilic substitution.

6.3.1 Side-chain chlorination of toluene

The degree of substitution in side-chain chlorination is dependent on the chlorine/toluene molar ratio, as shown in Figure 6.4. For the manufacture of benzyl chloride and benzal chloride, the reaction has to be restricted to a low conversion to suppress the production of benzotrichloride.

The purity of the toluene is critical and traces of water are generally removed by distillation before chlorination.

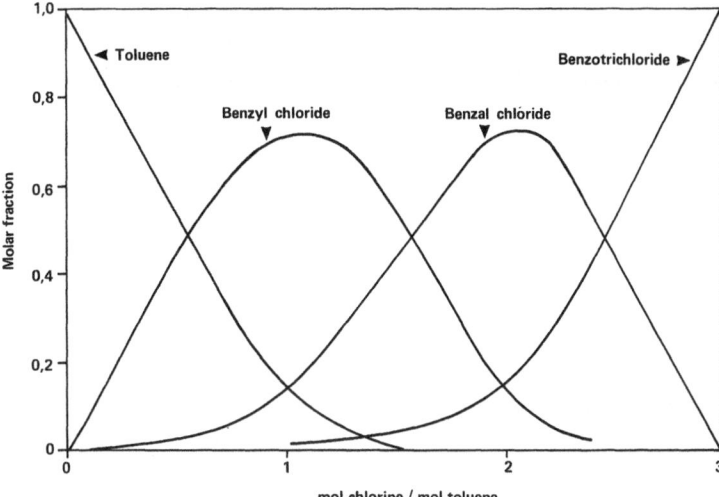

Figure 6.4: Product distribution in the side-chain chlorination of toluene versus the molar chlorine/toluene ratio

6.3.1.1 Benzyl chloride

To avoid nuclear chlorination due to Friedel-Crafts catalysis by iron contaminants, benzyl chloride production is carried out in vessels made from pure nickel, or in enamel or borosilicate glass apparatus. The chain reaction takes place with photochemically-produced chlorine radicals, using mercury arc lamps; the reaction temperature ranges from 80 to 160 °C. The process is carried out with a chlorine deficiency, so that a high yield of benzyl chloride is achieved with low conversion. The reaction mixture is separated by distillation.

The most important product from benzyl chloride, production of which was 35,000 t in the USA in 1985, is butyl benzyl phthalate, which is manufactured by the reaction of monosodium butyl phthalate with benzyl chloride. Butyl benzyl phthalate is used as a plasticizer in the manufacture of PVC-plastics.

Benzyl alcohol is the second most important product from benzyl chloride. Stanislao Cannizzaro deduced its structure in 1853 from the reaction of benzaldehyde with potassium hydroxide.

252 Production and uses of toluene derivatives

Benzyl alcohol is obtained in a 70% yield from benzyl chloride and soda solution at 90 °C in a reaction lasting several hours. The reaction product is separated by distillation. Dibenzyl ether, a by-product, is used as a plasticizer.

Benzyl alcohol

Dibenzyl ether

The third most significant application for benzyl chloride is the production of benzyl cyanide. Benzyl cyanide is obtained from benzyl chloride by the reaction with sodium cyanide. The most important product derived from benzyl cyanide is phenylacetic acid, which is easily obtained by hydrolysis.

Phenylacetic acid is especially used in the production of benzyl penicillin (penicillin G) by a biotechnical process.

Penicillin G

In 1928/29 Alexander Fleming discovered that the growth of a staphylococcus was inhibited by a substance produced by a mold. Since the mold in question was identified as Penicillium notatum, Fleming named the substance penicillin; however, he did not make any further systematic attempts to isolate it.

The full significance of penicillin as an effective antibiotic was recognised by Howard W. Florey and Ernst B. Chain (1940). This discovery stimulated intensive research toward the biotechnical production of penicillin. It was possible to increase production from a few milligrams of penicillin per liter of culture broth to a level of 35 g/l by mutation of the strain and improvements in biochemical process technology. Figure 6.5 shows the production flow diagram for penicillin G.

Production and uses of toluene derivatives   253

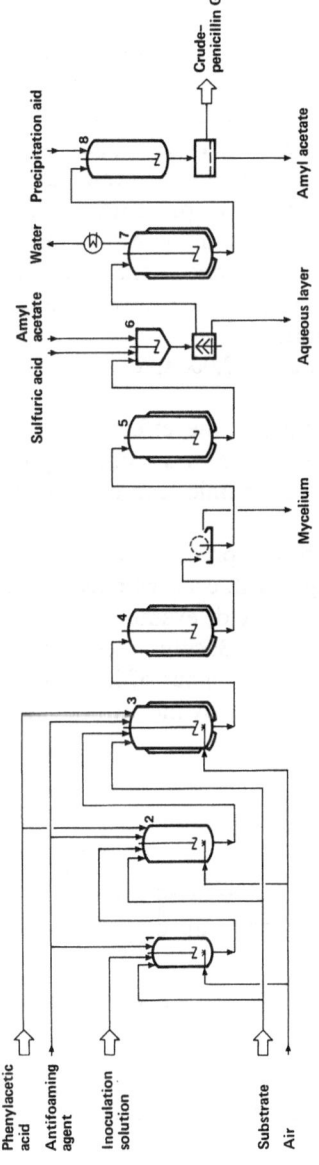

1 Prefermenter I; **2** Prefermenter II; **3** Production fermenter; **4** Cooling vessel; **5** Storage tank for penicillin concentrate; **6** Mixing vessel; **7** Dewatering vessel; **8** Precipitation vessel

Figure 6.5: Flow diagram for the production of penicillin G

In a multistage reaction penicillin G is nowadays produced exclusively by the aerobic submerged process, whereby oxygen is introduced by aerating the base of the reactor (fermenter); strains of the mold Penicillium chrysogenum are the predominantly used microorganism.

Corn steep solution, which is a by-product from the recovery of starch from maize, is generally used as the fermentation substrate. In addition, other sources of carbon, in the form of lactose, glucose, saccharose, oils and fats, can be added to the fermentation medium. In the production of penicillin G, phenylacetic acid is added during the fermentation as potassium salt in neutral aqueous solution.

The fermentation, which is carried out batch-wise at 25 °C, lasts some 200 hours in total. In subsequent washing, the cell mass (mycelium) formed is filtered off on rotating drum filters, and the filtrate is acidified to pH 2 with sulfuric or phosphoric acid. Penicillin G is concentrated by extraction with solvents such as butanol or amyl acetate. When the acid has been converted to the potassium salt, penicillin G is recovered by filtration and drying as a 99.5% pure product. The mycelium can be used as an animal feedstuff.

World production of penicillin G is around 12,000 tpa, the largest producers being *Gist-Brocades, Glaxo* and *Beecham*.

In connection with investigations to improve the effectiveness of penicillin antibiotics, it has been established that substitution in the phenyl group, or other types of substituents in the thiazolidine-ß-lactam unit, lead to an increased effectiveness over penicillin G.

Production of modified penicillins is based predominantly on penicillin G (semi-synthetic penicillin). Penicillin G is split with carrier-linked acylase enzymes into phenylacetic acid and 6-aminopenicillanic acid; examples of enzyme sources for the penicillinacylase are Escherichia coli or Bacillus megaterium. 6-Aminopenicillanic acid may be converted into a large number of highly-effective semi-synthetic penicillin antibiotics, such as amoxycillin, by the introduction of suitable side-chains.

6-Aminopenicillanic acid

Amoxycillin

The most important penicillin derivative in terms of quantity is ampicillin, which is obtained by the reaction of 6-aminopenicillanic acid with the protected amine-group of D-phenylglycine (Dane process).

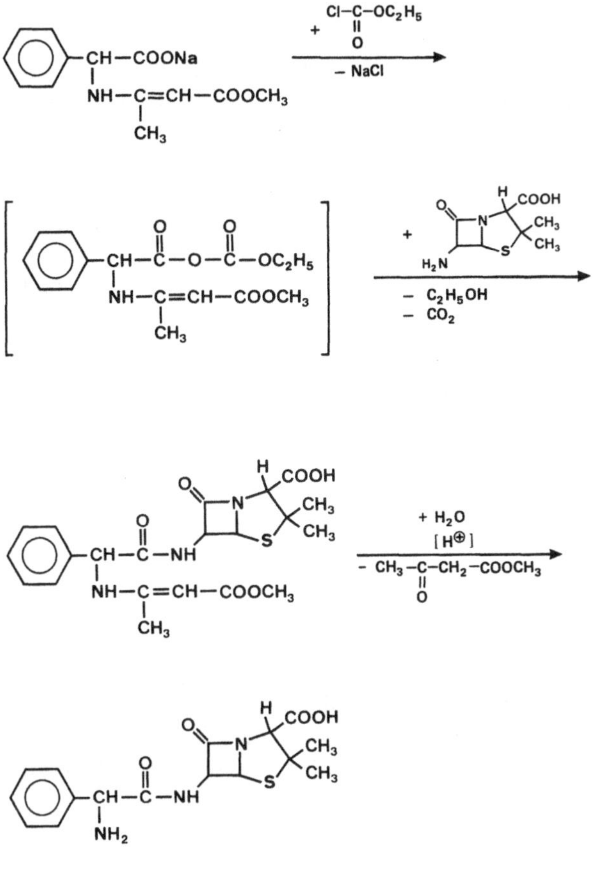

Production and uses of toluene derivatives

[Scheme showing synthesis of ampicillin:]

Ph–CH(COONa)–NH–C(CH₃)=CH–COOCH₃ + Cl–C(=O)–OC₂H₅ → (−NaCl)

[Ph–CH(–C(=O)–O–C(=O)–OC₂H₅)–NH–C(CH₃)=CH–COOCH₃] + 6-APA (H₂N-β-lactam-thiazolidine with COOH, CH₃, CH₃) → (−C₂H₅OH, −CO₂)

Ph–CH(–C(=O)–NH–β-lactam with COOH, CH₃, CH₃)–NH–C(CH₃)=CH–COOCH₃ + H₂O [H⁺] → (−CH₃–C(=O)–CH₂–COOCH₃)

Ph–CH(NH₂)–C(=O)–NH–β-lactam (COOH, CH₃, CH₃)

Ampicillin

Phenylacetic acid offers interesting prospects for the production of pure stereoisomeric amine compounds. This involves splitting the mixed amines in the form of DL-amides of phenylacetic acid with enzymes, when only the L-form is affected by the enzymatic cleavage, so that the stereoisomeric amines can be separated.

Other important products based on benzyl alcohol are the esters formed with acetic acid, salicylic acid and benzoic acid which are used as fragrances. In addition, benzyl alcohol is used as a thinner in epoxy resin hardeners.

A new application for benzyl alcohol has been found in the production of the sweetener aspartame, via the intermediate phenylalanine.

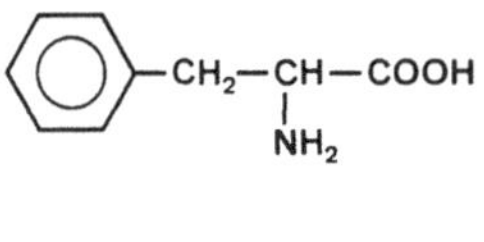

Ph–CH₂–CH(NH₂)–COOH

Phenylalanine

Ph–CH₂–CH(–C(=O)–OCH₃)–NH–C(=O)–CH(NH₂)–CH₂–COOH

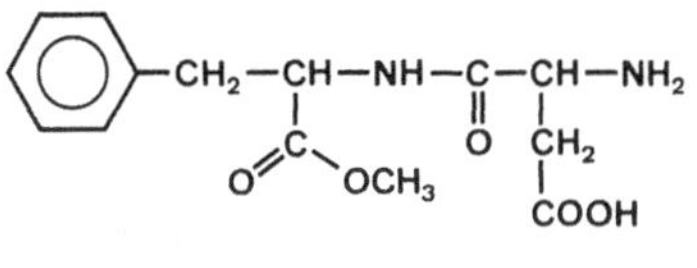

Aspartame

6.3.1.2 Benzal chloride

The production of benzal chloride is relatively insignificant in comparison with that of benzyl chloride. Benzal chloride is mainly used to make benzaldehyde, obtained by alkaline hydrolysis (Na_2CO_3) or in an acidic medium ($ZnCl_2$) at 120 to 130 °C. Benzaldehyde however can also be recovered as a byproduct of the oxidation of toluene to benzoic acid, occurring in quantities of between 5 and 8%; this route leads to benzaldehyde which is distinguished by the lack of chlorine impurities. The main uses for benzaldehyde are the production of benzyl alcohol (see Chapter 6.3.1.1), chloramphenicol, 2-phenylglycine and cinnamaldehyde.

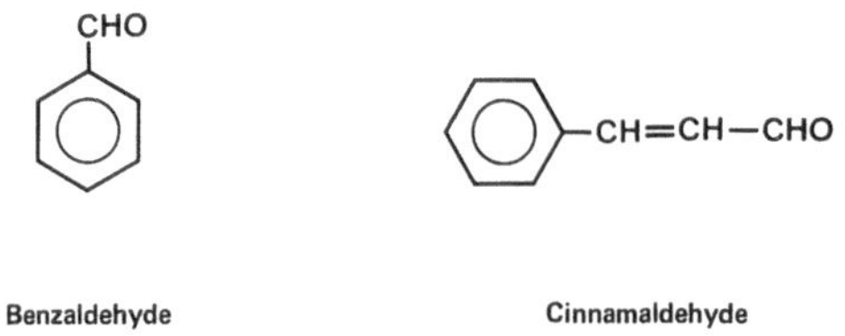

Benzaldehyde

Cinnamaldehyde

The antibiotic chloramphenicol, which is mainly used in veterinary medicine, is produced by the *Boehringer Mannheim* process from cinnamic alcohol, which in turn is obtained from benzaldehyde through aldol condensation with acetaldehyde and reduction of the intermediate cinnamaldehyde.

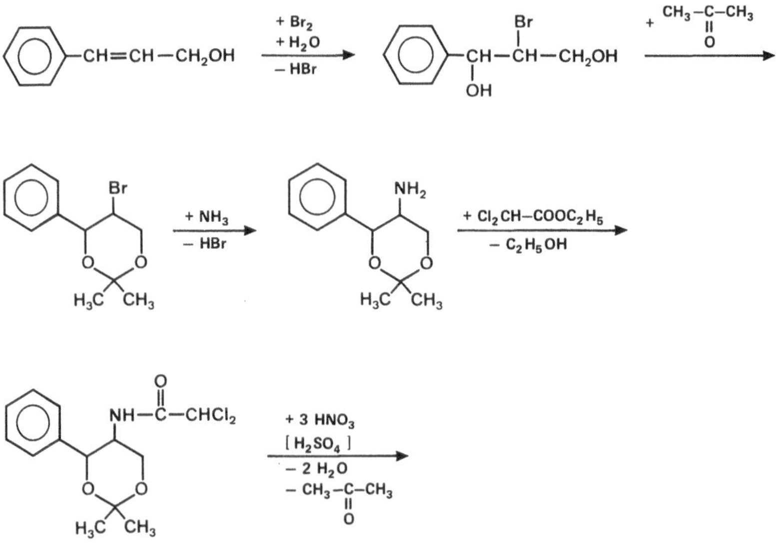

$$\text{O}_2\text{N}-\underset{\underset{\text{O}-\text{NO}_2}{|}}{\bigcirc}-\text{CH}-\underset{}{\overset{\overset{\text{NH}-\overset{\text{O}}{\overset{\|}{\text{C}}}-\text{CHCl}_2}{|}}{\text{CH}}}-\text{CH}_2-\text{O}-\text{NO}_2 \quad \xrightarrow[-2\,\text{HNO}_3]{+2\,\text{H}_2\text{O}} \quad \text{O}_2\text{N}-\bigcirc-\underset{\underset{\text{OH}}{|}}{\text{CH}}-\underset{\underset{\text{CH}_2\text{OH}}{|}}{\text{CH}}-\text{NH}-\overset{\overset{\text{O}}{\|}}{\text{C}}-\text{CHCl}_2$$

Chloramphenicol

2-Phenylglycine is produced as a DL-amino-acid mixture from benzaldehyde, sodium cyanide and ammonium chloride. The D-form of 2-phenylglycine is of particular importance and is used in the manufacture of a wide range of semi-synthetic penicillins, such as ampicillin, which is produced worldwide in a quantity of approx. 3,000 t.

D-(2-Phenylglycine)

Cinnamaldehyde is obtained by the aldol condensation of acetaldehyde with benzaldehyde; when reacted with long chain n-aldehydes (C_7/C_8), it is used to produce fragrances and perfumes.

Self-condensation of benzaldehyde produces benzoin, which is used as a photoinitiator for UV hardening of paints. Benzil, which is obtained by oxidation of benzoin, has the same application.

Benzoin   Benzil

The sympathomimetic ephedrine is produced using benzaldehyde as the starting material. Benzaldehyde is converted biotechnically using Carl Neuberg's synthesis (1921) with acetaldehyde which arises from fermentation of molasses to produce optically active phenylacetylcarbinol ((−)-1-hydroxy-1-phenylacetone), which is transformed into L-(−)-ephedrine by catalytic hydrogenation in the presence of methylamine.

Phenylacetylcarbinol

L-(−)-Ephedrine

6.3.1.3 Benzotrichloride

Benzotrichloride is produced by the side-chain chlorination of toluene with excess chlorine at 100 to 140 °C. West European production in 1985 was 15,000 t. The most important application for benzotrichloride is the production of benzoyl chloride (see Chapter 6.2) by hydrolysis with equimolar quantities of water at 120 to 140 °C, with the addition of $ZnCl_2$; the benzoyl chloride is purified by distillation.

The reaction of benzotrichloride with anhydrous HF at 120 °C in a nickel or stainless-steel reactor produces benzotrifluoride, from which 3-trifluoromethylaniline is obtained by selective nitration and subsequent catalytic reduction of the 3-trifluoromethylnitrobenzene.

Benzotrifluoride

3-Trifluoromethylaniline

3-Trifluoromethylaniline (3-aminobenzotrifluoride) is a very versatile intermediate, used predominantly in the production of plant protection agents and, to a lesser extent, pharmaceutical products.

An important herbicide based on 3-aminobenzotrifluoride is *Ciba Geigy's* fluometuron, which is obtained by reaction of the related isocyanate with dimethylamine.

Fluometuron

Norflurazon, developed by *Sandoz,* is also produced from 3-aminobenzotrifluoride; in the final stage of this process, the reaction product from 3-trifluoromethylphenylhydrazine and mucochloric acid reacts with methylamine.

Benzotrichloride is also used as a feedstock in the production of dyestuffs such as Vat Yellow 2 (a bis-thiazole compound), which is obtained from benzotrichloride and 2,6-diaminoanthraquinone with sulfur.

6.3.2 Nuclear chlorination of toluene

Chlorination of the aromatic nucleus is generally carried out at relatively low temperatures (ca. 50 °C) under atmospheric pressure, yielding ca. 95% monochlorotoluene and 5% dichlorotoluenes; the proportions of 2- and 4-chlorotoluene can be controlled by the choice of catalyst. Separation of the 2-/4-isomers is possible by distillation. Capacity for nuclear chlorination of toluene in Western Europe is approx. 100,000 tpa.

4-Chlorotoluene is of particular importance, with a production in Western Europe in 1985 of around 20,000 t. The most important product from 4-chlorotoluene is 4-chlorobenzotrichloride, which is produced by photochlorination.

4-Chlorotoluene   4-Chlorobenzotrichloride   2-Chlorotoluene

It serves as an intermediate in, e.g., the production of 4-chlorobenzotrifluoride, which is further processed to manufacture the plant protection agent trifluralin. Nucleophilic substitution of 4-chloro-3,5-dinitrobenzotrifluoride with di-n-propylamine leads to trifluralin.

Trifluralin

World production of this pre-emergence herbicide, developed by *Eli Lilly*, is around 25,000 tpa.

Another plant protection agent based on 4-chlorotoluene is fenvalerate (see Chapter 5.3.4.3.1), which works in a similar way as pyrethroid insecticides.

For the production of the necessary 'chrysanthemum acid substituent', 4-chlorotoluene is chlorinated photochemically and the 4-chlorobenzyl chloride converted into the nitrile with sodium cyanide. Base-catalysed introduction of the isopropyl group and subsequent hydrolysis of the nitrile, followed by chlorination, yields 2-isopropyl-(4-chlorophenyl)-acetyl chloride as an intermediate component for the production of fenvalerate.

2-Chlorotoluene, which is manufactured in Western Europe in quantities of ca. 20,000 tpa, is used mainly in the production of cresols (see Chapter 5.3.4.3) by hydrolysis. In addition, it is used, via the intermediate 2-chlorobenzal chloride, as a raw material for the manufacture of optical brighteners, such as benzaldehyde-2-sulfonic acid.

6.4 Sulfonic acid derivatives of toluene

Toluenesulfonic acids are produced by the reaction of toluene with oleum, SO_3 or chlorosulfonic acid. The position of the sulfonic acid group can be controlled by the choice of reaction conditions. Toluene-4-sulfonic acid is obtained by the reaction of toluene with 90 to 95% sulfuric acid at 95 to 100 °C. The crude mixture consists of 75 to 85% toluene-4-sulfonic acid, 10 to 20% toluene-2-sulfonic acid, and 2 to 5% toluene-3-sulfonic acid. Toluene-4-sulfonic acid can be purified by crystallization from 66 to 70% sulfuric acid.

The formation of toluene-2-sulfonic acid is favored by low temperatures. The reaction of toluene with 96% sulfuric acid at 40 °C yields predominantly toluene-2-sulfonic acid.

Toluene-3-sulfonic acid is produced by the isomerization of the toluene monosulfonic acid mixture at 140 to 200 °C.

Figure 6.6 shows the flow diagram for the batch-wise sulfonation of toluene with oleum; the water produced during the reaction can be separated by azeotropic distillation with excess toluene.

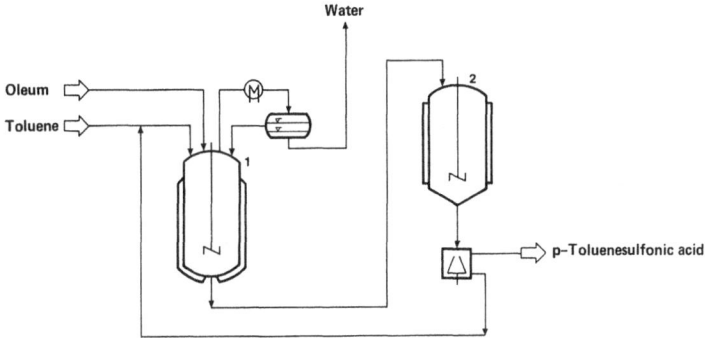

1 Sulfonation reactor; **2** Crystallizer

Figure 6.6: Process scheme for the sulfonation of toluene

In continuous processes particularly, sulfonation is carried out with SO_3. The exothermic reaction can be controlled by vaporizing excess toluene.

Toluene-4-sulfonic acid is used in 65% solution as an accelerator for hardening furan and phenolic resins, mainly used in foundries. The salts, especially the sodium salt in 45% solution, are base materials for detergents, principally to reduce the viscosity of alkylbenzene sulfonates.

Toluene-4-sulfonic acid is used as a starting material in the production of 2-chloro-5-aminotoluene-4-sulfonic acid (CLT-acid). CLT-acid is manufactured by chlorination of toluene-4-sulfonic acid, transformation of 2-chlorotoluene-4-sulfonic acid into 2-chloro-5-nitrotoluene-4-sulfonic acid followed by the reduction of the nitro group with iron/HCl. CLT-acid is used, for example, as a raw material in the production of Pigment Red 52:1, Pigment Red 53:1 and other metal complexes.

2-Chlorotoluene-4-sulfonic acid   2-Chloro-5-nitrotoluene-4-sulfonic acid   CLT-acid

Pigment Red 52:1   Pigment Red 53:1

Alkali fusion of toluene-4-sulfonic acid can produce p-cresol (see Chapter 5.3.4.3); this route is carried out by *Synthetic Chemicals* (UK) and *Sherwin-Williams* (USA).

Total consumption of aromatic sulfonic acids, i.e. benzene-, toluene-, xylene- and cumenesulfonic acids in Western Europe was around 40,000 t in 1985.

Toluene polysulfonic acids are of little importance in comparison with monosulfonic acid derivatives.

6.5 Toluenesulfonyl chloride

The reaction of toluene and excess chlorosulfonic acid produces a mixture of toluene-2- and toluene-4-sulfonyl chloride. At low temperatures (approx. 0 °C) the formation of the 2-isomer is favored, whereas at high temperatures a concentration of the 4-isomer of around 80% can be obtained. The isomers can be separated by distillation or by crystallization.

Toluene-2-sulfonyl chloride is converted into the corresponding sulfonamide, which, after an oxidation step, is used in the production of the sweetener saccharin.

Toluene-2-sulfonyl chloride   Toluene-4-sulfonyl chloride

The content of 2-sulfonamide, a suspected carcinogen, is limited in commercial saccharin to 10–25 ppm, varying from country to country.

Saccharin can also be produced from anthranilic acid by way of dithiosodiumsalicylate.

Dithiosodiumsalicylate   Saccharin

Saccharin is a possible starting material for the antiphlogistic piroxicam (*Pfizer*), a derivative of 2-aminopyridine.

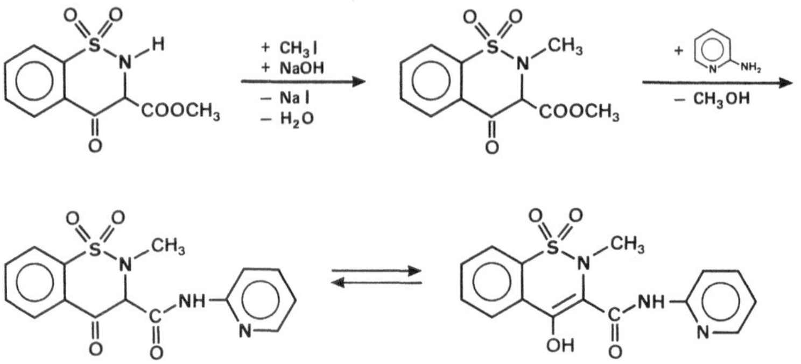

264 Production and uses of toluene derivatives

Piroxicam

Consumption of saccharin in Western Europe is around 2,200 tpa.

Toluene-4-sulfonyl chloride is used in the production of chloramine-T, a chlorination agent.

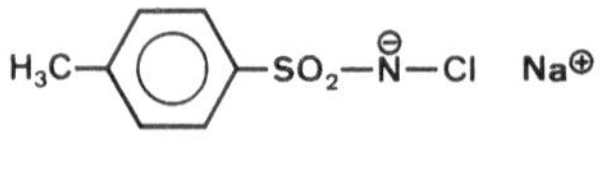

Chloramine T

The mixed isomers of toluene sulfonamide are used as plasticizers for melamine-formaldehyde resins.

6.6 Other toluene derivatives

A further important derivative of toluene is 4-tert-butyltoluene, which is obtained by the alkylation of toluene with isobutene. 4-tert-Butyltoluene is the starting material in the production of 4-tert-butylbenzoic acid, which is used as a corrosion inhibitor and modifier for alkyd resins.

Mild oxidation of 4-tert-butyltoluene produces 4-tert-butylbenzaldehyde, which finds application as a perfumery component.

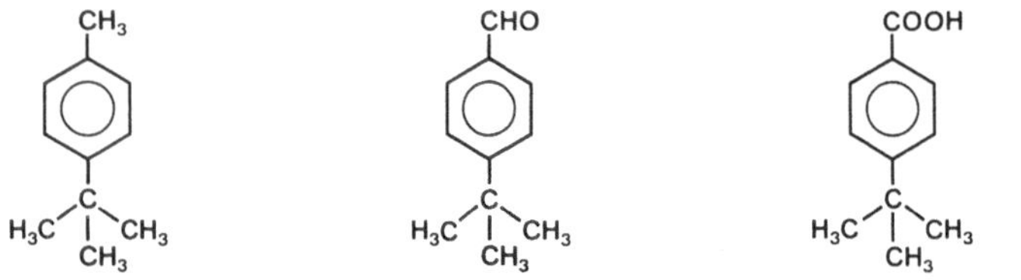

4-tert-Butyltoluene   4-tert-Butylbenzaldehyde   4-tert-Butylbenzoic acid

7 Production and uses of xylene derivatives

Whereas benzene and toluene serve as the raw materials for a wide range of products, applications for the three xylene isomers, o-, m- and p-xylene, are basically limited to chemicals arising through oxidation, i.e. phthalic anhydride (PA) from o-xylene, isophthalic acid from m-xylene and terephthalic acid from p-xylene.

Phthalic anhydride (PA)   Isophthalic acid   Terephthalic acid

7.1 o-Xylene and its derivatives

7.1.1 Oxidation of o-xylene to phthalic anhydride

Phthalic acid was discovered by the French chemist Auguste Laurent in 1836. During experiments with naphthalene, he discovered an acidic substance, which he called 'naphthalene acid'. In 1869, Carl Graebe established that this naphthalene acid was, in fact, o-benzenedicarboxylic acid (phthalic acid).

Commercial production of phthalic anhydride (PA) was taken up by *BASF* in 1872, by the oxidation of naphthalene with manganese dioxide and hydrochloric acid, to obtain the required base material (PA) for the manufacture of the dyestuffs fluorescein and eosine, and later for phenolphthalein; however, the yield was only 5 to 7%.

Despite considerable headway in the process, which involved the use of chromic acid and later oleum, and which increased the yield to around 15%, non-catalytic methods remained highly unsatisfactory.

Fluorescein

Eosine

Phenolphthalein

During research into the development of a commercial synthesis of indigo in 1891, Eugen Sapper discovered an important improvement by the use of a mercury sulfate catalyst; the Sapper process remained in use until 1925. During World War I, Alfred Wohl in Germany and Harry D. Gibbs and Courtney Conover in the USA, independently discovered the catalytic gas-phase oxidation of naphthalene with vanadium pentoxide.

As a consequence of the lengthy patent dispute arising from this parallel development, and restrictions on the importation of German PA during the war, a high temperature process for gas-phase oxidation of naphthalene, using mercury for heat transfer, was perfected in the USA in 1917.

Concurrently, *BASF* developed a durable catalyst for a process operating at low temperature, by which naphthalene could be converted into PA in high yields, initially 73.5% and later rising to 87%. The cylindrical *BASF* catalyst contained 10% vanadium pentoxide, 20 to 30% potassium sulfate and 60 to 70% porous silica. The reaction was carried out in a multitubular reactor at 380 to 390 °C; heat was removed by a salt bath.

PA production received a further innovative impetus from the use of an alternative feedstock. In 1944/45, for the first time, *Oronite Chemical Co.* (*Standard Oil (CA)*) in the USA oxidized o-xylene in a salt-bath reactor to produce PA. The catalyst was a low-porosity carrier (silica quartz or silicon carbide), which was coated with 7 to 8% molten V_2O_5. The reaction temperatures of 450 to 600 °C were considerably higher than those used in the *BASF* process.

At the present time, phthalic anhydride is produced from both naphthalene and o-xylene. The use of naphthalene is especially high in Japan, at just 40%, whereas in the USA and some West European countries it is relatively low, since in these countries naphthalene is most commonly used in the manufacture of dyes and agricultural chemicals. Currently, 80% of world phthalic anhydride production is based on o-xylene.

Oxidation of aromatics, including o-xylene and naphthalene, to carboxylic acids can be carried out in both the liquid- and the gas-phase. Liquid-phase oxidation is generally distinguished by high selectivity at high conversion rates. The disadvantage is that down-stream processing of the reaction products requires separation of the reaction solvent.

In gas-phase oxidation, the cost of purifying the reaction products is lower, since there is no solvent recovery involved; a drawback is the reduced selectivity, because of the necessarily higher temperatures.

Only gas-phase oxidation has proved commercially viable for the oxidation of o-xylene and naphthalene. A high yield with high gas throughput, controlled removal of the heat of reaction with optimal use of process steam, the highest possible precipitation and recovery of the crude product, and efficient purification of the product are essential features of the industrial processes.

The reaction diagram overleaf, showing o-xylene oxidation with possible competing side-reactions to other organic oxygen compounds, illustrates the complex reaction nature of the gas-phase oxidation.

Naphthoquinone is obtained as an additional by-product in the oxidation of naphthalene.

A distinction is made between low temperature processes (*BASF* catalyst, *von Heyden* catalyst) and high temperature processes, which are operated at around 450 °C.

The commercial oxidation of o-xylene in a fixed bed is nowadays carried out exclusively in multitubular reactors with a capacity of up to 50,000 tpa; large-scale reactors are fitted with 25,000 tubes, each with a diameter of 25 mm. The most common catalyst is vanadium pentoxide, the reactivity of which is modified by the addition of salts (e.g., K_2SO_4) and oxides (e.g., TiO_2) and by the shape of the surface-coated catalyst.

In place of the earlier cylindrical catalyst pellets, spherical pellets with smooth or porous surfaces made from porcelain, magnesium silicate, quartz and silicon carbide are used to obtain higher space velocities. A very thin layer of V_2O_5/TiO_2 is applied to the carrier to manufacture the catalyst; a ratio of TiO_2 to V_2O_5 with 2 to 15% V_2O_5 in the active mass has been found to be very efficient. Contact time is between 0.15 to 0.6 sec. Current service life of the catalyst is of the order of 2 to 4 years.

In the *BASF* process, a typical low-temperature process, o-xylene is fed into the multitubular reactor with air which has been preheated to around 150 °C, and the reaction gas flows downwards through the catalyst bed (Figure 7.1).

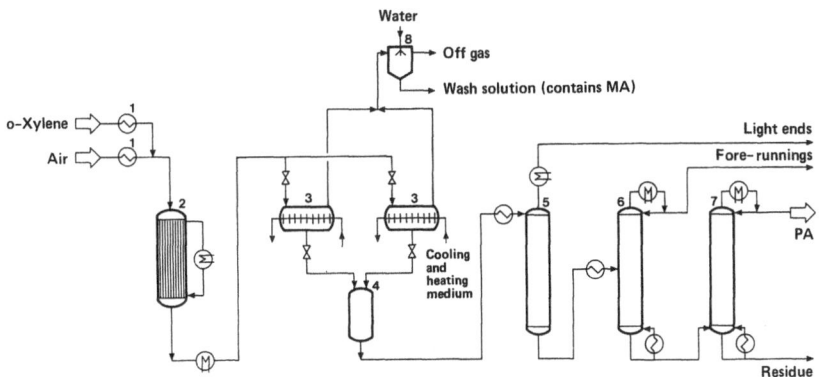

1 Preheater; **2** Multitubular reactor; **3** Switch condenser; **4** Crude PA storage vessel; **5** Thermal pretreatment; **6** Fore-runnings column; **7** PA-column; **8** Off gas purification

Figure 7.1: Low-temperature process for the oxidation of o-xylene

268 Production and uses of xylene derivatives

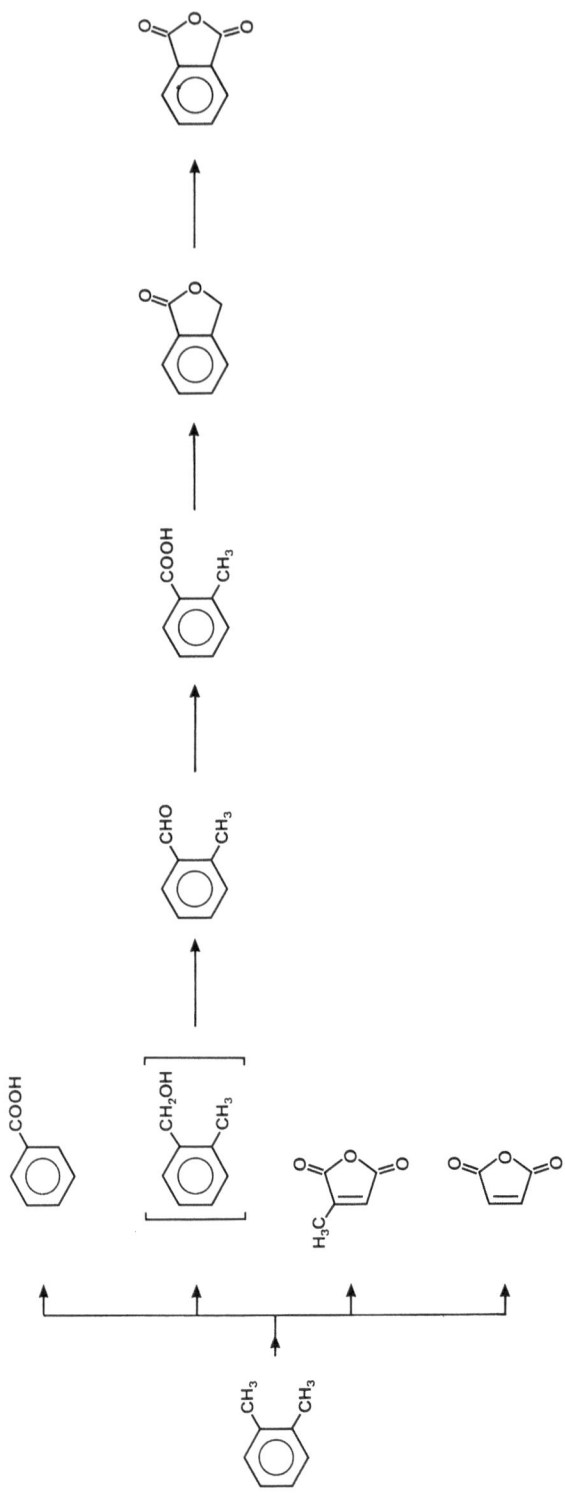

The concentration of o-xylene (or naphthalene) in the air, 60–70 g/Nm³, is in the explosive range, which for o-xylene extends from 44 g/Nm³ to 335 g/Nm³. High loading is particularly desirable to improve the energy efficiency.

The heat of the reaction is removed by means of an eutectic salt bath of potassium nitrate/sodium nitrite (59%/41%) with a melting point of around 141 °C and is used to generate high pressure steam.

The hot reaction gases leave the reactor at a temperature of 350 to 450 °C and are fed to the gas separator. In the fixed bed process condensing is carried out by desublimation. Desublimation of solid PA requires a rapid cooling below the dew point of 125 °C and is carried out in heat exchangers which are fitted with oval finned pipes. Cooling to separate the crude phthalic anhydride and heat for melting are provided by a heat-transfer oil.

Figure 7.2: Thermal pretreatment and distillation units of the PA plant of *Hüls/ Veba*, Bottrop/ West Germany

Before the final upgrading by distillation, the crude phthalic anhydride is thermally pretreated at temperatures of 230 to 300 °C in the pre-decomposer. This causes the by-products (maleic acid, maleic anhydride, o-tolualdehyde, benzoic acid, phthalide, among others) to be partly decomposed, resinified or driven off. The heat-treated crude phthalic anhydride is fractionated by continuous distillation into fore-runnings, pure phthalic anhydride and a residue.

The yield achieved is around 80% of the possible theoretical value, corresponding to 108 kg phthalic anhydride from 100 kg of pure o-xylene. By the application of suitable processing conditions, maleic anhydride can be obtained as a by-product in levels up to 4%.

Figure 7.2 shows part of the distillation plant and thermal pretreatment section of the PA plant operated by *Hüls/Veba*, Bottrop/West Germany, which has a total capacity of 90,000 tpa.

The high temperature process has been developed particularly in the USA. The process operated at 450 °C permits a greater space velocity resulting in shorter contact time; a disadvantage, however, is the increased risk of explosion, since the flash point of a mixture of air/o-xylene is 456 °C, and that of air/PA 580 °C.

Although mercury was initially used in high temperature processes as a coolant, from 1940 onwards the eutectic salt mixture of potassium nitrate and sodium nitrite has been used as a substitute.

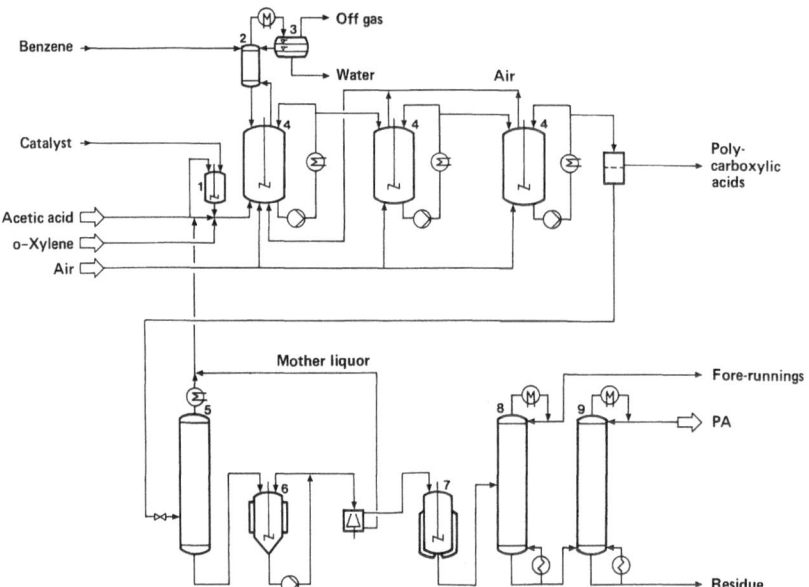

1 Catalyst mixing vessel; 2 Distillation column; 3 Water separator; 4 Reactor cascade;
5 Flash column; 6 Crystallizer; 7 Decomposer; 8 and 9 Vacuum columns

Figure 7.3: Flow diagram of the liquid-phase oxidation of o-xylene

The liquid-phase oxidation of o-xylene was of little comparable technical significance following the introduction of gas-phase oxidation, and was used only in rare cases. In the early 1970's, *Rhône Poulenc* (Chauny/France) started a plant with a capacity of 19,000 tpa. In a process developed by *Rhône Poulenc/Progil*, o-xylene was fed into a three-stage cascade reactor together with acetic acid and the solution of the catalyst (Co-salts). In the first reactor a conversion of between 50 and 60% was achieved, in the second between 35 and 40% and in the third between 8 and 10%. The reaction was carried out under a pressure of 6 bar and at temperatures ranging from 150 to 165 °C.

In order to avoid explosions, the off gas must be largely free from oxygen. The reaction is therefore controlled to a residual content of 2% oxygen in the off gas. The gas is cooled and the reaction water separated; for this purpose, unconverted o-xylene can be used as an entrainer. The crude phthalic anhydride is removed from the solution in a flash depressurizing stage followed by crystallization and rinsing with o-xylene. After further purification with acetic acid, the PA is melted and distilled under vacuum to obtain the desired purity.

Figure 7.3 shows the flow diagram for the *Rhône Poulenc/Progil* process.

Liquid-phase oxidation cannot compete with gas-phase oxidation, so at the present time there is no large-scale plant in operation using liquid-phase oxidation for PA production.

The manufacture of phthalic anhydride from naphthalene by the fluidized-bed process is outlined in Chapter 9.3.1.

Commercial refined PA exhibits a crystallizing point of 130.8 °C with a purity of 99.8%; the content of maleic anhydride is 0.05% maximum, the content of benzoic acid is 0.1% maximum.

Table 7.1 summarizes the production figures for the most important PA-producer countries.

Table 7.1: Production of the major PA-producer countries (1985)

	(1,000 t)
USA	400
Canada	20
Brazil	70
France	75
West Germany	210
Italy	90
Great Britain	70
Austria	30
Spain	30
Soviet Union	220
Yugoslavia	40
Japan	280
Korea (South)	60
Australia	20
Others	485
Total production	2,100

The leading producers are *Exxon, Koppers* (USA), *BASF, Hüls/Veba* (West Germany) and *Alusuisse Italia* (Italy).

In Western Europe and Japan, over 60% of the phthalic anhydride is used in the production of phthalic esters; in the United States this figure is around 55%.

Other applications for phthalic anhydride are the production of alkyd resins and unsaturated polyester resins, each representing around 20%, complemented by the production of dyes and pigments.

7.1.2 Production of phthalic esters

The main application for phthalic anhydride is the production of phthalic esters for use as plasticizers.

Plasticizers are auxiliary agents used in the industrial processing of plastics. By lowering the intermolecular forces between the molecular chains, they endow high polymeric substances with certain desirable physical characteristics, e.g. reduced brittleness, higher plasticity, increased elastic properties, lower hardness and, where necessary, increased adhesion.

James A. Cutting was the first to discover, in 1854, that camphor was a suitable plasticizer for nitrocellulose.

Camphor

This discovery was first applied commercially in 1868, when celluloid was produced by John Wesley Hyatt. Phthalic esters were recognised as plasticizers from 1880 when they were used in celluloid in place of camphor.

The phthalic esters are the most important industrial plasticizers and, in practice, almost all commercially available aliphatic or cyclic alcohols are used in the production of phthalates. The standard plasticizer is diisooctyl phthalate (DOP), which represents a compromise between solvent effect, volatility and overall performance. Next to 2-ethylhexanol, isononyl alcohol and straight-chained C_7 to C_9-alcohols are used as standard components for phthalic esters.

Table 7.2 summarizes the properties of various phthalic esters.

Table 7.2: Characteristic properties of phthalic esters

Ester	M	d_{20}^{20}	Distillation range at 6.66 mbar (T_1–T_{95})	Pensky-Martens flash point (°C)
Di-2-ethylhexyl phthalate (DOP)	390	0.982–0.984	230–233	ca. 200
Diisononyl phthalate (DINP)	418	0.976–0.980	244–252	ca. 200
Diisobutyl phthalate (DIBP)	278	1.039–1.042	171–177	ca. 172
Diphthalate of straight-chain C_9–C_{11} alcohols		0.964–0.967	264–279	ca. 200
Diphthalate of straight-chain C_6–C_{10} alcohols		0.973–0.976	235–270	ca. 200

The reaction equation for esterification shows the two stages of this reaction.

$$\text{phthalic anhydride} \xrightarrow{+\,R-OH} \text{monoester (COOR, COOH)} \xrightarrow{+\,R-OH,\ -H_2O} \text{diester (COOR, COOR)}$$

Figure 7.4 depicts a flow diagram (*Hüls/Veba*) for the batchwise production of diisooctyl phthalate.

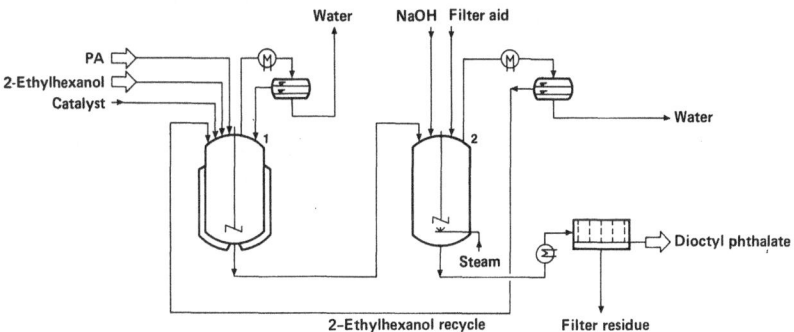

1 Esterification reactor; **2** Finishing reactor

Figure 7.4: Flow diagram for the production of diisooctyl phthalate

The formation of the diester from the monoester can be catalyzed by acids or amphoteric substances. Sulfuric acid, sodium aluminate, as well as titanium and zirconium esters are particularly important catalysts. Esterification can also be essentially autocatalytic, with the phthalic monoester acting as an acid catalyst. However, because of the higher pH-value, this process requires longer esterification times.

The acid-catalysed esterification reaction is carried out in a stirred reactor. Phthalic anhydride and 2-ethylhexanol are fed into the reactor in the molar ratio 1:2.5; the resulting reaction water is distilled off azeotropically, with the alcohol acting as the entrainer. On completion of the reaction, traces of unconverted monoesters are neutralized with alkali; this is followed by separation of the surplus alcohol by steam stripping. A filter aid is added, then the ester is filtered for final refining.

Other alcohols, produced predominantly by the Oxo-process, can be esterified by this method using phthalic anhydride.

85% of the production of phthalic esters is employed as plasticizers for PVC. The remaining 15% is used in auxiliaries for paints, dispersions, cellulose, polystyrene and other polymers.

In addition to phthalic esters, the most important applications for phthalic anhydride are in unsaturated polyester resins, together with alkyd resins produced by reaction with polyhydric alcohols. These polymers are used principally as raw materials in paint manufacture.

7.1.3 Other products from phthalic anhydride

Other uses for phthalic anhydride, but of less importance in terms of quantity, are in the production of pigments, dyes and phthalimide, which is used as a raw material in the production of anthranilic acid, pesticides and pharmaceutical products.

The reaction of chlorobenzene with phthalic anhydride to yield 2-chloroanthraquinone, which is used as an intermediate in the production of indanthrone (see Chapter 11.3.2), was of great industrial importance for the early tar-based dyestuffs industry. Reaction of phthalic anhydride with quinaldine yields quinophthalone, which is the basis of the quinoline yellow dyes.

Quinophthalone

Heating PA with phenol in the presence of concentrated sulfuric acid leads to phenolphthalein (see Chapter 7.1.1), which was discovered by Adolf von Baeyer in 1871.

The phthalocyanines, as copper, cobalt and nickel complexes, discovered in the 1920's, form a versatile group of pigments; they display high color-fastness with blue to green shades, depending on the degree of halogen substitution. Phthalocyanines can be produced by the widely-used synthesis from phthalic anhydride and urea as well as from phthalodinitrile.

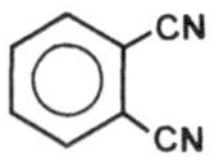

Phthalodinitrile

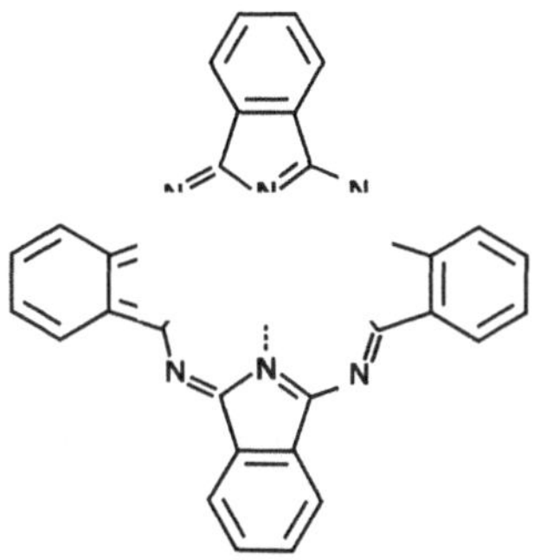

Copper phthalocyanine

The reaction of urea, CuCl and phthalic anhydride is carried out at ca. 200 °C in a process lasting 2 to 3 hours, using ammonium molybdate as a catalyst; nitrobenzene, trichlorobenzene or kerosene can be used as solvents. When phthalodinitrile is used, the reaction is carried out without catalyst at ca. 200 °C as a baking process or in solution. When the crude phthalocyanine has been produced, it is conditioned to suitable crystal modifications, e.g. by dissolving in sulfuric acid and subsequent hydrolysis, or treatment with organic solvents.

Phthalocyanines are used as pigments for green-blue printing inks, plastics and paints.

Phthalimide is obtained by the reaction of phthalic anhydride and ammonia at temperatures from 250 to 280 °C, in yields of 98% and with a purity of 99%.

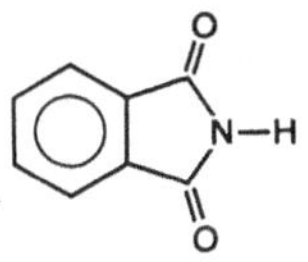

Phthalimide

Possible alternative routes through the reaction with urea or by the oxidative ammonolysis of o-xylene are of limited industrial significance.

The reaction of potassium phthalimide with trichloromethanesulfenyl chloride yields folpet (N-trichloromethylthiophthalimide), a fungicide developed by *Chevron,* which has a world production of some 6,000 tpa. Future production, however, will probably decline, because of local bans (e.g. West Germany 1986).

A further important product from phthalimide is anthranilic acid. Hofmann-degradation of phthalimide with sodium hypochlorite produces the sodium salt of anthranilic acid in an exothermic reaction. This salt is transformed into free anthranilic acid by the reaction with sulfuric acid.

Production of anthranilic acid in Western Europe in 1985 amounted to around 8,000 tpa; it is mainly used in the production of plant protection chemicals such as bentazone (*BASF*), which is obtained by the reaction of anthranilic acid with isopropylamine and subsequent sulfonation with SO_3 and ring closure with $POCl_3$.

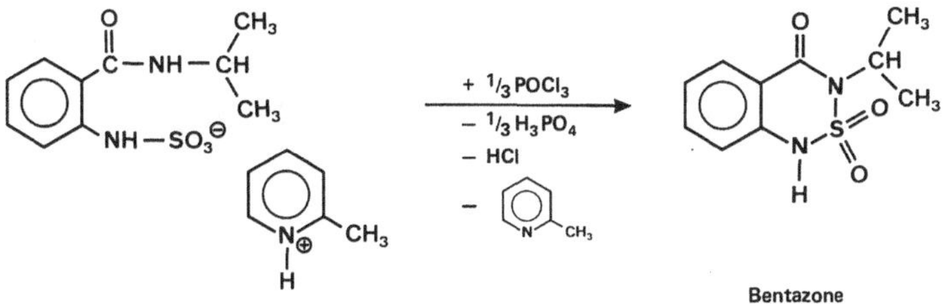

Bentazone

The reaction of anthranilic acid amide with nitrous acid yields 1,2,3-benzotriazin-4-one, from which the insecticide azinphos-methyl (*Bayer*) is made with formaldehyde and O,O-dimethyldithiophosphoric acid.

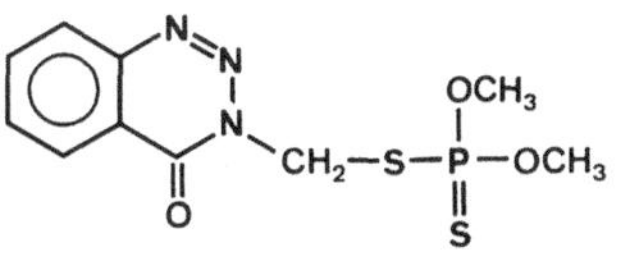

Azinphos–methyl

Thalidomide, marketed under the trade name 'Contergan', is also a phthalimide derivative, which, because of its extremely teratogenic effects, was used for only a short time as a sedative.

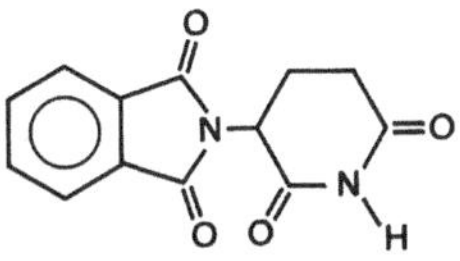

Thalidomide

7.1.4 Nitration of o-xylene

The nitration of o-xylene at around 30 °C and separation of the product isomers by distillation and crystallization yields 3,4-dimethylnitrobenzene, which is transformed into 3,4-xylidine by catalytic reduction.

3,4-Dimethylnitrobenzene

3,4-Xylidine

3,4-Xylidine is used as a raw material in the production of riboflavin (vitamin B$_2$), the synthesis of which is based on the condensation of 3,4-dimethylaniline with D-ribose to give the Schiff's base. By hydrogenation with Raney nickel and coupling the amine with benzenediazonium chloride followed by condensation with barbituric acid, riboflavin is obtained.

Production of vitamin B$_2$ worldwide is around 2,500 tpa.

7.2 m-Xylene and its derivatives

Unlike o-xylene and p-xylene, the application of m-xylene is limited to the production of isophthalic acid, nitrated xylenes and m-xylylenediamine. Production of 3,5-dimethylphenol, which was carried out in earlier times by alkali fusion, has now been replaced by gas-phase aromatization of isophorone (see Chapter 5.3.4.3.2).

7.2.1 Production of isophthalic acid

The oxidation process for m-xylene is based on the liquid-phase oxidation for the production of terephthalic acid. Air oxidation in acetic acid is catalysed by cobalt and manganese salts and bromine and conducted at temperatures ranging from 170 to 230 °C and pressures of 20 to 25 bar. Following the reaction, isophthalic acid is separated by crystallization and the mother liquor is recycled.

Isophthalic acid is produced by only a small number of companies in the USA (*Amoco*), Japan (*Mitsubishi Gas Chemical*) and Italy (*Sisas*). Its main use is in the production of unsaturated polyester resins. It yields polyesters of greater strength and higher resistance to corrosion than those derived from phthalic acid.

Other uses are in the production of alkyd resins, although isophthalic acid has no noticeable advantage over PA. However, during the 1960's, there was a shortage of PA in the USA for a time, and isophthalic acid was used as a substitute, despite its higher cost.

Isophthaloyl chloride is an industrially important derivative of isophthalic acid. It is produced by chlorination ($SOCl_2/Cl_2$) of m-xylene via the intermediate 1,3-bis-(trichloromethyl)-benzene and its reaction with isophthalic acid. Isophthaloylchloride, along with m-phenylenediamine, serves as a monomer component for the production of the high-strength and heat-resistant aramid fiber Nomex (*Du Pont*).

7.2.2 Other products from m-xylene

m-Xylene can be converted into isophthalodinitrile with ammonia and air; isophthalodinitrile can be transformed into m-xylylenediamine by hydrogenation.

Figure 7.5 shows the process developed by *Showa Denko* for ammoxidation and subsequent hydrogenation, which can also be used for m-/p-xylene mixtures.

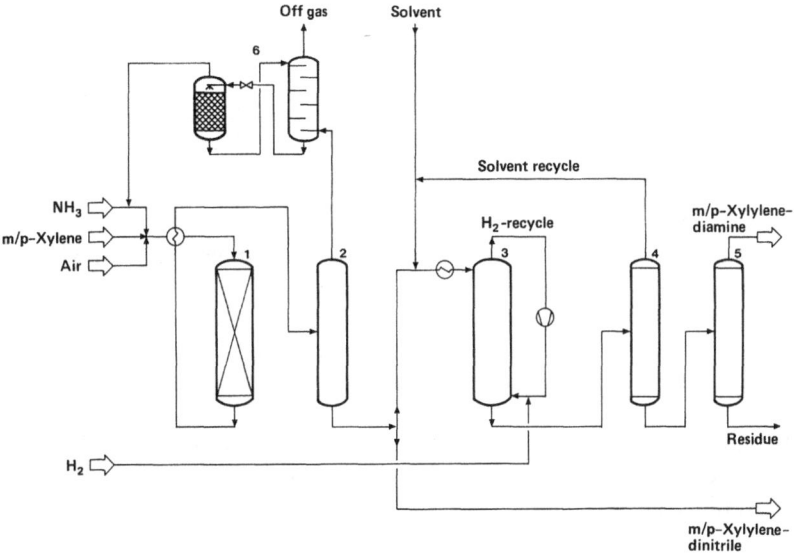

1 Ammoxidation reactor; **2** Product separator; **3** Hydrogenation reactor; **4** Solvent recovery; **5** Distillation column; **6** NH$_3$-recovery

Figure 7.5: Flow diagram of ammoxidation and subsequent hydrogenation of m-/p-xylene

m-Xylylenediamine is used as a curing agent for epoxy resins and as a monomer component in the production of special polyamides.

Chlorination of isophthalodinitrile leads to the fungicide tetrachloroisophthalodinitrile (chlorothalonil), developed by *Diamond Shamrock*.

Chlorothalonil

The high temperature sulfonation of m-xylene followed by alkali fusion yields 3,5-dimethylphenol, but this is no longer an economical process.

Production and uses of xylene derivatives

[Reaction scheme: m-xylene + H_2SO_4 → 2,4-dimethylbenzenesulfonic acid (−H_2O); + NaOH → sodium salt (−H_2O); + 2 NaOH → sodium 3,5-dimethylphenoxide (−Na_2SO_3, −H_2O); + $H^{\oplus}$ → 3,5-dimethylphenol (−$Na^{\oplus}$)]

The nitration of m-xylene produces a mixture of isomers of nitroxylene, which consists of 80 to 85% 2,4-dimethylnitrobenzene and 15 to 20% 2,6-dimethylnitrobenzene; the proportion of 3,5-dimethylnitrobenzene is minor. Separation of this mixture of isomers is effected by distillation and melt crystallization. The corresponding xylidines are produced by reduction of the nitro compounds with Raney nickel.

2,4-Xylidine is used in the preparation of N-acetoacetyl-2,4-xylidide (AAX), an intermediate in pigment manufacture; it is accessible by the reaction of 2,4-xylidine with diketene.

[Reaction scheme: 2,4-xylidine + diketene ($H_2C=C-CH_2$, $O-C=O$) → N-acetoacetyl-2,4-xylidide ($H_3C-C_6H_3(CH_3)-NH-C(O)-CH_2-C(O)-CH_3$)]

One of the most important pigments based on AAX is Pigment Yellow 13 (see Chapter 5.5.3.4).

2,6-Xylidine can be produced by the reduction of 2,6-dimethylnitrobenzene, which is formed during nitration of m- and p-xylenes, and also by ammonolysis of 2,6-xylenol. It is used in the production of plant protection agents such as the fungicide metalaxyl (*Ciba Geigy*) (see page 171) and the herbicide metazachlor (*BASF*).

[Structure of Metazachlor]

Metazachlor

7.3 p-Xylene and its derivatives

7.3.1 Terephthalic acid

In line with the o- and m-xylene isomers, the industrial importance of p-xylene is based virtually exclusively on the production of the respective dicarboxylic acid, namely terephthalic acid. Terephthalic acid and its esters are used in the production of polyester fibers, which, alongside polyamide and acrylic fibers, are the major commercial fibers.

Until the end of World War II, terephthalic acid and its esters were of no great technical significance. Wallace H. Carothers, the discoverer of nylon fiber, had already produced a polyester from phthalic acid and glycol in 1930, but high-melting fibers could not be manufactured from these polymers. This was only achieved when the sterically more bulky phthalic acid was replaced by terephthalic acid. The development of polyester fibers, which followed their discovery by John Rex Whinfield and James Tennant Dickson in 1939 at the *Calico Printer's Association* in Great Britain, and the technical application of the process by *ICI* (1949) and *Du Pont* (1953) caused an upsurge of interest in technologies of p-xylene conversion.

Initially, the development of fibers based on terephthalic acid met with extraordinary difficulties. Terephthalic acid is a white powder, which is virtually insoluble in almost all solvents, does not melt and cannot be distilled. These properties render refining of crude terephthalic acid very intricate. Since high purity of the monomer feedstock material is an absolute necessity for the production of synthetic fibers, an alternative purification route via the dimethyl ester was developed. Dimethyl terephthalate (DMT) is a crystallizable substance which can also be distilled; it is therefore relatively easy to produce in pure form.

The development of methods for the production of terephthalic acid progressed differently in Germany, Japan and the USA. In Germany, the process using the methyl ester was developed, whereas in the USA, efforts were concentrated on purifying the crude terephthalic acid. In Japan, on the other hand, the rearrangement processes (*Henkel* I and II) were developed to large-scale maturity, since in Japan, in particular, the rapid growth of the steel industry meant that PA based on coal-derived naphthalene was available as a raw material in increasing quantities. Up to the early 1960's, the dimethyl ester production route dominated in terms of volume; later there was increasing use – especially in the USA and Japan – of the direct method for refining terephthalic acid.

Initially, oxidation of p-xylene was carried out with dilute nitric acid (30 to 40%) at 165 °C and a pressure of 10 bar; the NO formed was recirculated. The terephthalic acid so produced contained nitrogen compounds and therefore warrented an improvement in its purity.

Nowadays, the production of technical grade terephthalic acid (TPA) is carried out by liquid-phase oxidation of p-xylene with air. Figure 7.6 shows the flow diagram for the *Amoco* process, an advanced version of the *Mid Century/Amoco* method, which has found widest application.

284   Production and uses of xylene derivatives

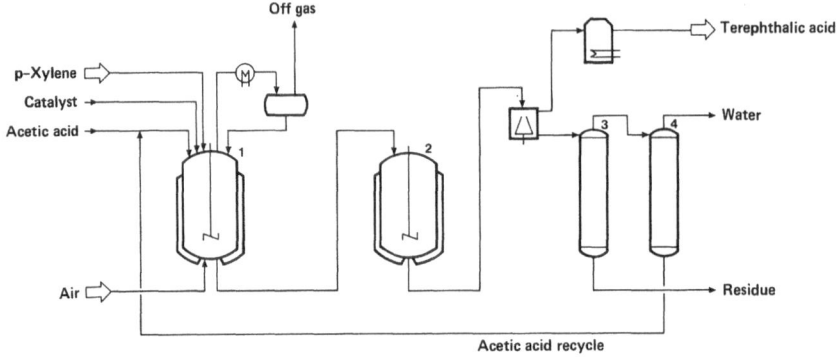

1 Oxidation reactor; 2 Storage vessel; 3 Residue distillation; 4 Dewatering column

Figure 7.6: Flow diagram of the *Amoco* process for the production of terephthalic acid (TPA)

p-Xylene, acetic acid, air and the catalyst (e.g. cobalt acetate, NaBr, CBr$_4$) are fed continuously into the reactor; oxidation occurs at temperatures ranging from 175 to 230 °C and a pressure of 15 to 35 bar. The air is added in excess of the stoichiometric ratio to minimize the formation of by-products. The heat of reaction is removed by vaporizing the acetic acid, which condenses and is fed back into the process. Residence time in the reactor is between 30 minutes and 3 hours, depending on process conditions. Conversion is over 95%, resulting in a yield of around 90 mol%. The reaction mixture is taken to a depressurizing vessel, and the terephthalic acid is recovered by crystallization. The mother liquor is refined by distillation.

The *Eastman-Kodak* process is operated in a similar way to the *Amoco* process, adding acetaldehyde as an oxidation accelerator. A further variant is the *Toray* process, which works with paraldehyde as an oxidation accelerator.

When bromine compounds are used as catalysts, only expensive reactor materials such as Hastelloy or titanium can be used, since the bromine compounds are strongly corrosive.

The crude terephthalic acid (CTA) contains a number of by-products; compounds with only one functional group, such as benzoic acid and methylbenzoic acid, can retard the polymerization process and reduce the degree of polymerization. Other compounds, such as 4-carboxybenzaldehyde, cause discoloration of the crude terephthalic acid.

The main upgrading step in refining pure terephthalic acid (PTA) consists of a catalytic hydrogenation. Crude terephthalic acid is slurried with water and, after being heated to around 250 °C, the slurry is pumped into a hydrogenation reactor, which contains a noble-metal catalyst (e.g. palladium) on a carbon carrier. The liquid-phase hydrogenation removes the impurities which have a tendency to discolor; 4-carboxybenzaldehyde is transformed into p-methylbenzoic acid. Subsequent upgrading is carried out by crystallization.

Figure 7.7 shows the diagram for the *Amoco* refining process.

Production and uses of xylene derivatives   285

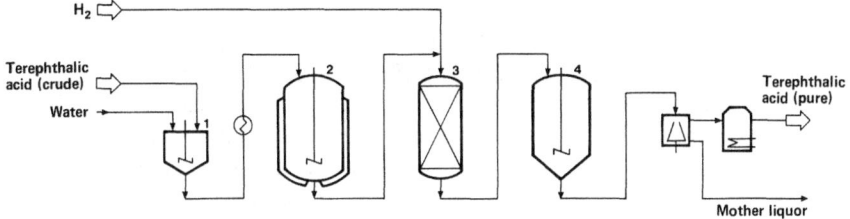

1 Slurry preparation vessel; **2** Dissolving vessel; **3** Hydrogenation reactor; **4** Crystallizer

Figure 7.7: Flow diagram of the purification of terephthalic acid

Figure 7.8 shows the hydrogenation stage of the production plant for pure terephthalic acid (PTA) operated by *Mitsui Petrochemical,* Iwakuni-Ohtake/Japan, which has a capacity of 150,000 tpa.

Figure 7.8: Part of the purification plant for the production of pure terephthalic acid of *Mitsui Petrochemical,* Iwakuni-Ohtake, Japan

To comply with specifications for 'polymer grade' terephthalic acid, the 4-carboxybenzaldehyde content must be less than 25 ppm. A further important quality criterion is the acid number; it must be 675 ± 2 mg KOH/g acid.

Because of the difficulty in producing pure terephthalic acid suitable for polymerization, in the early 1950's processes were developed to produce dimethyl terephthalate. This involves first oxidizing p-xylene to p-methylbenzoic acid, followed by esterification of the first acid group with methanol. The subsequent oxidation of the second methyl group can then be carried out more easily, producing terephthalic acid monomethyl ester, which is converted into the diester by further addition of methanol.

Figure 7.9 shows the flow diagram for the *Dynamit Nobel/ Witten* process.

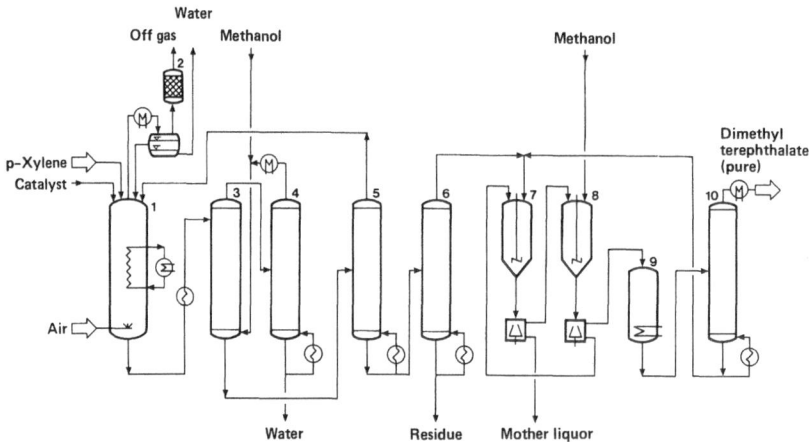

1 Oxidation reactor; **2** Off gas purification; **3** Esterification reactor; **4** Methanol-dewatering column; **5** p-Methylbenzoic acid methyl ester column; **6** Dimethyl terephthalate column; **7** and **8** Crystallizers; **9** Melting pot; **10** Distillation column

Figure 7.9: Flow diagram of the production of terephthalic acid dimethyl ester

In the *Chemische Werke Witten* process, which was further developed by *Dynamit Nobel* and *Hercules*, p-xylene, air and the catalyst are fed continuously into the oxidation reactor, to which recirculated p-methylbenzoic acid methyl ester is also added. Oxidation is effected at a temperature of 140 to 170 °C and a pressure of 4 to 7 bar. The heat of reaction is removed by the vaporization of water and excess p-xylene. The further reaction with methanol is carried out at 200 to 250 °C under slightly raised pressure (20 bar) in the esterification reactor, to keep the reaction mixture in the liquid-phase. The esterification products flow to the crude ester column, where p-methylbenzoic acid methyl ester is separated from the crude dimethyl terephthalate. p-Methylbenzoic acid methyl ester is recycled to the oxidation reactor, where oxidation of the second methyl group occurs. The crude dimethyl terephthalate is purified to 'fiber grade' quality by distillation and crystallization from methanol, and subsequent redistillation in a column with around 30 trays. The yield of dimethyl terephthalate (m.p. 141 °C) is generally about 87 mol%.

In plants of a capacity up to 150,000 tpa, the reactor system is arranged as a three-stage cascade.

In addition to the use of p-xylene as a raw material in the production of terephthalic acid, processes were also operated in the past to produce terephthalic acid from toluene and phthalic anhydride.

In the *Henkel* process (*Henkel* I-process), which was used on a large scale particularly in Japan until the late 1970's, phthalic anhydride was converted into the dipotassium salt of phthalic acid, which isomerized to dipotassium terephthalate under a pressure of carbon dioxide of 10 to 50 bar at 350 to 400 °C.

The *Henkel* II-process was also preferably operated in Japan. In this process, toluene is oxidized with air over cobalt catalysts to yield benzoic acid which is then transformed into potassium benzoate by subsequent neutralization. In the presence of cadmium oxide or zinc oxide, at temperatures of 450 °C and under CO_2 pressure, disproportionation to dipotassium terephthalate occurs; this is then converted into terephthalic acid. Benzene is a by-product of the disproportionation.

Production and uses of xylene derivatives

[Reaction scheme: 2 toluene (CH₃) + 3 O₂ → 2 benzoic acid (COOH) + 2 H₂O; + 2 KOH → 2 potassium benzoate (COOK) − 2 H₂O; Disproportionation → dipotassium terephthalate − benzene]

[Reaction scheme: dipotassium terephthalate + H₂SO₄ → terephthalic acid − K₂SO₄]

Terephthalic acid and dimethyl terephthalate are used almost exclusively in the production of terephthalic acid diglycol ester and other esters, which are transformed into polyester condensation and processed to fibers (e.g. Diolen, Enka; Terylene, ICI) and films.

[Reaction scheme: dimethyl terephthalate + 2 HO−CH₂−CH₂−OH → HO−CH₂−CH₂−O−CO−C₆H₄−CO−O−CH₂−CH₂−OH + 2 CH₃OH]

[Reaction scheme: n HO−CH₂−CH₂−O−CO−C₆H₄−CO−O−CH₂−CH₂−OH → [−CO−C₆H₄−CO−O−CH₂−CH₂−O−]ₙ + n HO−CH₂−CH₂−OH]

Since the transesterification of dimethyl terephthalate (DMT) requires a methanol recovery plant, and the esterification of terephthalic acid (TPA) with ethylene glycol renders higher yields possible, processes to produce TPA are gaining a greater share of the manufacturing capacity, at the expense of DMT.

Table 7.3 shows a compilation of production figures for the most important terephthalic acid (TPA) producer countries.

Table 7.3: Output of major terephthalic acid (TPA) producing countries (1985)

	(1,000 t)
USA	1,200
Mexico	260
Brazil	75
Benelux	80
Italy	85
Great Britain	340
Spain	90
Japan	850
Korea (South)	175
Taiwan	480
West Germany	–
Others	65
Total production in Western world	3,700

By far the largest producer of TPA is *Amoco* (USA), followed by *ICI* (England). To complement the production figures for terephthalic acid (TPA), the following table compiles production figures for dimethyl terephthalate (DMT); this comparison shows the varying regional importance of the production of DMT and TPA.

Table 7.4: Production of dimethyl terephthalate (DMT) (1985)

	(1,000 t)
USA	1,500
Mexico	180
Brazil	60
Benelux	100
France	60
West Germany	610
Italy	115
Spain	60
Japan	330
India	50
Others	635
Total production in Western world	3,700

The most important US-producer of DMT is *Du Pont;* in Western Europe *Dynamit Nobel* operates the largest capacities.

Owing to the importance of terephthalic acid as a large tonnage dicarboxylic acid, there has been no shortage of attempts to find alternative processes to p-xylene oxidation.

Lummus has developed an ammoxidation process, in which p-xylene is transformed into terephthalic acid dinitrile with ammonia on vanadium catalysts. The terephthalic acid dinitrile is then hydrolysed into the free acid.

Another alternative to the production of terephthalic acid is the process developed by *Mitsubishi Gas Chemical,* in which a complex generated from toluene, HF and BF_3 is made to react under pressure with carbon monoxide (Gattermann-Koch). After decomposition of the complex, p-methylbenzaldehyde can be recovered by crystallization and may then be oxidized to terephthalic acid.

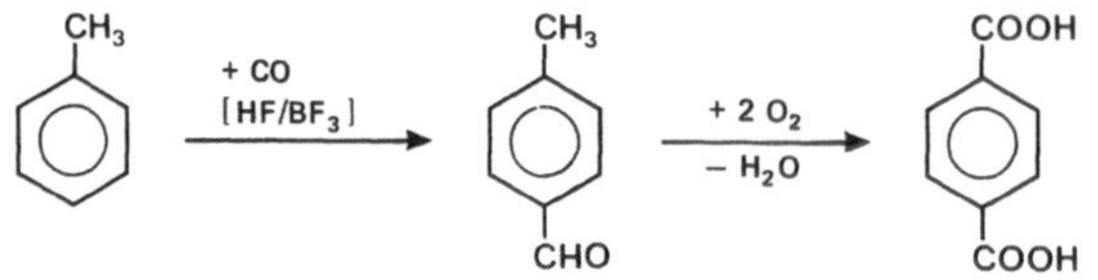

Neither alternative process has yet been able to replace p-xylene oxidation.

Of less importance than polyester in terms of quantity, but with great potential for growth, is the condensation product of terephthaloyl chloride and p-phenylenediamine, which can be spun to high tensile strength fibers (Kevlar, *Du Pont;* Twaron, *Akzo*).

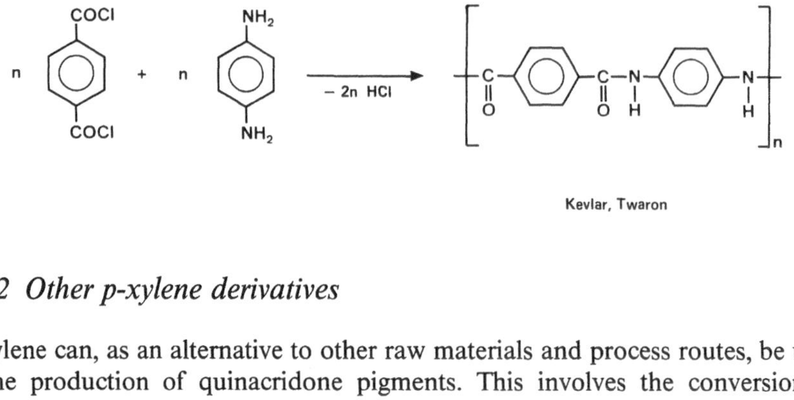

Kevlar, Twaron

7.3.2 Other p-xylene derivatives

p-Xylene can, as an alternative to other raw materials and process routes, be used in the production of quinacridone pigments. This involves the conversion of p-xylene to 2,5-dibromo-1,4-xylene, followed by the oxidation to 2,5-dibromoterephthalic acid. Substitution with arylamines, catalysed by copper acetate, and condensation in an acid medium produces quinacridones such as Pigment Violet 19.

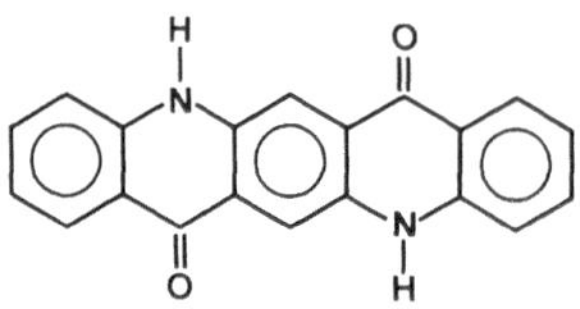

Pigment Violet 19

8 Polyalkylated benzenes – production and uses

Petroleum- and coal-derived heavy gasoline fractions with a boiling range of around 160 to 220 °C contain polymethylated benzenes, such as trimethylbenzenes (pseudocumene, mesitylene and hemimellitene), together with the tetramethylated benzenes durene, isodurene and prehnitene. Indane and indene compounds, penta- and hexamethylbenzene and cumene, are also present in these heavy gasoline fraction. (Cumene is predominantly converted to phenol as described in Chapter 5.2).

Pseudocumene   Mesitylene   Hemimellitene

Durene   Isodurene   Prehnitene

Pentamethylbenzene   Hexamethylbenzene

Table 8.1 shows the composition of C_9-aromatics fractions from pyrolysis gasoline and catalytic reforming.

Table 8.1: Composition of C_9-aromatics from pyrolysis gasoline and catalytic reforming (in percentage)

C_9-aromatics	from pyrolysis benzene	from catalytic reformer
Cumene	4.2	0.6
n-Propylbenzene	12.3	5.2
o-Ethyltoluene	11.8	9.1
m-Ethyltoluene	24.0	17.4
p-Ethyltoluene	11.5	8.6
Mesitylene	5.6	7.4
Pseudocumene	14.6	41.3
Hemimellitene	3.3	8.2
Indane	12.7	2.0

Of the polymethyl benzenes, only pseudocumene, mesitylene and durene have any noteworthy industrial significance.

8.1 Pseudocumene

Pseudocumene (1,2,4-trimethylbenzene) is produced by fractional distillation of the trimethylbenzene cut from the heavy gasoline residues of catalytic reforming; because of the very slight differences in boiling point of their constituents, efficient fractionation is required in distillation columns with up to 300 trays.

The most important derivatives of pseudocumene are trimellitic anhydride and 2,3,5-trimethylaniline, an intermediate in the production of vitamin E (see Chapter 5.3.4.3.2).

In addition, durene can be produced by methylation of pseudocumene.

The oxidation of pseudocumene to trimellitic anhydride can be carried out in the liquid-phase using cobalt/manganese salts and bromine compounds as catalysts in acetic acid (*Amoco* process).

Trimellitic anhydride

A process developed by *Bergbau-Forschung* recommends dilute nitric acid as oxidizing agent (Figure 8.1).

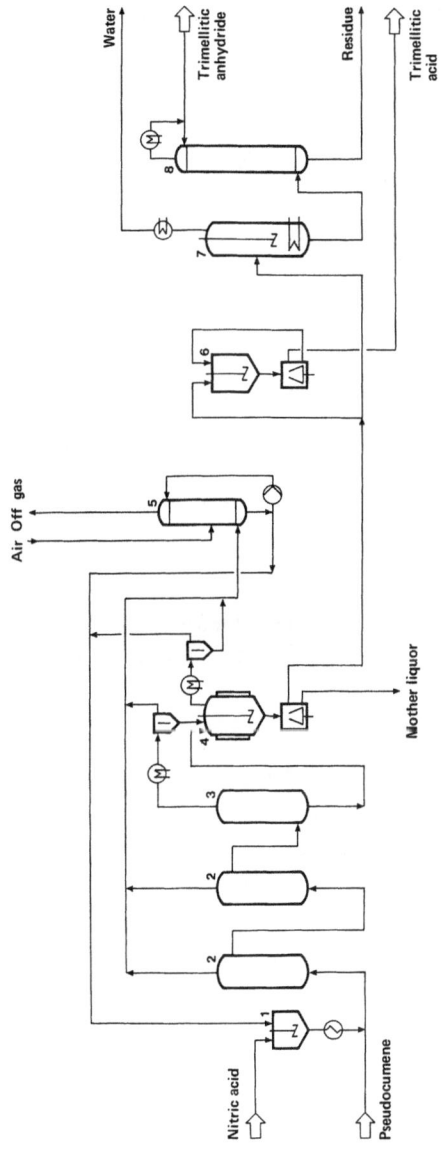

Figure 8.1: Flow diagram of pseudocumene oxidation with dilute nitric acid

1 Mixing vessel; **2** Reactors; **3** Finishing reactor; **4** Crystallizer; **5** Absorption column; **6** Mixing vessel; **7** Dehydration vessel; **8** Distillation column

The reaction is carried out at 170 to 190 °C, under a pressure of 20 bar with 7% nitric acid.

Trimellitic anhydride is used as a raw material for plasticizers, as a component in polyesterimides and as a hardener for epoxy resins. High-temperature resistant and high-strength polyimides are produced by the reaction of trimellitic acid chloride with an aromatic diamine, such as 4,4'-diaminodiphenylmethane (e.g. Torlan, Amoco).

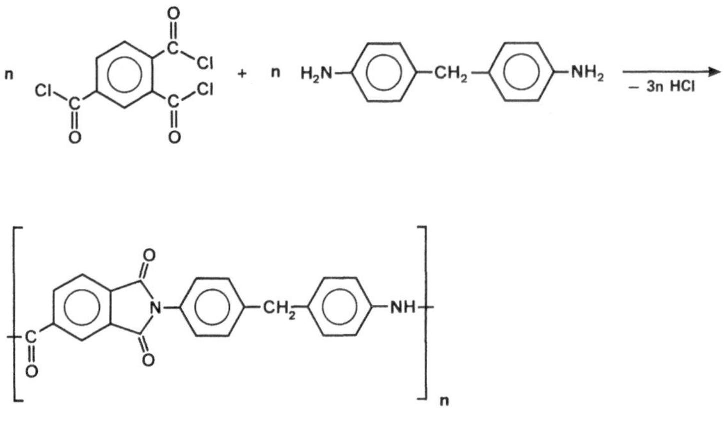

Torlan

Nitration and reduction of pseudocumene yields 2,3,5-trimethylaniline, which is used as a raw material in the production of vitamin E, via the corresponding trimethylhydroquinone.

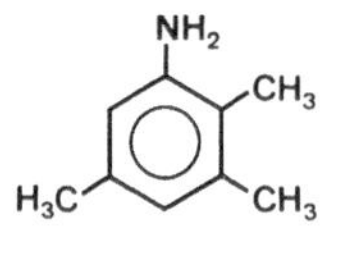

2,3,5-Trimethylaniline

8.2 Mesitylene

Mesitylene can be recovered by distillation of the C_9-aromatics of reformer residues; separation from the co-boiling component, o-ethyltoluene, is, however, extremely intricate. Mesitylene is oxidized in small quantities to trimesic acid (1,3,5-benzenetricarboxylic acid). The oxidation can be carried out in either the gas-phase or the liquid-phase.

The nitration and reduction of mesitylene yields mesidine (2,4,6-trimethylaniline), which is used as an intermediate in the production of dyestuffs.

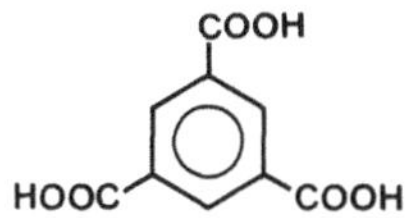

Trimesic acid

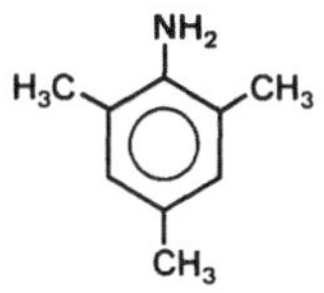

Mesidine

8.3 Durene

Durene is recovered from reformer residues by low-temperature crystallization; recovery by distillation is not possible, because of the virtually identical boiling point of isodurene. Furthermore, durene occurs in gasoline produced by the recently developed *Mobil* process (see Chapter 3.4.1); high concentrations in these methanol-derived gasolines can lead to blockages in the carburetor, a result of the tendency of durene to crystallize.

Durene is predominantly oxidized to pyromellitic dianhydride; this anhydride can also be produced by oxidation of the corresponding triisopropyltoluenes and diisopropylxylenes. The favored process is gas-phase oxidation with V_2O_5 as a catalyst, at temperatures from 400 to 600 °C.

Pyromellitic dianhydride is mainly used in the production of polyimides, e.g. by reaction with an aromatic diamine, such as 4,4'-diaminodiphenyl ether; the high-temperature resistant plastic Kapton (*Du Pont*) is obtained by this method.

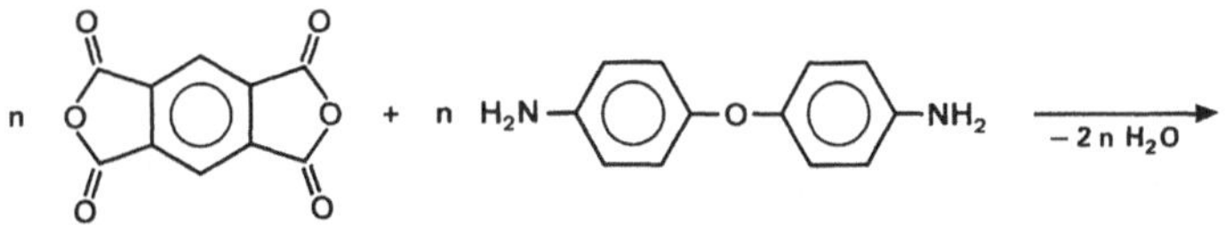

Pyromellitic dianhydride

4,4'-Diaminodiphenyl ether

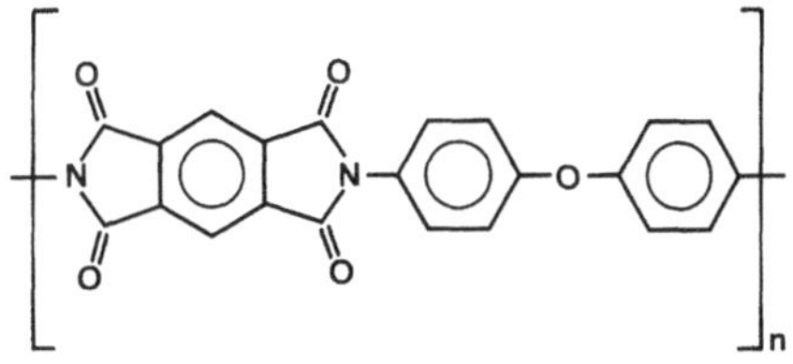

Kapton

8.4 Other cumene derivatives

8.4.1 Nitrocumene and isoproturon

Next to phenol the herbicide isoproturon is one of the few cumene derivatives with any large-scale importance. Nitration of cumene yields 2-/4-nitrocumene in the ratio 35:65. The 4-isomer is recovered by vacuum distillation and is then reduced to cumidine (4-isopropylaniline). (Cumidine can also be produced by ammonolysis of 4-isopropylphenol, which arises as a by-product during the oxidation of 1,4-diisopropylbenzene to produce hydroquinone). Cumidine is reacted with phosgene to give 4-isopropylphenyl isocyanate. Reaction of 4-isopropylphenyl isocyanate with dimethylamine yields isoproturon, which in Western Europe is produced in quantities of around 6,000 tpa.

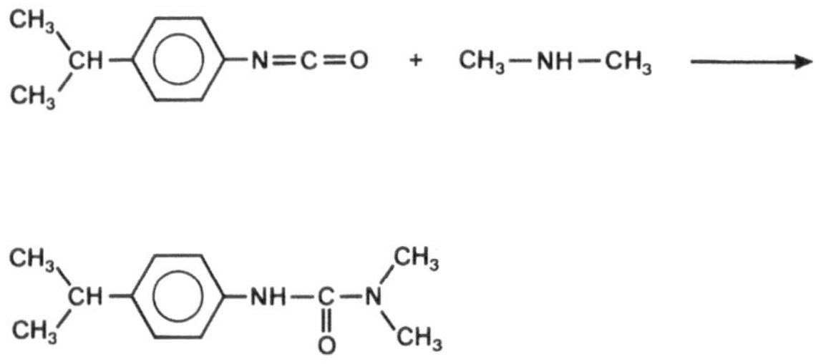

Isoproturon

8.4.2 Cumenesulfonic acid

By the reaction of cumene with sulfuric acid, in a manner similar to toluene sulfonation (see Chapter 6.4), cumenesulfonic acid is obtained, which, after neutralization with sodium hydroxide in aqueous solution, is used extensively as a surfactant (hydrotrope).

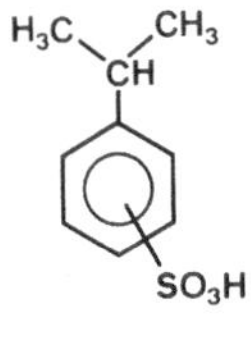

Cumenesulfonic acid

8.5 Indan and indene

Indan can be recovered by distillation of the heavy gasoline from coal tar refining. Indene, which is present in pyrolysis gasoline and in coal tar heavy gasoline, is of particular technical importance. It is polymerized with coumarone and other olefins to produce indene/coumarone resins.

Indan   Indene   Coumarone

Indene/coumarone resins find extensive application, especially in the production of adhesives, as reinforcers and tackifiers in the production of commerical rubber products, and in paint manufacture. Production of indene-derived resins in Western Europe is around 110,000 tpa.

9 Naphthalene – production and uses

9.1 History

Naphthalene was first isolated from coal tar in 1819 by Alexander Garden; it represents about 10% of this complex mixture of aromatics. The industrial importance of naphthalene dates from the latter half of the last century, owing mainly to the ease with which it can be converted into sulfonic acids and thence also to the naphthols, for use as dyestuffs intermediates. However, the first synthetic naphthalene-based dye was a nitro-derivative, Martius Yellow (Acid Yellow 24), which was patented in 1864 by Carl Alexander Martius.

Acid Yellow 24

The development of synthetic dyestuffs provided further impetus for advances in naphthalene chemistry. Towards the end of the last century, its oxidation into phthalic anhydride (PA) attained particular importance as a step in a successful route for the economical synthesis of indigo. In the 20th century, the classic uses of naphthalene have been extended into new areas, for example, the development of the naphthol AS dyes from the parent naphthols, and the production of PA-based plasticizers and pesticides. The most recent developments in the field of industrial naphthalene chemistry concern the production of alkylnaphthalene-derivatives as solvents for use in carbonless copy papers, as well as the manufacture of naphthoquinone for the synthesis of anthraquinone.

There are potential future uses for naphthalene derivatives, especially dihydroxy- and dicarboxy compounds, in the production of liquid-crystal polymers for the manufacture of high-value engineering plastics (see Chapter 10.2).

9.2 Naphthalene recovery

Coal tar is the traditional feedstock for the recovery of naphthalene. The introduction of new cracking processes in the last decades has given rise to the availability of petroleum-based raw materials. In addition to coal-derived naphthalene, petroleum-derived naphthalene has been recovered since 1961 – predominantly in the USA – by dealkylation of the residues of ethylene production and of reforming- and cat-cracker fractions which are rich in aromatics. The worldwide production of petroleum-derived naphthalene has been declining in the last decade, representing presently only around 5% of the total, since adequate supplies of coal-based naphthalene are available.

Because of its high aromaticity, coal tar offers the most suitable source of naphthalene. Next to coal tar in terms of aromaticity comes pyrolysis tar (see Chapter 3.6, Figure 3.60), with an aromaticity of around 0.8. It is possible to recover naphthalene from both of these raw materials, and from naphthalene oil from crude-oil cracking (see Chapter 3.3.2.5.2), by distillation and crystallization.

In other petroleum-derived naphthalene feedstocks of lower aromaticity, such as reforming residues and cat-cracker recycling oils, there is, as a result of their lower aromaticity, a higher proportion of alkylnaphthalenes; thus the naphthalene content is not sufficient for direct recovery in an economic manner. Dealkylation processes must therefore be used to increase the concentration of naphthalene.

9.2.1 Naphthalene from coal tar

In coal tar refining, rectification is used to produce a naphthalene fraction which, depending on the separation efficiency of the naphthalene column, can have a naphthalene content rising to over 90%. Table 9.1 shows the composition of a close-cut coal tar naphthalene fraction.

Table 9.1: Composition of a representative coal tar naphthalene fraction

Indan	0.1 wt%
Indene	1.0 wt%
Methylindenes	2.0 wt%
Naphthalene	89.0 wt%
Thianaphthene	2.5 wt%
Phenols	3.5 wt%
2-Methylnaphthalene	1.3 wt%
1-Methylnaphthalene	0.6 wt%
Total sulfur content	6,000 ppm
Total nitrogen content	750 ppm

In industrial practice, it is not the naphthalene content which is used to denote the grade of the oil, but the crystallization point. Figure 9.1 shows the dependence of the crystallization point on the naphthalene content of coal-based naphthalene fractions.

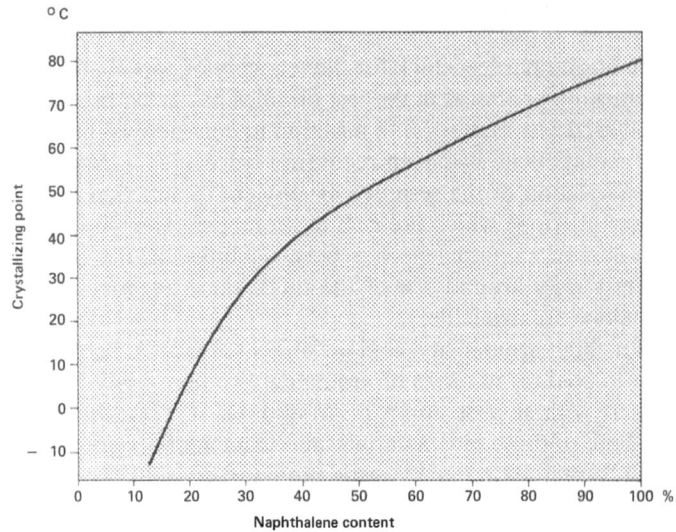

Figure 9.1: Plot of crystallizing point versus naphthalene content of coal tar-based naphthalene fractions

The crystallizing point of a 'technical' naphthalene fraction from the primary distillation may be as high as 76 °C, corresponding to a naphthalene content above 90%. To achieve further purification, it is necessary to separate the naphthalene from other components having similar boiling points. The proposed end-use for the product determines the extent of refining which is required. A distinction is made between 'technical' grades of naphthalene, which have crystallization points of 78.0, 78.5 and 79.0 °C (corresponding to a naphthalene content of 95 to 98%) and 'refined' naphthalene, with a crystallization point of over 79.6 °C, representing a naphthalene content in excess of 98.7%.

For the production of 'technical' naphthalene (crystallization point 78.5 °C) a redistillation of the naphthalene oil using columns with approx. 80 trays and reflux ratios of from 7 to 10:1 is sufficient.

Phenols are generally extracted from the naphthalene oil before distillation, using a 15 to 40% sodium hydroxide solution. Figure 9.2 shows the process diagram for the distillation of a naphthalene oil fraction to recover naphthalene with a crystallization point of 78.5 °C. The pre-column is operated under atmospheric conditions, while the main column operates at a slight overpressure, to achieve optimum heat economy.

Refined naphthalene, with a crystallization point above 79.6 °C, is usually obtained by crystallization or hydrogenation. The choice of process is made on economic grounds, especially the price of hydrogen, and on the required level of sulfur in the refined naphthalene. Crystallization is characterized by high naphthalene yields and favorable energy costs; high yields are also favored by a high concentration of naphthalene in the feedstock. However, the sulfur content of coal-derived naphthalene cannot be reduced below 300–400 ppm by crystallization

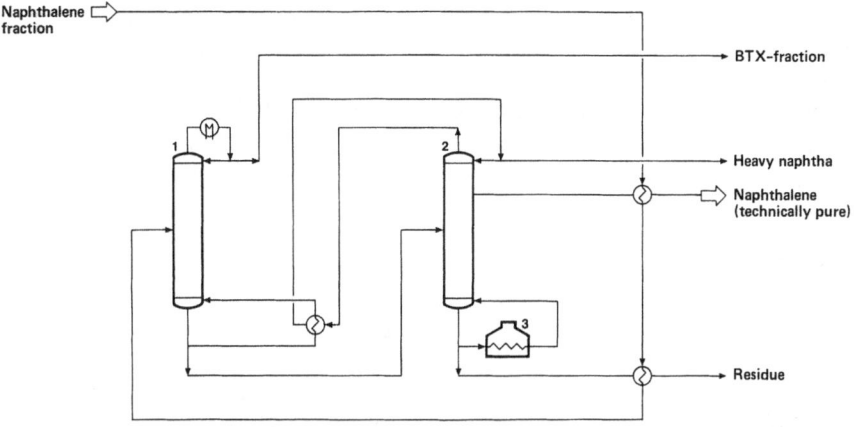

1 Pre-column; **2** Main column; **3** Tubular furnace

Figure 9.2: Production of 'technical' naphthalene (crystallizing point ca. 78.5 °C) by distillation

processes within the framework of reasonable economics. A sulfur content below 100 ppm may be obtained by hydrorefining, but this process is associated with the disadvantage of loss of naphthalene through the formation of 1,2,3,4-tetrahydronaphthalene (tetralin) and pyrolysis products such as gases and pitch-like residues resulting from the application of elevated temperatures.

Figure 9.3 shows the flow-sheet for the hydrogenation of coal tar-derived naphthalene (Unionfining); the naphthalene product can be further purified by crystallization or distillation.

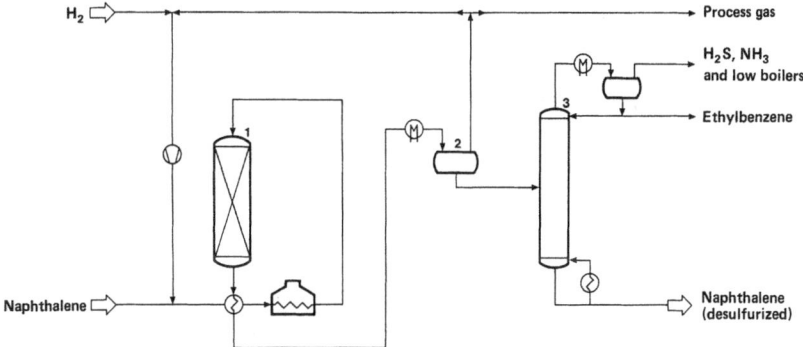

1 Reactor; **2** Gas separator; **3** Stripping column

Figure 9.3: Flow diagram for the hydrogenation of coal-derived naphthalene (Unionfining)

Hydrogenation is carried out at 400 °C under a pressure of about 14 bar with cobalt/molybdenum catalysts. Thianaphthene is thereby converted into H_2S and ethylbenzene; other naphthalene co-boiling materials are broken down to lower-boiling hydrocarbons, which can be separated by distillation.

Naphthalene purified by hydrogenation has a sulfur content of 10 to 100 ppm. The crystallization point ranges between 77.5 and 79 °C. Tetralin is present as a by-product at a concentration of ca. 1%.

Hydrorefining of naphthalene is especially used in the production of low-sulfur naphthalene for phthalic anhydride production by the fluid-bed process, since sulfur compounds can poison the fluid-bed catalyst.

The recovery of naphthalene by crystallization has grown in importance, as a result of the advances in crystallization technologies and increased fuel costs. From the energy point of view, the marked advantage of crystallization consists in the fact that crystallization is carried out in the temperature range between 20 and 95 °C. A further asset is that the latent heat of fusion of naphthalene is lower than the latent heat of vaporization by a factor of 2.5.

The main aim of naphthalene crystallization is the separation of thianaphthene, which, with a boiling point of 219.9 °C, boils just 1.9 °C higher than naphthalene. However, the large difference in melting points (48 °C) facilitates separation to be carried out by crystallization, although this is complicated by the formation of mixed crystals of naphthalene and thianaphthene, rendering it necessary to carry out the process in several stages.

Figure 9.4 shows the naphthalene/thianaphthene phase diagram.

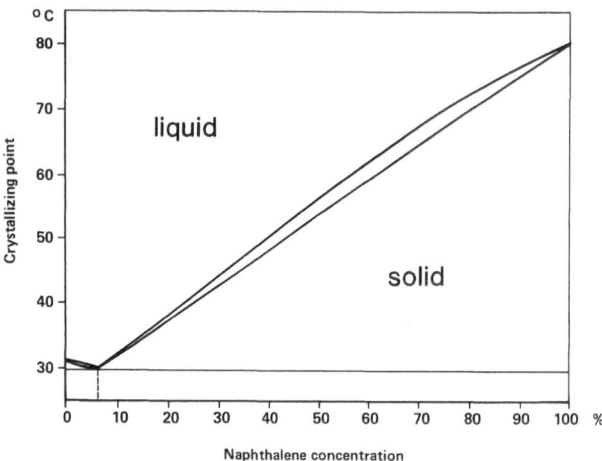

Figure 9.4: Phase diagram for naphthalene/thianaphthene

The *Sulzer-MWB* and the Brodie crystallization processes are most commonly used to recover naphthalene by crystallization. Figure 9.5 shows the *Sulzer-MWB* flow sheet, which is operated in a modified form in plants with a naphthalene production of up to 60,000 tpa.

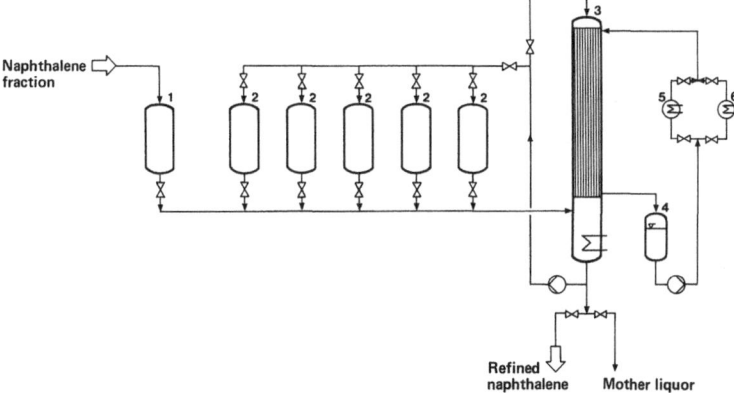

1 Feed tank; **2** Storage vessels (intermediate fractions); **3** Multitubular crystallizer; **4** Recycle water vessel; **5** Cooling system; **6** Heating system

Figure 9.5: Flow diagram of the *Sulzer-MWB* process for the production of naphthalene by crystallization from the melt

The naphthalene fractions produced by the crystallization process are stored in tanks and fed alternately into the crystallizer. The crystallizer contains around 1,100 cooling tubes (diameter: 25 mm), through which the naphthalene fraction trickles downwards in turbulent flow and partly crystallizes out on the tube walls. The remaining melt is recycled and pumped into a storage tank at the end of the crystallization process. The crystals which have been deposited on the tube walls are then partly melted for further purification. Following the removal of the drained liquid, the purified naphthalene is melted. Four to six crystallization stages are required to obtain refined naphthalene with a crystallization point of 80 °C, depending on the quality of the feedstock. The yield is typically between 88 and 94%, depending on the concentration of the feedstock fraction.

Figure 9.6 shows the *Rütgerswerke* naphthalene crystallization plant at Castrop-Rauxel/West Germany, which has a capacity of 60,000 tpa and applies a modified *Sulzer-MWB* process.

304 Naphthalene – production and uses

Figure 9.6: *Rütgerswerke*'s crystallization plant at Castrop-Rauxel/West Germany for the production of refined naphthalene

The continuous Brodie crystallizer (Figure 9.7) consists of three horizontal, scraped-wall chillers, in cascade formation, and one vertical purification column. The crystals are produced in the upper part of the crystallizer (recovery section); they then pass through the middle enrichment zone (refining section) and arrive at the base of the apparatus as molten product. Brodie crystallization is applied for the purification of naphthalene, for example, by *Nippon Steel Chemical,* Tobata/ Japan.

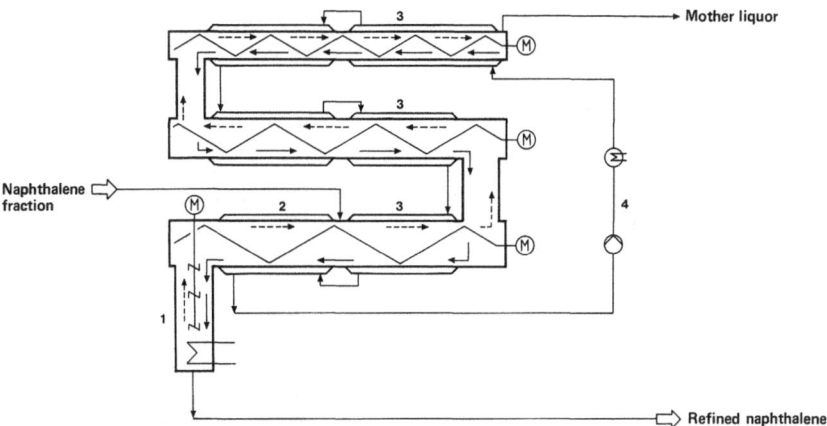

1 Purifying section; **2** Refining section; **3** Recovery section; **4** Cooling and heating system

Figure 9.7: Flow diagram of the Brodie crystallizer

Table 9.2 shows the comparative quality of the feedstock fraction and the pure naphthalene recovered in the Brodie crystallizer.

Table 9.2: Composition of feedstock material and refined naphthalene recovered by the Brodie crystallizer

	Feedstock	Refined naphthalene
Crystallization point, °C	67.0	80.2
Naphthalene, %	71.0	99.9
Thianaphthene, %	2.2	0.1
1-Methylnaphthalene, %	0.5	Traces
2-Methylnaphthalene, %	1.1	Traces
Sulfur, ppm	5500	310
Nitrogen, ppm	840	10

9.2.2 Naphthalene from petroleum-derived raw materials

Naphthalene is present, in varying amounts, in all petroleum-derived pyrolysis products which have been exposed to a temperature of over 500 °C or to catalytic processes under relatively severe conditions. It can thus be found in the appropriate distillation fractions of the liquid products of steam cracking for the production of ethylene, in by-products of crude oil cracking and in residues of catalytic gasoline reforming as well as catalytic cracking of gas oils. The extraction of kerosene fractions provides a further source of naphthalene, after the separation of aliphatics.

Table 9.3 compares the typical composition of petroleum-derived naphthalene fractions in the boiling range 210 to 295 °C with a coal tar-derived naphthalene fraction.

Table 9.3: Composition of fractions containing naphthalene

	Reformer residue	Light catal. cycle oil	Pyrolysis-tar fraction	Broad-cut coal-tar naphthalene fraction
Total aromatics content, %wt	90–95	45–65	70–95	95-100
Composition of aromatics, %wt				
Alkylbenzenes	20	25	20	5
Indanes and Tetralins	15	25	10	5
Alkylindenes	2	7	18	3
Naphthalene and Alkylnaphthalenes	55	35	45	75
Biphenyls and Acenaphthenes	6	6	5	10
Anthracene and Phenanthrene	2	2	2	2

The naphthalene content is relatively high in the pyrolysis-tar naphthalene oil fraction from steam cracking of naphtha. Naphthalene can be recovered from this fraction by distillation and crystallization. The naphthalene content of a close-cut fraction from pyrolysis tar can reach 80%, with methylindenes as common co-products which must be separated by further refining.

Table 9.4 shows the composition of a naphthalene fraction with a crystallization point of 69.5 °C, recovered from pyrolysis tar by distillation.

Table 9.4: Composition of a naphthalene fraction from pyrolysis tar

Indene	0.7%
Methylindenes	17.0%
Naphthalene	78.5%
2-Methylnaphthalene	1.2%
1-Methylnaphthalene	0.3%
Dimethylnaphthalenes	0.15%
Sulfur content	500 ppm

If the naphthalene content of petroleum-derived naphthalene fractions is not sufficient for direct recovery, naphthalene enrichment is carried out by dealkylation of methylnaphthalenes, by analogy with the production of benzene by dealkylation of toluene.

Figure 9.8 shows the *Union Oil* (CA) Unidak process for the manufacture of naphthalene from reformer residues.

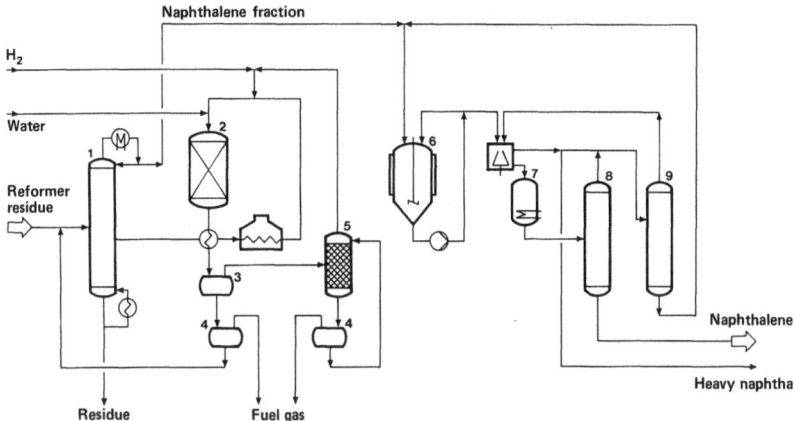

1 Naphthalene column; **2** Dealkylation reactor; **3** High-pressure gas separator; **4** Low-pressure gas separator; **5** Methane scrubber; **6** Crystallizer; **7** Melting vessel; **8** Stripping column; **9** Solvent recovery column

Figure 9.8: Flow diagram for the hydrodealkylation of alkylnaphthalenes

In the first process stage, the naphthalene homologs are further concentrated in a pre-column. Dealkylation of alkylnaphthalenes takes place at 35 bar and at a temperature of 540 to 620 °C. The molar ratio of hydrogen to feedstock material is around 4–10:1. The conversion is restricted to about 60%, in order to limit the hydrogenation of naphthalene.

The crude naphthalene fraction is then further purified by crystallization. The largest producer of petroleum-derived naphthalene is *Texaco*, Delaware City/USA.

Table 9.5 shows the most important naphthalene-producer countries.

Table 9.5: Naphthalene production (1985)

	(1,000 t)
France	30
Italy	15
West Germany	120
Spain	10
Japan	175
Korea (South)	20
USA	170
Canada	20
Poland	50
USSR	140
Others	200
Total production	950

308 Naphthalene – production and uses

The largest producers of refined naphthalene are *Rütgerswerke* (West Germany), *Nippon Steel Chemical* (Japan), *CdF-Chimie/HGD* (France) and *Recordchem* (Canada).

The main uses of naphthalene in Western Europe, Japan and the USA are shown in Figure 9.9.

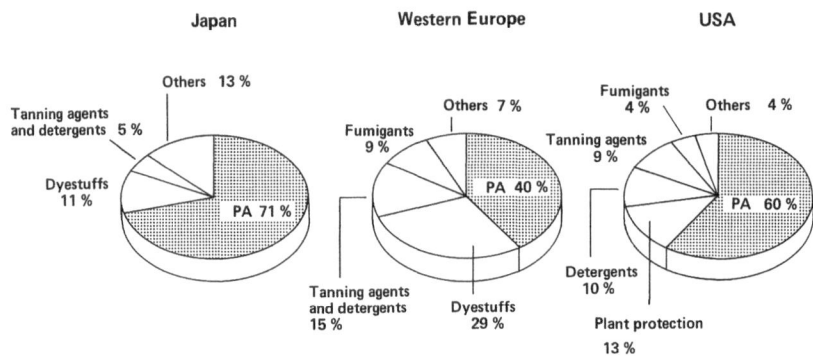

Figure 9.9: Uses of naphthalene in Japan, USA and Western Europe (1985)

Although in Japan and the USA the main use of naphthalene is the production of phthalic anhydride, in Western Europe naphthalene is used in large quantities for the production of dyestuffs and tanning agents, as well as for surface-active compounds.

Refined coal-derived naphthalene (crystallization point above 80 °C), produced by crystallization or sublimation, has a worldwide application for some 30,000 t as fumigants (moth balls). In order to obtain color-stability, this grade of naphthalene requires a particularly low content of unsaturated co-products (indenes).

Although formerly commonly used as solvents, transformer oils and chemical intermediates, the chloronaphthalenes are now of little commercial significance, because of their toxicity and environmental persistence.

9.3 Naphthalene derivatives

Naphthalene chemistry differs fundamentally from benzene chemistry in that two isomers can be produced by monosubstitution, facilitating the formation of by-products. Furthermore, in comparison with benzene, naphthalene is more reactive in electrophilic substitution, so that this type of reaction can be carried out under relatively mild conditions. Naphthalene chemistry is therefore characterized by especially-optimized reaction procedures and complex refining processes. Increased reactivity and substitution capacity offer advantages which can be used, for example, in dyestuffs production, where the naphthalene moiety serves as a carrier for a variety of auxochromic groups.

9.3.1 Production of phthalic anhydride (PA) from naphthalene

The main application for naphthalene is traditionally the production of phthalic anhydride. The oxidation of naphthalene to phthalic anhydride has undergone noticeable changes in recent years. Important stages in this development were mercury catalysis and the introduction of fixed-bed and fluidized-bed processes (see Chapter 7.1.1). Since there are no basic differences between the gas-phase oxidation of naphthalene and of o-xylene to produce phthalic anhydride, it is possible to use one plant with suitable modifications for both feedstock materials (dual-feed plants).

Since naphthalene and o-xylene have different boiling points, when using naphthalene it is necessary to install a vaporizer. In addition, account must be taken of the different explosive limits, which range from 0.9 vol% (45 g naphthalene/Nm^3) to 5.9 vol% (320 g naphthalene/Nm^3) for naphthalene and from 1.0 vol% (44 g o-xylene/Nm^3) to 7.6 vol% (335 g o-xylene/Nm^3) for o-xylene. The ignition temperature of naphthalene is 520 °C and thus substantially above the normal operating range for low-temperature processes; the enthalpy of reaction (ΔH_{298} = -1790 kJ/mol) is of the same order of magnitude as that for o-xylene oxidation.

The development of fluidized-bed technology was of particular importance for chemical process engineering. It was first used to produce phthalic anhydride by oxidation of naphthalene in the USA in 1944/45, using the fluidized-bed technology introduced by *Standard Oil* (CA) in Baton Rouge/Louisiana in 1941 for the catalytic cracking of petroleum fractions. The advantage of fluidized-bed technology consists in the uniformity of the reaction temperature, which allows better control of reaction conditions.

In contrast to the fixed-bed reactor, where the contact time in high-temperature processes is 0.2 sec. and in low-temperature processes 3 sec., the contact time in the fluidized bed is 10 to 20 sec. A further difference is the lower air/naphthalene ratio, which is 10:1 to 12:1 in the fluidized bed, whereas in fixed-bed processes values of 15:1 to 30:1 are usual.

Figure 9.10 shows the flow diagram for the *Sherwin-Williams/Badger* process for the production of PA from naphthalene, using the fluidized-bed process.

In contrast to the fixed-bed process, naphthalene is fed to the reactor in liquid form at the base of the catalyst bed; spontaneous evaporation then takes place by contact with the hot catalyst particles. Air is introduced below the naphthalene feed through a distribution plate. The temperature in the reaction zone is between 340 and 380 °C and the operating pressure is around 4 bar.

Any catalyst which becomes entrained in the reaction gases is separated by special filters and returned to the reactor. Up to 60% of the phthalic anhydride can be separated from the reaction product in liquid form by cooling. The remainder is condensed as solid. The uncondensable portions of the reaction product are partially removed by washing; the tail gases are incinerated.

The crude phthalic anhydride is purified by vacuum distillation. The yield is around 87-88 wt%, based on the naphthalene feed, and has a crystallizing point of 130.5-130.8 °C corresponding to a purity of 96-97%.

Vanadium pentoxide, on silica gel carriers, is used as a catalyst in the fluidized-bed process. Potassium hydrogen sulfate is added as a modifier, especially to pre-

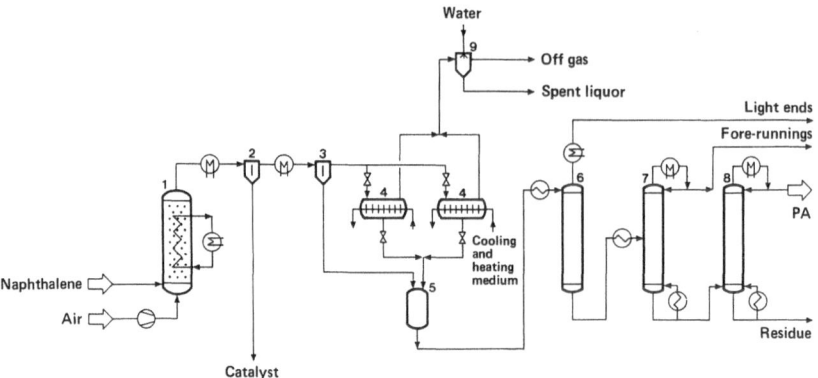

1 Fluid-bed reactor; **2** Catalyst separator; **3** Liquid-PA condenser; **4** Desublimation switch condenser; **5** PA storage tank; **6** Thermal pretreatment; **7** Light-ends column; **8** PA-column; **9** Gas purification

Figure 9.10: Fluidized-bed process for the production of PA from naphthalene

vent overoxidation of the naphthalene by reducing the catalyst activity. The particle size of the catalyst in the fluidized bed is between 40 and 300 µm.

The first catalyst used for fluidized-bed processes in the 1940's had poor resistance to sulfur, so that naphthalene had to be desulfurized beforehand. The fluidized-bed process was therefore most commonly used to oxidize petroleum-derived naphthalene from reformer residues.

Since that time, sulfur-resistant catalysts have been developed. However, the importance of the fluidized-bed phthalic anhydride process is declining; since the proportion of raw material costs is about 70% of total production costs, the feedstocks o-xylene and coal tar naphthalene can be more cost-effectively oxidized in fixed-bed reactors.

Applications for phthalic anhydride are described in Chapter 7.1.2.

9.3.2 Production and uses of naphthoquinone

The oxidation of naphthalene to naphthoquinone is closely related to its oxidation for the production of phthalic anhydride, as naphthoquinone is formed as an undesirable by-product in the latter process. The use for naphthoquinone is almost entirely limited to its reaction with butadiene to produce anthraquinone/tetrahydroanthraquinone.

$$\text{naphthalene} \xrightarrow[-H_2O]{+3/2\ O_2} \text{naphthoquinone}$$

The large-scale production of naphthoquinone by gas-phase oxidation is fundamentally more complex than the production of phthalic anhydride for two main reasons. First, the process conditions and the selected catalyst must assure the minimized formation of co-produced phthalic anhydride and second, as a result, the purification of the crude product is more complex and involves separation of unconverted naphthalene, phthalic anhydride and naphthoquinone, and the purification of the waste gases.

By analogy with the production of phthalic anhydride, the reaction is performed in a multitubular reactor. The catalyst consists of vanadium pentoxide on a silica support, modified by potassium and ammonium sulfates or potassium hydrogen sulfate.

The reaction is carried out at a temperature of 400 °C and a concentration of around 40 g naphthalene/Nm³ air. Based on the naphthalene feed, at 94% conversion about 50% naphthoquinone and about 50% phthalic anhydride are obtained.

To purify the crude naphthoquinone, the phthalic anhydride is hydrolyzed to phthalic acid and extracted by washing with water. The naphthoquinone is then extracted with an aromatic solvent such as toluene, and further purified by an alkali wash to remove residual phthalic acid.

The PA co-product is recovered after hydrolysis by crystallization as phthalic acid and then converted to phthalic anhydride by dehydration.

Next to the *Kawasaki Kasei* process, which is used commercially and has a high naphthalene conversion rate, mention should be made especially of the *Bayer* process, which involves only partial conversion of the naphthalene through oxidation, and further reaction of the crude oxidation product with butadiene.

Naphthoquinone is converted into tetrahydroanthraquinone by treatment with excess butadiene (molar ratio 1:3) in a Diels-Alder reaction, typically at 120 °C and at a pressure of 20 bar.

$$\text{naphthoquinone} + \text{butadiene} \longrightarrow$$

$$\text{tetrahydroanthraquinone} \xrightarrow[-2\ H_2O]{+O_2} \text{Anthraquinone}$$

312 Naphthalene – production and uses

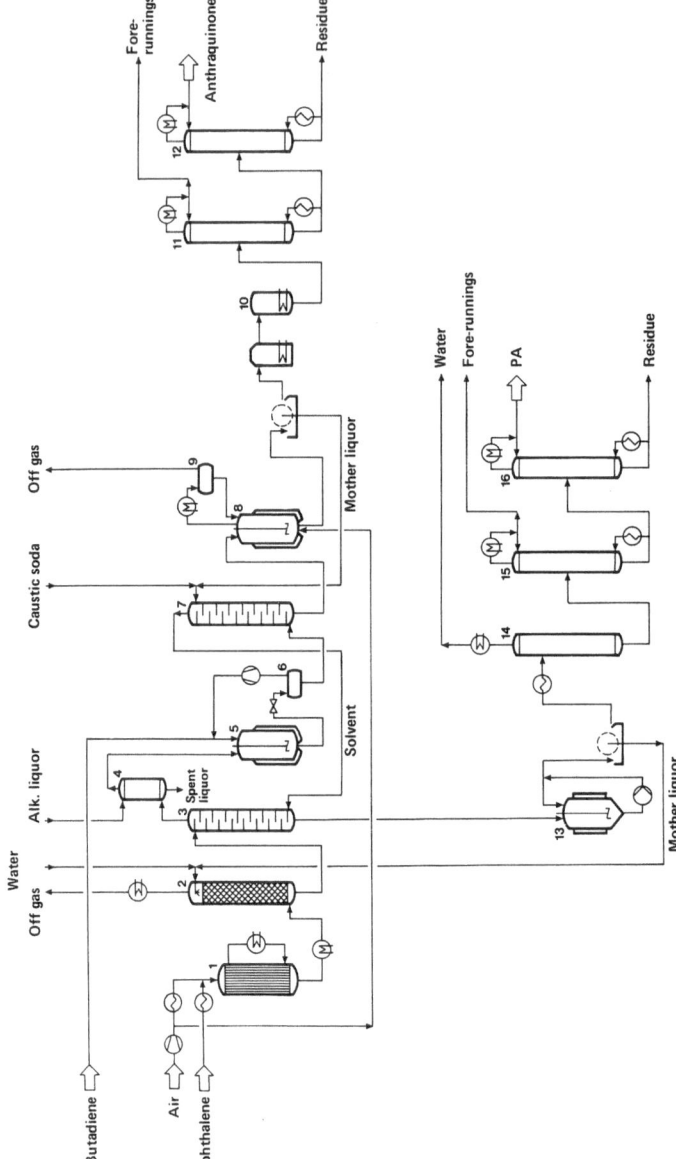

Figure 9.11: Flow diagram of a continuous synthesis of naphthoquinone and further reaction with butadiene to anthraquinone

1 Oxidation reactor; **2** Purification column; **3** Extraction column; **4** Post treatment; **5** Diels-Alder reactor; **6** Pressure let-down vessel; **7** Extraction column; **8** Oxidation reactor; **9** Gas separator; **10** Melting pot; **11** and **12** Distillation columns; **13** Crystallizer; **14** Dehydration column; **15** and **16** Distillation columns

Tetrahydroanthraquinone is used as a catalyst in wood pulping (see Chapter 11.6) and for the production of anthraquinone by dehydrogenation.

Oxidative dehydrogenation of tetrahydroanthraquinone to anthraquinone takes place in the liquid phase with a conversion of almost 100%; selectivity is also practically quantitative. The crude anthraquinone is removed from the base of the oxidation reactor and further purified by distillation.

A flow diagram of continuous naphthoquinone production with a high conversion of naphthalene and subsequent anthraquinone production is depicted in Figure 9.11.

Naphthoquinone/tetrahydroanthraquinone production is presently carried out exclusively by *Kawasaki Kasei* (*Mitsubishi Chemical Ind.*) in a 7,000 tpa plant, using a continuous process.

The naphthoquinone process complements the other production routes to anthraquinone, described in Chapter 11.

9.3.3 Production and uses of naphthols

The most important naphthalene-based compounds which retain the two-ring system are the two naphthol isomers, commonly used as intermediates for azo dyes, pigments and insecticides.

9.3.3.1 1-Naphthol and its derivatives

In comparison with the 2-isomer, 1-naphthol is of relatively little importance as a dyestuff component. Its main use is in the production of the insecticide 1-naphthyl N-methylcarbamate (carbaryl), which was one of the most important organic pesticides used in the USA in the 1970's.

The following syntheses can be used for the production of 1-naphthol:

1. Alkali fusion of naphthalene-1-sulfonic acid

2. Hydrolysis of 1-chloronaphthalene

3. Pressure hydrolysis of 1-naphthylamine

[Naphthalene-NH$_2$] + H$_2$O → [Naphthalene-OH], −NH$_3$

4. Dehydrogenation of 1-tetralone to 1-naphthol

[1-tetralone] −H$_2$ → [1-naphthol]

The most important process to obtain 1-naphthol has been developed by *Union Carbide* via tetralin to 1-naphthol (see Chapter 9.3.6). Processes applying alkali fusion of naphthalene-1-sulfonic acid, manufactured at 180 °C by selective sulfonation, and by pressure hydrolysis of 1-naphthylamine at 185 °C with 20% sulfuric acid are of less importance in terms of quantities.

1-Naphthol production in the Western world is around 15,000 tpa, with the USA as the largest producer.

The main derivative of 1-naphthol is carbaryl, which is produced by the reaction of 1-naphthol with methyl isocyanate. Methyl isocyanate, a toxic liquid boiling at 38 °C, is obtained, among other methods, by the reaction of phosgene with methylamine. Because of its toxicity, it should only be stored for short periods to avoid possible risks during storage (Bhopal accident).

CH_3-NH_2 + $COCl_2$ −HCl → $CH_3-NH-C(=O)Cl$ −HCl → $CH_3-N=C=O$

[1-naphthol] + $CH_3-N=C=O$ → [Carbaryl]

Carbaryl

Another important derivative of 1-naphthol is propranolol, a β-receptor blocker developed by *ICI* and with a worldwide production of around 500 tpa. Synthesis takes place by the reaction of 1-naphthol with epichlorohydrin, followed by substitution of the chlorine in 1-chloro-3-(1-naphthoxy)-2-propanol with isopropylamine to yield propranolol.

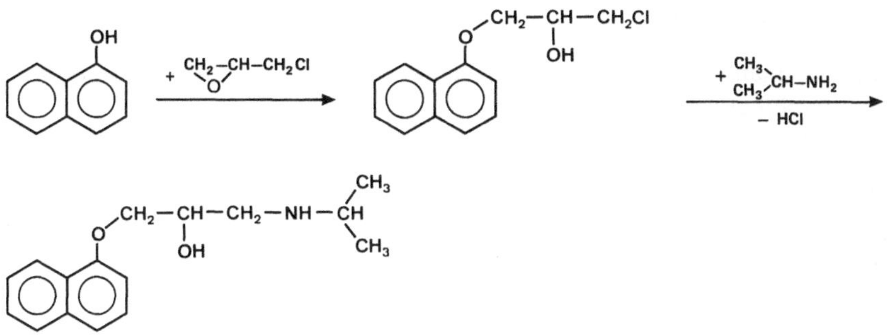

[Propranolol synthesis scheme]

Propranolol

Sulfonation of 1-naphthol with sulfuric acid at 65 °C in nitrobenzene leads to 1-hydroxynaphthalene-4-sulfonic acid (Nevile-Winther acid), which serves as a coupling component for azo dyestuffs such as Acid Orange 19.

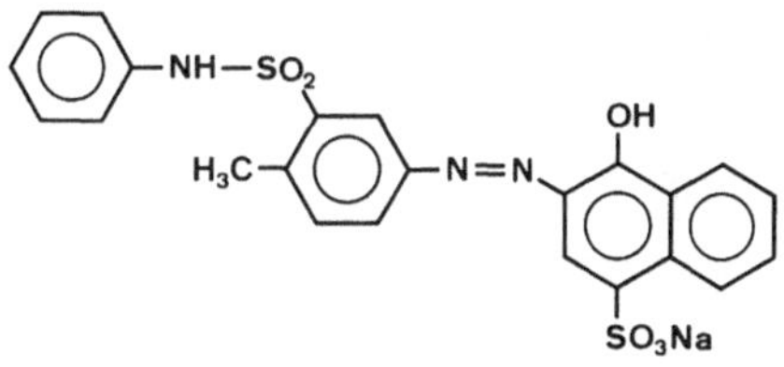

Acid Orange 19

Nevile-Winther acid can also be obtained by a reversed Bucherer reaction from 1-aminonaphthalene-4-sulfonic acid with aqueous sodium hydrogen sulfite solution and SO_2 at 95 °C and a pressure of 2.5 bar.

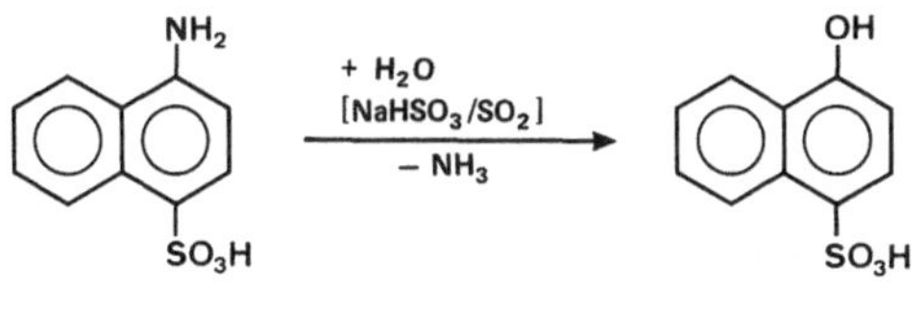

Nevile-Winther acid

316   Naphthalene – production and uses

9.3.3.2 2-Naphthol and its derivatives

In terms of quantity and scope of application, 2-naphthol is the more important of the two naphthol isomers.

Two routes have gained industrial prominence to produce 2-naphthol, namely alkali fusion of sodium naphthalene-2-sulfonate and the oxidative cleavage of 2-isopropylnaphthalene.

Figure 9.12 shows the process diagram for the alkali fusion process which is used exclusively today to make 2-naphthol.

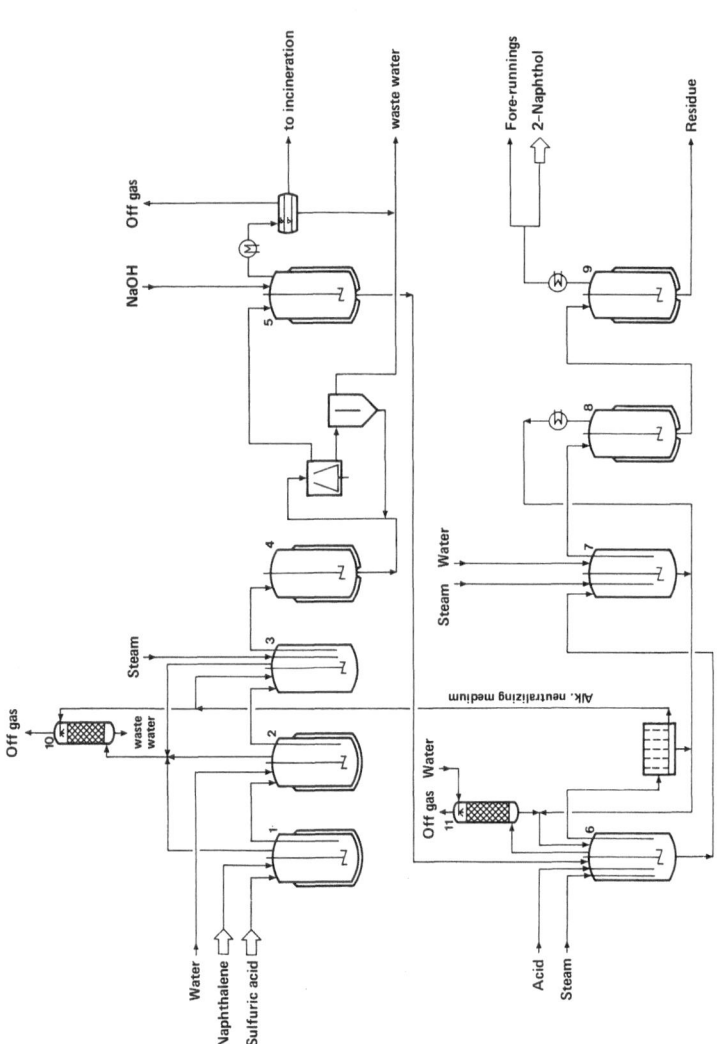

1 Sulfonation reactor; **2** Hydrolysis reactor; **3** Neutralizing vessel; **4** Crystallizer; **5** Melt reactor; **6** Dissolving and neutralizing vessel; **7** Washing vessel; **8** Dewatering reactor; **9** Discontinuous vacuum distillation; **10** and **11** Gas purification

Figure 9.12: Production of 2-naphthol by alkali fusion

In the first stage of the process, naphthalene is transformed into naphthalene-2-sulfonic acid, using sulfuric acid under thermodynamically-controlled reaction conditions. This is followed by conversion of the naphthalene-2-sulfonic acid to the sodium salt with approximately 2.5 mol sodium hydroxide at 300 to 320 °C, with a yield of some 75%.

Purification of the crude 2-naphthol is carried out by vacuum distillation. The high sulfate content in the primary effluent from 2-naphthol production is greatly reduced in modern production plants by the recovery of sodium sulfate.

American Cyanamid operated a plant with a capacity of 14,000 tpa 2-naphthol, for some years prior to 1982 for the oxidation of 2-isopropylnaphthalene. 2-Isopropylnaphthalene can be obtained from propylene and naphthalene at 150 to 240 °C and 10 bar with a phosphoric acid catalyst using a large excess of naphthalene, followed by isomerization to a mixture of 1- and 2-isopropylnaphthalenes (5:95). The introduction of air (oxygen) at 110 °C produces the α-hydroperoxide of the 2-isomer. The hydroperoxide is cleaved with sulfuric acid, in a manner analogous to the Hock synthesis of phenol; the 2-naphthol yield is around 95%.

The 2-naphthol capacity in the Western world is approximately 50,000 tpa, with *Acna* (Italy) and *Hoechst* (West Germany) operating the largest plants; China's capacity is around 7,000 tpa. Other important producing countries are Poland, Romania and Czechoslovakia.

Examples of technically important colorants based on 2-naphthol are Pigment Orange 5, Pigment Red 3 (see Chapter 6.1.1) and the azo dye Acid Orange 7.

Acid Orange 7

Pigment Orange 5

The most important derivatives of 2-naphthol are 2-hydroxy-3-naphthoic acid (BON) and its anilide, naphthol AS, which are used as coupling components in the production of azo dyes. To produce BON, 2-naphthol is first transformed into sodium 2-naphtholate with a 50% sodium hydroxide solution. The 2-naphtholate is obtained as a fine powder. This step is followed by Kolbe-Schmitt carboxylation with CO_2 at temperatures of 230 to 260 °C and a pressure of 15 bar; the yield is over 90%.

West European production of 2-hydroxy-3-naphthoic acid was approximately 8,000 t in 1985.

When BON reacts with aniline at 70 to 80 °C in toluene in the presence of PCl_3, the corresponding amide of 2-naphthol-3-carboxylic acid is produced; this anilide (Azoic Coupling Component 2) is the base compound of the naphthol AS dyestuffs.

Azoic Coupling Component 2

Naphthol AS dyestuffs were invented as early as 1912 by Adolph Winther, August Leopold Laska and Arthur Zitscher of *Hoechst;* 'AS' is the abbreviation of 'Amid einer Säure' (amide of an acid). Naphthol AS dyestuff components are distinguished from the unsubstituted 2-naphthol derivatives by their increased affinity for the dyed substrate and by higher chemical stability in the atmosphere. Nitro-, chloro-, methyl- and methoxybenzene derivatives can serve as aromatic moieties of the amide group. The coupling reaction with the diazo salt takes place in the 1-position. Examples of early important naphthol AS dyestuffs are Griesheim Red, discovered in 1912, and Indra Red.

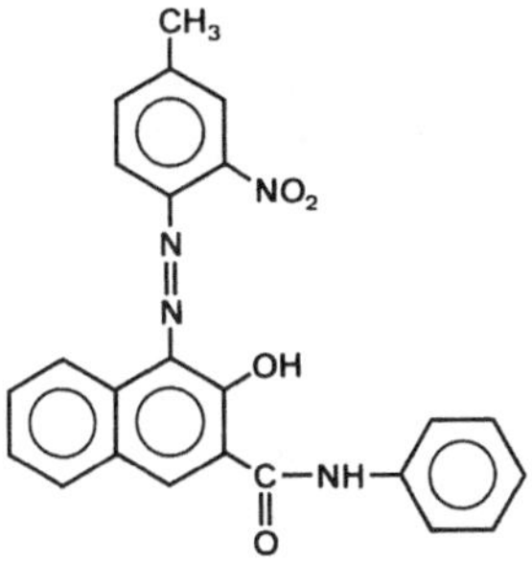

Griesheim Red

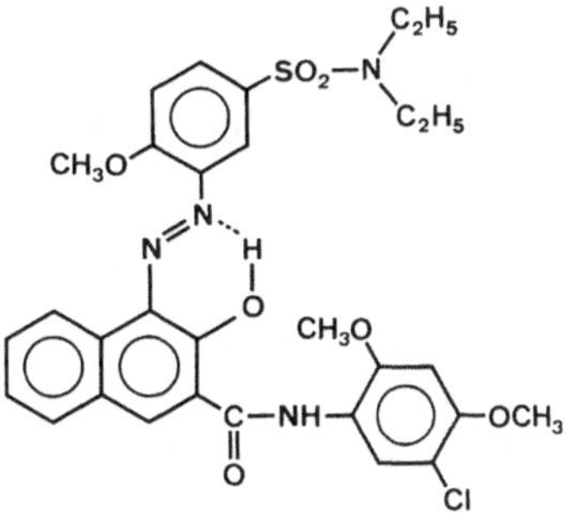

Indra Red

Important naphthol AS pigments are Pigment Red 7 and Pigment Red 112.

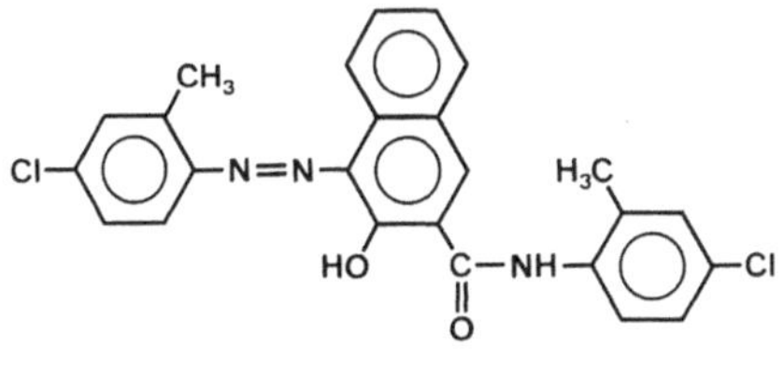

Pigment Red 7

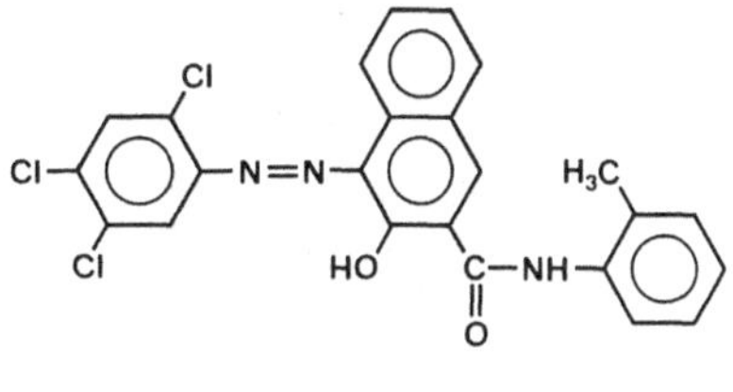

Pigment Red 112

The production of 2-naphthol AS dyestuffs worldwide is approximately 25,000 tpa.

Another important derivative of 2-naphthol is Schäffer acid, which is produced by sulfonation of 2-naphthol with the addition of sodium sulfate at 85 to 105 °C.

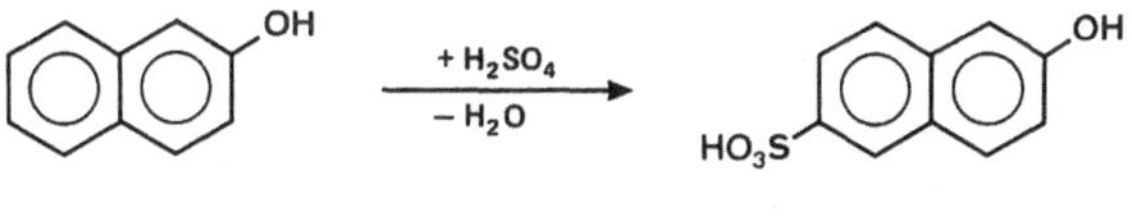

Schäffer acid

Schäffer acid is used as a coupling component in the production of azo dyes such as Acid Black 26.

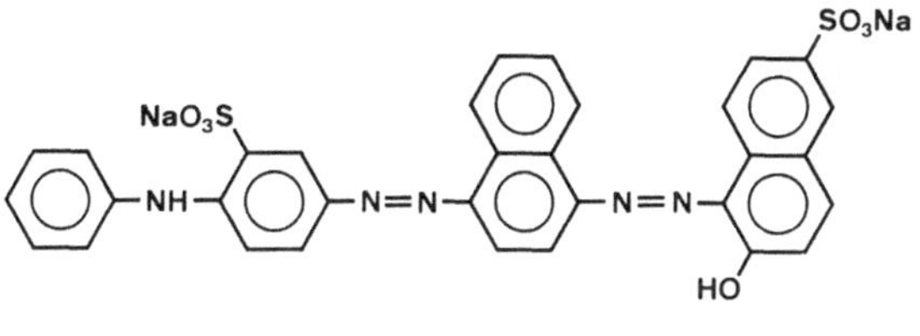

Acid Black 26

Schäffer acid can be converted by alkali fusion at 295 °C into 2,6-dihydroxynaphthalene, which is used as a component in the manufacture of aromatic polyesters which, in common with the corresponding amides, display liquid-crystal characteristics.

2-Hydroxynaphthalene-1-sulfonic acid can be prepared by reaction of 2-naphthol with chlorosulfonic acid at 0 °C in inert solvents such as o-dichlorobenzene or chloroethane. It is not available in satisfactory yields by the direct sulfonation of 2-naphthol with sulfuric acid. Ammonolysis at 150 °C transforms 2-hydroxynaphthalene-1-sulfonic acid into 2-aminonaphthalene-1-sulfonic acid (Tobias acid), which, in quantity terms, is one of the most important derivatives of 2-naphthol.

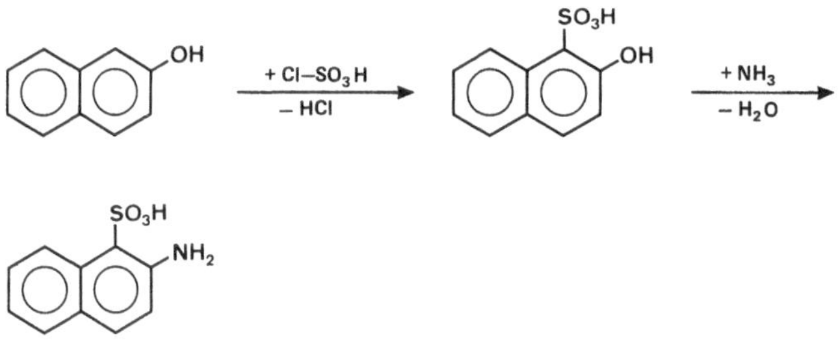

Tobias acid

Examples of pigments widely used – especially in the USA – for printing inks and dyes for plastics based on Tobias acid, are Pigment Red 63:1 and Pigment Red 49:1.

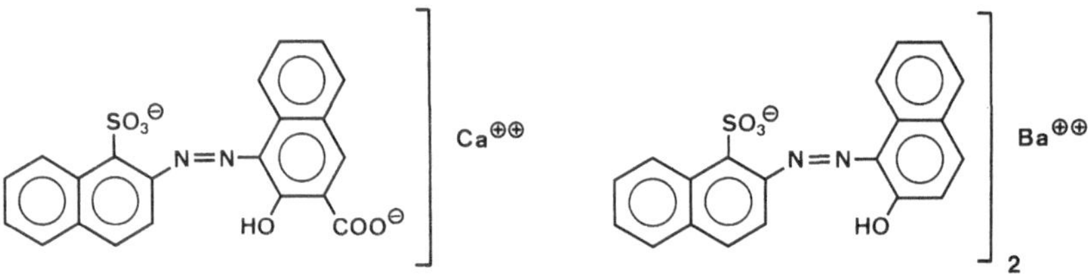

Pigment Red 63:1

Pigment Red 49:1

Around half of the Tobias acid produced in Western Europe is used to produce J-acid, an intermediate for dyes, e.g., Direct Blue 71.

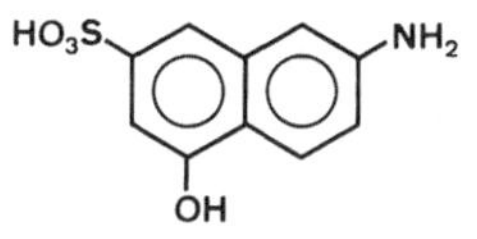

J-acid

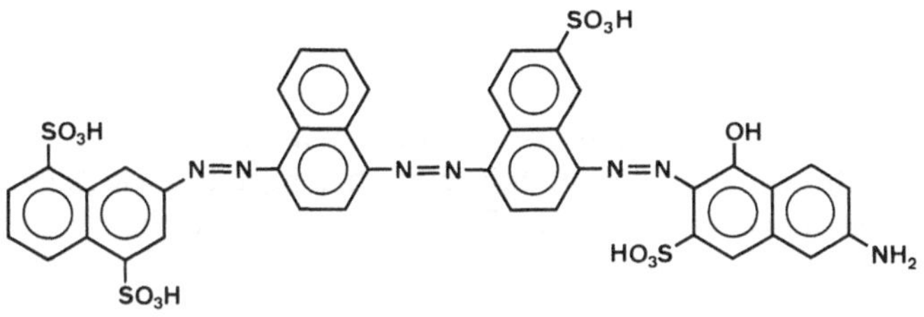

Direct Blue 71

J-acid can be obtained from Tobias acid by sulfonation with 30% oleum at 100 °C to 2-aminonaphthalene-1,5,7-trisulfonic acid, hydrolytic removal of the sulfonic acid group in the 1-position, and alkali fusion.

N-phenyl-2-naphthylamine (PBN), which is used as an antioxidant in rubber processing, is obtained by the reaction of aniline with 2-naphthol at 180 °C.

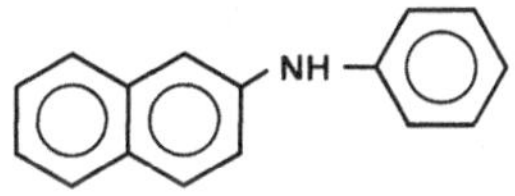

N-Phenyl-2-naphthylamine

The disulfonation of 2-naphthol leads to 2-hydroxynaphthalene-3,6-disulfonic acid (R-acid) and 2-hydroxynaphthalene-6,8-disulfonic acid (G-acid), which are used in the production of dyestuffs such as Acid Red 26 and Acid Red 18.

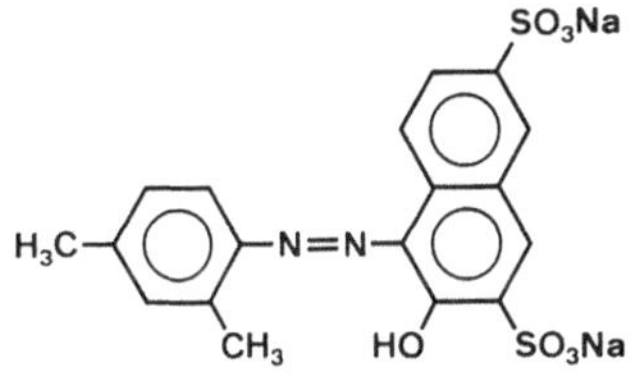

Acid Red 26

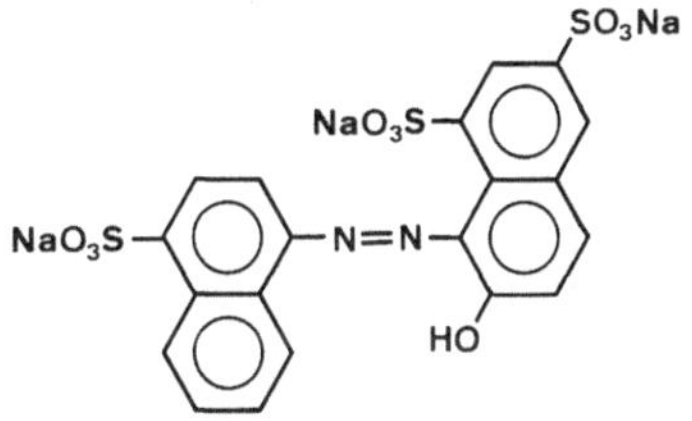

Acid Red 18

Other derivatives of 2-naphthol, less significant in terms of quantity, are the semi-synthetic penicillin nafcillin, and the antimycotic tolnaftate, which is produced from 2-naphthol, thiophosgene and N-methyl-m-toluidine (*Nippon Soda*).

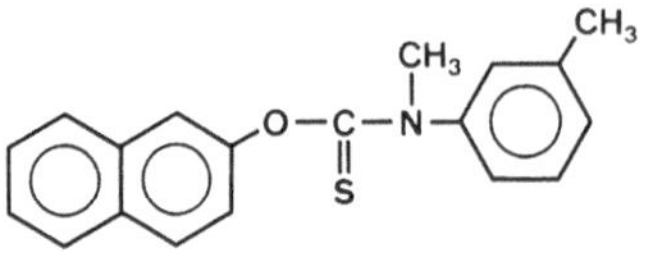

322 Naphthalene – production and uses

Tolnaftate

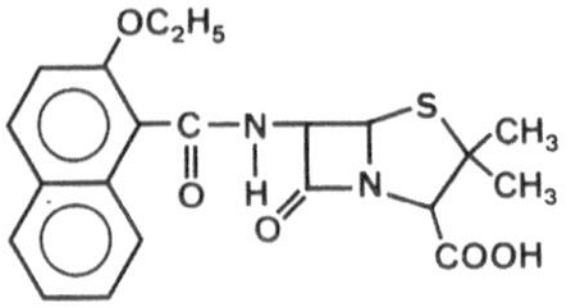

Nafcillin

The anti-rheumatic naproxen is also derived from 2-naphthol; it is produced by the Friedel-Crafts acylation of 2-methoxynaphthalene and subsequent Willgerodt-Kindler reaction. Since only the *S*-configuration is effective, the racemic mixture is resolved with the alkaloid cinchonidine.

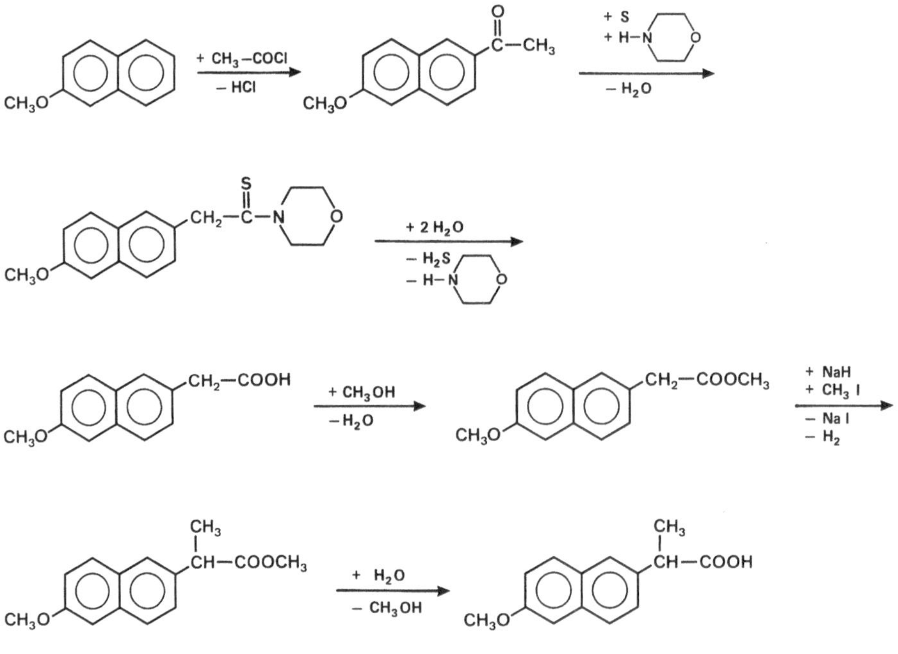

Naproxen

9.3.4 Sulfonic acid derivatives of naphthalene

Sulfonation followed by nitration produces further naphthalene derivatives which are used predominantly as dyestuff intermediates.

Technical sulfonation is carried out by the classic method with sulfuric acid in agitated reactors made from cast iron or coated with enamel. More modern processes operate in organic solvents with chlorosulfonic acid or sulfur trioxide at temperatures below 10 °C. Reaction conditions for the sulfonation of naphthalene to yield defined derivatives are summarized in Figure 9.13.

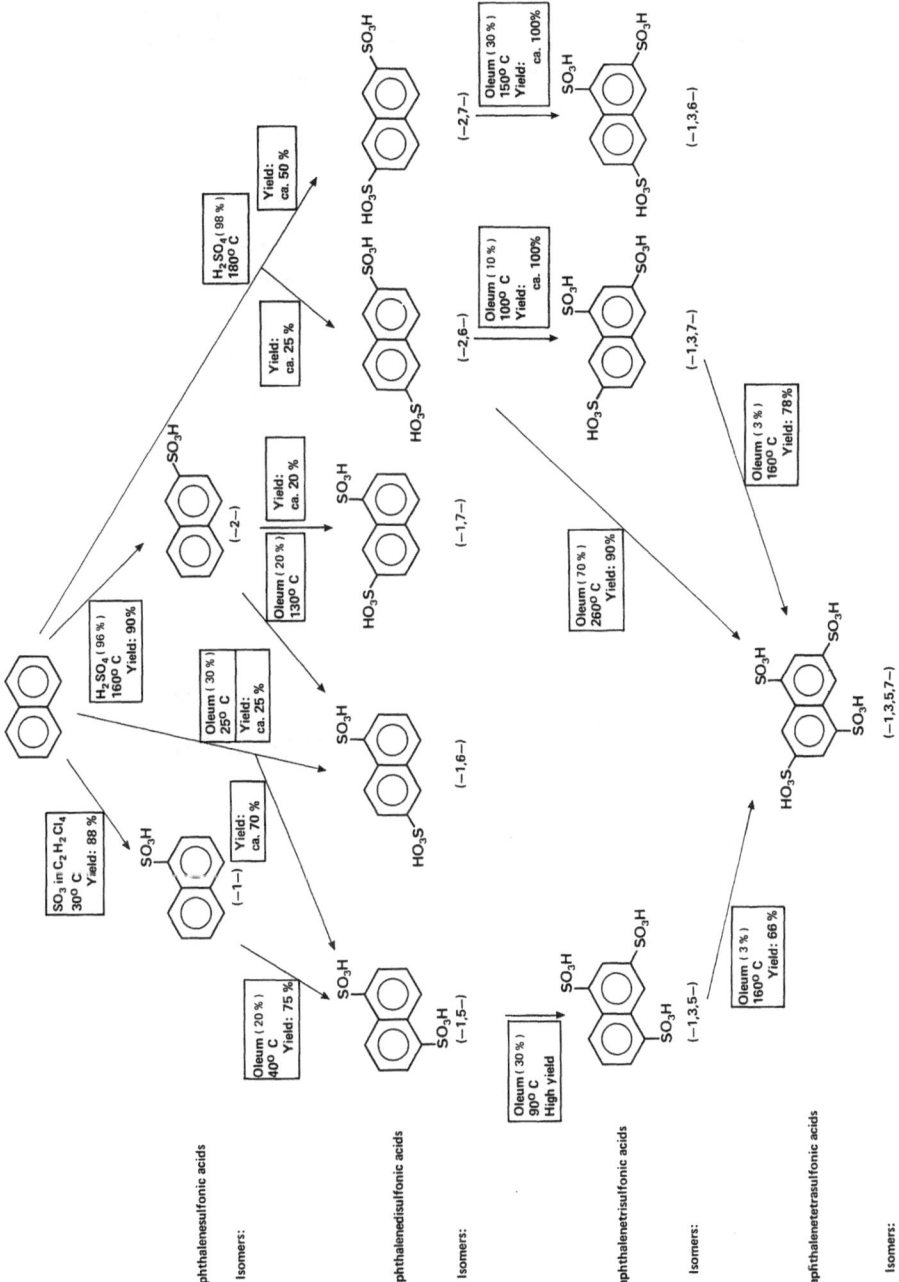

Figure 9.13: Selected reaction conditions for the sulfonation of naphthalene

Among the naphthalenesulfonic acid derivatives, particular note should be taken of the so-called letter acids. Table 9.6 shows some important letter acids, which are used as coupling components in the production of azo dyes.

Table 9.6: Letter acids

Name	Chemical name	Structure
G-acid	2-hydroxynaphthalene-6,8-disulfonic acid	6-SO_3H, 8-HO_3S, 2-OH on naphthalene
H-acid	1-amino-8-hydroxy-naphthalene-3,6-disulfonic acid	8-OH, 1-NH_2, 3-HO_3S, 6-SO_3H on naphthalene
J-acid	2-amino-5-hydroxy-naphthalene-7-sulfonic acid	5-HO_3S, 2-NH_2, 7-OH on naphthalene
R-acid	2-hydroxynaphthalene-3,6-disulfonic acid	2-OH, 3-SO_3H, 6-HO_3S on naphthalene
δ-acid	2-amino-8-hydroxy-naphthalene-6-sulfonic acid	8-OH, 1-NH_2, 6-HO_3S on naphthalene
T-acid (Koch-acid)	1-aminonaphthalene-3,6,8-trisulfonic acid	1-NH_2, 8-SO_3H, 3-HO_3S, 6-SO_3H on naphthalene

Figure 9.14 shows as an example the process diagram for the production of H-acid.

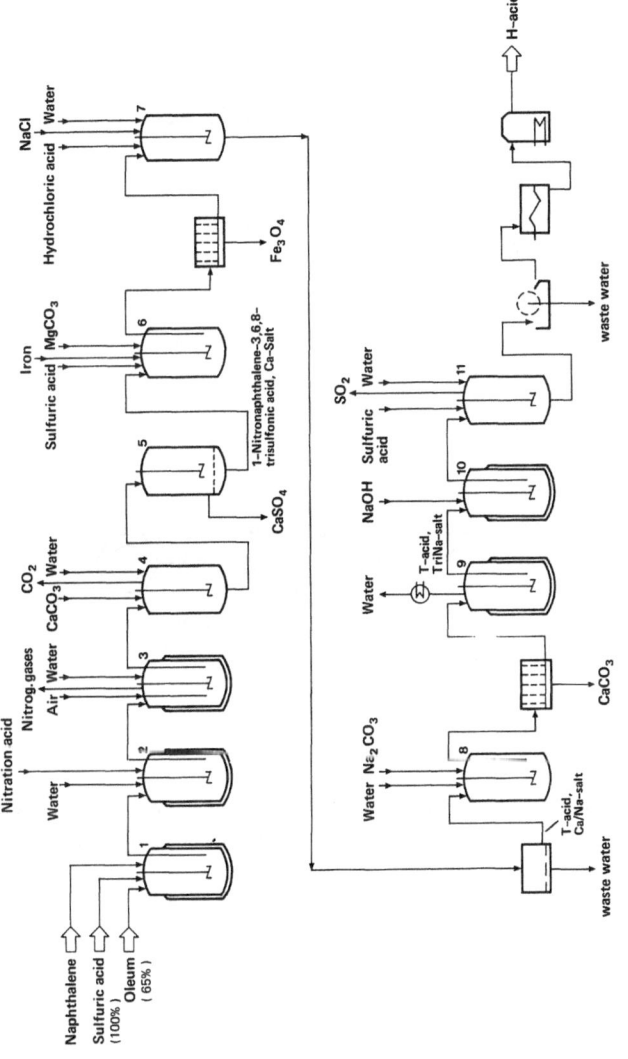

1 Sulfonation reactor; **2** Nitration reactor; **3** NO_x-separation; **4** Neutralizing vessel; **5** Filtration vessel; **6** Reduction reactor; **7** Precipitation vessel; **8** Tri-Na-salt formation; **9** Concentration vessel; **10** Melt reactor; **11** Precipitation vessel

Figure 9.14: Production of H-acid

In the first reaction stage, naphthalene is sulfonated with sulfuric acid. Subsequent nitration and neutralization of the naphthalene-trisulfonic acid with lime yields the calcium salt of 1-nitronaphthalene-3,6,8-trisulfonic acid, which is traditionally reduced to T-acid (Koch-acid) with iron filings and hydrochloric acid; modern processes use continuous catalytic hydrogenation with a nickel catalyst. The H-acid is obtained by subsequent fusion of the T-acid with sodium hydroxide and neutralization with sulfuric acid. It serves as a coupling component in the production of azo dyes such as Direct Blue 15 or Acid Black 1.

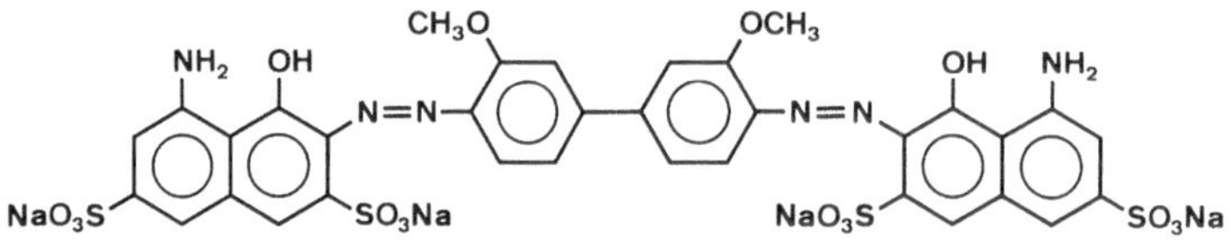

Direct Blue 15

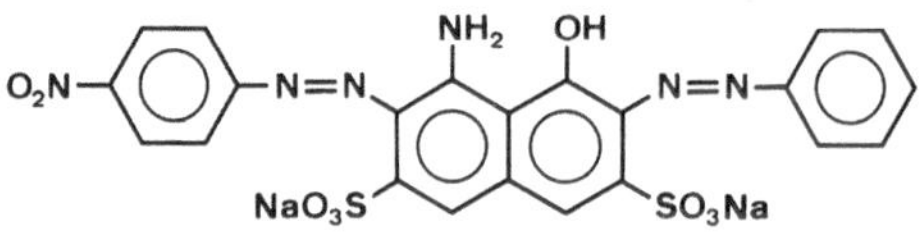

Acid Black 1

An example of an azo dyestuff in which γ-acid is the coupling component is Acid Red 337, which is obtained through the reaction of diazotized o-trifluoromethylaniline and γ-acid.

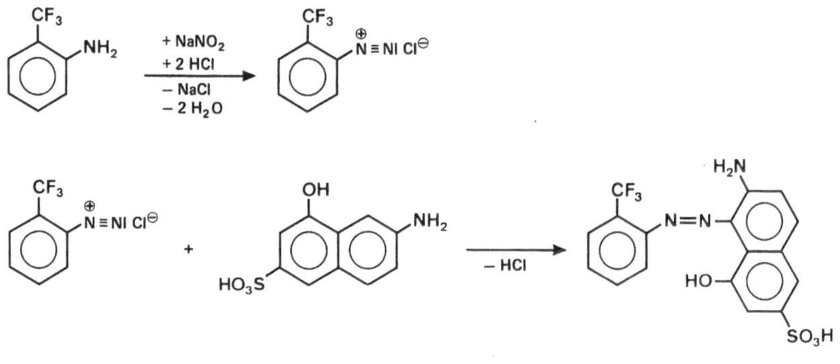

Acid Red 337

The o-trifluoromethylaniline is diazotized in aqueous hydrochloric acid with sodium nitrite at 0 to 5 °C. Filtration is followed by coupling with γ-acid to produce the dyestuff, which is principally used to color polyamide fibers. Figure 9.15 shows the process diagram for diazotization to produce Acid Red 337.

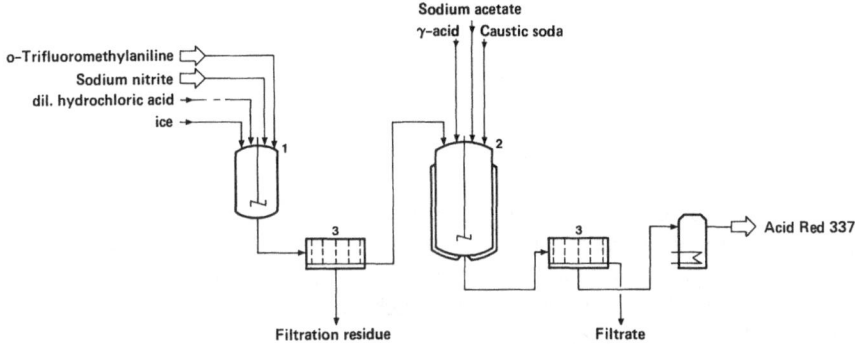

1 Diazotization reactor; 2 Coupling reactor; 3 Filtration

Figure 9.15: Flow diagram for diazotization for the production of Acid Red 337

9.3.5 Nitro- and aminonaphthalenes

Nitronaphthalenes are obtained by treating naphthalene with nitric acid; they form the basis for the production of naphthylamine. Since 2-naphthylamine has lost its earlier large-scale commercial importance, especially as a dye intermediate, because of its carcinogenic properties, mononitration of naphthalene is used exclusively today to produce 1-nitronaphthalene.

When the nitration is carried out at 40 to 55 °C, the selectivity of the reaction can be well governed so that the 2-nitronaphthalene forms only 5% of the reaction product, and dinitronaphthalenes are formed to an extent less than 0.3%. The technical grade 1-nitronaphthalene is recovered from the reaction mixture by distillation.

1-Nitronaphthalene is reduced with iron or hydrogen with nickel catalysts; it is purified by distillation. In modern processes, the 2-naphthylamine content can be reduced to below 10 ppm.

1-Naphthylamine is a significant intermediate for azo dyes and it is also used as the amino component in the manufacture of the herbicide naptalam (*Uniroyal*), which is obtained by reaction with phthalic anhydride in xylene at 25 °C.

Naptalam

9.3.6 Production of tetralin and tetralone

The hydrogenation of naphthalene to tetralin is analogous to the hydrogenation of benzene to produce cyclohexane. The reaction can be carried out in the liquid phase with nickel catalysts at 200 °C and 10 to 15 bar hydrogen pressure. The feedstock is desulfurized naphthalene, which can be obtained by treatment of crude naphthalene with sodium.

World production of tetralin is around 25,000 tpa. In addition to the main application already described – the production of 1-tetralone (see Chapter 9.3.3.1), – tetralin is also used as a solvent. An earlier application as a hydrogen donor in coal hydrogenation is of no technical significance nowadays, but it is frequently carried out as a model reaction.

The oxidation of tetralin leads to the hydroperoxide, which can be split into 1-tetralone and 1-tetralol. Dehydrogenation of tetralone and tetralol takes place in two stages, to avoid the possible dehydration of tetralol on the platinum catalyst. In the first stage, tetralol is dehydrogenated to tetralone in the gas phase in the presence of hydrogen at 200 to 325 °C and nickel or copper catalysts; in the second stage, dehydrogenation on the platinum catalyst at 350 to 400 °C produces 1-naphthol.

9.3.7 Production of isopropylnaphthalene derivatives

A new application for naphthalene with considerable growth potential is the production of alkylated naphthalenes, especially that of diisopropylnaphthalenes. Resulting from the mutual depression of melting-points, the mixture of diisopropylnaphthalene is liquid and can therefore be used as a high-value solvent for dyes in the production of carbonless copy papers. The solvents for dyes must meet certain preconditions, i.e. exhibit high solvent power, be free from odor, be toxico-

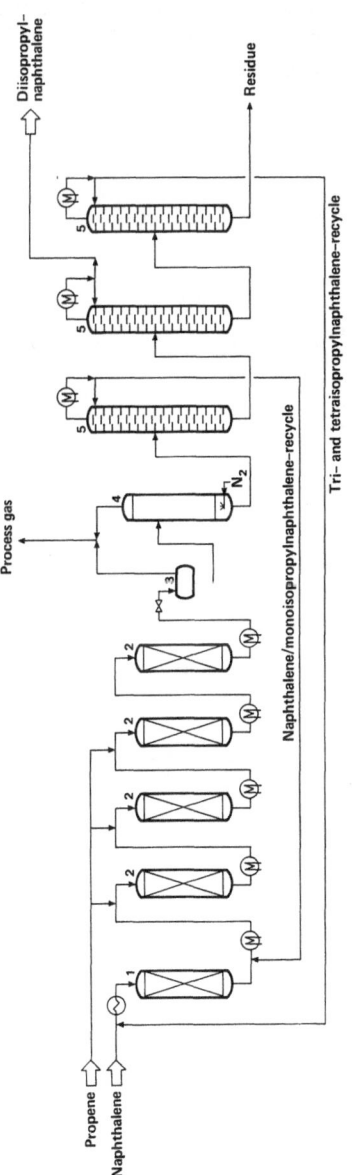

1 Transalkylation reactor; **2** Propylation reactor; **3** Pressure let-down vessel; **4** Strip column; **5** Vacuum-distillation columns

Figure 9.16: Flow diagram of the propylation of naphthalene

330 Naphthalene – production and uses

logically safe, and have suitable viscosity and satisfactory biodegradability. Isopropylated naphthalenes with a high proportion of diisopropylnaphthalenes fulfill all these quality criteria.

The process developed by *Kureha Chemical* to produce diisopropylnaphthalene mixtures operates in a manner similar to that of benzene alkylation; it comprises three stages consisting of transalkylation, alkylation and distillation of the reaction product. Naphthalene and recycled tri- and tetra-isopropylnaphthalenes first react in a transalkylation stage; the product is then fed together with recycled mono-isopropylnaphthalenes to a second stage where the reaction with propene produces mainly diisopropylnaphthalenes. The reaction mixture is split by vacuum distillation. The reaction is performed at 7 bar and 200 °C using a silica-alumina catalyst. Figure 9.16 shows the flow sheet for the *Kureha* process.

Figure 9.17: Production plant of *Rütgers Kureha Solvents* for the manufacture of KMC solvent, Duisburg-Meiderich/West Germany

Capacity for the production of diisopropylnaphthalene, sold under the trade name KMC, totals each 10,000 tpa in Japan and West Germany. Figure 9.17 shows the *Rütgers Kureha Solvents* production plant in Duisburg-Meiderich/West Germany; the erection of a similiar plant in the USA has been announced.

Other possible applications for isopropylnaphthalenes are the production of 2,6-naphthalenedicarboxylic acid (see Chapter 10.2), and the manufacture of 2-isopropenylnaphthalene, which can be used as a vinylogous monomer to produce polymers with a high glass-transition temperature.

2,6-Naphthalene-
dicarboxylic acid

2-Isopropenyl-
naphthalene

9.3.8 Other alkylnaphthalenes from naphthalene

The alkaline earth salts of sulfonated dinonylnaphthalenes have commercial importance as lube-oil additives. They are produced by Friedel-Crafts alkylation of naphthalene with nonene at 60 to 80 °C using aluminum chloride and concentrating the dinonylnaphthalene by distillation and then sulfonating it with SO_3 or oleum at 20 to 40 °C.

Barium dinonylnaphthalene
sulfonate

Through the reaction of the naphthalenesulfonic acids with formaldehyde, surface-active substances are produced which are used as wetting agents and tanning substances. These products belong to the syntans, which complement the range of natural tanning and wetting agents and, to some extend, the inorganic tanning agents based on chromium (III).

Condensation products from naphthalenesulfonic acids and formaldehyde are also used as additives to improve the flow behavior of concrete. The degree of condensation (n) of these products, with a worldwide production of around 40,000 tpa, ranges from 5 to 8.

9.4 Process review

Table 9.7 summarizes the most important processes of the industrial chemistry of naphthalene and its derivatives.

Table 9.7: Summary of the most important processes in industrial naphthalene chemistry

Process	Objective of process	Process conditions				Other characteristics
		Pressure (bar)	Temp. (°C)	Catalyst	Reaction comp.	
1. Hydrogenation						
Unionfining process	Naphthalene refining	14	400	Co/Mo	H_2	Desulfurization by conversion of thianaphthene in H_2S and ethylbenzene
Alkylnaphthalene dealkylation (UNIDAK process)	Naphthalene recovery	35	540–620	Co/Mo	H_2	Process for recovering petroleum-derived naphthalene
Naphthalene hydrogenation	Tetralin	10–15	200	Ni	H_2	Liquid-phase reaction analogous to benzene hydrogenation
2. Oxidation						
Naphthalene oxidation (*BASF* process/ *von Heyden* process)	PA	1–1.5	350–380	V_2O_5	O_2	Fixed-bed process
Naphthalene oxidation (*Sherwin-Williams/ Badger*)	PA	4	340–380	V_2O_5	O_2	Fluid-bed process; little used today
Naphthalene oxidation	Naphthoquinone	atmos.	400	V_2O_5	O_2	Fixed-bed process
3. Alkali fusion						
Hoechst process	2-Naphthol	atmos.	300–320	–	Naphthalene-2-sulfonic acid/ NaOH	Only route to 2-naphthol in use today

Naphthalene – production and uses

Table 9.7 (continued)

Process	Purpose of process	Process conditions				Other characteristics
		Pressure (bar)	Temp. (°C)	Catalyst	Reaction comp.	
4. Sulfonation						
Naphthalene sulfonation	Naphthalene-1-sulfonic acid	atmos.	10	–	SO_3	Reaction occurs in organic solvents (e.g. $C_2H_2Cl_4$)
Naphthalene sulfonation	Naphthalene-2-sulfonic acid	atmos.	160	–	H_2SO_4	Sulfonation under thermodynamically controlled conditions
Dinonylnaphthalene sulfonation	Production of additives for lubricants	atmos.	20–40	–	Oleum	Sulfonation in organic solvents (e.g. n-heptane)
5. Nitration						
Naphthalene nitration	1-Nitronaphthalene	atmos.	40–55	–	HNO_3	Raw material for 1-naphthylamine
6. Alkylation						
Naphthalene propylation	Diisopropylnaphthalenes	7	200	SiO_2/Al_2O_3	Propene	Multi-stage liquid phase reaction

10 Alkylnaphthalenes and other bicyclic aromatics – production and uses

Apart from naphthalene, other commercially important bicyclic aromatics are especially biphenyl (also referred to as diphenyl), acenaphthene and acenaphthylene. Like the methylnaphthalenes, they are present in varying concentrations in coal tar. Where the concentration is sufficiently high, as is the case for methylnaphthalenes and acenaphthene (or acenaphthylene), then they are recovered from coal tar; for compounds present in low concentrations, on the other hand, such as biphenyl, production is normally via a synthetic route.

10.1 Biphenyl

Biphenyl, first described by Rudolph Fittig in 1862, is present in coal tar in a concentration of around 0.4%. Because of this low concentration, biphenyl is manufactured by thermal dehydrogenation of benzene, as well as by co-production in the dealkylation of toluene.

Biphenyl

Thermal dehydrogenation of benzene occurs under atmospheric pressure at 600 to 800 °C. Reaction times are kept below 3 sec to avoid excessive build-up of pyrolysis products; at very short reaction times (less than 0.1 sec) the yield and the conversion rates fall noticeably.

In addition to biphenyl, an isomeric mixture of terphenyls is produced, which are separated from biphenyl as high-boiling compounds by distillation. As a rule, the terphenyl fraction contains around 8% o-, 49% m- and 23% p-terphenyl, as well as around 20% triphenylene and small amounts of quaterphenyl.

o-Terphenyl   m-Terphenyl   p-Terphenyl

Triphenylene   Quaterphenyl

Partial hydrogenation of the terphenyl isomers with Raney nickel yields hydrogenated terphenyls (vide infra), which are used as solvents for dyes, e.g. for the manufacture of carbonless copy papers.

The main source of biphenyl is the dealkylation of toluene, through which biphenyl occurs as a co-product in yields of around 1%; it can be produced in a technically pure form (93 to 95%), from the residue of benzene recovery, by distillation.

Methylbiphenyl

Dimethylbiphenyl

In addition to biphenyl, methylbiphenyls and dimethylbiphenyls also occur as by-products of the dealkylation of toluene.

Biphenyl is commonly used in eutectic mixture with diphenyl oxide (diphenyl ether) as a heat transfer oil, generally in mixtures of 26.5% biphenyl and 73.5% diphenyl ether.

Through Friedel-Crafts chlorination of biphenyl with $FeCl_3$ as a catalyst, a total of 209 chlorobiphenyls can be obtained. The earlier importance of biphenyl for the production of polychlorinated biphenyls (PCB), with a chlorine content of around 50 to 55%, for plasticizers and transformer oils no longer exists today, since it was recognized in the late 1960's that these compounds are not readily biodegradable and that their combustion at temperatures between 600 and 900 °C can release toxic dioxins. Production of polychlorinated biphenyls was therefore banned in a number of industrialized countries, e.g. in the USA in 1976.

10.2 Methylnaphthalenes

Both methylnaphthalene isomers are present in coal tar, at levels of 0.7% (1-methylnaphthalene) and 1.5% (2-methylnaphthalene).

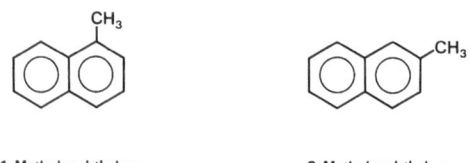

1-Methylnaphthalene

2-Methylnaphthalene

They are concentrated by distillation in the methylnaphthalene fraction, boiling between 230 and 245 °C; further refining involves removal of phenols and bases. The isomers are separated by their different crystallizing points. 2-Methylnaphthalene, with a crystallizing point of 34.6 °C, precipitates by cooling the concentrated methylnaphthalene fraction, while 1-methylnaphthalene, with a crystallizing point of −30.5 °C, is concentrated in the mother liquor. 1-Methylnaphthalene is recovered by redistilling the mother liquor. The crystallized product, rich in 2-methylnaphthalene, which has a crystallizing point between 25 and 28 °C, is washed with sulfuric acid, neutralized and then redistilled. Technical grade 2-methylnaphthalene is 95 to 98% pure.

Methylnaphthalenes are also found in petroleum-derived feedstocks, such as pyrolysis tar from ethylene production and cat-cracker residues (see Chapter 9.2.2). The 1-/2-methylnaphthalene isomer ratio, which is around 1:2.5 for methylnaphthalenes in coal tar, is higher in this case, because of the lower temperature exposition of the raw material, and is close to 1:1. Recovery of methylnaphthalenes from these sources is generally only carried out on a small scale, since it can be extremely intricate to separate co-boiling compounds; petroleum-derived methylnaphthalenes has served as a feedstock for naphthalene production by dealkylation, especially in the USA in the 1960's and 70's.

Worldwide production of 2-methylnaphthalene is around 1,500 tpa. In addition, 1-/2-methylnaphthalene mixtures are used in roughly the same amount as solvents and heat transfer oils.

Applications for 1-methylnaphthalene are relatively sparse; it is primarily used as a feedstock for production of 1-naphthylacetic acid (NAA), one of the first plant growth regulators produced in 1936 by *ICI*, which is obtained by side-chain chlorination, followed by reaction of the naphthyl chloride with potassium cyanide and subsequent hydrolysis.

1-Naphthylacetic acid (NAA)

2-Methylnaphthalene is commonly used as a feedstock in the production of vitamin K_3 (menadione). Methylnaphthalene is oxidized to menadione with chromic or nitric acid, in a similar method to anthracene oxidation. Menadione is used as an intermediate in the production of vitamin K_1. To produce vitamin K_1, menadione is reduced with hydrogen on Pd/activated-carbon catalysts to menadiol. Esterification of the two hydroxyl groups with acetic anhydride yields menadiol diacetate, which is converted into the 1-monoacetate with ammonia. Vitamin K_1 (phytomenadione) is produced by the reaction of 1-menadiol monoacetate with phytol, using a BF_3/ether complex as catalyst, followed by hydrolysis and dehydrogenation.

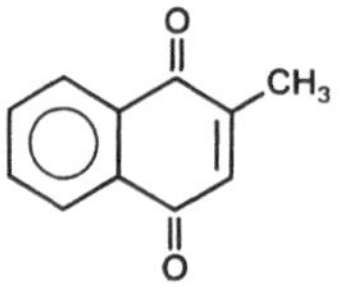

Menadione

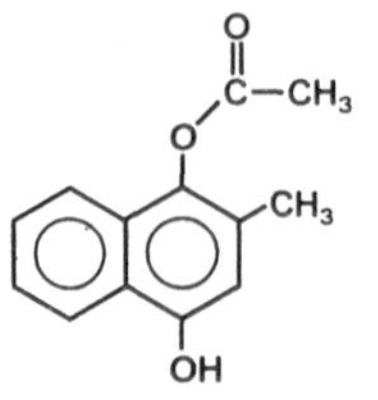

1-Menadiol monoacetate

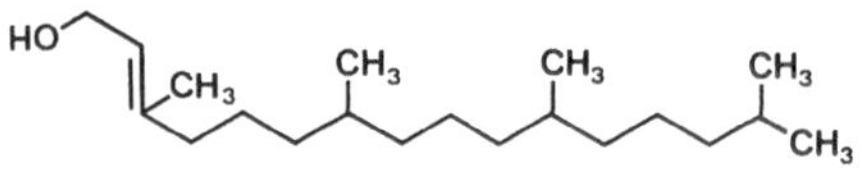

Phytol

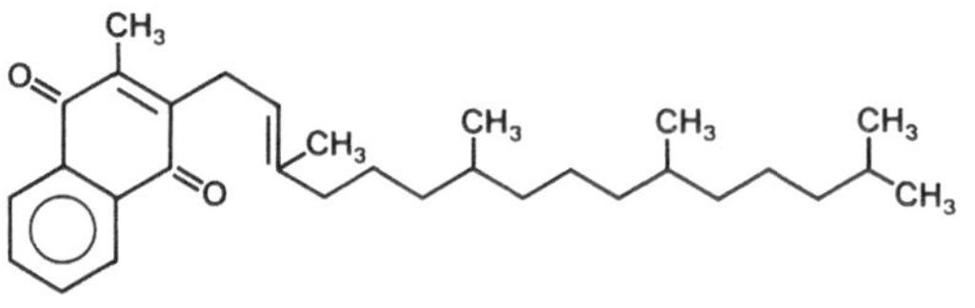

Vitamin K$_1$ (Phytomenadione)

2-Methylnaphthalene can be converted into 2-methyl-6-acetylnaphthalene by acylation with BF$_3$/acetic anhydride, with high selectivity and conversion rates. 2-Methyl-6-acetylnaphthalene, when oxidized at 130 °C and at a pressure of 6 to 8 bar using cobalt/bromine catalysts in acetic acid produces naphthalene-2,6-dicarboxylic acid.

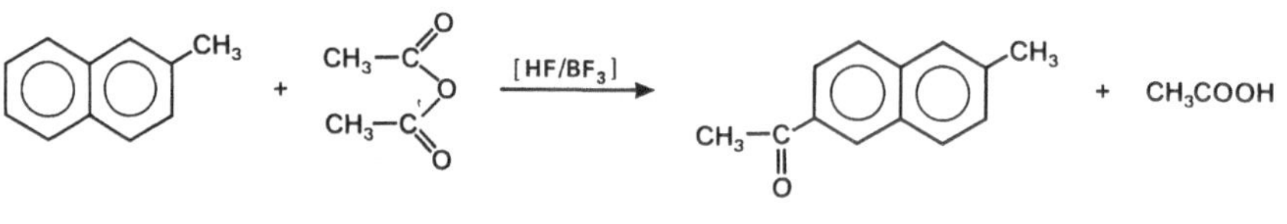

A further method for the synthesis of naphthalene-2,6-dicarboxylic acid is based on 2-methylnaphthalene or mixtures of 1-/2-methylnaphthalene. The methylnaphthalenes are oxidized to carboxylic acids; subsequent *Henkel* rearrangement, which was carried out a number of years ago on a large scale in Japan (*Teijin*) at 400 to 500 °C under CO$_2$ pressure of around 100 bar, converts the monocarboxylic acid to dicarboxylic acid.

[Reaction scheme: 2-methylnaphthalene + 3/2 O₂ → 2-naphthoic acid (−H₂O); + KOH → potassium 2-naphthoate (−H₂O); 2 equivalents of potassium 2-naphthoate → (−naphthalene) dipotassium naphthalene-2,6-dicarboxylate (KOOC–C₁₀H₆–COOK); + H₂SO₄ → naphthalene-2,6-dicarboxylic acid (HOOC–C₁₀H₆–COOH) (−K₂SO₄)]

Small amounts of 2-methylnaphthalene are used in sulfonated form as textile auxiliaries, surfactants and emulsifiers.

The two isomeric methylnaphthalenes can also be interconverted, e.g. on zeolites or with HF/BF$_3$.

[Equilibrium: 1-methylnaphthalene ⇌ 2-methylnaphthalene, catalyzed by [HF/BF$_3$]]

Among the di-substituted alkylnaphthalene derivatives, 2,6-dimethylnaphthalene is important as it can also be used as a base material for the production of naphthalene-2,6-dicarboxylic acid. It is produced from the coal tar fraction boiling between 255 and 265 °C, by redistillation and crystallization. Oxidation of 2,6-dimethylnaphthalene is carried out in the same way as the oxidation of p-xylene, in the liquid phase with Co/Br-catalysts. Naphthalene-2,6-dicarboxylic acid, or its acid chloride and ester, are components in the production of high-value polymer films and fibers, with high temperature resistance; these polymers can display liquid-crystalline character in the molten state.

10.3 Acenaphthene/acenaphthylene

Acenaphthene was discovered in coal tar by Pierre E. M. Berthelot in 1873. Its concentration level in coal tar is only 0.2%; in addition, however, coal tar contains around 2.5% of acenaphthylene, the dehydrogenated form of acenaphthene.

Acenaphthene   Acenaphthylene

A large portion of the acenaphthylene is hydrogenated to acenaphthene (see Chapter 3.2.3) by hydrogen transfer which occurs during the distillation of coal tar.

Acenaphthene is recovered by distilling the wash-oil fraction of coal tar, which contains 25% acenaphthene. Redistillation of this tar oil fraction produces an acenaphthene concentrate, boiling between 265 and 285 °C, from which 95 to 99% pure acenaphthene can be recovered by crystallization. In 1985, production of acenaphthene was 2,500 t.

The main use of acenaphthene is the production of naphthalic anhydride by gas-phase oxidation, carried out with vanadium oxide catalysts at 330 to 400 °C. Liquid-phase oxidation with chromates is also used industrially, but on a much smaller scale.

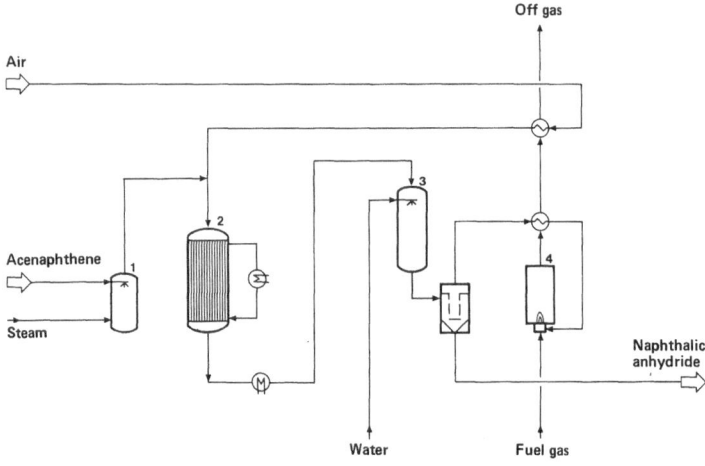

1 Vaporizer; 2 Multitubular reactor; 3 Desublimation tower; 4 Waste-gas incineration

Figure 10.1: Flow diagram of the gas-phase oxidation of acenaphthene

Figure 10.1 shows the flow diagram for gas-phase oxidation, which operates like that for naphthalene oxidation.

Technically pure (around 95%) naphthalic anhydride (NA) is used as an intermediate in the production of perylenetetracarboxylic acid pigments, which are distinguished by high color-fastness and strength.

In this case, NA is first made to react with ammonia to form the acid imide, from which perylenetetracarboxylic-diimide is produced by alkali fusion and oxidation; this is followed by hydrolysis with concentrated sulfuric acid, to perylenetetracarboxylic acid.

3,4,9,10-Perylenetatracarboxylic acid

The most important pigments based on perylenetetracarboxylic acid are Pigment Red 179, which is commonly used to color automobile paints, and the bluish-red pigment, Pigment Red 149, used e.g. in coloring plastics.

Pigment Red 149

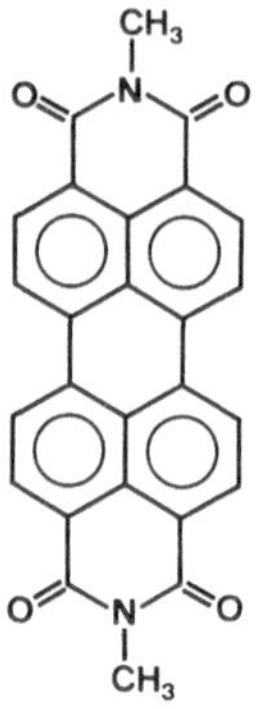

Pigment Red 179

If justified by demand, acenaphthylene is produced not by recovery from the corresponding coal-tar fraction, but by the catalytic dehydrogenation of acenaphthene in the gas phase.

Acenaphthylene can be used as a raw material in the production of resins, particularly in mixtures with other vinylaromatics; these resins are distinguished by high softening points.

11 Anthracene – production and uses

Anthracene was first discovered in coal tar by Jean B.A. Dumas and Auguste Laurent in 1832. The importance of anthracene for industrial aromatic chemistry began with the synthesis of the dyestuff alizarin by Carl Graebe and Carl Th. Liebermann, as well as by William H. Perkin in 1868, replacing the natural dye produced from madder. Anthraquinone dyestuffs have remained the most important class of dyes, alongside azo-dyes, since the beginning of the chemistry of synthetic dyestuffs.

From the time of the first alizarin synthesis, anthraquinone has been the most versatile anthracene derivative. This applies both for dye production and for the more recent applications of anthraquinone, as an additive (redox-catalyst) in wood pulping, and as a hydrogen carrier in H_2O_2 production.

11.1 Production of anthracene

While it is possible to recover the bicyclic naphthalene from pyrolysis products of coal and petroleum-derived raw materials, the tricyclic anthracene can be isolated only from high-temperature coal tar, since anthracene is present only in small amounts in other pyrolysis products and in the residues of catalytic hydrocarbon conversion.

Synthetic production of anthracene has gained no industrial significance so far, since sufficient quantities are present in coal tar; in theory, supplies of around 150,000 tpa of anthracene are available to meet a demand of some 20,000 tpa.

The feedstock for anthracene production from coal tar is the anthracene fraction ('anthracene oil'), which boils between 300 and 400 °C and contains around 6% anthracene (Table 11.1).

Table 11.1: Composition of anthracene oil from coal tar (in %)

Dimethylnaphthalenes	0.7	Phenanthrene	18.8
Acenaphthene	3.1	Carbazole	3.8
Dibenzofuran	4.0	Methylphenanthrenes	12.3
Fluorene	7.7	Fluoranthene	8.0
Methylfluorene	10.6	Pyrene	4.1
Dibenzothiophene	1.9	Other aromatics	19.2
Anthracene	5.8		

The first stage in the production of anthracene is the recovery of a 25 to 30% anthracene concentrate by crystallization, which can be carried out in two stages to increase the yield. The crystallizate, known as 'anthracene cake', is generally concentrated to around 50% by vacuum distillation. The main co-boiling compound of '50's anthracene' is phenanthrene while carbazole is reduced to below 2%. Subsequent refining to yield 'pure anthracene', containing over 95%, is normally achieved by recrystallization in polar solvents, such as acetophenone, mixtures of cyclohexanol/cyclohexanone or N-methylpyrrolidone; in addition, distillation or azeotropic distillation with ethylene glycol can be used for purification.

The quality of anthracene obtained by this method is sufficient for the manufacture of anthraquinone by the usual processes. Because of hydrogen transfer during coal tar refining, the fore-runnings which arise during the distillation of the anthracene cake are enriched with 9,10-dihydroanthracene, which can be converted into anthracene by oxidation with air. This anthracene, recoverable by subsequent crystallization, is distinguished by an exceptionally low nitrogen content.

Figure 11.1 shows the flow diagram for recovery of pure anthracene from anthracene oil, by the combined application of crystallization and distillation.

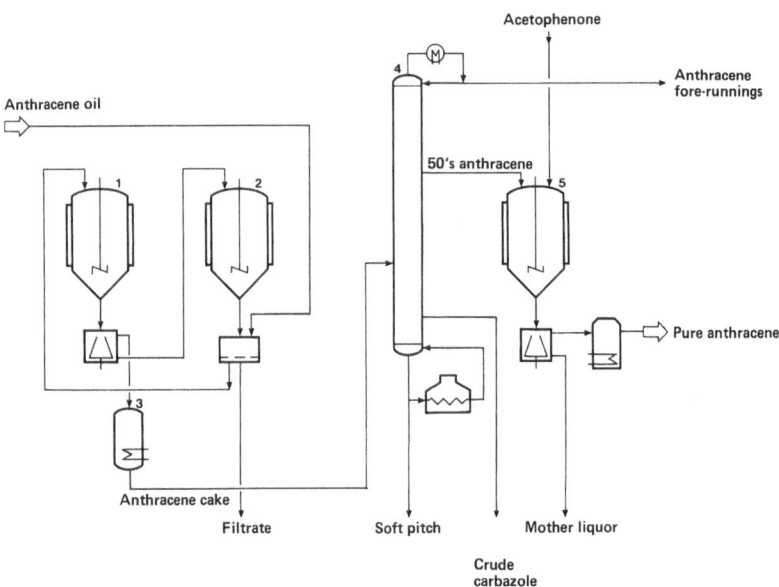

1 Crystallizer (1st stage); **2** Crystallizer (2nd stage); **3** Melting vessel; **4** Distillation column; **5** Crystallizer

Figure 11.1: Flow diagram for production of anthracene from anthracene oil

In spite of the adequate availability of anthracene from coal tar, processes have been developed to synthesize anthracene from phenanthrene, which is present in coal tar in higher concentrations, or from diphenylmethane.

In the *Bergbau-Forschung* process, phenanthrene is first hydrogenated to sym-octahydrophenanthrene, then converted to sym-octahydroanthracene, and finally dehydrogenated.

This synthesis, in common with the production of anthracene based on diphenylmethane by Friedel-Crafts reaction and dehydrogenation, has had no industrial significance to date.

World production of anthracene from coal tar is around 20,000 tpa. *Rütgerswerke*/West Germany, with around 8,000 tpa and *Nippon Steel Chemical*/Japan, with around 3,000 tpa are the major producers.

The industrial importance of anthracene consists mainly in its use as a raw material in the production of anthraquinone; small amounts are used in production of the tranquilizer benzoctamine, which is produced from anthracene and acrolein in a Diels-Alder reaction, followed by reductive amination of the aldehyde with methylamine.

Benzoctamine

A new application for anthracene could develop with the increasing importance of hydrogen as a fuel, in the form of magnesium hydride. Magnesium hydride contains 150 g hydrogen per liter, which is released on heating to 250 to 300 °C. A metalorganic complex of magnesium with anthracene is an intermediate stage in the synthesis of magnesium hydride. This complex is produced by the reaction of pure magnesium with anthracene in tetrahydrofuran, at ambient temperature.

11.2 Production of anthraquinone

Two methods are available to the production of anthraquinone and anthraquinone-derivatives:

1. oxidation of anthracene,
2. synthesis of anthraquinone from smaller moieties

Because of adequate availability of anthracene, oxidation is preferred to synthesis, but the latter can, however, be advantageous when substitution is required in specific positions.

11.2.1 Oxidation of anthracene to anthraquinone

The most elegant method of oxidizing anthracene is gas-phase oxidation, which was first described by Alfred Wohl in 1916, in the context of an investigation into gas-phase oxidation of polynuclear aromatics, particularly for the synthesis of phthalic anhydride (PA). In addition, the oxidation of anthracene is also carried out with dichromate and, to a lesser extent, with nitric acid.

Industrial gas-phase oxidation of anthracene is performed in a way comparable to the oxidation of naphthalene to PA or naphthoquinone.

Technically-pure anthracene, containing around 95%, is used as feedstock for anthracene oxidation; it is vaporized with the injection of steam and pre-heated air and then passed together with additional air into a fixed-bed multi-tubular reactor. As for naphthalene oxidation, vanadium pentoxide, doped with alkaline metal oxides, serves as the catalyst. The reaction temperature is between 380 and 420 °C. The anthraquinone is precipitated in powder form through condensation of the reaction product by quenching with cold air. After melting and redistillation, 99% pure anthraquinone is produced.

If the catalyst is modified with titanium and phosphorus compounds it is possible under suitable reaction conditions, to use technical anthracene with a purity lower than 95%. The impurities in the technical anthracene are oxidized to the corresponding carboxylic acids, which can be separated from the anthraquinone with aqueous alkali (*Nihon Iyoryu Kogyo*).

Figure 11.2 shows the flow diagram for gas-phase oxidation of anthracene. Depending on the catalyst, the yield is 80 to 95%.

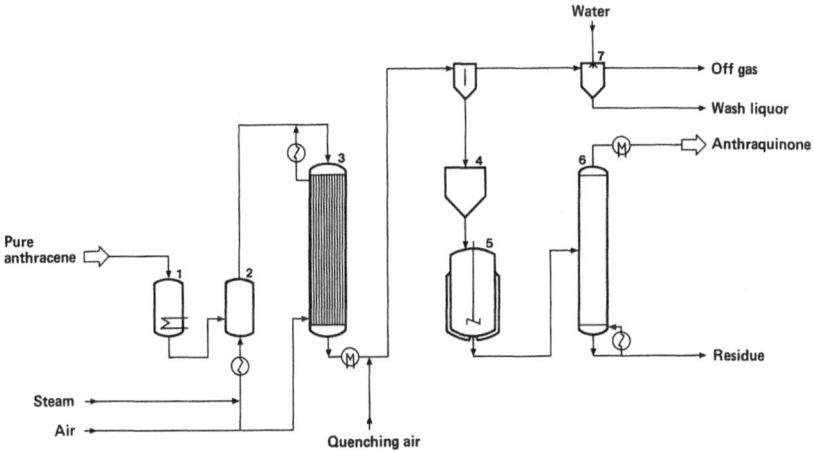

1 Melting vessel; 2 Vaporizer; 3 Multi-tubular reactor; 4 Crude anthraquinone silo; 5 Melting and polymerization vessel; 6 Vacuum distillation; 7 Gas scrubber

Figure 11.2: Flow diagram of gas-phase oxidation of anthracene

Apart from gas-phase oxidation, liquid-phase oxidation with dichromate is an expedient process if the chromium-(III)-sulfate by-product can be used as a tanning agent. The largest anthraquinone plant (*Bayer*/West Germany) with a capacity of 11,000 tpa applies this process. Batch oxidation is carried out with pulverized 94-95% anthracene in a stirred tank reactor with the addition of sodium dichromate and sulfuric acid at 60 to 105 °C. The reaction takes between 25 and 30 hours. The yield of anthraquinone, which is produced in purities of around 95%, is over 90%. The purity can be increased to over 99% by recrystallization from nitrobenzene or cyclohexanol/cyclohexanone.

If there is no possibility of using the chromium-(III)-sulfate lye as a tanning agent, the dichromates can be regenerated by electrochemical oxidation (12 to 16 V).

Oxidation of anthracene with nitric acid is of less significance, but it is applied on a large scale in England (*ICI*).

11.2.2 Synthetic production of anthraquinone

Synthetic production of anthraquinone is based on Friedel-Crafts acylation of benzene with phthalic anhydride, and the Diels-Alder reaction of naphthoquinone with butadiene.

The synthesis of anthraquinone from benzene and PA is applied particularly in developing countries, such as India. In this process, PA, benzene and $AlCl_3$, in the ratio 1:1:2 react together at temperatures below 45 °C. Much foaming occurs and the reaction gives around 95% yield of o-benzoylbenzoic acid. o-Benzoylbenzoic acid is then condensed with sulfuric acid at 115 to 140 °C. The disadvantage of this method consists in the basic problem of Friedel-Crafts acylation, i.e. high consumption of catalyst, linked with the production of heavily contaminated effluent. An advantage is the high yield of anthraquinone, which is over 95%.

The Diels-Alder reaction of naphthoquinone with butadiene is described in Chapter 9.3.2.

A method has been developed by *BASF* to produce anthraquinone by dimerization of styrene followed by oxidation of the 1-methyl-3-phenylindane.

This route, like the Diels-Alder reaction of benzoquinone with butadiene, has found no large-scale application to date.

The same applies to the *American Cyanamid* process, in which anthraquinone is produced in 90% yield from benzene by way of benzophenone by pressure carbonylation at 40 bar with $CuCl_2/PdCl_2$ as catalyst, at 220 to 250 °C.

World capacity for anthraquinone is currently around 30,000 tpa. Chromic acid oxidation processes, which account for around 40%, have the largest share. They are followed by gas-phase oxidation, with around 30%, and Friedel-Crafts acylation of benzene with phthalic anhydride, and nitric acid oxidation, with a combined share of around 30%.

The most important use of anthraquinone is in the production of dyes; a more recent application is in redox catalysis in wood pulping to produce cellulose.

11.3 Anthraquinone derivatives

11.3.1 Production of anthraquinone derivatives from anthraquinone

Alongside azo-dyes, dyes produced from anthraquinone represent one of the most important classes of dyestuffs. As to their color, anthraquinone dyes, with their blue and turquoise shades, complement the yellow to red azo dyes. In spite of their relatively complicated methods of synthesis, anthraquinone dyes find wide application, if their color-fastness and other qualities are superior to those of the competing azo dyes.

Before the introduction of synthetic anthraquinone derivatives, natural anthraquinone dyes have been used for centuries. Apart from alizarin, madder also contains e.g. purpurin, pseudopurpurin, rubiadin and munjistin.

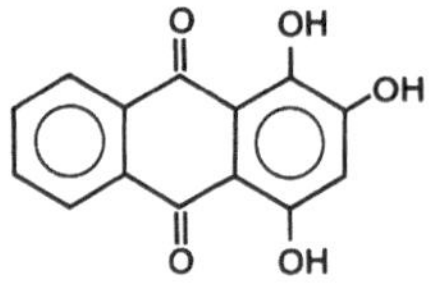

Purpurin

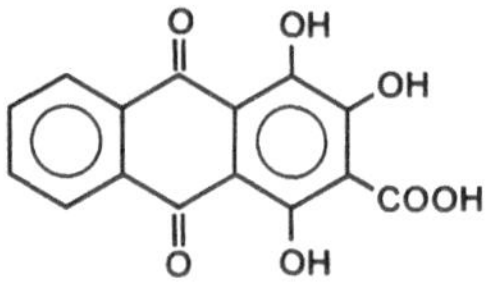

Pseudopurpurin

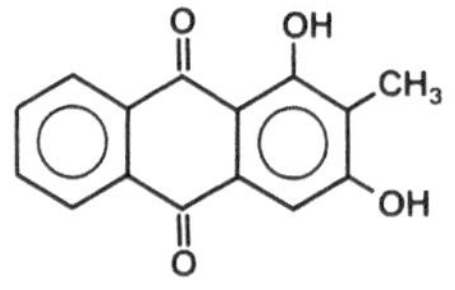

Rubiadin

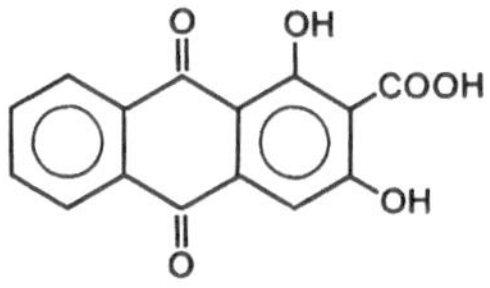

Munjistin

Not only plants, but also insects served as a basis for the production of anthraquinone dyes. A dyestuff known in America for centuries, carminic acid, was produced from cochineal beetles; this dye was first imported into Spain in 1530.

[Carminic acid structure]

Carminic acid

Kermesic acid, a dyestuff produced from kermococcus ilicis (Ilex-scale) was also used in Europe in the 14th and 15th centuries to dye carpets.

[Kermesic acid structure]

Kermesic acid

By far the most versatile synthetic anthraquinone dyestuffs are based on anthraquinone derivatives which are substituted in the 1-, 1,4-, 1,5- and 1,8-positions; 1,2-substituted derivatives, as in the case of alizarin, are also significant.

Production of dyestuff intermediates based on anthraquinone commonly involves the exchange of ring substituents. The exchange of sulfonic acid groups is of particular importance, as used in the last century in the classic alizarin synthesis.

One of the major objectives of anthraquinone chemistry has been the development of selective substitution processes. A milestone in the history of anthraquinone dyes was the discovery, published in 1903 by Michail Iljinski, that anthraquinone can be sulfonated with high selectivity to yield anthraquinone-1-sulfonic acid when mercury is used as a catalyst. Up to a few years ago, this route remained by far the most important reaction in anthraquinone chemistry.

The sulfonation of anthraquinone to produce anthraquinone-1-sulfonic acid is carried out under relatively mild conditions in the presence of mercury (0.5%) with 20% oleum. To avoid disulfonation, the reaction is restricted to a conversion of only 50%, and at a temperature of 120 °C. Anthraquinone is soluble in oleum, pro-

[Reaction scheme: anthraquinone + H_2SO_4 [Hg], $-H_2O$ → anthraquinone-1-sulfonic acid]

Anthraquinone-1-sulfonic acid

ducing a red color. Unconverted anthraquinone is separated together with most of the mercury, which is present as sulfide, by dilution; it is then washed, and fed back into the process. Anthraquinone-1-sulfonic acid is precipitated as the potassium salt (diamond salt).

The main use of anthraquinone-1-sulfonic acid is in the production of 1-aminoanthraquinone. Production of 1-aminoanthraquinone by way of sulfonic acid is diminishing, because of the involved use of mercury. In its place, nitration of anthraquinone, followed by the reduction of 1-nitroanthraquinone, is becoming increasingly common, as is pressure ammonolysis.

1-Nitroanthraquinone   1-Aminoanthraquinone

1-Aminoanthraquinone is produced by the classic route from the potassium salt of anthraquinone-1-sulfonic acid with ammonia and sodium 3-nitrobenzenesulfonate at 175 °C and a pressure of 25 to 30 bar. The sodium 3-nitrobenzenesulfonate serves as an oxidizing agent for the sulfite which is released.

1-Nitroanthraquinone is reduced to 1-aminoanthraquinone with sodium sulfide or sodium hydrogen sulfide; more recently, reduction with Raney nickel has been developed. Alongside the reduction, it is also possible to replace the nitro group with the amino group by reaction with ammonia at 140 to 170 °C, under pressure.

Bromamine acid (1-amino-4-bromoanthraquinone-2-sulfonic acid) is the most important intermediate for the production of acid anthraquinone dyestuffs.

Bromamine acid

It is manufactured by the reaction of 1-aminoanthraquinone with chlorosulfonic acid to yield 1-aminoanthraquinone-2-sulfonic acid; from this, 1-amino-4-bromoanthraquinone-2-sulfonic acid can be obtained by the reaction with bromine, with ice cooling. Examples of dyes from bromamine acid are Alizarine Brilliant Blue R (Acid Blue 62), which is obtained from bromamine acid by reaction with cyclohexylamine in the presence of sodium hydroxide; if aniline is used in place of cyclohexylamine, then Alizarine Saphirol A (Acid Blue 25) is obtained.

Acid Blue 62

Acid Blue 25

Figure 11.3 shows the flow diagram for production of the dyestuff Reactive Blue 19, which is also obtained from bromamine acid.

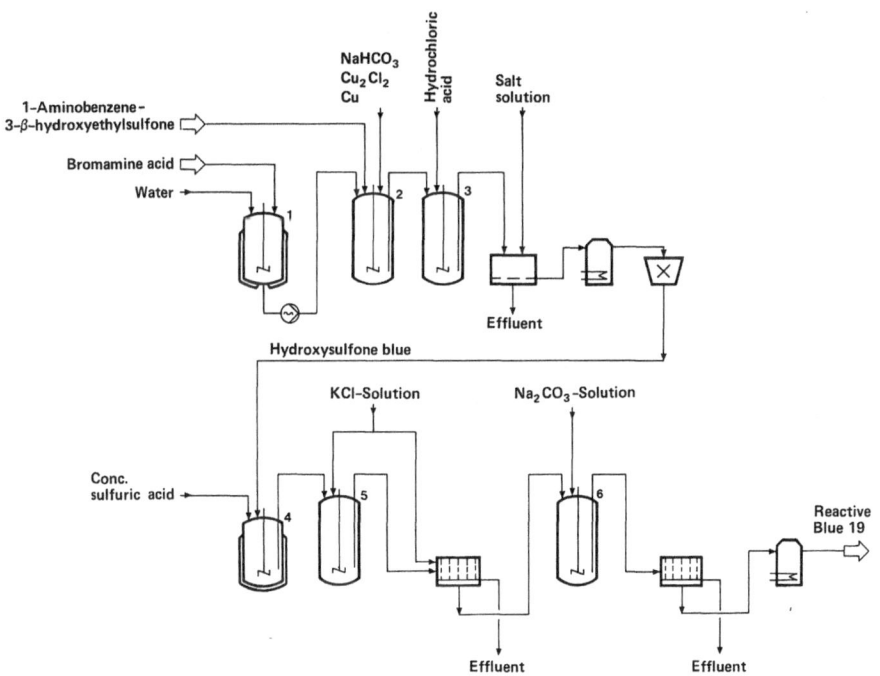

1 Mixing vessel; 2 Reactor; 3 Acidification vessel; 4 Esterification reactor; 5 Precipitation vessel; 6 Reactor

Figure 11.3: Flow diagram of production of Reactive Blue 19

Bromamine acid is made to react with 1-aminobenzene-3-ß-hydroxyethylsulfone at 70 °C; after acidification, the precipitated hydroxysulfone blue is filtered off and, after refining, is then esterified with sulfuric acid. Subsequent refining takes place with soda solution.

[Reaction scheme showing bromamine acid (anthraquinone with NH₂, SO₃Na, Br substituents) reacting with 1-aminobenzene-3-β-hydroxyethylsulfone (NH₂-C₆H₄-SO₂-CH₂-CH₂-OH), with loss of HBr, to form an intermediate anthraquinone with NH₂, SO₃Na, and NH-C₆H₄-SO₂-CH₂-CH₂-OH substituents. This is then treated with H₂SO₄/NaOH, with loss of 2 H₂O, to give Reactive Blue 19, with the side chain SO₂-CH₂-CH₂-O-SO₃Na.]

Reactive Blue 19

11.3.2 Synthetic production of anthraquinone derivatives

Where selective introduction of substituents into the anthraquinone molecule is difficult, synthesis from smaller components is frequently used. The most important reaction component in this context is phthalic anhydride, which undergoes Friedel-Crafts reaction with other substituted aromatics such as chlorobenzene, chlorophenol or toluene to yield the desired anthraquinone derivative.

Production of 1,4-dihydroxyanthraquinone (quinizarin) is a particularly important example. It is obtained by the reaction of phthalic anhydride with p-chlorophenol at 170 to 190 °C.

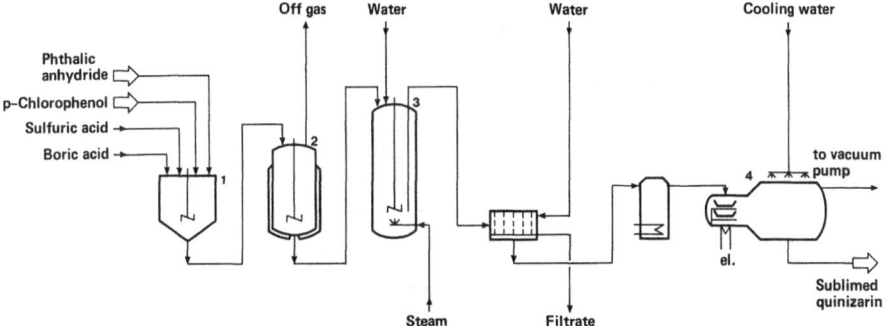

Quinizarin

Figure 11.4 shows the flow diagram for the synthesis of quinizarin.

1 Mixing vessel; **2** Reactor; **3** Hydrolysis reactor; **4** Vacuum sublimation chamber

Figure 11.4: Flow diagram for the synthesis of quinizarin

The boric acid ester, formed as an intermediate with quinizarin, prevents further oxidation, and is hydrolysed with steam at the boiling point. Crude quinizarin is subsequently refined by sublimation.

Acid Green 25

356 Anthracene – production and uses

Quinizarin is used as a raw material in the production of 1,4-diaminoanthraquinone dyestuffs, such as e.g. Alizarine Cyanine Green G (Acid Green 25), which is obtained from quinizarin and p-toluidine, with subsequent sulfonation and neutralization.

The reaction of chlorobenzene with phthalic anhydride to yield 2-chloroanthraquinone is also industrially important. The flow diagram for this Friedel-Crafts reaction is shown in Figure 11.5.

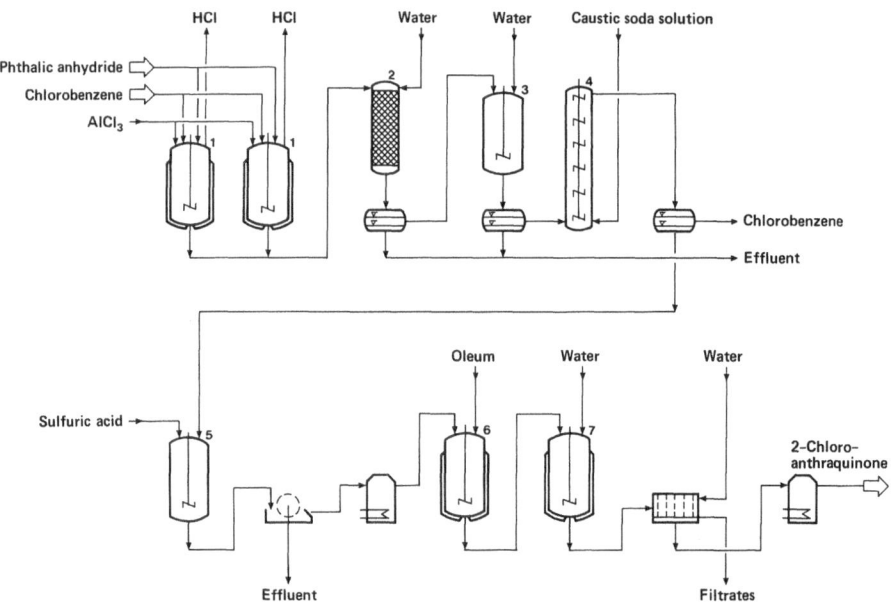

1 Reactors; 2 Hydrolysis column; 3 Washing vessel; 4 Extractor; 5 Precipitation vessel; 6 Reactor; 7 Dilution vessel

Figure 11.5: Production of 2-chloroanthraquinone from chlorobenzene and PA

The reaction is carried out at 75 to 80 °C. Sulfuric acid can be used as a catalyst in place of aluminum chloride. 2-(4'-Chlorobenzoyl)-benzoic acid is obtained as an intermediate and is recovered by vacuum filtration. Subsequent condensation to 2-chloroanthraquinone is effected with oleum.

2-(4'-Chlorobenzoyl)
benzoic acid

2-Chloroanthraquinone

2-Chloroanthraquinone is especially useful as a base material for the production of 2-aminoanthraquinone, from which the vat dye indanthrene (Vat Blue 4) is produced by a method discovered by René Bohn in 1901, involving a reaction in alkaline solution with atmospheric oxygen in the presence of potassium nitrate at 150 to 200 °C. At higher temperatures and using antimony-(V)-chloride or aluminum chloride, flavanthrone (Vat Yellow 1) is obtained; this is nowadays produced from 1-chloro-2-aminoanthraquinone and PA.

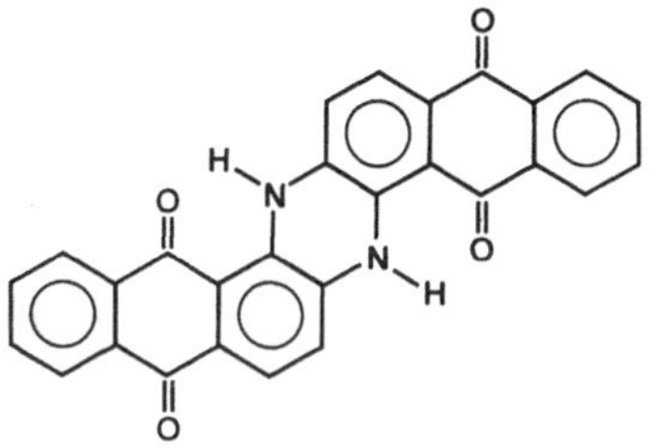

Vat Blue 4

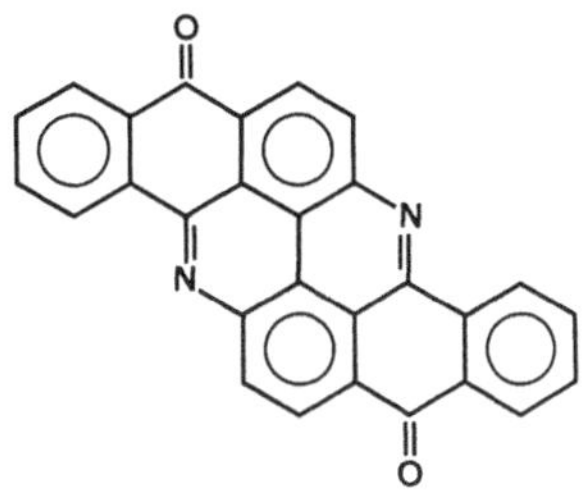

Vat Yellow 1

11.4 Higher condensed dyes from anthraquinone

Alongside the production of dyes based on the anthraquinone structure, anthraquinone also serves as a base material for the production of more highly condensed aromatics, especially benzanthrones. Benzanthrone is produced from anthraquinone by reaction with glycerol, sulfuric acid and iron. Anthrone occurs as an intermediate in the reaction and condenses at its methylene group with acrolein, which is produced by dehydration from the glycerol; this leads to a ring closure. Subsequent dehydrogenation of the dihydrobenzanthrone can be achieved by its oxidation in the presence of copper-II-sulfate.

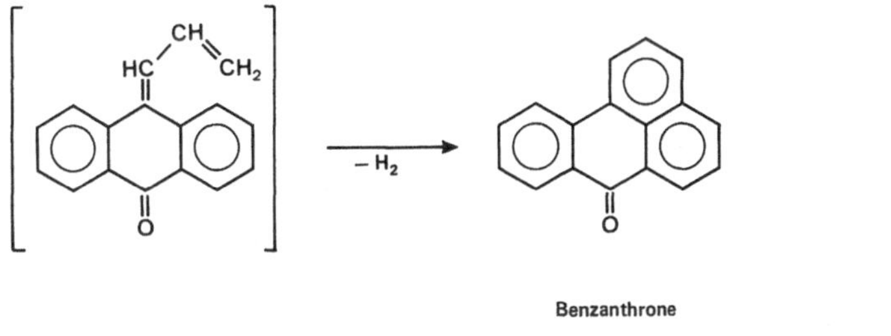

Benzanthrone

Benzanthrone per se is a vat dye, but it is more commonly used to produce Indanthren Brilliant Green B (Vat Green 1). In this reaction, benzanthrone reacts with air in the presence of potassium hydroxide in isopropanol, to form 4,4'-dibenzanthronyl. On further heating, the 4,4'-dibenzanthronyl undergoes ring closure to yield violanthrone. 4,4'-Dibenzanthronyl can also be converted into 16,17-violanthronequinone with manganese dioxide and sulfuric acid at 35 °C. Reduction with sodium hydrogen sulfite leads to 16,17-dihydroxyviolanthrone, from which Indanthren Brilliant Green B (Vat Green 1) is produced by methylation.

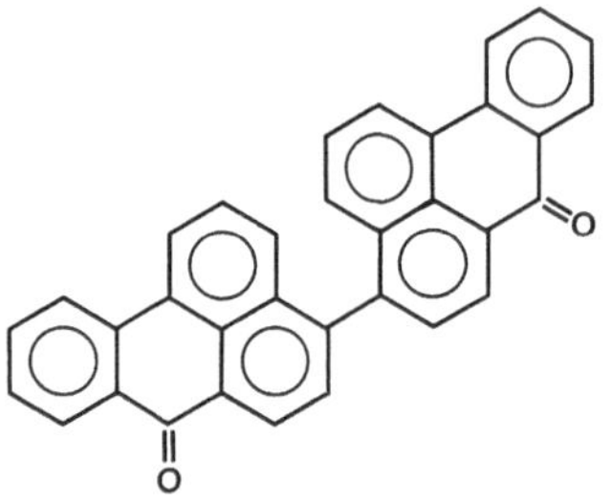

4,4'-Dibenzanthronyl

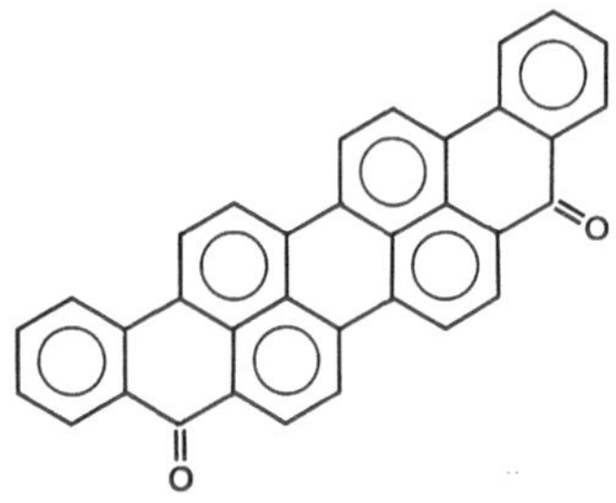

Dibenzanthrone/Vat blue 20
(Violanthrone)

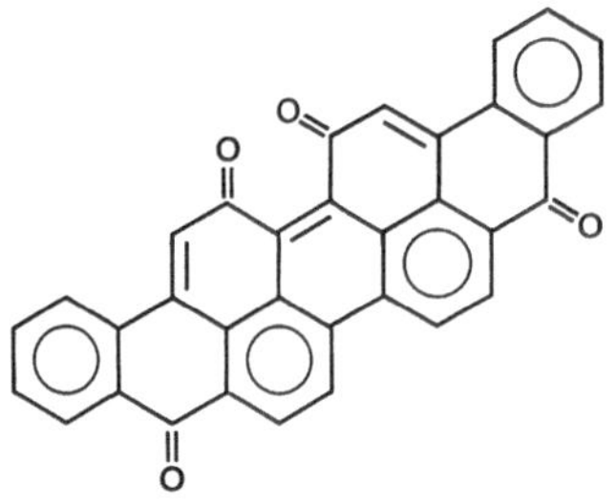

16,17-Violanthronequinone

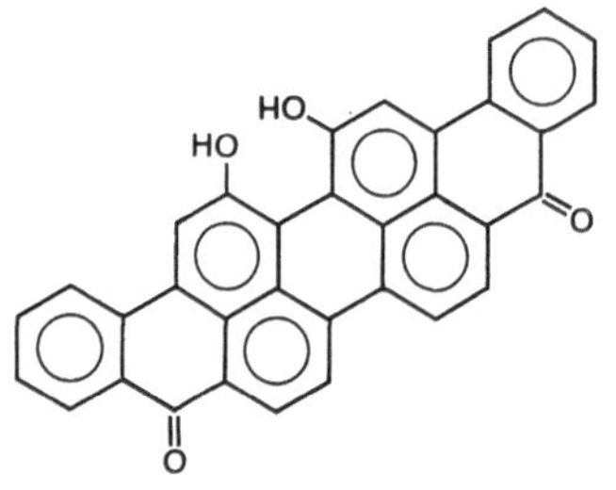

16,17-Dihydroxyviolanthrone

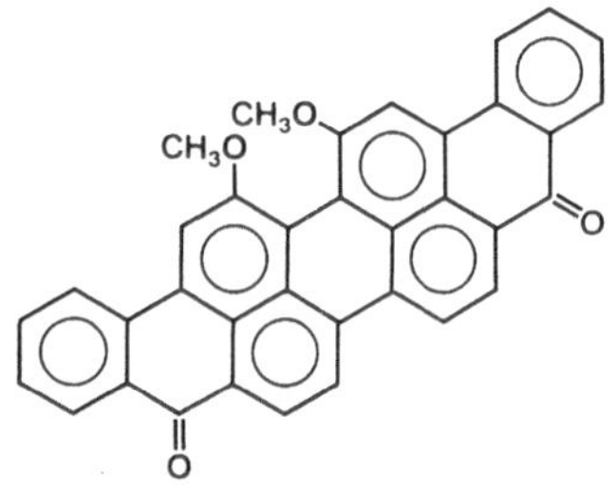

Vat Green 1

11.5 Anthraquinone as a catalyst in the production of hydrogen peroxide

Hydrogen peroxide, which is produced worldwide in quantities of 800,000 tpa is manufactured by wet-chemical processes based on barium peroxide, electrochemical processes and organic autoxidation processes. Hydrogen peroxide production by autoxidation was developed to industrial viability by *BASF* in the 1930's, in the form of the hydrazobenzene process. The hydrazobenzene process, however, was characterized by the disadvantage that sodium amalgam had to be used to reduce azobenzene to hydrazobenzene. Georg Pfleiderer and Hans-Joachim Riedl then used alkylated anthraquinones in place of azobenzene. The first plant to use this process to produce H_2O_2 was started in Memphis, Tenn. by *Du Pont* in 1953.

In the anthraquinone process, an alkylanthraquinone is catalytically reduced to the corresponding hydroquinone with hydrogen; the hydroquinone reacts with oxygen (air), becoming re-oxidized to the anthraquinone derivative, with the formation of hydrogen peroxide.

2-Ethylanthraquinone, 2-tert-butylanthraquinone and 2-pentylanthraquinone are commonly used as anthraquinone derivatives; these alkylanthraquinones are obtained by Friedel-Crafts acylation of PA and the corresponding alkylbenzenes, and are then dissolved in an aromatic solvent (alkylated benzenes, methylnaphthalene). The alkylanthraquinones are hydrogenated with a catalyst (palladium, Raney nickel) under low hydrogen pressure and at low temperature. The hydrogenation not only converts the quinone to the hydroquinone, but may also hydro-

2-Ethyl-5,6,7,8-tetrahydroanthraquinone epoxide

genate the unsubstituted anthracene ring. Reaction of this tetrahydroanthrahydroquinone with oxygen can form an epoxide, which does not participate in the formation of H_2O_2, and therefore leads to a loss of active quinone.

Reaction conditions in hydroquinone production are selected to ensure the most selective hydrogenation possible of the oxygen of quinone to the hydroquinone.

Figure 11.6 shows the scheme of alkylanthraquinone hydrogenation using the *Degussa* process. The reaction is carried out in a loop reactor; by varying the diameter of the reactor tubes different reactant flow velocities are achievable.

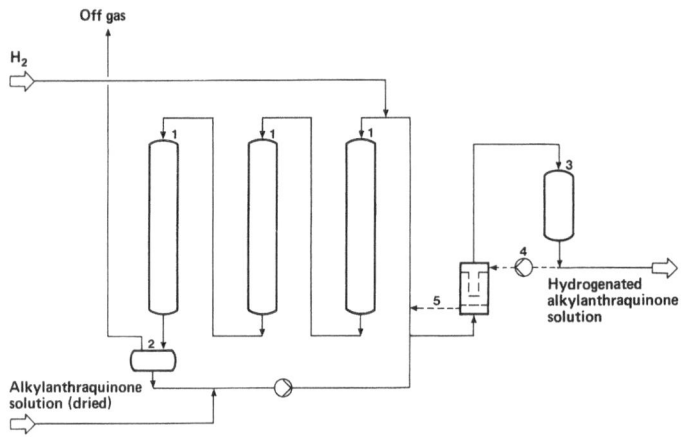

1 Hydrogenation reactors; 2 Gas separator; 3 Storage vessel; 4 Backwash pump; 5 Backwash circuit

Figure 11.6: Flow diagram for the hydrogenation of alkylanthraquinone

11.6 Wood pulping with anthraquinone

In 1977, a very promising new application for anthraquinone, with exiting future prospects was introduced in wood pulping.

A primary objective of wood pulping is to separate lignin from cellulose. Two chemical processes are commonly used to achieve this, namely the sulfite and the sulfate methods (Kraft-process). In the sulfite process, the lignin in wood is sulfonated with aqueous solutions of sulfites or hydrogen sulfites, with added SO_2, at high temperatures, so that soluble lignin sulfonic acids are produced, which can be dissolved out of the wood. The sulfate method uses sodium hydroxide, sodium sulfide and sodium carbonate as basic pulping chemicals. It derived its name from the small amounts of sodium sulfate which are added to balance out alkali losses.

The addition of less than 0.1% of anthraquinone can increase the cellulose yield and accelerate the delignification process, leading to better utilization of the capacity. Furthermore, with alkaline pulping, less sodium sulfide has to be added, so that pollution of the atmosphere and contamination of the waste water are reduced.

The reaction of lignin occurs as a result of redox catalysis by the anthraquinone. The mechanism of the reaction between anthraquinone and lignin involves a cleavage in the ß-position.

In addition to anthraquinone, tetrahydroanthraquinone can also be used to accelerate wood pulping.

The anthraquinone pulping process is especially used in Japan, Scandinavia, USA and Canada.

12 Additional polynuclear aromatics – production and uses

Next to anthracene, the higher condensed aromatics which are of commercial importance are phenanthrene, fluorene, fluoranthene and pyrene. Worldwide production of these polynuclear aromatics is around 2,000 tpa.

12.1 Phenanthrene

Phenanthrene is present in coal tar at a concentration of about 5%; it was discovered by Rudolph Fittig and Eugen Ostermayer in 1872.

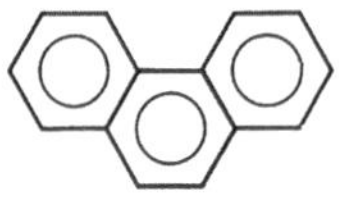

Phenanthrene

In nature, phenanthrene compounds occur in alkaloids, such as morphine.

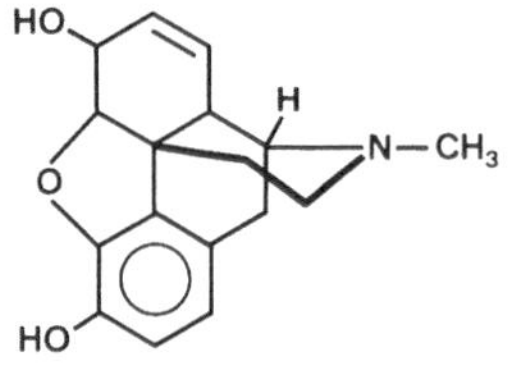

Morphine

As a feedstock for the production of phenanthrene, the filtrate from pure anthracene production is particularly suitable (see Chapter 11.1), containing around 65% phenanthrene with an anthracene content below 10%. Distillation produces 90% pure phenanthrene; the major contaminants, in terms of quantity, of technically pure phenanthrene are anthracene and diphenylene sulfide (dibenzothiophene).

Diphenylene sulfide

Anthracene can be separated from the other polycyclics by adduct formation (Diels-Alder reaction) with maleic anhydride.

The most important product from phenanthrene is 9,10-phenanthrenequinone, which is obtained in 85 to 90% yield by liquid-phase oxidation (or in suspension) of phenanthrene with dichromate in sulfuric acid at 80 to 85 °C. Gas-phase oxidation gives poorer yields.

9,10-Phenanthrenequinone

9,10-Phenanthrenequinone is used as a base material for the plant growth regulator flurenol-n-butyl ester, which is produced from phenanthrenequinone by benzilic acid rearrangement to yield 9-hydroxyfluorene-9-carboxylic acid, and subsequent esterification with n-butanol. It is applied in mixtures with other herbicides, such as e.g. MCPA.

Flurenol-n-butyl ester

Chloroflurenol-methyl ester, likewise a growth regulator, is related to flurenol-n-butyl ester.

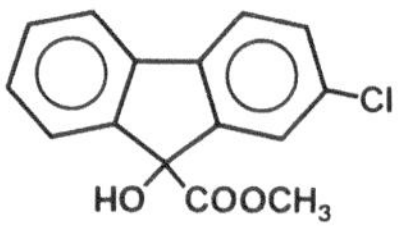

Chloroflurenol-methyl ester

Another derivative of phenanthrene is 2,2'-diphenic acid, which can be used in the production of polyesters and alkyd resins.

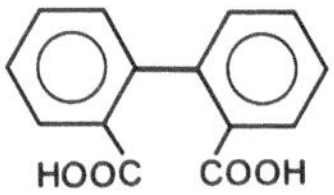

2,2'-Diphenic acid

12.2 Fluorene

Fluorene was discovered in coal tar, in which it is present in amounts of around 2%, by Pierre M. Berthelot in 1867.

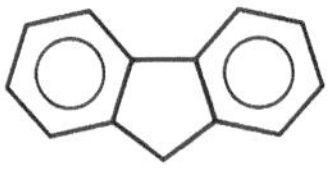

Fluorene

It is recovered by redistillation of the fluorene oil fraction, which boils between 290 and 305 °C (or from the distillation fore-runnings in anthracene production), followed by recrystallization, for example, from solvent naphtha. Technical fluorene, 95% pure, is commonly used to produce fluorenone by liquid-phase oxidation with air/oxygen at around 100 °C. Fluorenone can principally be used as a mild oxidant for Oppenauer oxidation, particularly in steroid chemistry.

Fluorenone

12.3 Fluoranthene

Next to naphthalene and phenanthrene fluoranthene is one of the main constituents of coal tar, being present in excess of 3%. It was discovered in coal tar by Rudolph Fittig and Ferdinand Gebhard in 1878.

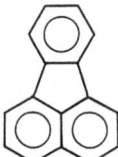

Fluoranthene

Fluoranthene is recovered by the distillation of high-boiling anthracene oil fractions, or from pitch distillates followed by recrystallization of the fluoranthene fraction, which boils between 375 and 385 °C. The technical product is around 95% pure.

Fluoranthene is especially used to produce fluorescent dyes.

12.4 Pyrene

Pyrene was discovered in coal tar by Carl Graebe in 1871. It is present in concentrations of nearly 2%.

Pyrene

The recovery of pyrene is based on the distillates of coal tar pitch, which arise during hard-pitch production and pitch coking. The pyrene fraction boils between 320 and 420 °C and contains 5 to 7% pyrene. It is concentrated to 50% by distillation, then further refined to 95% technical pyrene by recrystallization from solvent naphtha or acetophenone.

Pyrene is mainly used in the production of 1,4,5,8-naphthalenetetracarboxylic acid, a starting material for perinone pigments, which are characterized e.g. by their good heat stability.

In the *Hoechst* process, pyrene reacts with bromine to yield 1,3,6,8-tetrabromopyrene; this is oxidized with sulfuric acid, to produce the intermediate 2,7-dibromo-1,2,3,6,7,8-hexahydro-1,3,6,8-tetraoxopyrene; subsequent oxidation in an alkaline medium yields the tetra-sodium salt of 1,4,5,8-naphthalenetetracarboxylic acid.

Reaction with 1,2-diaminobenzene in glacial acetic acid produces a mixture of 'cis'- and 'trans'-isomer perinones; the isomers can be separated by fractional precipitation with ethanol/KOH, the trans-isomer (Pigment Orange 43) precipitating as an orange-colored addition compound. Pigment Orange 43 is particularly used to color plastics such as PVC or polyethylene, and for PAN dope-dyeing. The cis-isomer (Pigment Red 194) is mainly used to pigment paints.

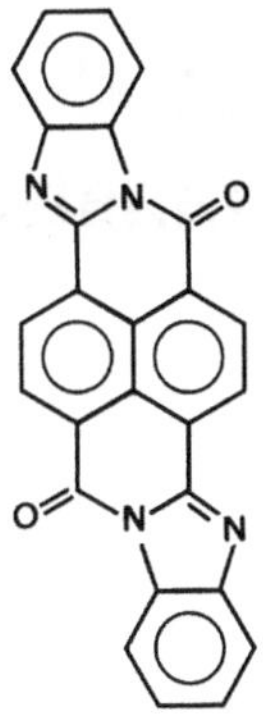

Additional polynuclear aromatics - production and uses

Pigment Orange 43

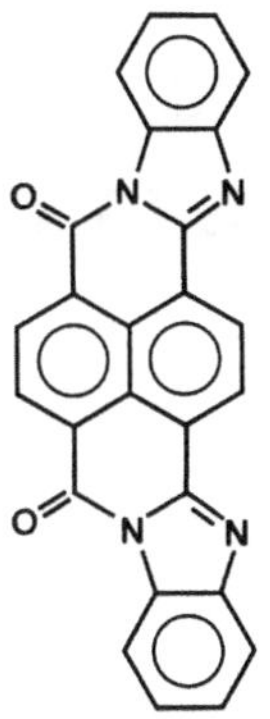

Pigment Red 194

13 Production and uses of carbon products from mixtures of condensed aromatics

Aromatic residue fractions produced in coal tar refining, naphtha pyrolysis and thermal as well as catalytic cracking (see Chapter 3) are suitable feedstocks for the manufacture of carbon products, especially because of their high C/H ratio. Graphitic carbon products include synthetic graphite and graphitic carbon fibers which are distinguished by both the anisotropy in the structure of the carbon lattice, and a lattice plane distance which is very close to the theoretical value for graphite. In addition, carbon products with a high proportion of isotropic regions are used on a large scale, especially for the manufacture of anodes for the aluminum industry.

The conversion of carbon-rich refining residues, which consist mainly of three-, four- and five-ring aromatics with partial alkyl substitution, into special cokes, graphite and carbon fibers occurs predominantly in the liquid phase. Carbon black, another important carbon product, on the other hand is obtained by pyrolysis in the gas phase.

The graphitic materials produced from aromatic mixtures are complemented by natural graphite, which is produced in particular in China, the Soviet Union, Sri Lanka, West Germany, Austria and Mexico, in quantities totalling around 500,000 tpa. It is recovered by mining and the raw graphite is cleaned by flotation. The level of ash-forming agents can be reduced by treatment with hydrofluoric acid and by alkali fusion. Natural graphite is used in the iron and steel industry to produce smelting crucibles and carburizers for adjusting the carbon content of steel. Recently, the production of graphite foils as a replacement for asbestos in packings has gained increasing importance.

13.1 Pyrolysis of aromatic hydrocarbon mixtures in the liquid phase

13.1.1 Formation of mesophase

There is a large number of stages of orientation in the production of carbon products with a graphitic crystalline structure from high-boiling, aromatic residues with the concurrent elimination of hydrogen.

The mesophase stage is of particular importance. It occurs in the temperature range between 300 and 500 °C and is characterized by the close association of large-area aromatics with the resultant formation of an optically-visible liquid-crystal phase.

After the semi-coke phase, which extends up to around 700 °C, further increase in temperature brings about the formation of anisotropic coke, which can be used as the feedstock for the production of graphite, especially graphite electrodes.

Insufficient pre-orientation in the mesophase stage, as a result of low aromaticity of the respective residue, or through too rapid a heating to temperatures in excess of 500 °C, leads to isotropic cokes, which are principally used for the manufacture of anodes for aluminum production by electrolysis.

The anisotropic cokes and carbon blacks partially consist of para-crystalline structures. The degree of orientation of the graphitic carbons can be determined by the spacing of the graphite layers; for the ideal crystal, it is 0.3354 nm, for coke 0.34 to 0.35 nm, and for carbon blacks 0.36 nm. Figure 13.1 shows the crystal lattice of hexagonal graphite.

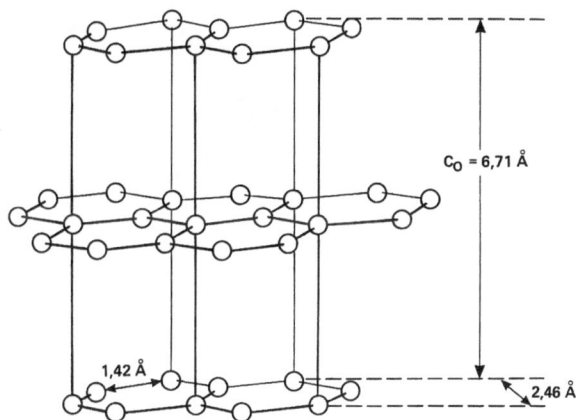

Figure 13.1: Lattice structure of the hexagonal graphite

During the course of the pyrolysis of a petroleum-derived residue (pyrolysis tar, cat-cracker residue) or a filtered coal tar pitch it is possible to observe, under the polarisation microscope and at a certain temperature, the formation of anisotropic spherules, which grow as the reaction time lengthens and the temperature increases, coalesce and, at around 500 to 600 °C, are transformed into a semi-coke phase with marked anisotropy. Figure 13.2 shows photomicrographs of a filtered coal tar pitch pyrolyzed at 400 °C with the formation of spherulitic mesophases after reaction times of 2, 6, 10 and 16 hours.

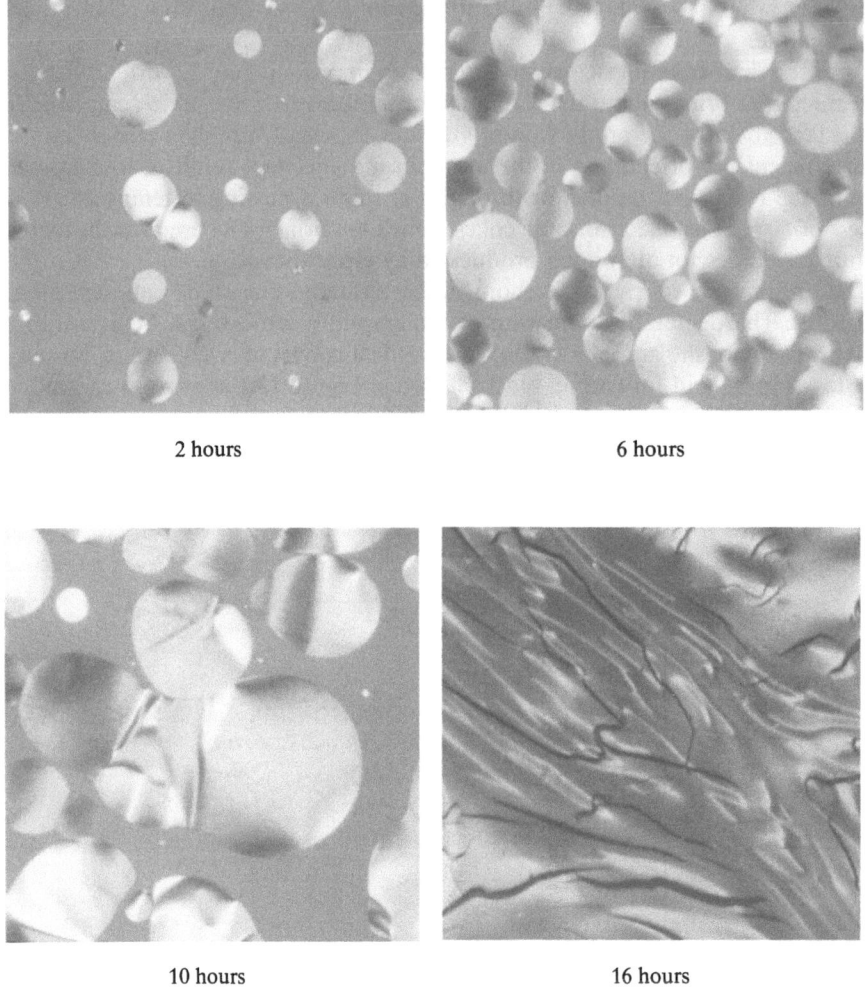

2 hours	6 hours
10 hours	16 hours

Figure 13.2: Progression of mesophase formation and coalescence during pyrolysis of filtered coal-tar pitch at 400 °C

During nucleation, growth and coalescence, the pitch phase exhibits a liquid-crystalline character. The optically-visible liquid-crystal spherical mesophases have a diameter of around 1 to 30 μm. The isotropic pitch matrix initially contains none or few of those aromatics which are capable of forming the liquid-crystal phase (mesogens). Only by thermally-induced reactions are larger mesogen molecules of different nature formed. The composition of the mixture of mesogens and non-mesogens undergoes constant change because of the thermal reaction in progress.

Unlike liquid-crystal formation in synthetically produced mesogens, such as e. g. 4′-octylbiphenyl-4-carbonitrile, reversible liquid-crystal formation is generally not observed in pitches, since, in the latter case, the formation of liquid crystals is accompanied by a chemical reaction (dehydrogenation).

4'-Octylbiphenyl-4-carbonitrile

A large number of conditions must be fulfilled for the formation of liquid crystals, which particularly concern the geometry of the potential mesogen. Using the example of the hexaphenyls, it can be shown that a high length/diameter ratio is necessary for mesophase formation; thus only the linear hexaphenyl is capable of forming the liquid-crystal phase, as was postulated by Paul J. Flory in 1956, on the basis of theoretical considerations.

(liquid-crystal forming)

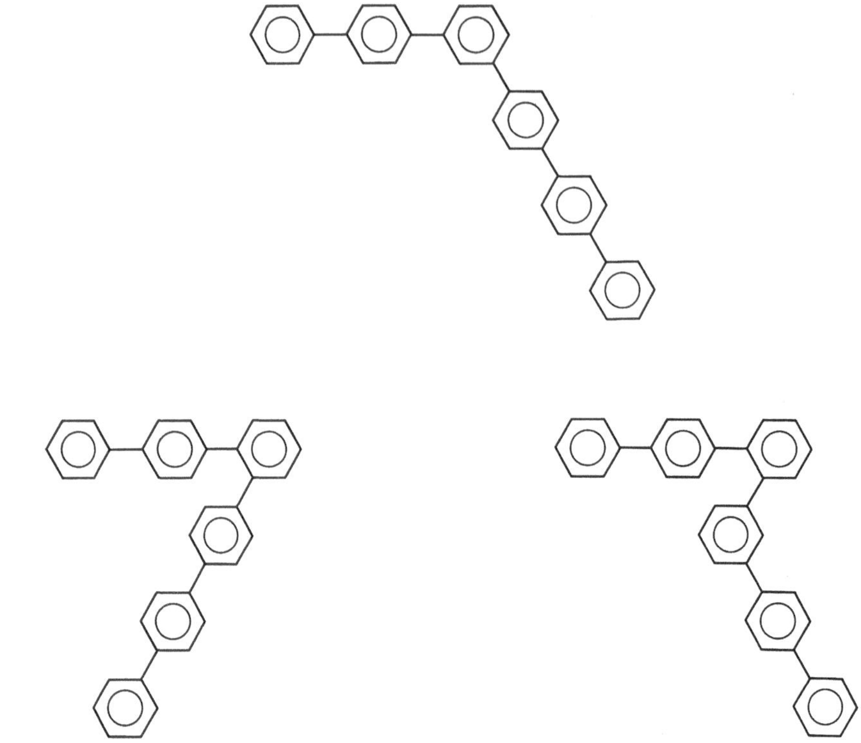

Figure 13.3: Structure of hexaphenyls

The mesophases are formed from aromatics which are produced from smaller molecules by condensation. Since the large-area molecules behave like discs, the liquid-crystal phases of pitch aromatics are referred to as discotic liquid crystals or, when arranged along a preferred axis, as discotic nematic phases.

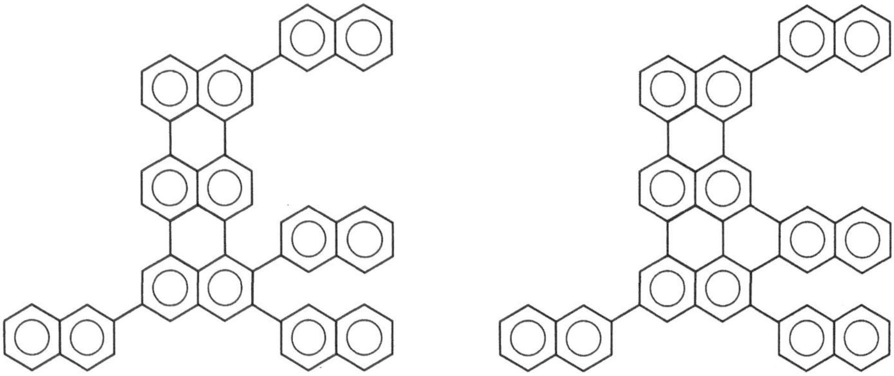

Heptamers

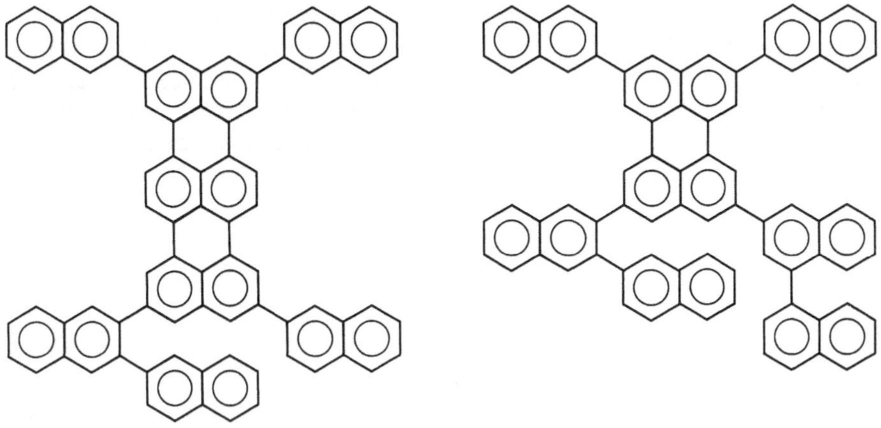

Production and uses of carbon products 373

Octamers

Figure 13.4: Condensation products from the pyrolysis of naphthalene

For a pitch from the pyrolysis of naphthalene, heptameric and octameric naphthalene condensation products have been postulated as model mesogens; the corresponding molecular weights are 878/876 and 1006/1008 (Figure 13.4).

Figure 13.5 shows the polymer structure suggested for a synthetically-produced pitch from phenanthrene.

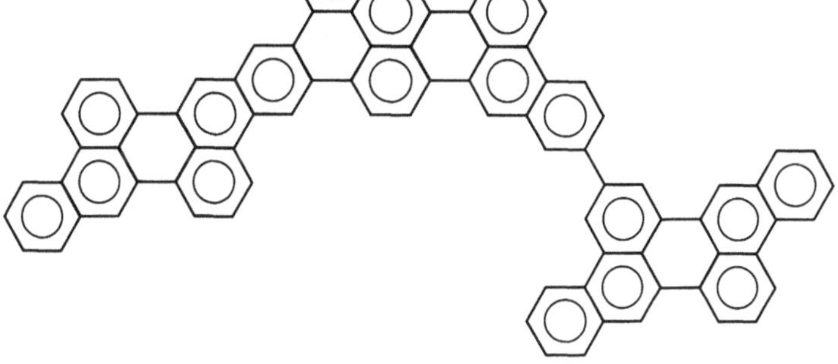

Figure 13.5: Phenanthrene polymer structure

The molecular weight of isotropic pitch and mesophase pitch differ especially in the width of their distributions. Figure 13.6 shows the molecular weight distribution of isotropic and mesophase pitch from the pyrolysis of naphthalene. The mesophase pitch displays a broader molecular weight distribution and, in addition, the maximum of the distribution curve is found at higher molecular weights than for isotropic pitch.

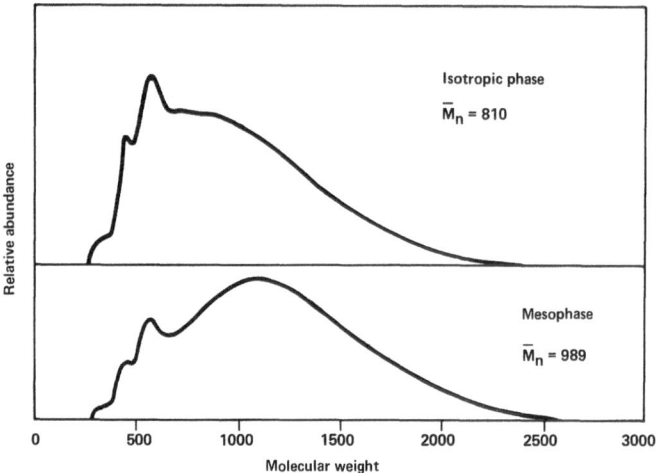

Figure 13.6: Molecular-weight distribution of mesophase pitch and isotropic pitch from the pyrolysis of naphthalene

Mesophase formation is also detectable in the viscosity behavior of pitch. Figure 13.7 shows the viscosity curves obtained on heating a coal tar pitch and measuring the viscosity under different shear stresses in a rotation viscometer. Mesophase formation is characterized by the occurrence of a minimum in the viscosity, which is particularly evident with the application of low shear stress.

The mesophase phenomenon, discovered by James D. Brooks and Geoffrey H. Taylor in 1965, made a decisive contribution to the understanding of the production of high-value carbon products from mesophase pitch, such as premium coke by delayed coking and carbon fibers by spinning.

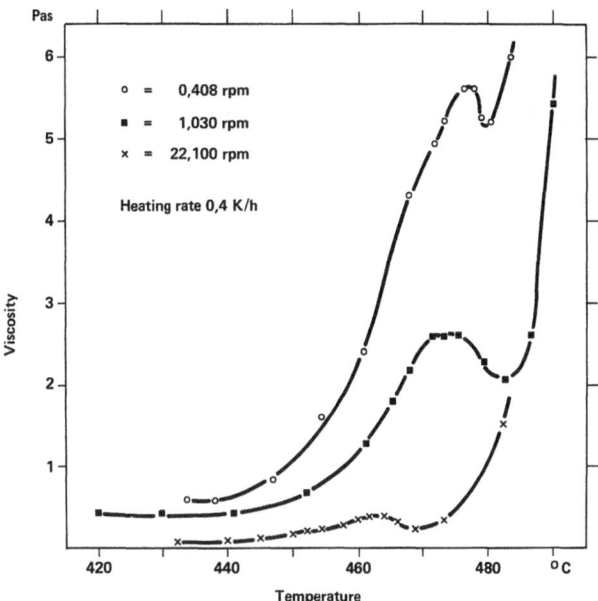

Figure 13.7: Viscosity pattern on heating a coal tar pitch, with mesophase formation

13.1.2 Production and uses of coke from aromatic residues by the delayed coking process

Hydrocarbon carbonization is carried out today with the following objectives:

1. Production of low-molecular weight liquid hydrocarbons with the concurrent formation of coke as an unavoidable by-product, i.e. the conversion of petroleum residues for producing middle distillates and gasoline.
2. Manufacture of high-value coke (premium coke) with simultaneous production of minor amounts of liquid and gaseous hydrocarbons, i.e. the production of premium coke from highly-aromatic residues (coal-tar pitch, pyrolysis residues from naphtha cracking, residues from thermal and catalytic cracking).

Coking is a disproportionation of hydrocarbons into highly-enriched carbon and hydrogen-enriched hydrocarbon compounds; it occurs predominantly in the liquid phase. The most important process for carbonizing liquid hydrocarbons (i.e. not the production of metallurgical coke from coal) is the delayed coking process, which was developed in the USA in the 1930's to convert petroleum residues to gasoline. In 1968, the delayed coking process was likewise introduced for coal tar pitch by *Nippon Steel Chemical* (Tobata, Japan). Coke production in the Western world by the delayed coking process is around 18 Mtpa, with by far the greatest part being produced in the USA.

A delayed coker consists of a furnace to heat the feedstock to around 500 °C, two pressure vessels (coke drums) with a diameter of 4 to 7 m and 20 to 30 m high, together with a distillation system to separate the volatile components. Figure 13.8 shows the flow diagram for the delayed coking process.

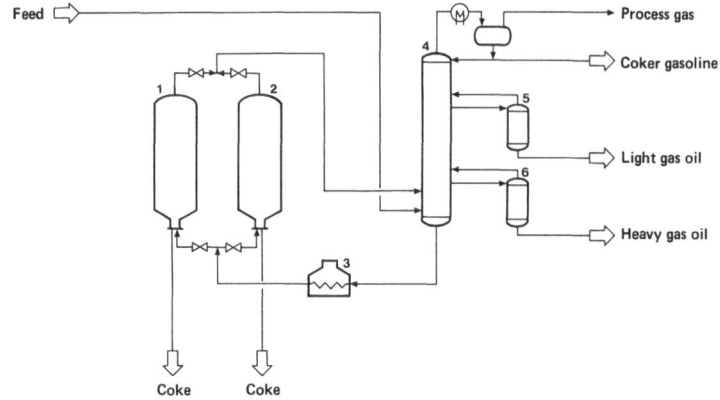

1 and 2 Coke drums; 3 Tubular furnace; 4 Distillation column; 5 and 6 Side columns

Figure 13.8: Flow diagram of the delayed coking process

The feedstock consists of the raw material and the recycled bottom product of the distillation. It is fed in with steam at a velocity of 2 m/sec and heated in chromium-alloy tubular furnaces to around 500 °C. Sufficient heat transfer is guaranteed by the high turbulence; in addition, the tendency of the coke to deposit on the heater walls is reduced. The coke drums are charged with the heated feedstock at the bottom. The filling height is normally monitored by radioactive probes, which are fitted into the side of the upper third of the drum. Foaming of the feedstock is suppressed by the addition of silicone oil. The coking pressure is around 2 to 7 bar; high pressures and high recycling ratio result in increased coke yield.

After an average coking time of around 12 hours, during which a slightly endothermic reaction transforms the hydrocarbons to high-molecular reaction products, the feed is switched to the second drum and the full drum is purged with steam (steaming phase), to drive off the volatile components which have not been converted into coke, until a green coke of a residual volatile content of below 15% remains. In the steaming phase, the coke is cooled to around 300 °C; further cooling is effected with water.

After de-pressurizing and opening the coke drum, a central hole with a diameter of 600 mm is drilled through the coke bed with a high pressure water jet, through which further horizontal cutting of the coke is carried out. Using a water drill which is lowered down, the coke is cut out in discs under water pressures up to 300 bar. The coke pieces, some of which have a diameter over 40 cm, are subsequently broken down and fed into a bin to remove water. When the coke has been cut and removed from the drum, the coke drum is checked, closed and then preheated with steam and coke vapors from the operating chamber. A new cycle begins every 48 hours. The volatile components are refined by distillation.

To produce gasoline and middle distillates by delayed coking, atmospheric or vacuum residues from petroleum refining are used. When the desired end-product is high-value coke, then highly-aromatic pyrolysis residues from ethylene production, thermal-cracker residues, cat-cracker residues and coal-tar pitch which has been freed of quinoline-insoluble components and ash-forming compounds are the usual feedstocks. The use of these raw materials in the delayed coking process leads to the formation of mesophases by the association of planar large-area aromatics; the mesophases are oriented in the direction of flow by the shearing action of the rising gas bubbles.

Figure 13.9: Delayed coker operated by *Conoco*, Immingham/ England

Figure 13.9 shows the four delayed coker units operated by *Conoco, Immingham/England*, which have a production capacity for 300,000 tpa of premium coke.

The green coke obtained with a highly-aromatic feedstock displays needle-like structures and is therefore known as needle coke. If, however, an atmospheric or vacuum residue from mineral-oil refining is used as feedstock, then a foam-like coke (sponge coke) is produced.

The coke yield is particularly dependent on the coking value of the raw material; for atmospheric residues, it is around 20%, for pyrolysis residues around 35% and for coal tar pitch over 50%. Table 13.1 compares the yields from delayed coking of residues of different origins.

Table 13.1: Yields from different feedstocks used in delayed coking

Feedstock	Brega	Light Arabian	Thermal-cracker tar	Heavy Arabian	Coal-tar pitch
Density (g/cm^3)	0.984	1.019	1.030	1.040	1.223
Conradson carbon residue (%)	14.6	15.4	8.0	24.2	31.2
Sulfur content (%)	1.06	4.1	0.7	5.25	0.48
Yields:					
Gas to C$_4$ fraction, (wt%)	7.0	11.1	15.0	13.2	3.0
Gasoline fraction (up to 195 °C, wt%)	18.6	16.1	10.5	13.5	10.7
Gas oil (over 195 °C, wt%)	52.4	45.8	32.5	40.4	25.4
Coke (wt%)	22.0	27.0	42.0	33.0	60.9

Green coke has a carbon content of around 92%. To produce carbon products, the green coke, which contains up to 15% volatiles, is calcined, reducing the volatile content to around 0.1%. Calcination takes place in rotary hearth or in rotary kiln calciners at temperatures from 1250 to 1450 °C; the volatile components are used as fuel.

Table 13.2 compares the elementary composition of green coke and calcined coke.

Table 13.2: Composition of green coke and calcined coke

	Green coke	Calcined coke
Carbon (%)	91.80	98.40
Hydrogen (%)	3.82	0.10
Oxygen (%)	1.30	0.02
Nitrogen (%)	0.95	0.22
Sulfur (%)	1.29	1.20
Ash (%)	0.35	0.35
C/H ratio (Atomic)	2.00	82.00

Green coke with a high sulfur and metal content is used as fuel e.g. in the cement industry. Green petroleum cokes also serve as modifiers for coking coal in the manufacture of metallurgical coke.

Calcined cokes are used, depending on their quality, to produce shaped graphite products (e.g. electrodes), anodes for aluminum electrolysis or as reduction agents (e.g. in titanium production).

To produce large graphite electrodes (Ultra High Power (UHP) electrodes), of a diameter up to 600 mm, the high-value premium coke is mixed with around 20% coal-tar binder pitch and extruded e.g. in a tilting extrusion press. The extrusion ram entering the mud cylinder forces the mix through the die at a pressure of 70 to 150 bar. The green electrodes are then baked in a ring furnace up to 1300 °C. The coal-tar pitch used as binder is made from tars especially selected on the basis of their QI content. To increase the carbon density of the baked carbon product, it is impregnated with low-QI pitches in a vacuum/pressure process.

Table 13.3 summarizes the major characteristics of binder and impregnating pitches. Eletrode binders are produced exclusively from coal tar whereas impregnating pitches can be produced from low-QI tar or by air blowing and heat-treatment of cat-cracker residues.

Table 13.3: Typical data for binder pitch and impregnating pitch used in the production of graphite electrodes

	Binder pitch	Impregnating pitch
Softening point, °C (Kraemer-Sarnow)	95	60
Quinoline insoluble, % (QI)	10	2
Toluene insoluble, % (TI)	39	18
Coking residue (Conradson)	56	38
Viscosity (150 °C, mPas)	1200	55

Graphitization of the baked and impregnated carbon products occurs at temperatures from 2,500 to 3,000 °C, using the Acheson process (transverse graphitization) or the Castner process (lengthwise graphitization).

Graphite electrodes are used in the electric arc production of steel, whereby 3 to 5 kg of electrode material is consumed to produce 1 t of steel. A further wide field of application for graphite is the electrical industry, where carbon/graphite is used e.g as material for brushes. Graphite is also used to manufacture apparatus (e.g. heat exchangers) and, most recently, for heating elements to manufacture pure silicon for the production of wafers.

World graphite production in 1985 was around 1 Mt. The main producer countries are the USA (e.g. *Union Carbide, Great Lakes*), Japan (e.g. *Showa Denko*) and West Germany (e.g. *Sigri, Conradty*).

The sponge-type 'regular coke' from the delayed coking of petroleum residues (petroleum coke) and coal-tar pitch (pitch coke), along with pitch coke from the horizontal-chamber coking process, are principally used in the manufacture of anodes for the aluminum industry (see Chapter 13.1.3).

13.1.3 Pitch coking in horizontal chamber ovens

In West Germany and the USSR pitch coke is also produced by horizontal-chamber coking in modified coke ovens, as used for coal coking.

To increase the density of the carbon, 'normal' (straight-run) pitch (softening point 70 °C), arising from the distillation of coal tar, is first blown with air in a batch process, or brought to a softening point of around 160 °C by flash distillation. This liquid hard pitch is then fed into the chambers at the top of the coke oven. After 14 to 24 hours coking at a temperature of up to 1200 °C an isotropic coke is obtained which is used to make anodes for the production of aluminum and as a feedstock for special graphites (e.g. reactor graphite).

The electrochemical production of aluminum utilizes pre-baked or Söderberg anodes. Pre-baked anodes are manufactured by mixing petroleum or pitch coke with around 20% of electrode binder (see Chapter 13.1.2), followed by molding, as is also usual for the production of graphite electrodes. The green electrodes are baked in a ring furnace at a temperature of 1200 °C.

Söderberg anodes, use of which is declining, are baked directly in the aluminum reduction cell. In this case, petroleum coke is mixed with 25 to 30% binder; the mixture is carbonized by the heat of the electrolysis bath (940 to 980 °C).

The world production of around 15.5 Mt of aluminum in 1986 required a total consumption of around 7 Mt of pre-baked and Söderberg anode material.

13.1.4 Production of carbon fibers

Carbon fibers were first made by Thomas Alva Edison in 1879 from cellulose for lamp filaments. In Great Britain in 1961 the Royal Air Force produced a high-value carbon fiber from polyacrylonitrile (PAN).

Production of pitch fibers was first investigated in Japan. In 1963, Sugio Otani obtained pitch fibers by the pyrolysis of lignin and later of PVC pitch. The first commercial product was the fiber from a pitch derived from crude oil pyrolysis, produced by *Kureha*.

Industrial production of high-modulus carbon fibers based on pitch started in 1982, based on the fundamental work of Leonard Sidney Singer in the USA (*Union Carbide*).

Dependent on their carbon content, carbon fibers are divided into the following categories:

- Carbonized fibers: C-content <90%
- Carbon fibers: C-content 91–99%
- Graphite fibers: C-content >99%

Whereas carbon fibers with a low carbon content are formed predominantly from aliphatic raw materials (rayon), carbon fibers with a high carbon content are produced from aromatic feedstocks or easy-to-aromatize base materials. The most important raw materials for the manufacture of high-carbon fibers are polyacrylonitrile and mesophase pitch.

Because of the different orientability of the highly aromatic molecules in the formation of mesophase pitch, the carbon fibers have a wide range of qualities.

Table 13.4 shows the properties of carbon fibers from different raw materials.

Table 13.4: Properties of carbon fibers from different origins

Raw material	Fiber type	Density (g/cm^3)	Tensile strength (GPa)	Elastic modulus (GPa)	Extension at break (%)	Electr. resistance ($\mu \Omega$ m)
Rayon[1]	50 S	1.67	1.9	390	0.5	10
	75 S	1.82	2.5	520	0.5	-
Polyacrylo-[2]	T800	1.80	5.6	290	1.9	13
nitrile	M50	1.91	2.4	490	0.4	7.6
Isotropic[3]	T101 F	1.65	0.8	33	2.4	150
pitch	T201 F	1.57	0.7	33	2.1	50
Mesophase[4]	P25	1.90	1.4	160	0.9	13
Pitch	P120	2.18	2.2	830	0.3	2.2
Graphite single-crystal		2.25	-	1000	-	0.4

[1] *Union Carbide*, Thornel; [2] *Toray*, Torayca; [3] *Kureha*; [4] *Union Carbide*, Thornel

Carbon fibers from different sources complement each other in their properties. Pitch fibers usually display higher density, higher elasticity and higher electrical conductivity, while polyacrylonitrile produces exceptionally high strength fibers.

There are also large differences in yields: the yield of carbon fibers from rayon is between 20 and 25%, from polyacrylonitrile 45 to 50% and from pitch around 75 to 85 %. The high yield from pitch is a fundamental reason for the great efforts which are being made worldwide to bring about wider use of pitch as a precursor for carbon fibers.

In order to make high-modulus carbon fibers from pitch, petroleum- or coal-tar pitch is first filtered to remove solids which inhibit the growth of mesophase. This is followed by heat treatment, which results in a mesophase content of 70 to 95%; then the mesophase pitch is spun into monofilaments with a fiber diameter of around 5 µm. Optimum viscosity and minor gas liberation of the pitch are of fundamental importance. The pitch monofilament is oxidized in the second stage to make it infusible. Oxygen is used as the oxidation agent. Unlike in the production of PAN fibers, it is not necessary to stretch the fibers before oxidation.

The next stage is the carbonization of the oxidized fiber, which is carried out at temperatures from 1,000 to 1,500 °C. This may be followed by graphitization at 2,500 to 3,000 °C for the production of high-modulus fibers.

In addition to the process parameters, the quality of the fiber depends on the properties of the feedstock. In particular, mild hydrogenation of the pitches enables high mesophase content to be achieved at relatively low viscosity, facilitating the spinning process. Predominantly aliphatic pitches, like the pitch used by *Kureha* from crude oil cracking, lead, on the other hand, to isotropic pitch fibers with lower moduli.

The total production of carbon fibers in 1987 was 4,500 t, with polyacrylonitrile fibers currently by far the most important feedstock.

Carbon fibers are especially used in the production of highly stressed construction materials, in combination with a matrix (epoxy resin, phenolic resin) and for the production of high-strength sports equipment (golf clubs, tennis racquets, skis). If a further reduction can be achieved in production costs, then a whole new field of application could open up, especially in the automobile industry.

Carbon fibers, therefore, in part complement the high-value aromatic fibers, such as aramid fibers, which likewise are produced through a liquid-crystal phase.

13.2 Pyrolysis of mixtures of aromatics in the gas phase – Carbon black production

Unlike the production of coke and carbon fibers, the production of carbon black from aromatic mixtures occurs by gas-phase pyrolysis.

Carbon black was produced even in pre-industrial times, from oils and resins, for the manufacture of pigments. At the beginning of the 19th century, the lamp black and thermal black processes were introduced, which used mainly aliphatic hydrocarbons (natural gas) as raw material.

Outstanding pioneers in the development of carbon black production were, in particular, the American brothers Samuel and Godfrey Cabot, who, together with Joseph Binney, Edwin Drew and C. Harold Smith, founded the carbon black companies *Cabot* and *Columbian Carbon*, respectively, in 1882 following the discovery of crude oil and natural gas in Pennsylvania.

In the furnace process, which today dominates carbon black production, oils rich in aromatics from naphtha or gas oil pyrolysis, cat-cracker residues (decant oils) together with mixtures of aromatics from coal tar, are used as feedstock. Table 13.5 summarizes the characteristic data for decant oil, pyrolysis oil from naphtha cracking and a carbon black feedstock derived from coal tar.

The Bureau of Mines Correlation Index (BMCI) defined below, is a particularly important criterion of the quality of the carbon black feedstock:

$$BMCI = 473.7 \cdot d - 456.8 + 48640 \cdot K^{-1}$$

Here, d is the density of the hydrocarbon mixture in g/ml at 15.6 °C and K the average boiling temperature in Kelvin.

Hydrocarbon mixtures of predominantly aliphatic nature have a correlation index (CI) from 15 to 50; benzene, according to the definition, has a CI of 100, while high-boiling aromatic mixtures have a correlation index in excess of

Table 13.5: Characteristic data for coal- and petroleum-derived aromatic feedstocks for carbon black

	Decant oil	Pyrolysis oil	Coal-tar carbon black feedstock
Sulfur, %	0.7–4	0.5–1	0.5–0.6
Ash, %	0.02–1	<0.01	<0.01
Pentane-insoluble, %	1–8	10–15	1–3
BMCI	110–135	125–140	140–160
Density, g/ml	1–1.13	1.04–1.08	1.07–1.16
Distillation (at 1013 mbar) Initial boiling point, °C	250–280	160–200	200–250
50 vol% at °C	300–350	220–260	260–320
Carbon content, %	87–88	91.5–92	91–91.5
Hydrogen content, %	9–9.5	7.2–8.4	5.8–6.2

100. The yield of carbon black increases as the correlation index rises; the highest yields are obtained using three- and four-ring aromatic compounds.

Apart from suitable flow behavior (viscosity, pour point), a low alkali content is of great importance in assessing the quality of the carbon black feedstock, since alkali affects the structure of the carbon black. Because of their high CI-index, tar-derived oils generally give the highest yields of carbon black (up to 70%). The yield is also strongly influenced by process parameters, in particular the air/feedstock ratio. The production of very small particle carbon blacks requires a high air/feedstock ratio, and therefore results in a relatively low yield.

Figure 13.10 shows the flow diagram for the production of carbon black by the furnace process.

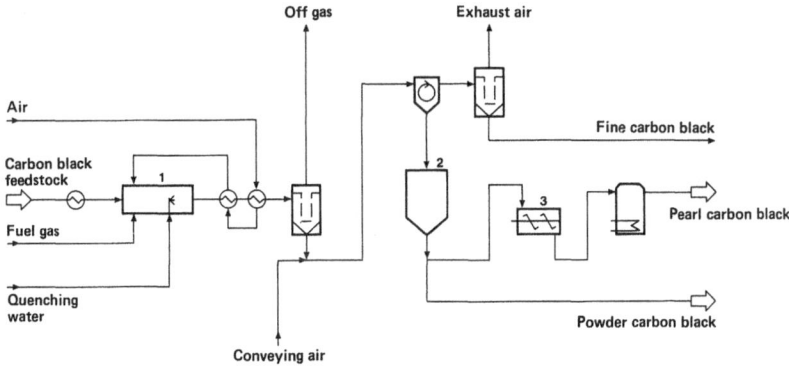

1 Carbon black reactor; **2** Silo; **3** Wet pearl machine

Figure 13.10: Flow diagram for the production of carbon black using the furnace process

The pre-heated carbon black oil is sprayed into the reactor, where it is cracked in a high temperature zone (1200 to 1800 °C). The reaction temperature is maintained by burning an additional energy carrier, such as natural gas, with an excess of atmospheric oxygen; the unused oxygen leads to partial combustion of the carbon black feedstock.

When carbon black has been formed, the pyrolysis reaction is stopped by quenching with water. Further cooling, down to 200 to 300 °C, is achieved by passing the gas/carbon black mixture into heat exchangers; the heat which is recovered is used to pre-heat the combustion air. The carbon black is separated in special filters. The residual gas is used in an incineration plant, to produce energy. The carbon black, which arises in powder form, is transferred to a pearl machine to convert it into pearl black for easier handling.

As to the formation mechanism of carbon black, there are indications that the carbon particles are formed by recombination of smaller hydrocarbons (acetylene, ethylene and their radicals as well as aromatic cracking products).

The properties of the carbon black are particularly dependent on the reaction conditions of the pyrolysis. The distribution of particle sizes from furnace carbon black extends from around 10 nm to 100 nm. (Smaller particles can be produced by the gas black method, larger particles by the thermal black process.) Small carbon black particles are obtained with high reaction temperatures and reaction times of around 10^{-2} sec., whereas the manufacture of carbon black with a particle diameter of 35 to 65 nm requires lower temperatures and reaction times of 1 to 2 sec.

Figure 13.11 shows the particle-size distributions for the major types of carbon black and their labelling according to the traditional ASTM standard.

SAF: super abrasion furnace black; *HAF:* high abrasion furnace black; *FEF:* fast extrusion furnace black; *SRF:* semi-reinforcing furnace black

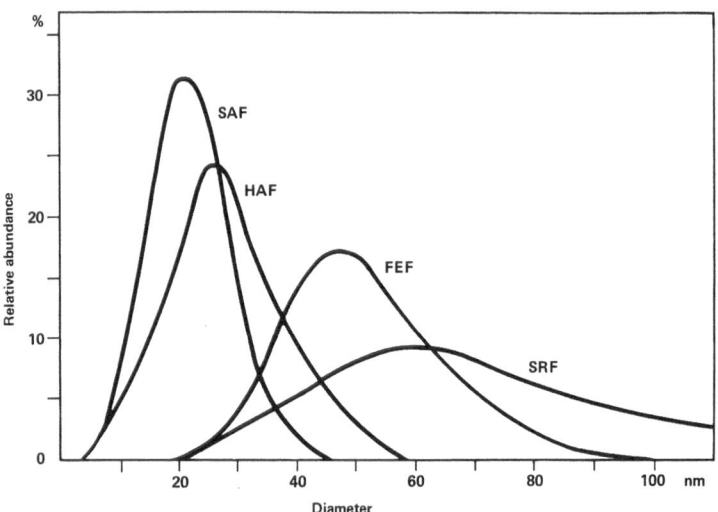

Figure 13.11: Spectrum of particle sizes of major carbon black types

As particle size grows, the specific surface area of the carbon blacks diminishes. A distinction is made between active rubber blacks with a specific area from 70 to 150 m^2/g, which are commonly used in the production of tire treads, and semi-active blacks, used for tire carcasses, which have a specific surface area of 15 to 70 m^2/g. Inactive carbon blacks have a specific surface area of <15 m^2/g; they are commonly used as fillers in the production of industrial rubber goods.

The most important type of carbon black in terms of quantity is *HAF*-black, which represents around 50% of total carbon black sales in Western Europe.

In West Germany, in addition to the furnace method, the gas black process developed by *Degussa* in 1935 is used for carbon black production, particularly for the manufacture of finely-textured pigment black for printing inks, as well as plastics and paints. The gas-black process uses in preference carbon black feedstocks derived from coal tar; previously, naphthalene and anthracene residues were used. To produce carbon black, the oil is vaporized in a carrier stream (hydrogen, coke-oven gas). Following partial oxidation of this mixture with atmospheric oxygen in the carbon black furnaces, the black is deposited on water-cooled rolls and collected in carbon black separators. Carbon black grades produced by the gas-black process complement furnace blacks with regard to their surface chemistry and particle size. In the gas black process carbon blacks with an average diameter of 8 to 25 nm are produced whereas the carbon black generated by the furnace process generally has a particle diamter ranging from 10 to 100 nm. Because it is manufactured in an oxidative atmosphere, the gas black, unlike the basic furnace black, is weakly acidic.

Carbon black production in the western world is around 4.0 Mtpa. The major producer countries are USA, Japan and the countries of Western Europe.

Table 13.6 summarizes the capacities of the major producer countries in the world.

Table 13.6: Capacities of the major carbon black producer countries (1985)

	(1,000 t)
USA	1,500
Mexico	75
Canada	170
Japan	660
India	125
Korea	75
Brazil	180
West Germany	400
United Kingdom	170
France	250
Holland	110
Italy	170
Spain	90
Australia	110
Other countries	415
Total capacity	4,500

Over 90% of carbon black production is consumed by the rubber processing industry as reinforcing black. The remaining 5 to 10% is used as pigments for paints, paper, printing inks and plastics.

14 Aromatic heterocyclics – production and uses

The industrial chemistry of the heterocyclic aromatics is extraordinarily diverse. A large number of mono- and polynuclear heterocyclics (vide infra) is present in coal tar and can be recovered from this source. Synthetic means of production, principally from petroleum-derived feedstocks have been added to the coal-derived raw materials since the fifties.

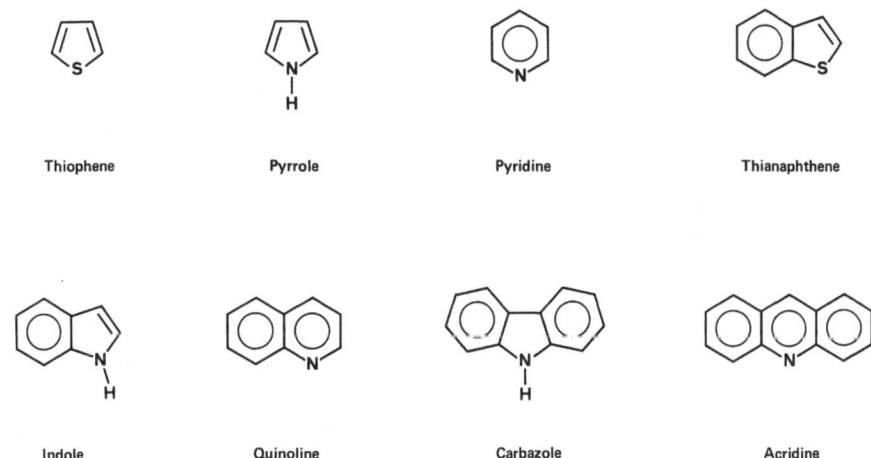

Thiophene Pyrrole Pyridine Thianaphthene

Indole Quinoline Carbazole Acridine

In addition to heterocyclics with one hetero-atom, aromatic compounds with several hetero-atoms are also commercially important. They are generally not produced by recovery from crude aromatic mixtures, but by synthesis from smaller building blocks. These heterocyclic compounds are used in substituted form predominantly as pesticides, dyes and pharmaceuticals.

14.1 Five-membered ring heterocyclics

14.1.1 Furan

Furan, which displays only limited aromatic character, is produced by decarbonylation of furfural (see Chapter 3.5.2.) on a ZnO/Cr_2O_3 catalyst at 400 °C. It is used principally to produce pyrrole, by reaction with ammonia.

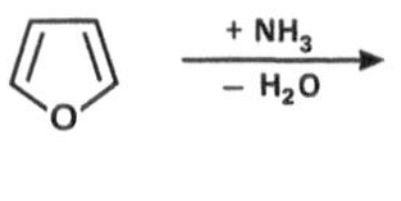

$$\text{[furan]} \xrightarrow[-H_2O]{+NH_3} \text{[pyrrole]}$$

The reaction of nitric acid and furfural in acetic anhydride produces 5-nitrofurfural diacetate, an important intermediate in the production of chemotherapeutic agents, such as nitrofurazone and nitrofurantoin.

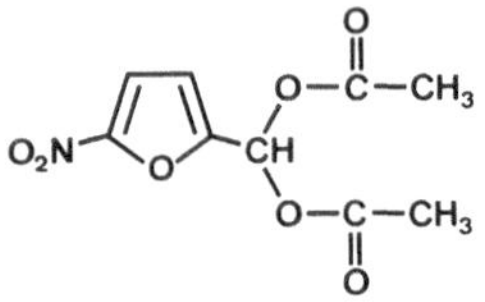

5-Nitrofurfural diacetate

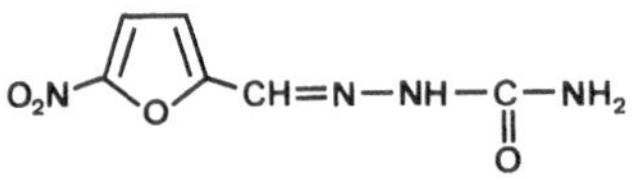

Nitrofurazone

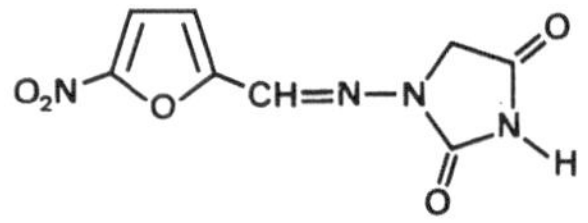

Nitrofurantoin

A recently-developed medicament for the treatment of ulcers, which is rapidly gaining an increasing share of the market, is ranitidine (*Glaxo*), a H_2-antagonist.

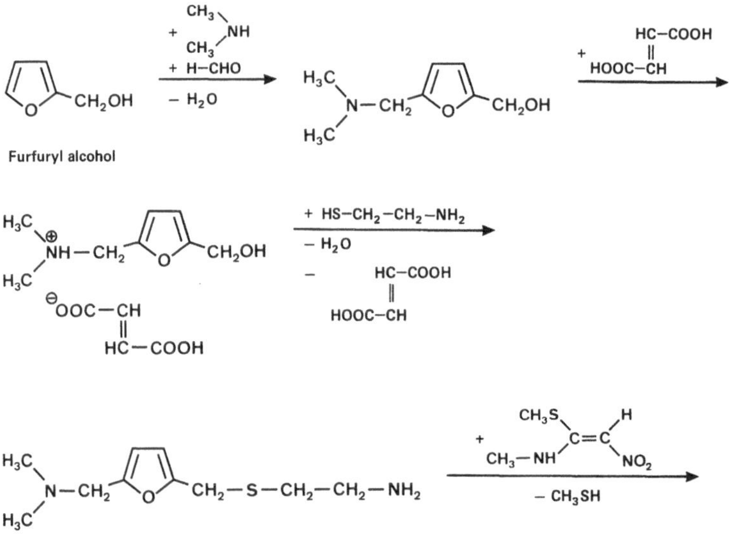

Furfuryl alcohol

[Ranitidine structure]

Ranitidine

It is produced from furfuryl alcohol by sequential reaction with dimethylamine hydrochloride/formaldehyde, cysteamine and N-methyl-1-(methylthio)-2-nitro-ethenamine. The production of ranitidine in 1986 amounted to 260 t.

Furfuryl alcohol is manufactured by catalytic reduction (modified copper catalyst) of furfural with hydrogen at a pressure of 1.5 to 2.5 bar and at a temperature of 120 °C. A multi-tubular fixed-bed reactor with 6,500 tubes of a diameter of 25 mm and a length of 1.8 m is used in industrial practice.

14.1.2 Thiophene

In 1883, Viktor Meyer first detected thiophene in coke-oven benzole, where it is present in concentrations of around 1%. (Because of the close relationship between thiophene and benzene, Meyer took the name 'kryptophen' ('hidden in benzene') into consideration.) Recovery of thiophene from coke-oven benzole, which is possible by the reaction of thiophene with concentrated sulfuric acid to thiophenesulfonic acid, is not used industrially since the synthesis is more economical.

The industrial synthesis of thiophene is based on butane and its reaction with sulfur or carbon disulfide. In comparison with benzene, thiophene can be more easily substituted electrophilically in the 2- or 5- position; this reaction is used to produce a wide range of pharmaceutical products. An example of this is the antihistamine thenalidine, which is obtained from thiophene by chloromethylation to 2-thienylmethyl chloride followed by reaction with 4-anilino-1-methylpiperidine.

[Reaction scheme producing Thenalidine]

Thenalidine

More important is the β-lactam antibiotic cefalotin (*Eli Lilly*), which was among the first cephalosporin preparations.

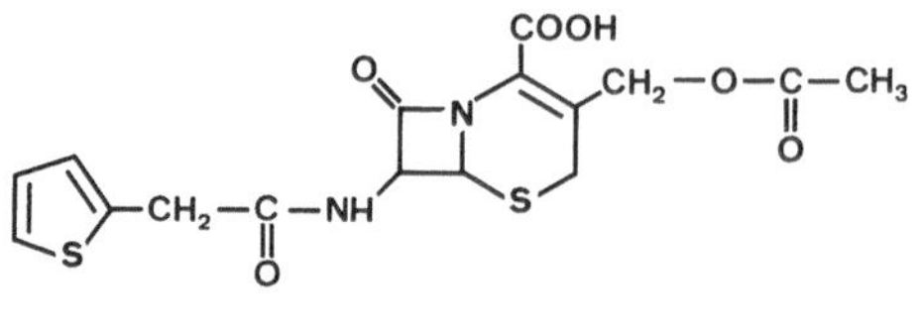

Cefalotin

14.1.3 Pyrrole

Pyrrole was discovered in 1834 by Friedlieb Ferdinand Runge in coal tar, where it is present in a concentration of less than 0.01%. Production from coal tar is, therefore, not economical. Pyrrole is synthesized by the reaction of furan with ammonia. Pyrrole is used in the production of polypyrrole and pharmaceuticals, such as the anti-inflammatory drug tolmetin. In tolmetin synthesis, the acidity of pyrrole is made use of. It is transformed into the potassium salt, which reacts with methyl chloride to form 1-methylpyrrole. The reaction of methylpyrrole with formaldehyde and dimethylamine produces the Mannich base, which is then quaternized

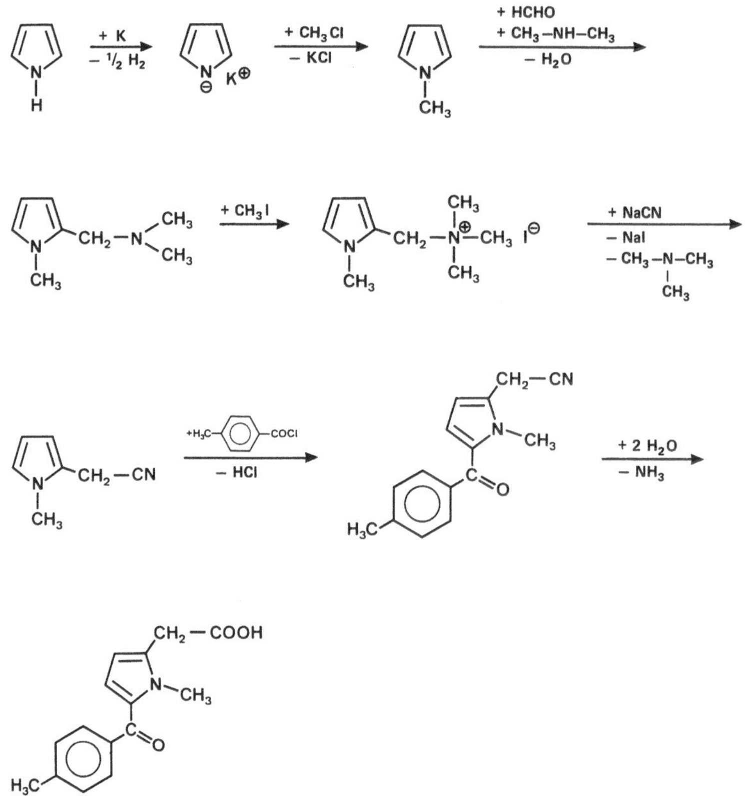

Tolmetin

with methyl iodide or dimethyl sulfate. By replacing the trimethylamino group with a cyanide group, the 2-acetonitrile derivative of 1-methylpyrrole is produced, which undergoes further reaction with p-methylbenzoyl chloride to yield tolmetin.

An alternative method of the synthesis of tolmetin, which is gaining in importance, is based on methylamine, acetonedicarboxylic acid and chloroacetaldehyde.

Pyrrole is also very important in natural substances, where, for example, it can be found in corrin, a basic building block for vitamin B_{12}.

Corrin

14.1.4 Five-membered ring heterocyclics with two or more hetero-atoms

Among the five-membered ring heterocyclics with more than one hetero-atom, the derivatives of imidazole, oxazole, thiazole and 1,2,4-triazole are particularly important.

Imidazole   1,3-Oxazole   1,3-Thiazole   1,2,4-Triazole

Oxazole and thiazole have virtually no significance as unsubstituted compounds. Imidazole is obtained in around 75% yield by the reaction of glyoxal, formaldehyde and ammonia at about 70 °C.

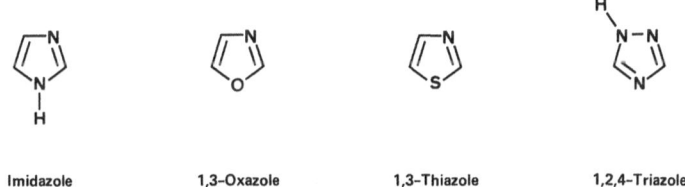

1,2,4-Triazole is produced in high yields from 4-amino-1,2,4-triazole by deamination or by condensation of hydrazine with formamide.

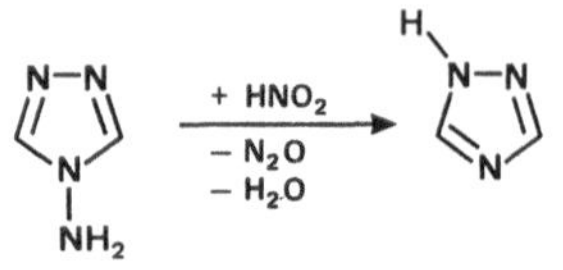

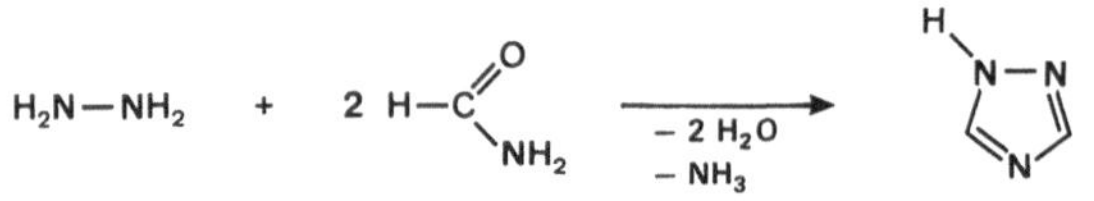

Examples of imidazole derivatives used as plant protection agents are the fungicides prochloraz (*Schering*) and imazalil (*Janssen-Pharmaceutica*).

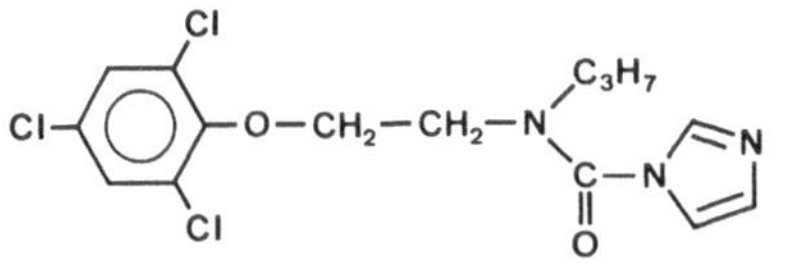

Prochloraz

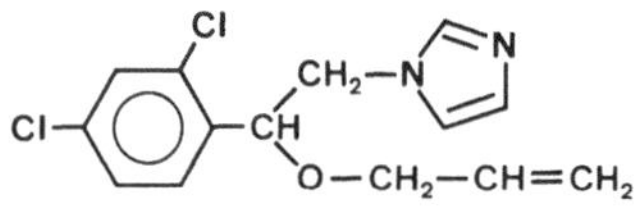

Imazalil

A wide range of pharmaceuticals is manufactured from imidazole compounds. An example is clotrimazole, which is produced from 2-chloro-triphenylchloromethane and imidazole; it is one of many imidazole-based broad-spectrum anti-

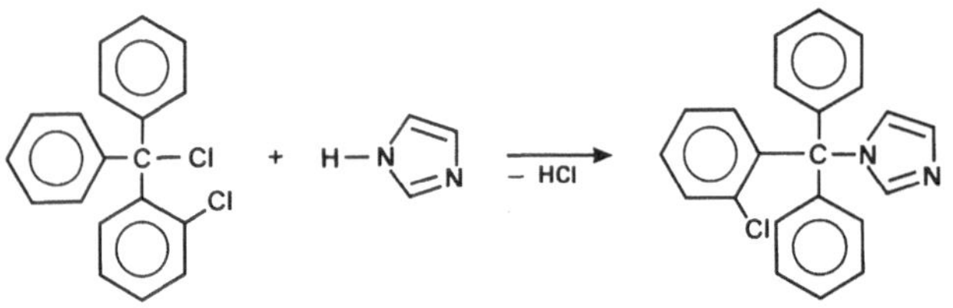

Clotrimazole

mycotics. Cimetidine, developed by *Smith Kline & French*, is also an imidazole derivative, and is used to treat ulcers. The imidazole is synthesized from formamide and 2-chloroethyl acetoacetate via 5-methylimidazole-4-carboxylic acid ethyl ester. The production of cimetidine in 1986 amounted to 800 t.

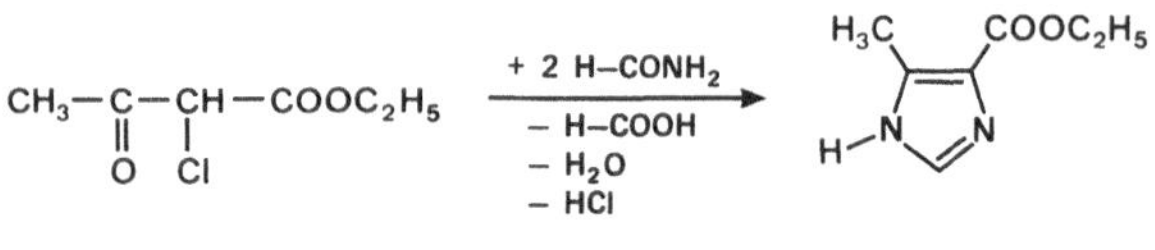

5-Methylimidazole-4-carboxylic acid ethyl ester

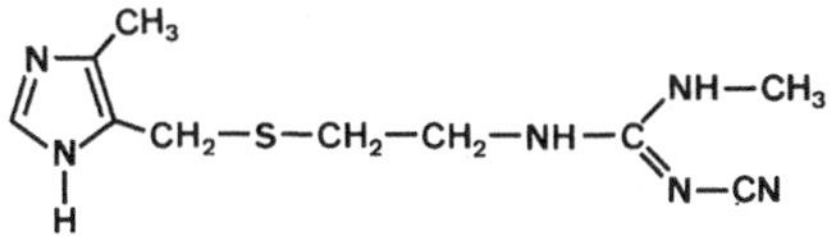

Cimetidine

Oxazole derivatives are used in the production of vitamin B_6 (see Chapter 14.2.1).

1,2,4-Triazoles serve as building blocks in the production of a wide range of fungicides, such as triadimefon (see Chapter 5.3.4.5), bitertanol (*Bayer*), propiconazole (*Ciba Geigy*) and diclobutrazol (*ICI*).

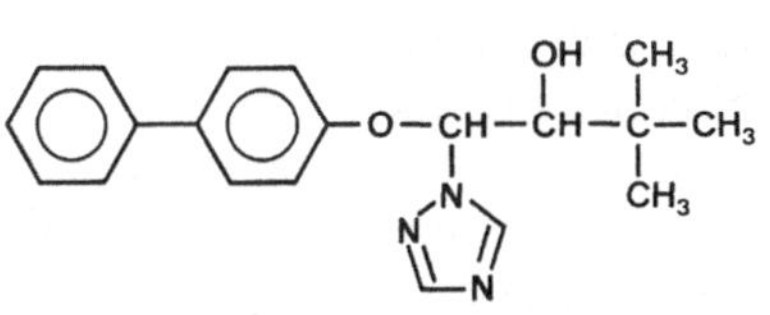

Bitertanol 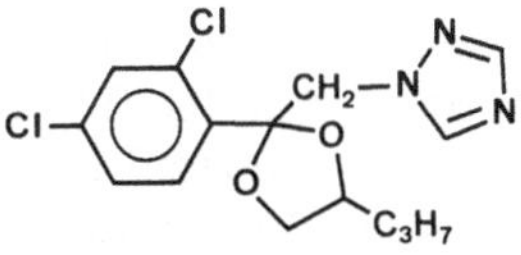   **Propiconazole**

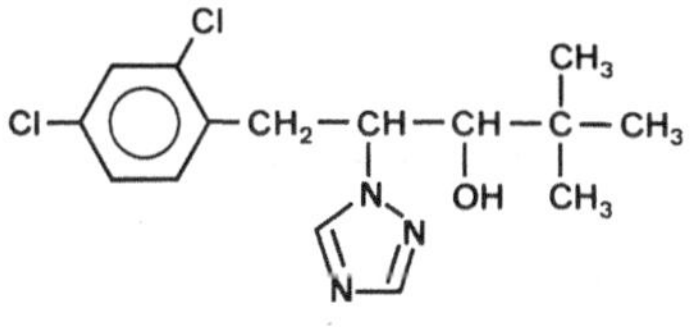

Diclobutrazol

Amitrole is a non-selective herbicide, which was introduced in the early 50's; it is obtained by condensation of formic acid and aminoguanidine.

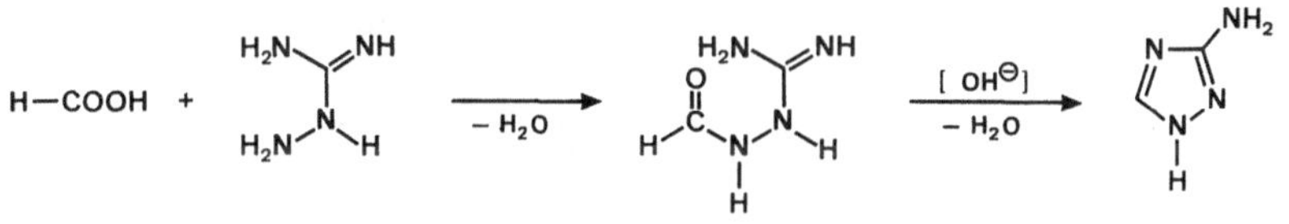

Amitrole

An example of an important thiazole derivative is the fungicide and antihelmintic thiabendazole, which is produced from o-phenylenediamine and 4-thiazole-carboxamide; 4-thiazole carboxamide is obtained from 4-methylthiazole by ammoxidation and partial hydrolysis.

Thiabendazole

14.2 Six-membered ring heterocyclics

14.2.1 Pyridine

The six-membered ring heterocyclic pyridine is of fundamentally greater commercial importance than the five-membered ring heterocyclics.

Pyridine was prepared by Thomas Anderson in 1851 from bone oil; he had already isolated 2-methylpyridine (α-picoline) from coal tar in 1846; this compound, together with β- and γ-picoline, is contained in coal pyrolysis products.

α–Picoline   β–Picoline   δ–Picoline

Like almost all aromatic compounds with one hetero-atom, up to the end of the 1950's, pyridine was produced exclusively from coal tar. Pyridine is contained mainly in the coal-tar light-oil fraction (see Chapter 3.2.3). In addition to the benzene hydrocarbons, phenols and polymerizable compounds, such as cyclopentadiene, light oil contains 2 to 7% of pyridine bases.

To recover pyridine bases, the light oil is first dephenolated. The pyridine bases are then extracted with sulfuric acid. 'Springing' with ammonia or caustic soda yields the free base mixture, which, after separation, is dewatered by azeotropic distillation with benzene.

Table 14.1 shows the composition of the crude pyridine bases.

Table 14.1: Composition of crude pyridine bases from light oil of coal tar

Pyridine	12%	2,5-Lutidine	4%
α-Picoline	10%	2,6-Lutidine	6%
β-Picoline	6%	2,3,6- and	
γ-Picoline	8%	2,4,6-Collidine	9%
2-Ethylpyridine	2%	Aniline	12%
2,3-Lutidine	2%	o-Toluidine	4%
2,4-Lutidine	11%	m- and p-Toluidine	8%
		Higher boiling bases	6%

The dewatered pyridine-base mixture is separated by distillation into pure pyridine and fractions containing higher boiling bases.

Figure 14.1 shows the flow diagram for the production of pyridine from light oil; in addition to pyridine, β-picoline and pyridine-base fractions with the indicated boiling ranges are recovered.

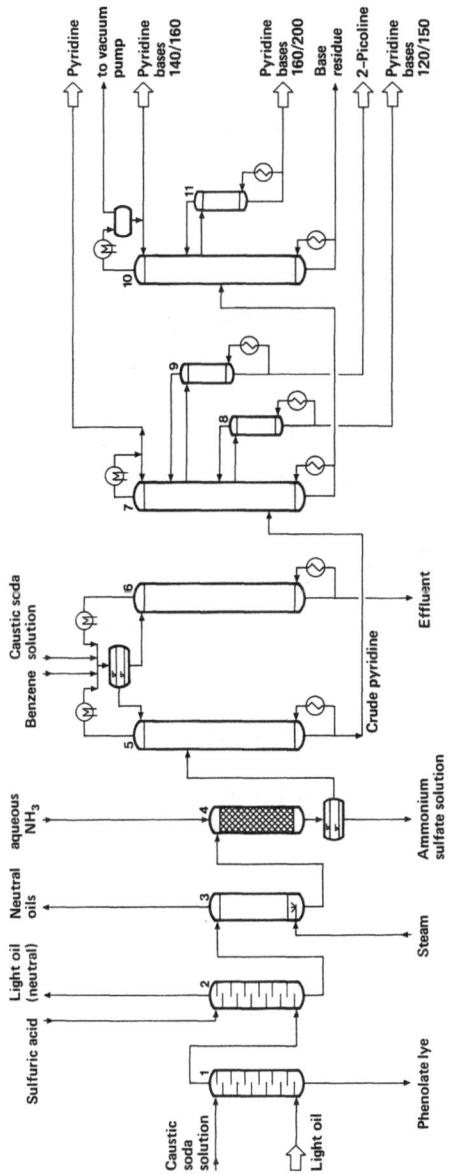

Figure 14.1: Flow diagram for the refining of pyridine bases

1 Extraction column (Phenols); **2** Extraction column (Bases); **3** Stripping column; **4** Base 'springing' column; **5** and **6** Dewatering columns; **7** Pyridine column; **8** and **9** Side columns; **10** Main vacuum column; **11** Vacuum side column

Current production of pyridine from coal tar in Western Europe is around 300 tpa.

The development of plant protection agents based on pyridine in the late 1950's (*ICI*) meant that coal-derived raw materials were insufficient to meet the increasing demand and methods of synthesizing pyridine were therefore introduced.

The main methods of the synthesis of pyridine are based on the reaction of aldehydes and ketones with ammonia; these methods produce alkylated pyridines in addition to pyridine.

Reaction conditions and feedstocks are selected in accordance with the desired product mix.

In industrial processes, formaldehyde, acetaldehyde and, to a lesser extent, acrolein are used as aldehydes; crotonaldehyde can also be employed for the synthesis of pyridines.

Figure 14.2 shows the typical dependence of the product composition on the feedstock components acetaldehyde/formaldehyde.

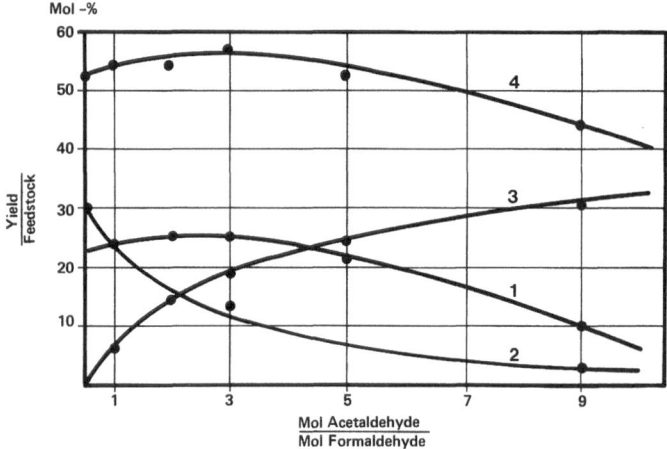

1 Pyridine; 2 3-Methylpyridine; 3 2- and 4-Methylpyridine; 4 Total yield

Figure 14.2: Dependence of the composition of pyridine bases on the ratio of acetaldehyde/formaldehyde feedstock

As the proportion of acetaldehyde increases, the yield of 3-methylpyridine drops; the proportion of pyridine also falls. Additionally, the total yield declines with an increased proportion of acetaldehyde.

The classic *Reilly* synthesis of pyridine is carried out on a SiO_2/Al_2O_3 catalyst activated by CdF_2 at temperatures of around 400 to 500 °C under atmospheric pressure.

The acetaldehyde/formaldehyde molar ratio is 1.4:1; methanol is added to increase the yield, which is around 40% pyridine and nearly 25% 3-methylpyridine. Figure 14.3 shows the flow diagram of the gas-phase reaction.

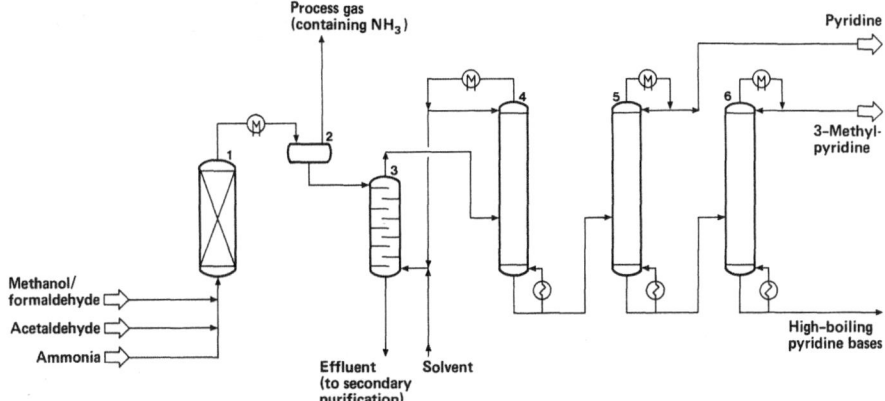

1 Reactor; **2** Gas separator; **3** Extraction column; **4** Solvent column; **5** Pyridine column; **6** Picoline column

Figure 14.3: Flow diagram of gas-phase reaction to produce pyridine and 3-methylpyridine from ammonia, formaldehyde and acetaldehyde

398 Aromatic heterocyclics – production and uses

Figure 14.4 shows part of the pyridine plant operated by *Reilly Tar & Chemical*, Tertre/Belgium. In the USA *Nepera* likewise operates large pyridine facilities.

Figure 14.4: Pyridine plant operated by *Reilly Tar & Chemical*, Tertre/Belgium

Apart from C_1 and C_2-aldehydes, C_3-aldehydes are also used in the production of pyridine. The process carried out by *Daicel* in Japan employes acrolein.

$$4\ CH_2{=}CH{-}CHO\ +\ 2\ NH_3\ \longrightarrow$$

$$\text{(3-methylpyridine)}\ +\ \text{(pyridine)}\ +\ HCHO\ +\ 3\ H_2O\ +\ H_2$$

Figure 14.5 shows a flow diagram for the synthesis of pyridine from ammonia and acrolein.

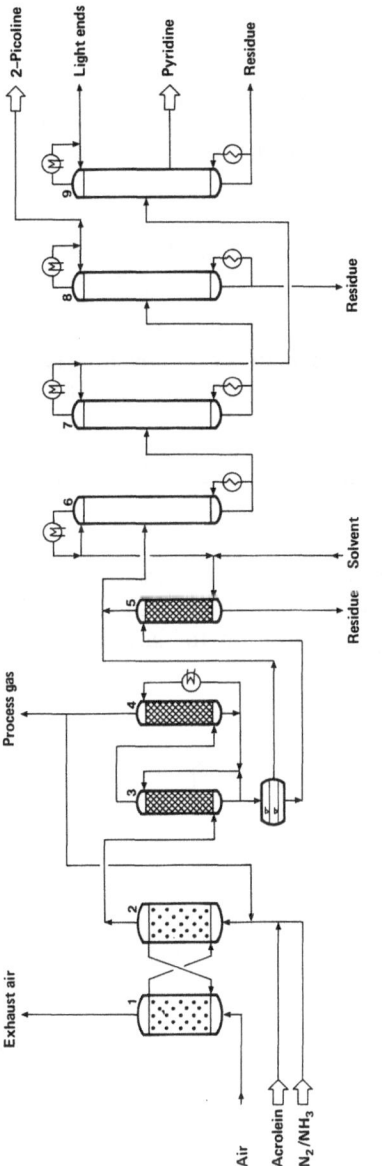

1 Catalyst regeneration; **2** Fluidized-bed reactor; **3** and **4** Absorption columns; **5** Extraction column; **6** Solvent column; **7** Light ends column; **8** α-Picoline column; **9** Pyridine column

Figure 14.5: Flow diagram of the synthesis of pyridine from ammonia and acrolein

Crotonaldehyde has also been suggested as a reaction component for the production of pyridine, but this process has not yet been applied for economic reasons.

$$CH_3-CH=CH-CHO + HCHO + NH_3 \longrightarrow \text{[pyridine]} + 2\ H_2O + H_2$$

Other possible methods of the production of pyridine are the ring expansion of cyclopentadiene, and the cyclization of pentene nitrile, obtained as a by-product in the dimerization of acrylonitrile.

Ammoxidation of cyclopentadiene (*Hoechst* process) is carried out with Al_2O_3 as a catalyst. The reaction is performed at atmospheric pressure and a temperature of around 300 °C. The yield is less than 50% for a conversion rate of 95 to 97%.

$$\text{[cyclopentadiene]} + NH_3 + O_2 \longrightarrow \text{[pyridine]} + 2\ H_2O$$

If cis-2-pentene nitrile is used, the yield is even lower, since the necessary higher temperatures cause fragmentation, with the concurrent formation of gas and char.

$$CH_3-CH=CH-CH_2-CN \longrightarrow \text{[pyridine]} + H_2$$

The synthesis of pyridine from renewable raw materials, i.e., tetrahydrofurfuryl alcohol has also been investigated several times. An intermediate of this process, which has found no industrial application to date, is piperidine, which is dehydrogenated to pyridine.

$$\text{[THF-CH}_2\text{OH]} \xrightarrow[-2\ H_2O]{+NH_3,\ +H_2} \text{[piperidine]} \xrightarrow{-3\ H_2} \text{[pyridine]}$$

World production of pyridine is around 30,000 tpa. The main producer countries are the USA with around 8,000 tpa, Japan with around 3,000 tpa and Western Europe with around 10,000 tpa.

14.2.2 Pyridine derivatives

In terms of its reactivity, particularly to electrophilic substitution, pyridine is comparable with nitrobenzene; the electrophilic introduction of substituents, therefore, requires a relatively high temperature. In comparison, nucleophilic substitution is more facile, especially after converting pyridine to the N-oxide.

Industrially, the most important derivatives are dimeric pyridines in the form of 4,4'-bipyridyl and 2,2'-bipyridyl. 2,2'-Bipyridyl can be synthesized by oxidative dimerization or by bromination to 2-bromopyridine and reaction with nonferrous metal catalysts such as copper.

4,4'-Bipyridyl is produced by the reaction of pyridine with sodium at $-45\ °C$ in dimethylformamide or NH_3 and oxidation of the intermediate, 4,4'-tetrahydrobipyridyl, with atmospheric oxygen.

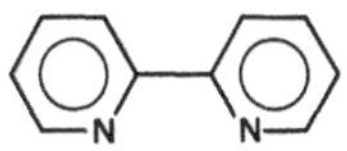

2,2'-Bipyridyl 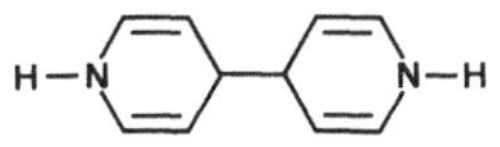   4,4'-Tetrahydrobipyridyl 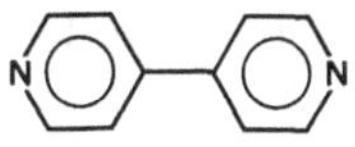   4,4'-Bipyridyl

Bipyridyls are converted by quaternization with dimethyl sulfate, methyl chloride or 1,2-dichloroethane into the herbicides paraquat and diquat, developed by William Boon (*ICI*). Paraquat is by far the more important of the two 'quat' herbicides.

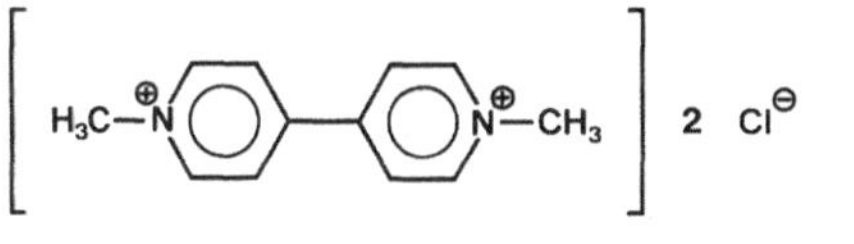

Paraquat 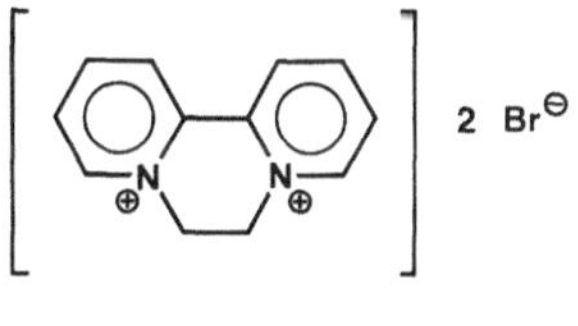   Diquat

One example of a pyridine product used as a pharmaceutical is cetylpyridinium chloride, which is obtained from pyridine and 1-chlorohexadecane; it is used as an antiseptic.

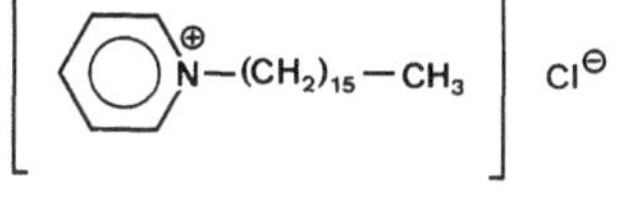

Cetylpyridinium chloride

Reaction of pyridine hydrochloride with stearamide and formaldehyde produces stearamido-methylpyridinium chloride, which is used for waterproofing textiles.

$$\left[\bigcirc\!\!\!\!\!\!{\overset{\oplus}{N}}\!-\!CH_2\!-\!NH\!-\!\underset{\underset{O}{\|}}{C}\!-\!(CH_2)_{16}\!-\!CH_3 \right] Cl^{\ominus}$$

Stearamido-methylpyridinium chloride

Chlorination of pyridine in the gas phase at temperatures around 350 °C produces 2-chloropyridine. Conversion is limited to 60 to 90%, to restrict polychlorination. Unconverted pyridine is precipitated as hydrochloride and is recycled.

2-Chloropyridine serves as a raw material in the production of pyridine-thiones; these are manufactured by the reaction of chloropyridine with H_2O_2 or other peroxides, followed by a nucleophilic reaction of the N-oxide with sodium hydrogen sulfide to produce the thiol. The zinc salt is used as an active ingredient (anti-dandruff agent) in hair shampoos.

2-Chloropyridine is also the starting point for a range of antihistamines, such as dexchlorpheniramine and its bromo-analog, brompheniramine.

Dexchlorpheniramine

Hydrogenation of pyridine yields piperidine, which serves as the base for the production of vulcanization accelerators, such as dipentamethylenethiuram tetrasulfide (DPTT).

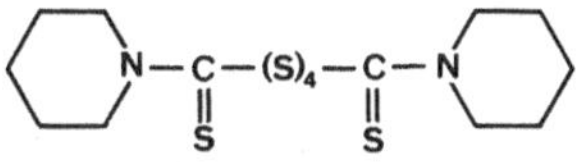

DPTT

The reduction can be carried out electrochemically or with Raney nickel at 200 °C.

Piperidine also arises as a by-product in the production of 4,4'-bipyridyl.

The insecticide chlorpyrifos, developed by *Dow*, is mainly produced not from pyridine, but by reaction of glutarimide and PCl_5 followed by nucleophilic chlorine exchange, via 3,5,6-trichloropyrid-2-one.

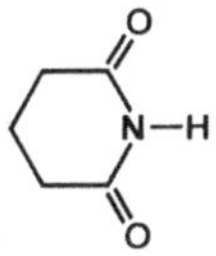

Glutarimide 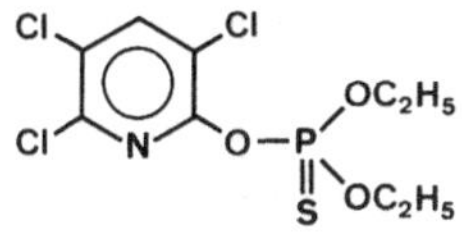 Chlorpyrifos

14.2.3 Alkylated pyridines

Alkylated pyridines play an important role in chemical technology as well as in nature. Nicotine is a natural alkylpyridine derivative, found in the tobacco plant; this alkaloid was used from the 17th century until the 1950's as a plant protection agent.

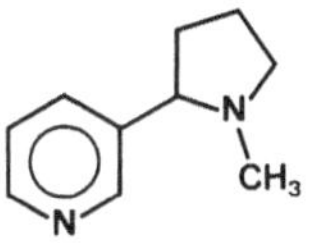

Nicotine

Among the synthetic alkylated pyridines, 5-ethyl-2-methylpyridine (MEP) is by far the most important; it is used to produce nicotinic acid. (Nicotinic acid (niacin, vitamin PP, vitamin B_3) as well as nicotinamide are both effective as vitamins.)

Production is carried out by liquid-phase synthesis from acetaldehyde and ammonia. Acetaldehyde is first trimerized to paraldehyde. The subsequent reaction with aqueous ammonia, which gives a 70% yield of 5-ethyl-2-methylpyridine, is performed in the liquid phase at 230 °C and at a pressure of around 150 bar with ammonium-salt catalysts.

Figure 14.6 shows a flow diagram for the production of 5-ethyl-2-methylpyridine.

World production of 5-ethyl-2-methylpyridine is currently around 15,000 tpa.

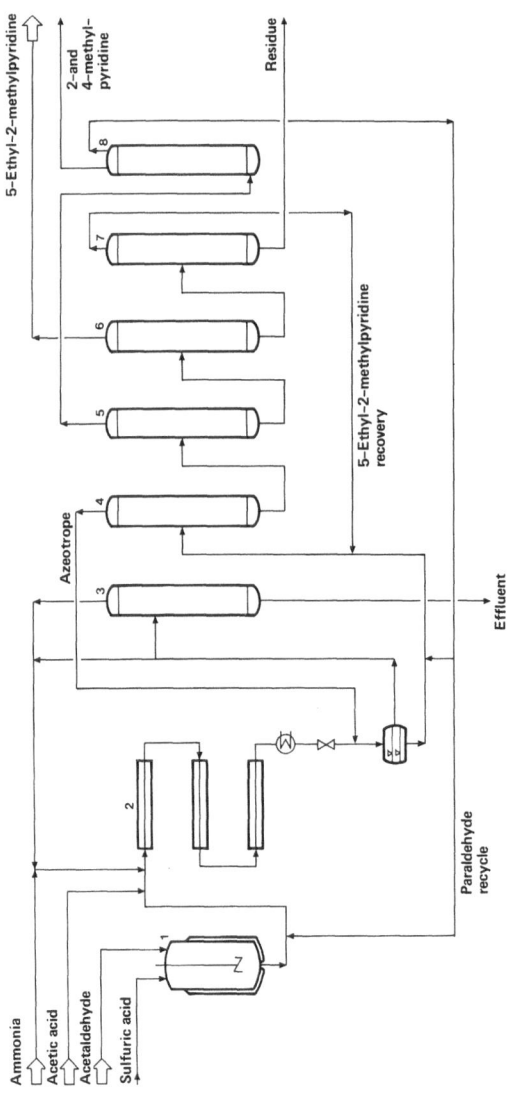

1 Reactor; **2** Tube reactor; **3** NH$_3$-stripping column; **4** Dewatering column; **5** Light ends column; **6** 5-Ethyl-2-methylpyridine column; **7** Residue column; **8** Batch fractionation column

Figure 14.6: Flow diagram of production of 5-ethyl-2-methylpyridine

Ethylmethylpyridine is converted into nicotinic acid by oxidation. The reaction takes place in titanium-lined tube reactors at around 290 bar and 330 °C. The yield is about 95%.

The process developed by *Lonza/Alusuisse* has the advantage of high selectivity and high molar yield; a disadvantage is the loss of two carbon atoms for each molecule of ethylmethylpyridine.

Figure 14.7 shows the niacin plant operated by *Lonza*, Visp/Switzerland.

Figure 14.7: Niacin plant operated by *Lonza*, Visp/Switzerland

406 Aromatic heterocyclics – production and uses

Figure 14.8 shows the flow diagram for the production of nicotinic acid by oxidation with nitric acid. This selective process makes it possible to produce nicotinic acid with a purity of over 99.5% and which can be increased even further by recrystallization.

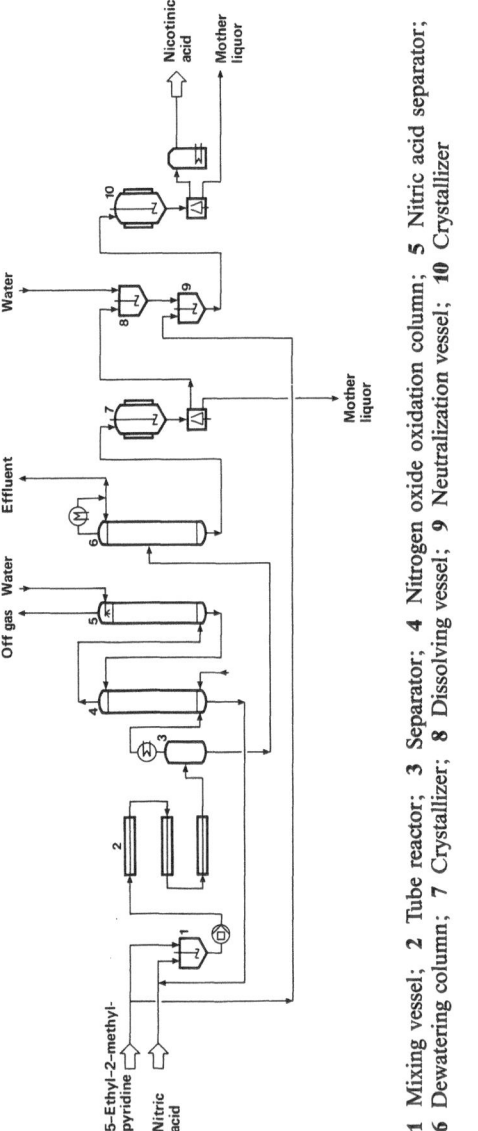

1 Mixing vessel; **2** Tube reactor; **3** Separator; **4** Nitrogen oxide oxidation column; **5** Nitric acid separator; **6** Dewatering column; **7** Crystallizer; **8** Dissolving vessel; **9** Neutralization vessel; **10** Crystallizer

Figure 14.8: Flow diagram of the oxidation of 5-ethyl-2-methylpyridine with nitric acid to produce nicotinic acid

In the USA, in particular, the preferred route for the production of niacin is based on 3-methylpyridine (β-picoline); oxidation is effected with air and ammonia.

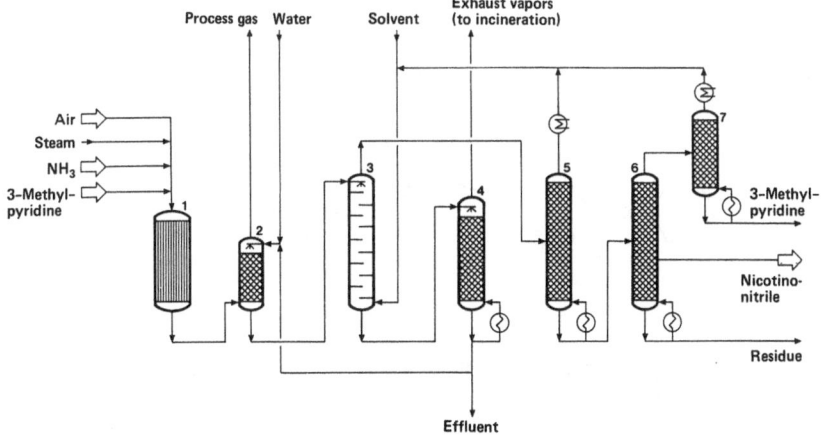

Figure 14.9 shows the flow diagram for ammoxidation of β-picoline.

1 Multitubular reactor; 2 Absorber; 3 Extraction column; 4 Process-water treatment; 5-7 Distillation columns

Figure 14.9: Flow diagram for ammoxidation of β-picoline

The gas-phase oxidation is carried out at around 350 °C on Al_2O_3/SiO_2 or TiO_2/V_2O_5 catalysts. The 3-cyanopyridine is then hydrolysed with caustic soda at 120 to 180 °C and 3 to 8 bar.

Nicotinic acid and nicotinamide are used as feed additives; in natural products, nicotinic acid is found primarily in yeasts, at levels of up to 500 ppm.

A newly-developed herbicide based on β-picoline is fluazifop-butyl (*ICI, Ishihara Sangyo Kaisha*). 2-Chloro-5-trichloromethylpyridine, which is obtained by chlorination, is used to produce this herbicide through fluorination with HF/SbF_3 to give 2-chloro-5-trifluoromethylpyridine, which then undergoes nucleophilic substitution with p-hydroxyphenoxypropionic acid butyl ester.

A new process has recently been introduced by *DSM* for the production of α-picoline (2-methylpyridine). Acetone serves as the raw material; reaction with a primary amine, such as cyclohexylamine or isopropylamine, and acrylonitrile leads to 1-cyanopentan-4-one. Subsequent cyclization at temperatures of around 200 °C gives α-picoline in almost quantitative yield.

Figure 14.10 shows the flow diagram for the *DSM* process.

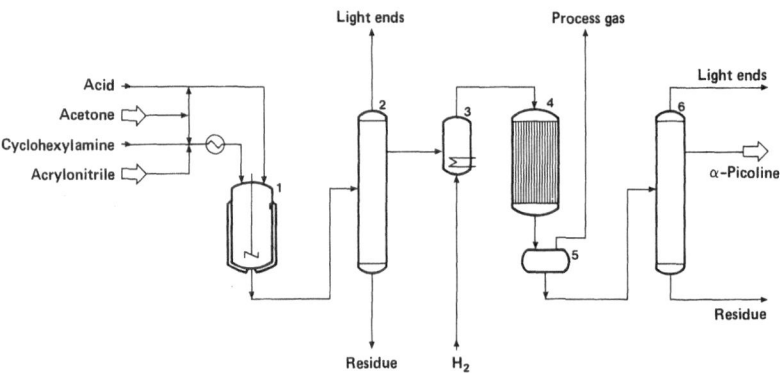

1 Reactor; 2 Distillation column; 3 Vaporizer; 4 Multi-tubular reactor; 5 Gas separator; 6 Distillation column

Figure 14.10: Flow diagram of α-picoline production via 1-cyanopentan-4-one

(If methyl ethyl ketone is used instead of acetone, 2,3-dimethylpyridine is obtained; this lutidine compound is used for the manufacture of 2,3-pyridine dicarboxylic acid (see Chapter 14.5.5))

The methyl group of α-picoline reacts, by virtue of its C-H acidity, with formaldehyde to form 2-pyridyl ethanol, which gives 2-vinylpyridine in the presence of bases. Vinylpyridine serves as a co-monomer in the production of modified styrene-butadiene rubber and special polyacrylic fibers to improve dye absorption.

$$\text{pyridine-CH}_3 \xrightarrow{+ \text{HCHO}} \text{pyridine-CH}_2\text{-CH}_2\text{OH} \xrightarrow{- H_2O} \text{pyridine-CH=CH}_2$$

α-Picoline is also used as a raw material in the manufacture of the herbicide picloram *(Dow)*, which is produced by chlorination, ammonolysis and oxidation.

Picloram

The application of γ-picoline (4-methylpyridine) is relatively limited. In the early 1950's, it was used in the synthesis of isoniazid, currently one of the most important anti-tuberculosis drugs; however, nowadays isoniazid is most commonly produced from the triammonium salt of citric acid by pyrolysis.

Citric acid triammonium salt $\xrightarrow[- 3 H_2O]{3 NH_4^{\oplus} \quad - 2 NH_3}$ 2,6-dihydroxypyridine-COOH $\xrightarrow[- 2/3 H_3PO_4]{+ 2/3 POCl_3}$ 2,6-dichloropyridine-COOH $\xrightarrow[- 2 HCl]{+ 2 H_2}$ pyridine-COOH $\xrightarrow[- H_2O]{+ C_2H_5OH}$

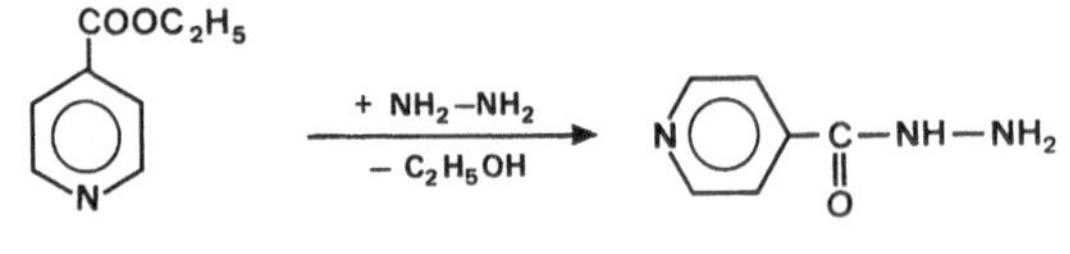

Isoniazid synthesis

[Structure: ethyl isonicotinate + NH$_2$–NH$_2$, − C$_2$H$_5$OH → isonicotinoyl hydrazide]

Isoniazid

The polymethylated pyridines (lutidines, collidines) are used as solvents or anti-corrosion agents, or can be converted into pyridine by dealkylation at temperatures between 700 and 800 °C.

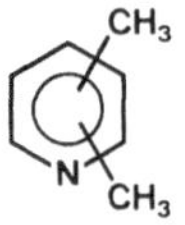

Lutidine 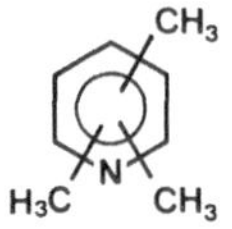   Collidine

An additional poly-substituted pyridine derivative is pyridoxol (pyridoxine/ vitamin B$_6$), which can be made, according to the classical method by Stanton

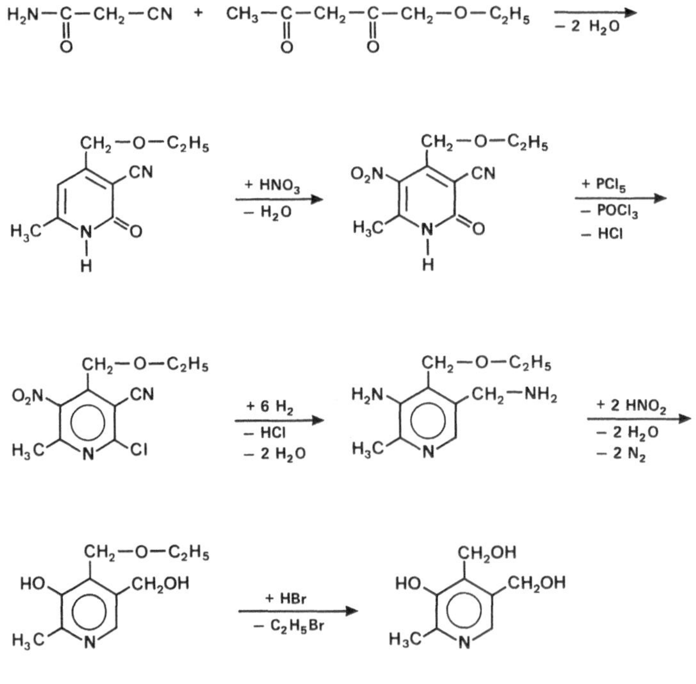

$$H_2N-C(=O)-CH_2-CN \; + \; CH_3-C(=O)-CH_2-C(=O)-CH_2-O-C_2H_5 \xrightarrow{-2\,H_2O}$$

[Reaction sequence:]

Step 1: pyridone intermediate + HNO$_3$, − H$_2$O → nitro-pyridone

Step 2: + PCl$_5$, − POCl$_3$, − HCl → chloro-nitropyridine

Step 3: + 6 H$_2$, − HCl, − 2 H$_2$O → diamino-pyridine derivative

Step 4: + 2 HNO$_2$, − 2 H$_2$O, − 2 N$_2$ → dihydroxymethyl pyridine

Step 5: + HBr, − C$_2$H$_5$Br → Pyridoxol

Pyridoxol

A. Harris and Karl Folkers, from cyanoacetamide and 1-ethoxypenta-2,4-dione. Newer processes include the use of 4-methyloxazoles.

In addition, γ-picoline is used in small amounts to produce 4-vinylpyridine.

14.3 Pyrimidine

Of the six-membered ring heterocyclics with two nitrogen atoms, pyridazine, pyrimidine and pyrazine, it is principally the pyrimidine compounds which are industrially important.

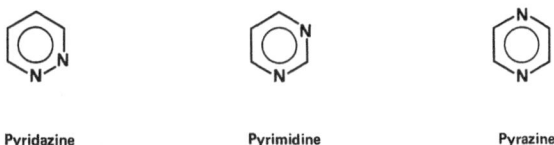

Pyridazine   Pyrimidine   Pyrazine

Pyrimidine occurs in natural substances, such as vitamin B_1 (thiamine), which is contained e.g. in grain husks, yeast or potatoes in a concentration up to 25 ppm.

Synthetic thiamine is produced by a multi-stage process.

The starting material is acetamidine, which is condensed with the formyl derivative of 2-ethoxypropionic acid ethyl ester to give 5-ethoxymethyl-4-hydroxy-2-methylpyrimidine.

The thiazole heterocyclic component is introduced in the form of 5-(2'-hydroxyethyl)-4-methyl thiazole.

Production of vitamin B_1 is about 2,000 tpa.

5-(2'-Hydroxyethyl)-
4-methylthiazole

5-Ethoxymethyl-4-hydroxy-
2-methylpyrimidine

Thiamine

Another route uses 2-methyl-4-aminocyanopyridine, CS_2 and 5-acetoxy-3-chloro-2-pentanone (made from butyrolactone).

Important insecticides such as diazinon (*Ciba Geigy*) and pirimicarb (*ICI*) are also based on pyrimidine compounds; for production of these compounds, the pyrimidine ring is synthesized from aliphatic building blocks.

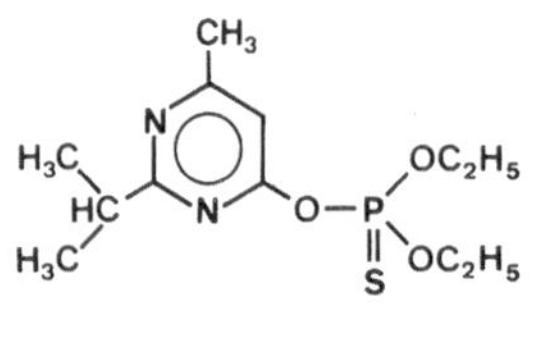

Diazinon

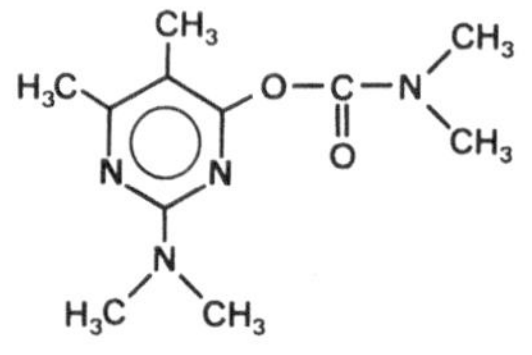

Pirimicarb

Other important pyrimidine compounds are derivatives of barbituric acid, substituted in the 5-position, which are produced by condensation of urea with malonic acid esters. However, these products are not typical aromatic compounds, since they occur predominantly in the keto form. Important representatives of the barbituric acids are the sedatives veronal and luminal.

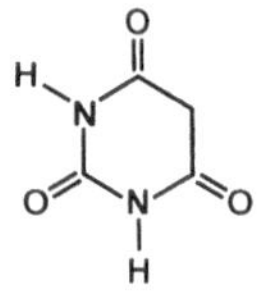

Barbituric acid

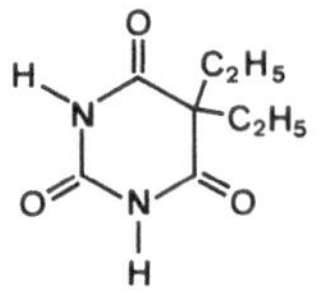

Veronal

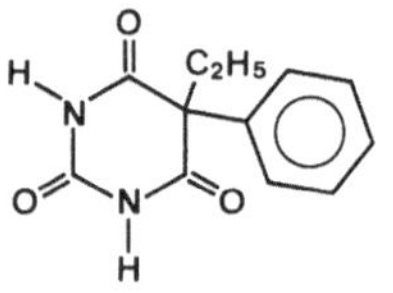

Luminal

The pyridazine compounds are particularly used as plant protection agents. A typical example is the *BASF* herbicide, pyrazon, which is produced from mucochloric anhydride with phenylhydrazine with replacement of one chlorine by ammonia (see Chapter 5.5.3.4).

14.4 Triazines

The derivatives of 1,3,5-triazine are by far the most important among the triazines.
2,4,6-Triamino-1,3,5-triazine (melamine) and 2,4,6-trichloro-1,3,5-triazine (cyanuric chloride) are important chemical intermediates.
Melamine can be obtained from dicyanodiamide by cyclization at 300 to 330 °C and 150 bar pressure.

Aromatic heterocyclics – production and uses   413

$$3 \quad \underset{H_2N}{\overset{H_2N}{>}}C{=}N{-}CN \longrightarrow 2 \quad \text{(Melamine)}$$

In addition, it is also possible to produce it from urea in an atmospheric pressure process (*BASF*) at temperatures of around 370 °C in a fluidized bed on an Al_2O_3 catalyst.

$$6\ NH_2{-}\underset{\underset{O}{\|}}{C}{-}NH_2 \xrightarrow[-6\ NH_3]{-3\ CO_2} \text{melamine}$$

Figure 14.11 shows the flow diagram for the manufacture of melamine from urea.

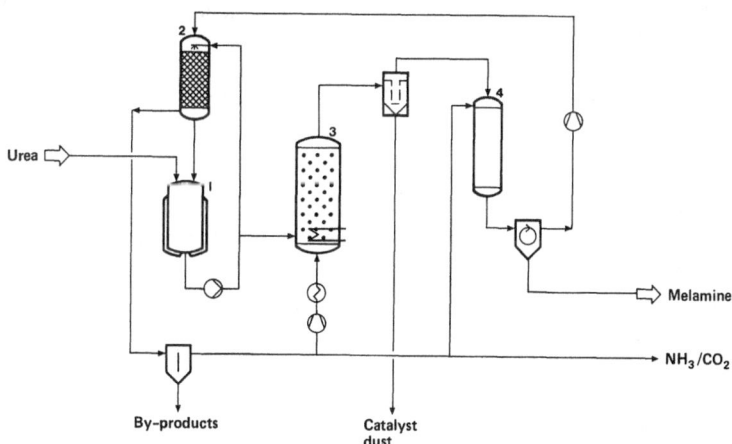

1 Melting vessel; 2 Gas scrubber; 3 Fluidized-bed reactor; 4 Desublimation chamber

Figure 14.11: Flow diagram for the production of melamine from urea

Melamine, by reaction with formaldehyde, is used in the production of resins for adhesives and binders. The largest US-producers are *Amel* and *Melamine Chemicals*.

2,4,6-Trichloro-1,3,5-triazine is an important intermediate, especially for the production of plant protection agents; it is made by the reaction of hydrogen cyanide with chlorine, in an aqueous medium, which gives cyanogen chloride. Trimerization of cyanogen chloride at temperatures in excess of 300 °C yields 2,4,6-trichloro-1,3,5-triazine, which is produced worldwide in quantities of around 100,000 tpa. Figure 14.12 shows a process for production of 2,4,6-trichloro-1,3,5-triazine.

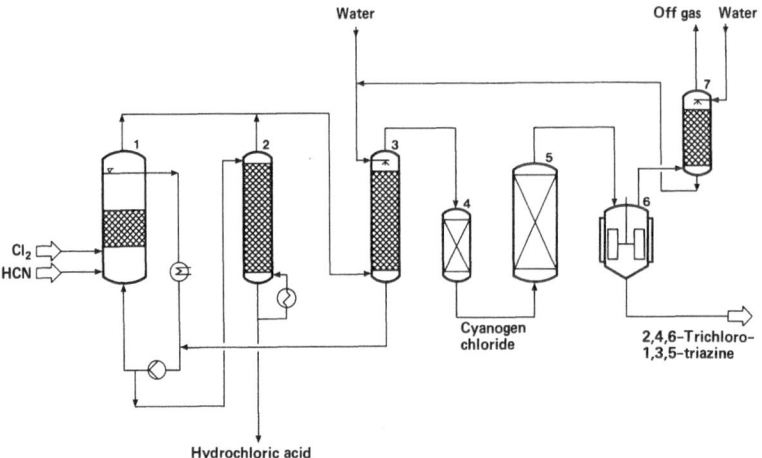

1 Cyanogen chloride reactor; **2** Degasifying column; **3** Scrubber; **4** Drier; **5** Trimerization reactor; **6** Desublimation chamber; **7** Scrubber

Figure 14.12: Flow diagram for the production of 2,4,6-trichloro-1,3,5-triazine

Nucleophilic reaction of 2,4,6-trichloro-1,3,5-triazine with ethylamine and further reaction with isopropylamine produces the plant protection agent atrazine (*Ciba Geigy*), which, with production of around 45,000 tpa, is one of the major pesticides.

Atrazine

Related to atrazine are the herbicides cyanazine (*Shell*), prometryn (*Ciba Geigy*) and prometon (*Ciba Geigy*).

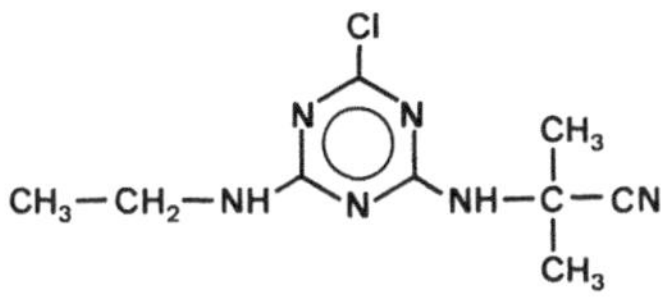

Cyanazine

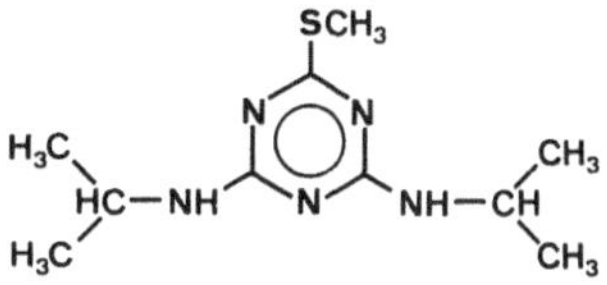

Prometryn

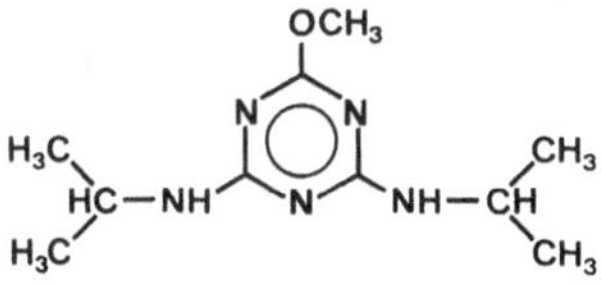

Prometon

The 1,3,5-triazine derivatives are synthesized by the introduction of a wide range of nucleophilic agents such as amines, alkoxides or mercaptides. The selectivity can be governed by temperature control. For the introduction of amines, the first chlorine is exchanged at 0 to 5 °C, the second at 30 to 50 °C and the third chlorine at 70 to 100 °C (Hans E. Fierz-David's rule).

2,4,6-Trichloro-1,3,5-triazine also serves as a component in the production of a wide range of reactive dyes and optical brighteners. An example of this is fluorescent brightener 28, produced by condensation of 2,4,6-trichloro-1,3,5-triazine with diaminostilbenedisulfonic acid, followed by replacement of the remaining chlorine atoms by the appropriate amino groups.

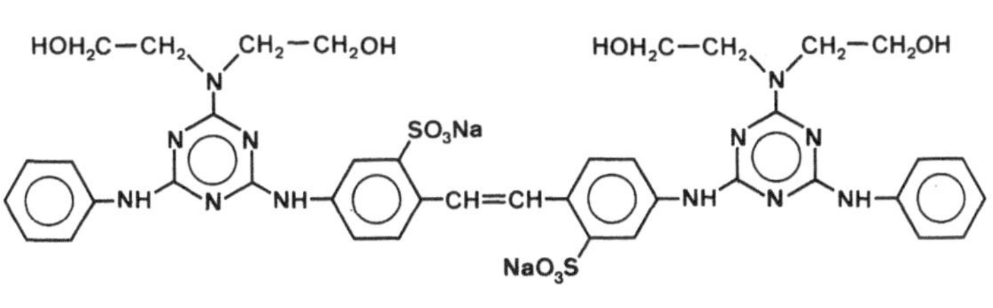

Fluorescent brightener 28

An example of a reactive dyestuff is Procion Brilliant Orange GS (Reactive Orange 1), which is a derivative of J-acid. The triazine group does not act as chromogen in this case, but as an anchor to fix the dye to cotton (triazine anchor).

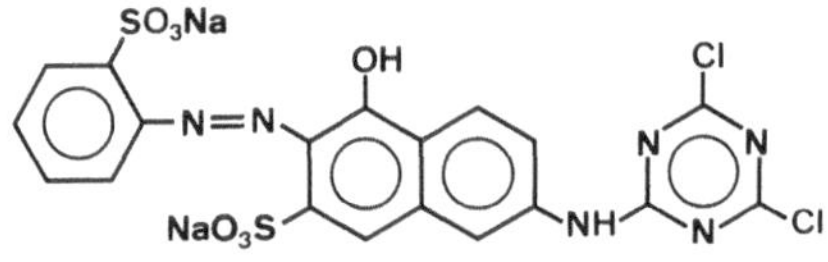

Reactive Orange 1

1,2,4-Triazine derivatives such as metribuzin also serve as plant protection agents; metribuzin is produced from butylglyoxylic acid and thiocarbohydrazide.

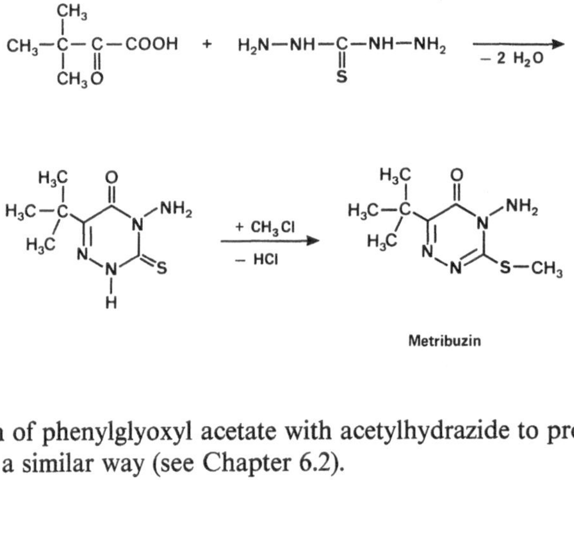

Metribuzin

The reaction of phenylglyoxyl acetate with acetylhydrazide to produce metamitron occurs in a similar way (see Chapter 6.2).

14.5 Condensed heterocyclics

14.5.1 Thianaphthene (2,3-Benzothiophene)

Thianaphthene was discovered in coal tar by Rudolf Weissgerber and Otto Kruber in 1920, after its presence had already been established in brown-coal tar by J. Boes in 1902.

Thianaphthene is present in coal tar in a concentration of 0.3%; although it has in terms of quantity considerable potential as a raw material, demand is currently very low. Thianaphthene can be isolated, after its concentration in the residue from the crystallization of naphthalene, by removal of phenols and bases, followed by a combination of crystallization and redistillation.

One possible application is the production of the anti-oestrogenic keoxifene (*Eli Lilly*).

Keoxifene

14.5.2 Indole

Of greater importance than thianaphthene is its nitrogen analogue, indole.

Indole is found in coal tar in a concentration of 0.2%; it is separated from the wash oil fraction by azeotropic distillation, extraction or alkali fusion.

In Japan, indole is synthesized (e.g *Mitsui Toatsu*) from aniline and ethylene glycol in a yield of around 70%.

In a purity of 99.9%, coal tar-derived indole is mainly used in perfumes as a fixative.

A further important application of indole is the manufacture of the amino acid tryptophan. Since tryptophan is needed as an essential amino acid in the L-form, it may be produced by biotechnical means. Possible methods include enzyme-catalyzed reaction of indole with L-serine and fermentation of nutrients (carbon source) containing anthranilic acid with mutants of Bacillus subtilis.

[Reaction scheme: anthranilic acid (o-aminobenzoic acid) and indole + serine (H₂N-CH(COOH)-CH₂OH) → Tryptophan]

Biotechnical production of tryptophan is analogous to the manufacture of penicillin G; methods of genetic recombination are currently being investigated as ways of increasing production.

Worldwide production of tryptophan is around 400 tpa; the largest production plants are operated in Japan (*Showa Denko, Mitsui Toatsu*).

14.5.3 Benzothiazole

The most important derivative of benzothiazole, commercially, is 2-mercaptobenzothiazole, which is obtained in 95% yield by the reaction of aniline with CS_2 and sulfur (see Chapter 5.5.3.2).

Benzothiazole

14.5.4 Benzotriazole

Reaction of o-phenylenediamine with nitrous acid produces benzotriazole, a corrosion inhibitor for copper alloys.

o-phenylenediamine + HNO_2 ⟶ Benzotriazole + 2 H_2O

14.5.5 Quinoline and isoquinoline

Quinoline was discovered in coal tar by Friedlieb Ferdinand Runge in 1834; it is present in concentrations of approximately 0.3%. Quinoline is recovered by extraction with sulfuric acid from the methylnaphthalene fraction of coal tar, followed by 'springing' with ammonia and rectification of the crude base mixture. Quinoline can be synthesized by the Skraup method, by the reaction of aniline with glycerol (or acrolein produced from glycerol) and catalytic gas-phase reaction of aniline with acetaldehyde. Since the supply of the tar-derived material has been adequate for a long time, synthetic production is not warrented.

Worldwide production of quinoline is around 2,100 tpa.

Skraup's quinoline synthesis

The main application for quinoline is the production of 8-hydroxyquinoline (oxine), which is obtained by alkali fusion of quinoline-8-sulfonic acid.

Sulfonation of quinoline to produce quinoline-8-sulfonic acid is carried out at 180 to 200 °C. Alkali fusion is performed at 250 °C and 45 bar; final purification of the crude 8-hydroxyquinoline is effected by distillation.

Apart from the production of 8-hydroxyquinoline (and its copper salt), which is mainly used as a fungicide and disinfectant, quinoline serves as a feedstock for production of pyridine-2,3-dicarboxylic acid (quinolinic acid), by oxidation. Quinolinic acid is used to manufacture the plant protection agent imazapyr (*American Cyanamid*).

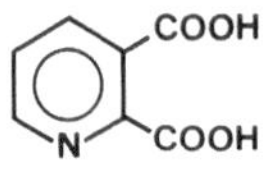

Quinolinic acid

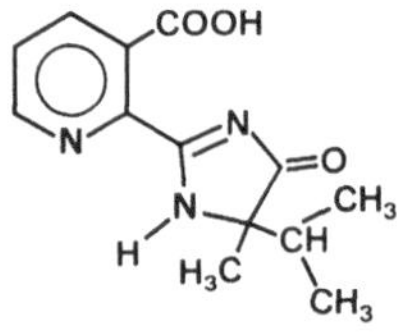

Imazapyr

Condensation of aniline with ketones, such as acetone or methyl isobutyl ketone, produces trimethyl-1,2-dihydroquinolines, such as 2,2,4-trimethyl-1,2-dihydroquinoline, which is used as an important anti-oxidant in rubber processing. West European production is around 10,000 tpa.

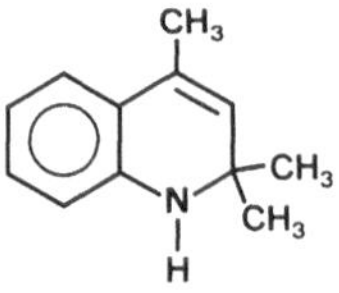

2,2,4-Trimethyl-1,2-dihydroquinoline

Chloroquine is an important anti-malarial, obtained from 4,7-dichloroquinoline by reaction with 4-amino-1-diethylamino-pentane. 4,7-Dichloroquinoline is produced from m-chloroaniline by cyclization with diethyloxalyl acetate and chlorination with $POCl_3$.

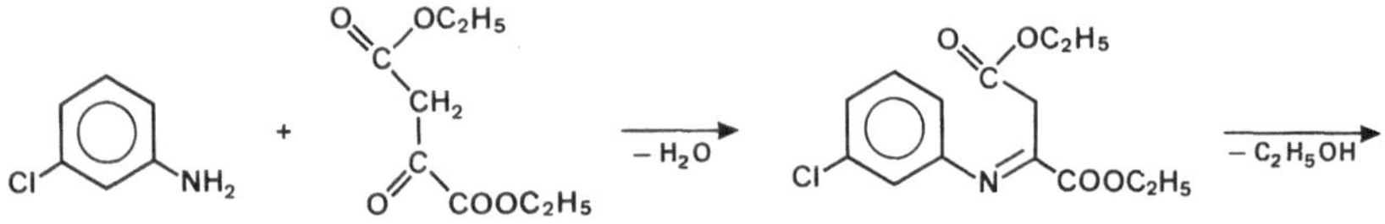

Aromatic heterocyclics – production and uses

In addition to quinoline, coal tar also contains alkylated quinolines, such as 2-methylquinoline and 4-methylquinoline; they are used in limited amounts to make dyestuffs.

Coal tar also contains 0.2% isoquinoline, which is recovered in the same manner as quinoline.

Isoquinoline is used predominantly in the production of medicaments such as the anthelmintic praziquantel, which is produced by the reaction of isoquinoline with cyclohexane carboxylic acid chloride and potassium cyanide, and three further reaction stages.

The alkaloid quinine is a quinoline derivative; Robert B. Woodward and William v. E. Doehring were the first to synthesize it successfully in 1944, almost 100 years after Perkin's unsuccessful attempts (see Chapter 1), using 7-hydroxy-isoquinoline as a base material.

7-Hydroxyisoquinoline

The most important pharmaceutical derivative of isoquinoline is papaverine, which is not, however, synthesized from isoquinoline, but from homoveratryl-amine and homoveratric acid, followed by isoquinoline ring closure by Bischler-Napieralski's method, with the aid of $POCl_3$, and subsequent reduction.

Papaverine

14.5.6 Carbazole

Carbazole was discovered in coal tar, where it is present in a concentration of around 1.5%, by Carl Graebe and Carl Glaser in 1872. Crude carbazole (70%) can be recovered as a side stream from the distillation to isolate anthracene; it can then be produced in pure form by crystallization from polar solvents such as acetophenone.

World production of carbazole is around 2,000 tpa; the largest producers are *Rütgerswerke* (West Germany) and *Nihon Iyoryu Kogyo/Nippon Steel Chemical* (Japan)

Apart from its use in the manufacture of pigments, carbazole is applied mainly to produce N-vinylcarbazole, which is obtained in a reaction with acetylene in the liquid phase at 150 to 250 °C and a pressure of around 25 bar, or in the gas phase at 350 °C in the presence of alkali hydroxides.

The reaction of carbazole with ethylene oxide, followed by dehydration, also leads to N-vinylcarbazole.

N-Vinylcarbazole is used as a co-monomer for the production of special plastics, particularly for the electrical industry.

Carbazole derivatives have been very important as pigments and dyes for a long time. They can be produced from carbazole or by cyclization of nitrogen compounds; examples of the latter route are the production of Vat Black 27 and Vat Orange 15.

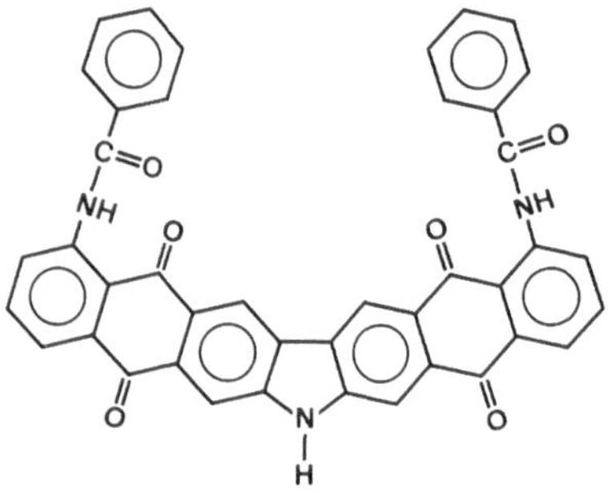

Vat Orange 15

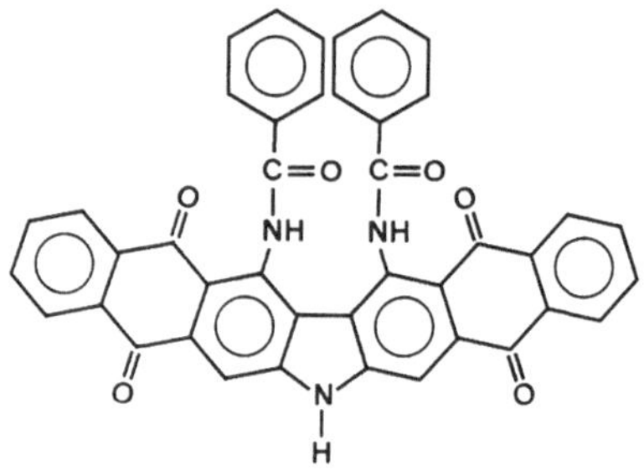

Vat Black 27

The universally applicable dioxazine pigment, especially Pigment Violet 23, which is distinguished by good color strength, is commercially important. It is produced by the reaction of 3-amino-N-ethylcarbazole and chloranil in o-dichlorobenzene at temperatures between 130 and 150 °C followed by oxidative ring closure to the dioxazine in the presence of benzenesulfonyl chloride. The 3-amino-N-ethylcarbazole required for the synthesis is produced from carbazole by N-alkylation, e.g. with diethyl sulfate or ethyl chloride, and nitration followed by the reduction to the amine.

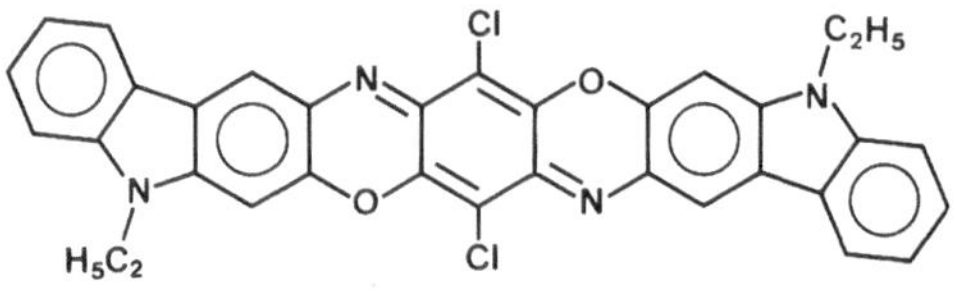

Pigment Violet 23

A traditionally important sulfur dye, based on carbazole is Hydron Blue R (Vat Blue 43), developed by *Cassella* in 1909. The reaction of p-nitrosophenol with carbazole in sulfuric acid yields 'indocarbazole' (N-(3-carbazole)-1,4-benzoquinonimine), which is transformed into the corresponding sulfur dye by reaction with sodium polysulfide. The structure of this dye, in common with the structure of other sulfur dyes, has not yet been fully elucidated, despite its economic significance.

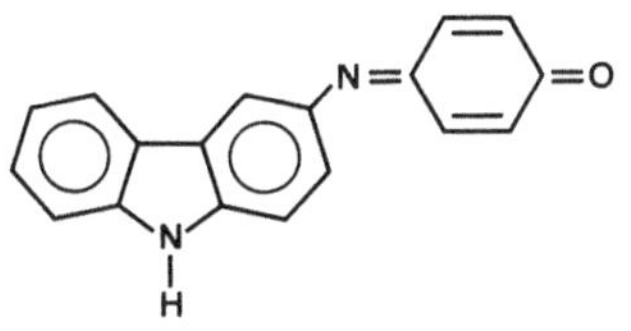

N-(3-carbazole)-1,4-benzoquinonimine

14.5.7 Dibenzofuran (Diphenylene oxide)

Diphenylene oxide is present in coal tar in a concentration of around 1%; it is recovered from the wash oil fraction by a combination of crystallization and distillation (see Chapter 3.2.3).

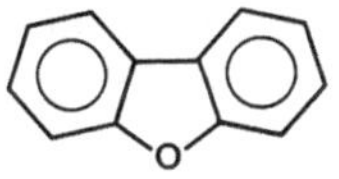

Diphenylene oxide

The most important product from diphenylene oxide is 2,2'-dihydroxybiphenyl, which is produced by alkali fusion (ether cleavage); it is used in small amounts as an intermediate in the production of plant protection agents and active ingredients for disinfectants.

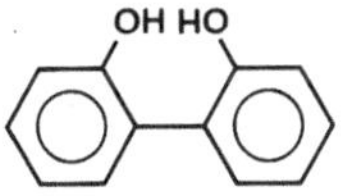

2,2'-Dihydroxybiphenyl

15 Toxicology / Environmental aspects

15.1 Basic toxicological considerations

Toxicology, the science concerning the studies of harmful effects of chemicals of both natural and industrial origin on living organisms, has the prime objective of investigating the possible risk to all forms of life from chemical substances.

The aim of toxicological investigations is to establish the risk of chemicals to human and animal health, in order to recognize possible hazards in time to take preventive steps.

In addition to acute and chronic toxic effects (especially carcinogenic, mutagenic, embryo-toxic and teratogenic effects), examination of the degradability or persistence of a chemical compound, together with the possibility of its bioaccumulation in the food chain are very much in the foreground of product environmental protection.

The acute toxic effect of a substance is generally determined by animal experiments, usually on the rat. Substances which cause death when administered in a maximum oral dose of 25 mg/kg of body weight (LD) are classed as highly toxic. Toxic substances are those which have a lethal effect in applications between 25 to 200 mg/kg. Less toxic, harmful substances cause death by doses between 200 and 2,000 mg/kg body weight. Substances with acute toxicity ratings above 2,000 mg/kg are classed as non-toxic.

Parallel to the examination of acute toxicity, the determination of chronic and sub-chronic toxicity (long-term effect) is playing an increasingly important role in the study of environmental effects.

The dosage level at which 50% of the experimental animals in a collective group die is known as the average lethal dose – LD_{50}.

The harmful effects of a chemical substance are determined by quantity, time of exposure and the mode of uptake (oral, inhalative, dermal). The words of Paracelsus from the 16th century still apply: 'All things are poison and nothing is without poison; only the dose determines that a thing is not poisonous'.

Table 15.1 summarizes the LD- and LD_{50}-values of some aromatics.

Table 15.1: LD- and LD_{50}-values of aromatics

Substance	Toxic values in mg/kg (oral, rat)	
Aniline	LD_{50}:	440
Anthracene	LD :	>16,000
Benzene	LD_{50}:	3,800
Benzenesulfonic acid	LD_{50}:	890
Chlorobenzene	LD_{50}:	2,910
2,4-D	LD_{50}:	500
DDT	LD_{50}:	500
Ethylbenzene	LD_{50}:	3,500
Fluoranthene	LD :	>16,000
o-Cresol	LD_{50}:	121
Naphthalene	LD :	>16,000
2-Naphthol	LD_{50}:	2,420
Phenanthrene	LD :	>16,000
Phenol	LD_{50}:	414
Pyrene	LD :	>16,000
Phthalic anhydride	LD_{50}:	4,020
Styrene	LD_{50}:	5,000
Terephthalic acid dimethyl ester	LD_{50}:	4,390
Toluene	LD_{50}:	5,000
o-Xylene	LD_{50}:	5,000
m-Xylene	LD_{50}:	5,000
p-Xylene	LD_{50}:	5,000

Unsubstituted aromatics and alkyl aromatics display a relatively low acute toxicity; substitution with reactive groups, such as e.g. amino and hydroxyl groups, generally causes a marked increase in acute toxicity.

15.2 Aspects of occupational medicine and legislation

Occupational medical investigation of industrial workers has a long tradition. In 1761, the London physician John Hill published a study in which he showed the cancer-producing effect of excessive snuff taking. This was followed in 1775 by a further publication on the high incidence of cancer among chimney sweeps. Percival Pott established that this increased cancer rate resulted from contact with soot.

The Japanese, Katsusaburo Yamagiwa and Koichi Ichikawa, also carried out experiments at the beginning of this century, which indicated that coal tar fractions could cause cancer in mice.

Dibenz[a,h]anthracene was the first carcinogenic aromatic compound to be identified, by Ernest L. Kennaway and Izrael Hieger in 1930. Kennaway also examined high-boiling tar fractions and isolated the isomeric benzo[a]pyrene and benzo[e]pyrene from 2 tons of pitch; he proved in animal experiments that only benzo[a]pyrene displays a strong carcinogenic effect.

In 1971, the International Agency for Research on Cancer (IARC) set up a program to determine the carcinogenic risks of chemicals to human beings.

When the carcinogenic potential has been identified, the industrial application of a carcinogenic substance has either been considerably restricted or appropriate measures have been introduced to enable the toxic compound to be handled more safely. There are examples of this as early as the 19th century. Because of the increased incidence of bladder cancer among workers involved in manufacture of the dye fuchsin (magenta), the aromatic amines in particular were investigated to study their carcinogenic effect. β-Naphthylamine and benzidine, together with auramine were the principal compounds recognized as being carcinogenic. The last plant for the manufacture of β-naphthylamine closed in Great Britain in 1949; enormous efforts in process engineering were made in α-naphthylamine production to minimize the β-naphthylamine content. Another example is Butter Yellow (Solvent Yellow 2) which was used to color margarine and butter; it has not been available commercially since 1938, because of its carcinogenic effect.

Solvent Yellow 2

In the development of new products, toxicological risks and, especially, carcinogenic effects are taken into account with extremely high priority. 2-Acetylaminofluorene, for example, was not introduced as an insecticide after a well-founded suspicion of carcinogenicity was established.

2-Acetylaminofluorene

In all major industrialized nations, laws are in force governing the handling of chemicals, namely the 'Gesetz zum Schutz vor gefährlichen Stoffen' (Chemicals law of September 16, 1980) in West Germany, the 'Health and Safety Act' in Great Britain, and in France, the 'decrét No. 85 á 217' of February, 1985 'portant sur le contrôle des produits chimiques'. In the USA, guidelines are laid down by the Occupational Safety and Health Administration (OSHA) and the Toxic Substance Control Act (TSCA) of 1976, and applied by the Environmental Protection Agency (EPA).

The objective of chemicals legislation is to protect man and the environment from the harmful effects of dangerous substances by making it obligatory to test and notify substances, and to classify, label and pack dangerous substances and preparations accordingly, as well as to impose bans and restrictions, and introduce regulations on the handling of toxic substances and on working practices.

Of particular importance within the framework of West German legislation is the 'Verordnung über gefährliche Stoffe' (Gefahrstoffverordnung of August 26, 1986) (Dangerous Substances Act). This Act, implementing the Chemicals Law, controls in detail the marketing, labelling and packaging, as well as the handling of dangerous substances.

Annex I of the Dangerous Substances Act specifies the classification and labelling of dangerous substances and preparations. Annex II sets forth special rules on handling carcinogenic substances and those causing damage to the fetus and leading to mutations. Annex VI contains the list of classified dangerous substances and preparations ('Liste eingestufter gefährlicher Stoffe und Zubereitungen').

Table 15.2 shows the classification of carcinogenic aromatic dangerous substances.

Table 15.2: Classification of dangerous aromatic substances in Annex VI of the dangerous substances act (Gefahrstoffverordnung); (Mass content of dangerous substance in %)

Dangerous carcinogenic substance	Groups		
	I (extremely harmful)	II (very harmful)	III (harmful)
o-Aminoazotoluene		≥ 0.1	$<0.1-0.01$
4-Aminobiphenyl	≥ 1	$<1-0.1$	$<0.1-0.01$
Benzidine	≥ 1	$<1-0.1$	$<0.1-0.01$
Benzidine salts	≥ 1	$<1-0.1$	$<0.1-0.01$
Benzene		≥ 1	
Benzo[a]pyrene*		≥ 0.1	$<0.1-0.005$
3,3'-Dichloro-benzidine		≥ 1	$<1 -0.1$
3,3'-Dimethyl-4,4'-diamino-diphenylmethane		≥ 1	$<1 -0.1$
2-Naphthylamine	≥ 1	$<1-0.1$	$<0.1-0.01$
5-Nitroacenaphthene		≥ 1	$<1 -0.1$
4-Nitrobiphenyl	≥ 1	$<1-0.1$	$<0.1-0.01$
2-Nitronaphthalene		≥ 1	$<1 -0.1$

* as an indicator substance of carcinogenic polycyclic aromatic hydrocarbons (PAH) in pyrolysis products from organic materials

The Senate commission of the 'Deutsche Forschungsgemeinschaft' (DFG – German Research Association) on investigations into harmful working substances publishes among other documents an annual listing of carcinogenic industrial chemical materials. The DFG commission also sets out the 'Technische Richtkonzentration' (TRK). The TRK value is the concentration of the substance as gas, vapor or aerosol in the atmosphere which serves as reference level for the relevant protective measures and the control of concentration in the workplace.

In addition, the Senate commission for the investigation of harmful working substances also sets out the maximum workplace concentration value (MAK), the highest permissible concentration of a substance as gas, vapor or aerosol in the workplace atmosphere, which, according to current knowledge, generally will not affect the health of the workers even with repeated (8 h) long-term exposure.

The most important aromatic compounds from the MAK list are given in Table 15.3. The MAK values may be altered, either up or down, in accordance with the progress of available toxicological knowledge.

Table 15.3: MAK values of aromatic compounds

Substance	Formula	MAK ml/m³ (ppm)	mg/m³
o-Aminoazotoluene			**
4-Aminobiphenyl			*
3-Amino-9-ethyl-carbazole			***
2-Amino-4-nitrotoluene			**
2-Aminopyridine		0.5	2
Aniline ***		2	8
ANTU 1-Naphthalenylthiourea			0.3
Atrazine			2
Auramine			**

Substance	Formula	MAK	
		ml/m³ (ppm)	mg/m³
Azinphos-methyl			0.2
Benzene			*
Benzidine and its salts			*
Benzo[a]pyrene			*****
p-Benzoquinone		0.1	0.4
Biphenyl		0.2	1
p-tert-Butyl-phenol		0.08	0.5
p-tert-Butyl-toluene		10	60

Substance	Formula	MAK	
		ml/m³ (ppm)	mg/m³
Carbaryl	(naphthyl-O-CO-NH-CH₃)		5
Chlorinated biphenyls: Chlorine content 42% ***		0.1	1
Chlorinated biphenyls: Chlorine content 54% ***		0.05	0.5
Chlorobenzene	(C₆H₅Cl)	50	230
1-Chloro-4-nitro-benzene	(4-Cl-C₆H₄-NO₂)		1
α-Chlorotoluene	(C₆H₅-CH₂Cl)	1	5
4-Chloro-o-toluidine			*
5-Chloro-o-toluidine			***
Chrysene			**

Substance	Formula	MAK	
		ml/m³ (ppm)	mg/m³
Cresols		5	22
2,4-D			10
DDT			1
2,4-Diamino-anisole			**
4,4'-Diamino-diphenyl ether			**
4,4'-Diamino-diphenylmethane			**
4,4'-Diamino-diphenyl sulfide			**
Diazinon			1
Dibenzoyl peroxide****			5

Substance	Formula	MAK	
		ml/m³ (ppm)	mg/m³
1,2-Dichloro-benzene		50	300
1,4-Dichloro-benzene		75	450
3,3'-Dichloro-benzidine**			0.1
α, α-Dichloro-toluene			***
1,4-Dihydroxy-benzene			2
3,3'-Dimethoxy-benzidine			**
N,N-Dimethylaniline		5	25
3,3'-Dimethyl-benzidine			**

Substance	Formula	MAK	
		ml/m³ (ppm)	mg/m³
α, α-Dimethyl-benzylhydroperoxide			****
3,3'-Dimethyl-4,4'-diamino-diphenylmethane			**
Dinitrobenzene		0.15	1
4,6-Dinitro-o-cresol			0.2
Dinitronaphthalenes			***
Dinitrotoluenes			**
Di-sec-octyl phthalate			10
Diphenyl ether		1	7

Substance	Formula	MAK	
		ml/m³ (ppm)	mg/m³
Diphenylmethane-4,4'-diisocyanate	OCN–C₆H₄–CH₂–C₆H₄–NCO	0.01	0.1
Ethylbenzene	C₆H₅–C₂H₅	100	440
O-Ethyl-O-(4-nitrophenyl)-phenylthiophosphonate (EPN)	C₆H₅–P(=S)(O–C₂H₅)–O–C₆H₄–NO₂		0.5
Fenthion	H₃C–S–C₆H₃(CH₃)–O–P(=S)(OCH₃)₂		0.2
Furfural	furan–CHO	5	20
Furfuryl alcohol	furan–CH₂OH	50	200
2-Methoxyaniline	C₆H₄(NH₂)(OCH₃)	0.1	0.5
4-Methoxyaniline	C₆H₄(NH₂)(OCH₃)	0.1	0.5
Methoxychlor	H₃CO–C₆H₄–CH(CCl₃)–C₆H₄–OCH₃		15

Substance	Formula	MAK	
		ml/m³ (ppm)	mg/m³
N-Methylaniline		2	9
4,4'-Methylene-bis-(2-chloroaniline)			**
4,4'-Methylene-bis-(N,N-dimethylaniline)			***
Methylstyrene		100	480
N-Methyl-2,4,6-N-tetranitroaniline			1.5
Michler's ketone			***
Naphthalene		10	50
1,5-Naphthalene diisocyanate		0.01	0.09
2-Naphthylamine			*

Substance	Formula	MAK	
		ml/m³ (ppm)	mg/m³
Nicotine		0.07	0.5
5-Nitroacenaphthene			**
2-Nitro-4-aminophenol			***
4-Nitroaniline		1	6
Nitrobenzene		1	5
4-Nitrobiphenyl			**
1-Nitronaphthalene			***
2-Nitronaphthalene			**
2-Nitro-p-phenylenediamine			***
Nitropyrenes			***

Substance	Formula	MAK	
		ml/m³ (ppm)	mg/m³
N-Nitrosoethylphenylamine	O=N–N(CH₂–CH₃)(C₆H₅)		**
N-Nitrosomethylphenylamine	O=N–N(CH₃)(C₆H₅)		**
Nitrotoluenes	4-NO₂-C₆H₄-CH₃	5	30
Paraquat-dichloride	[H₃C–N⁺(C₅H₄)–(C₅H₄)N⁺–CH₃] 2 Cl⁻		0.1
Parathion	(C₂H₅O)₂P(=S)–O–C₆H₄–NO₂		0.1
Pentachlorophenol	C₆Cl₅OH	0.05	0.5
Phenol	C₆H₅OH	5	19
p-Phenylene-diamine	H₂N–C₆H₄–NH₂		0.1
Phenylglycidyl-ether ***	C₆H₅–O–CH₂–CH(–O–)CH₂	1	6
Phenylhydrazine ***	C₆H₅–NH–NH₂	5	22

Substance	Formula	MAK	
		ml/m³ (ppm)	mg/m³

Substance	Formula	ml/m³ (ppm)	mg/m³
N-Phenyl-2-naphthylamine			***
Phthalic anhydride			5
Propenylbenzene (iso)		100	480
Propoxur			2
Propylbenzene (iso)		50	245
Pyridine		5	15
Pyrolysis products from org. material			*** *****
Rotenone			5

Substance	Formula	MAK	
		ml/m³ (ppm)	mg/m³
Styrene	CH=CH₂, phenyl ring	20	85
2,3,7,8-Tetrachloro-dibenzo-p-dioxin	dibenzodioxin with 4 Cl		**
Toluene	CH₃-phenyl	100	375
Toluene-2,4-diamine	CH₃-phenyl with 2 NH₂		**
Toluene-2,4-diisocyanate	CH₃-phenyl with 2 NCO	0.01	0.07
Toluene-2,6-diisocyanate	CH₃-phenyl with 2 NCO	0.01	0.07
2-Toluidine	NH₂, CH₃-phenyl		**
1,2,4-Trichloro-benzene	phenyl with 3 Cl	5	40
2,4,5-Trichloro-phenoxyacetic acid	Cl₃-phenyl-O-CH₂-COOH		10

Substance	Formula	MAK	
		ml/m³ (ppm)	mg/m³
α, α, α-Tri-chlorotoluene			***
Trimellitic anhydride		0.005	0.04
2,4,5-Trimethyl-aniline			**
2,4,7-Trinitro-fluorenone			***
2,4,6-Trinitro-phenol			0.1
2,4,6-Trinitro-toluene		0.15	1.5
Warfarin			0.5
Xylenes		100	440

Substance	Formula	MAK	
		ml/m³ (ppm)	mg/m³
Xylidines	NH₂ group on benzene ring with H₃C and CH₃ substituents	5	25
2,4-Xylidine***	NH₂, CH₃ at 2-position, CH₃ at 4-position on benzene ring	5	25

In addition, lignite tar, hard coal tar, coal tar pitch, coal tar oils with a carcinogenic potential and their mixtures (preparations) are included amongst industrial carcinogens.

The MAK list contains the following additional notes:

* III A$_1$ Carcinogenic substances; clearly proven and justifiably suspected
Substances which, according to experience, may cause malignant growths in humans
** III A$_2$ Substances, which, in the opinion of the Commission, have been shown to be carcinogenic only in animal experiments, but under conditions which are comparable with possible human exposure in the workplace or from which comparability may be deduced
*** III B   Substances justifiably suspected of carcinogenic potential; recent discoveries in cancer research require the consideration of further substances, which are suspected of having considerable carcinogenic potential and which urgently need further investigation. The concentration values given for carcinogenic dangerous substances are TRK values
**** Va   Organic peroxides
***** Vd   Pyrolysis products from organic materials

Particularly low MAK values are applied for aromatics with reactive substituents such as amines, phenols and isocyanates.

Animal experiments with unsubstituted polycyclic aromatic compounds have shown some of the compounds with four and more rings to be carcinogenic; with naphthalene, anthracene and phenanthrene, no carcinogenic effects have been established.

Although the toxic effects of industrially produced aromatics have generally been well investigated, relatively little is known in many instances about the numerous poisons which occur naturally. Highly toxic natural aromatic poisons are, for example, aflatoxin B_1, which is produced as a metabolic product of Aspergillus flavus, and colchicine, obtained from the meadow saffron.

Aflatoxin B₁

Colchicine

Bruce N. Ames has established that the main risk to man from mutagenic and carcinogenic substances results from the natural constituents of foodstuffs and from general (social) living habits; the influence of industrial chemicals is small in comparison.

15.3 Environmental aspects and biological degradation of aromatics

In the production of industrial chemicals in West Germany, the emission of harmful substances into the atmosphere is controlled by the technical guideline on clean air (Technische Anleitung zur Reinhaltung der Luft) of February 27, 1986, within the framework of the first general order for the Federal emissions law (Bundes-Immissionsschutzgesetz).

Permissible emissions of carcinogenic substances are divided within this guideline, into three classes of concentration in waste gas. Substances with the highest carcinogenic potential, such as benzo[a]pyrene, dibenz[a,h]anthracene and 2-naphthylamine are listed in Class I and must not exceed a mass concentration of 0.1 mg/m³ in the waste gas, at a mass flow of 0.5 g/h or above.

In Class II, where a maximum concentration of 1 mg/m³ is set for a mass flow of 5 g/h or above, the only compound included from the range of aromatics is 3,3'-dichlorobenzidine.

Benzene belongs to Class III, in which a concentration of 5 mg/m³ is the limit for a mass flow of 25 g/h or above.

Aromatics are degraded in water and soil principally by biological processes; degradation of aromatics which are adsorbed on particulate matter can also be accelerated by UV-radiation.

The first stage of biochemical decomposition of aromatics generally involves the introduction of free oxygen atoms into the aromatic molecule. This process is catalyzed by dioxygenases. The oxidation product from benzene is pyrocatechol. Although the reaction mechanism in mammalian cells is largely understood in all its stages, there is still uncertainty about the primary microbial oxidation stage. Since, unlike the reaction of mammals, in which trans-1,2-dihydroxydihydrobenzene is produced, cis-1,2-dihydroxydihydrobenzene is found in microbial decomposition of benzene, so it can be assumed that the first metabolite is a cyclic peroxide. 1,2-Epoxybenzene is found in mammalian cells.

The dihydroxylation of the aromatic ring is followed by its cleavage. In microbial metabolisms, there are three known ring-opening mechanisms:

1. the ortho-path, established for pyrocatechol and protocatechuic acid,
2. the meta-path, proven for pyrocatechol and substituted pyrocatechols, and
3. the homogentisic acid path.

$$\text{Pyrocatechol} \xrightarrow[O_2]{\text{(Pyrocatechase)}} \text{cis,cis-muconic acid} \longrightarrow$$

$$HOOC-CH_2-\underset{\underset{O}{\|}}{C}-CH_2-CH_2-COOH \longrightarrow HOOC-CH_2-CH_2-COOH + CH_3-COOH$$

$$\text{Pyrocatechol} \xrightarrow[O_2]{\text{(Metapyrocatechase)}} \text{2-hydroxymuconic semialdehyde}$$

$$\xrightarrow{\textit{Ps. arvilla}} CO_2 + CH_3-COOH + CH_3-\underset{\underset{O}{\|}}{C}-COOH$$

$$\xrightarrow{\textit{Ps. spec.}} H-COOH + CH_3-CHO + CH_3-\underset{\underset{O}{\|}}{C}-COOH$$

$$\text{Homogentisic acid} \xrightarrow[O_2]{\text{(Homogentisic acid oxygenase)}} \text{maleylacetoacetate} \longrightarrow$$

$$\longrightarrow \underset{HOOC}{\overset{H}{\diagdown}}C=C\underset{H}{\overset{COOH}{\diagup}} + 2\ CH_3-COOH$$

In principle, the bio-degradation of aromatics is more difficult than the breakdown of aliphatic compounds. However, in recent years considerable headway has been made in breaking down aromatic compounds, for example in the decomposition of lignin or brown coal.

The persistence of certain substances, combined with other effects such as bioaccumulation and toxic effects is undesirable. The combination of these characteristics has led, for example, to the banning of DDT in the major industrialized nations. Nevertheless, some resistance to biodegradation is necessary, especially in plant protection agents.

Chlorinated aromatics are often especially resistant to biological decomposition and become concentrated in the food chain; this is particularly true for para-substituted aromatics. This persistence of chlorinated aromatics, known as para-recalcitrance, is a result of the lack of suitable degrading enzyme systems or of enzyme inhibition in metabolic systems.

Rapid biological decomposition of aromatics is necessary for biological purification of effluents rich in aromatics.

In the Federal Republic of Germany, the discharge of contaminated waste water is controlled by the Water Resources Law (Wasserhaushaltsgesetz - WHG of October 16, 1976) and the law on discharging effluent into bodies of water (Abwasserabgaben-Gesetz - AbwAG of September 13, 1976), together with specific regulations in § 7a of the WHG, which lays down the state of purification which must be achieved when discharging into bodies of water.

16 The future of aromatic chemistry

The rise of industrial organic chemistry in the mid-19th century was initiated by pioneering innovations in the field of aromatic chemistry. It has since undergone continuous further development. Plastics and pesticides followed dyes and pharmaceuticals as the main areas of innovation.

The innovative impetus of industrial aromatic chemistry is likely to be maintained in the future, characterized by a qualitative growth.

The main sources of feedstock for the production of aromatics will continue to be the pyrolysis products of naphtha cracking and coal carbonization, and the catalytic reforming of gasoline fractions. These raw materials will be complemented by catalytic processes to provide aromatics from small aliphatic building blocks. The first attempts along these lines are the *Mobil*-MTG-process and the Cyclar process (*UOP/BP*). In principle, these processes are taking up attempts by Pierre E. M. Berthelot in 1866 to obtain benzene by trimerization of acetylene.

The raw material base for aromatic chemistry is sufficiently large to cope with wider future applications, since aromatics are used in vast quantities to produce fuels and industrial oils; in comparison the raw material requirement for the industrial aromatic chemistry is relatively small.

Among the methods used to recover pure aromatics, crystallization processes will gain further ground, since they generally are distinguished by a lower energy consumption than distillative processes.

In future, catalytic processes will become increasingly important in the development of aromatic conversion processes. The prime objectives will include the highly selective production of pure grades of the desired product and an improvement in environmental protection. Catalysts with corrosive properties will be replaced by less corrosive ones, such as the zeolites.

The range of processes to further upgrade aromatics will be extended by the inclusion of biotechnical processes. Aromatic molecules are often more difficult to convert by micro-organisms than aliphatic molecules, but developments in recent times have indicated that, for example, aromatics which were long considered as barely degradable can indeed be broken down. Typical of this development are the recent results in the field of biochemical degradation of lignin and brown coal.

The association of biotechnology with pharmaceutical chemistry will continue to provide significant impulses for the aromatic chemistry in the future. A large number of chemicals which play a dominant role in natural processes are of aromatic character, for example, tryptophan, the alkaloids quinine and morphine, and nucleic acids. Complex molecules with aromatic moieties, such as vitamin E, penicillins and more simple aromatics such as acetylsalicylic acid, paracetamol and ephedrine serve as traditional drugs. The increased use of biotechnical processes in the production of pharmaceutically effective compounds thus offers interesting prospects for the future.

There are constant new developments in the synthetic production of drugs using heterocyclics and mono- and polynuclear aromatics; one of the most recent examples is the antimycotic naftifine, an allylamine derivative of naphthalene.

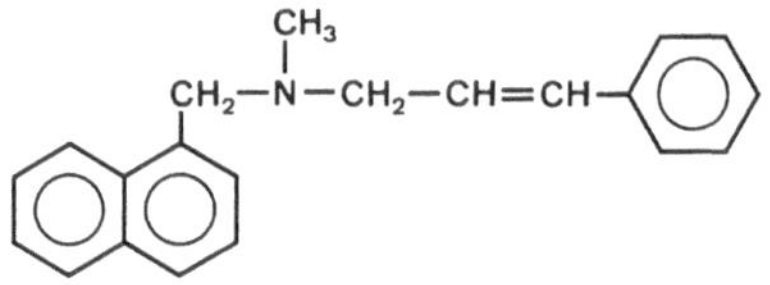

Naftifine

In the development of new plant protection agents, future pesticides will be distinguished by higher selectivity with lower dosages of application, together with adjusted biodegradability and the formation of non-toxic degradation products. Once again, aromatics have an important role to play, as evidenced by the example of synthetic pyrethroid insecticides or by the herbicides imazaquin (*American Cyanamid*), a derivative of quinoline, and chlorsulfuron (*Du Pont*).

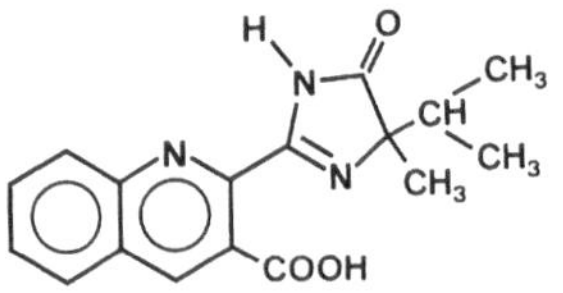

Imazaquin

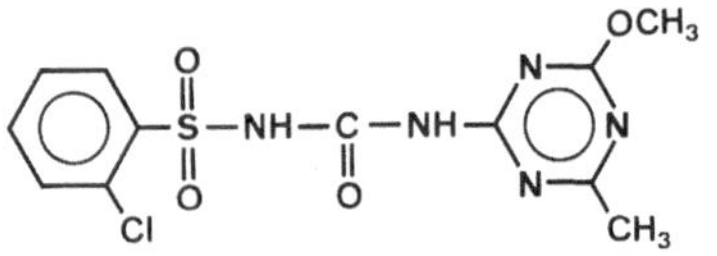

Chlorsulfuron

The potential to form pre-oriented phases (liquid crystals) offers additional areas of application for aromatics, especially for engineering plastics for high-value composites.

Industrial aromatic chemistry will therefore continue to provide a field of rich potential for the chemist and the process engineer.

Appendix

	Summation formula	Melting point (°C)	Boiling point (°C/1,013 mbar)
1. *Hydrocarbons*			
Benzene	C_6H_6	5.5	80.1
Indene	C_9H_8	−1.5	182.4
Indan	C_9H_{10}	−51.4	177.8
Naphthalene	$C_{10}H_8$	80.2	217.9
Acenaphthylene	$C_{12}H_8$	92.5	104 (4 mbar)
Acenaphthene	$C_{12}H_{10}$	95	279
Biphenyl	$C_{12}H_{10}$	69.2	254.9
Fluorene	$C_{13}H_{10}$	115	294
Anthracene	$C_{14}H_{10}$	218	340
Phenanthrene	$C_{14}H_{10}$	100	338.4
Pyrene	$C_{16}H_{10}$	150	393.5
Benzo[a]pyrene	$C_{20}H_{12}$	180	495.5
Benzo[e]pyrene	$C_{20}H_{12}$	178	492.9
2. *N-Heterocyclics*			
Pyrrole	C_4H_5N	−23.4	129.8
Pyridine	C_5H_5N	−42	115.3
Indole	C_8H_7N	52	254
Quinoline	C_9H_7N	−15	237.7
Isoquinoline	C_9H_7N	26.5	243.3
Carbazole	$C_{12}H_9N$	244.8	354.8
Acridine	$C_{13}H_9N$	111	343.9
Pyrimidine	$C_4H_4N_2$	21	124
1,3,5-Triazine	$C_3H_3N_3$	86	114
3. *O-Heterocyclics*			
Furan	C_4H_4O	−85.6	32
Coumarone	C_8H_6O	−28.9	171.4
Diphenylene oxide	$C_{12}H_8O$	85	285.1
Xanthene	$C_{13}H_{10}O$	101.5	311

	Summation formula	Melting point (°C)	Boiling point (°C/1,013 mbar)
4. S-Heterocyclics			
Thiophene	C_4H_4S	−38.2	84
Thianaphthene	C_8H_6S	31.3	219.9
Diphenylene sulfide	$C_{12}H_8S$	97	331.4
5. Phenols			
Phenol	C_6H_6O	40.9	181.8
Catechol	$C_6H_6O_2$	105	245.9
Resorcinol	$C_6H_6O_2$	110.7	276.5
Hydroquinone	$C_6H_6O_2$	170.3	285 (971 mbar)
1-Naphthol	$C_{10}H_8O$	96	288
2-Naphthol	$C_{10}H_8O$	122	295
1,5-Dihydroxynaphthalene	$C_{10}H_8O_2$	258	d.
1,8-Dihydroxynaphthalene	$C_{10}H_8O_2$	140	–
2-Aminophenol	C_6H_7ON	173	subl.
3-Aminophenol	C_6H_7ON	122.1	164 (15 mbar)
4-Aminophenol	C_6H_7ON	186	subl.
8-Hydroxyquinoline	C_9H_7ON	75.8	266.6 (1000 mbar)
6. Carboxylic acids and anhydrides			
Benzoic acid	$C_7H_6O_2$	121.7	249.2
Salicylic acid	$C_7H_6O_3$	159	211 (27 mbar)
Phthalic anhydride	$C_8H_4O_3$	130.8	284.5
Phthalic acid	$C_8H_6O_4$	208	d.
Isophthalic acid	$C_8H_6O_4$	330	subl.
Terephthalic acid	$C_8H_6O_4$	subl.	–
Naphthalic anhydride	$C_{12}H_6O_3$	274	subl.
1,4,5,8-Naphthalene-tetracarboxylic acid	$C_{14}H_8O_8$	320	–
Anthranilic acid	$C_7H_7O_2N$	146.1	subl.
7. Anilines			
Aniline	C_6H_7N	−6.2	184.4
1,2-Diaminobenzene	$C_6H_8N_2$	103.8	256
1,3-Diaminobenzene	$C_6H_8N_2$	62.8	287
1,4-Diaminobenzene	$C_6H_8N_2$	140	267
1-Aminonaphthalene	$C_{10}H_9N$	50	300.8
2-Aminonaphthalene	$C_{10}H_9N$	110.1	306.1

	Summation formula	Melting point (°C)	Boiling point (°C/1,013 mbar)
8. *Alkylderivatives*			
Toluene	C_7H_8	−95	110.8
Styrene	C_8H_8	−31	145.2
o-Xylene	C_8H_{10}	−25	144
m-Xylene	C_8H_{10}	−47.4	139.3
p-Xylene	C_8H_{10}	13.2	138.5
Ethylbenzene	C_8H_{10}	−94.4	136.2
Pseudocumene	C_9H_{12}	−43.8	169.4
Mesitylene	C_9H_{12}	−45	164.8
Cumene	C_9H_{12}	−96.9	152.5
Durene	$C_{10}H_{14}$	79.2	196.8
o-Cymene	$C_{10}H_{14}$	−71.5	177
m-Cymene	$C_{10}H_{14}$	−63.7	175.1
p-Cymene	$C_{10}H_{14}$	−73.5	177.1
p-Diisopropylbenzene	$C_{12}H_{18}$	−17	210.3
1-Methylnaphthalene	$C_{11}H_{10}$	−30.5	244.6
2-Methylnaphthalene	$C_{11}H_{10}$	34.6	241
1,4-Dimethylnaphthalene	$C_{12}H_{12}$	7.6	268
2,3-Dimethylnaphthalene	$C_{12}H_{12}$	104	265
2,6-Dimethylnaphthalene	$C_{12}H_{12}$	112	262
1-Isopropylnaphthalene	$C_{13}H_{14}$	−15.7	267.9
2-Isopropylnaphthalene	$C_{13}H_{14}$	12	268
2,6-Diisopropyl-naphthalene	$C_{16}H_{20}$	68.5	–
α-Picoline	C_6H_7N	−66.7	128.8
β-Picoline	C_6H_7N	−18.3	143.5
γ-Picoline	C_6H_7N	3.7	143.1
o-Cresol	C_7H_8O	30.8	191.0
m-Cresol	C_7H_8O	10.9	202.0
p-Cresol	C_7H_8O	34.7	201.9
2,6-Xylenol	$C_8H_{10}O$	45.8	201
3,5-Xylenol	$C_8H_{10}O$	63.3	221.7
4-tert-Butylphenol	$C_{10}H_{14}O$	99	239.5
2-Toluidine	C_7H_9N	−16.3	199.7
3-Toluidine	C_7H_9N	−31.5	203.3
4-Toluidine	C_7H_9N	43.5	200.3
Cumidine	$C_9H_{13}N$	−63	225
9. *Chlorine and bromine derivatives*			
Chlorobenzene	C_6H_5Cl	−45.2	132.1
1,2-Dichlorobenzene	$C_6H_4Cl_2$	−17.6	179
1,3-Dichlorobenzene	$C_6H_4Cl_2$	−24.8	172

	Summation formula	Melting point (°C)	Boiling point (°C/1,013 mbar)
1,4-Dichlorobenzene	$C_6H_4Cl_2$	53	174
1,3,5-Trichlorobenzene	$C_6H_3Cl_3$	63.5	208.5
1,2,4,5-Tetrachlorobenzene	$C_6H_2Cl_4$	138	245
Hexachlorobenzene	C_6Cl_6	230	309 (987 mbar)
Benzyl chloride	C_7H_7Cl	−39	179.4
Benzal chloride	$C_7H_6Cl_2$	−17.4	205.2
Benzotrichloride	$C_7H_5Cl_3$	− 4.7	220.7
2-Chlorotoluene	C_7H_7Cl	−34	159.5
3-Chlorotoluene	C_7H_7Cl	−47.8	161.6
4-Chlorotoluene	C_7H_7Cl	7.5	162.2
1-Chloronaphthalene	$C_{10}H_7Cl$	−20	259.3
2-Chloronaphthalene	$C_{10}H_7Cl$	56	264 (999 mbar)
1,4-Dichloronaphthalene	$C_{10}H_6Cl_2$	68	286.5 (984 mbar)
1,5-Dichloronaphthalene	$C_{10}H_6Cl_2$	107	subl.
2-Bromophenol	C_6H_5OBr	5.6	194.5
3-Bromophenol	C_6H_5OBr	33	236.5
4-Bromophenol	C_6H_5OBr	63.5	238
2-Chlorophenol	C_6H_5OCl	7	174.9
3-Chlorophenol	C_6H_5OCl	33	214
4-Chlorophenol	C_6H_5OCl	43	217
2,4-Dichlorophenol	$C_6H_4OCl_2$	45	210
1-Chloro-2-nitrobenzene	$C_6H_4O_2NCl$	32.5	245.5 (1002 mbar)
1-Chloro-3-nitrobenzene	$C_6H_4O_2NCl$	44.4	235.6
1-Chloro-4-nitrobenzene	$C_6H_4O_2NCl$	83.6	242
Cyanuric chloride	$C_3N_3Cl_3$	145.7	194

10. *Nitroderivatives*

	Summation formula	Melting point (°C)	Boiling point (°C/1,013 mbar)
Nitrobenzene	$C_6H_5O_2N$	5.7	210.9
1,3,5-Trinitrobenzene	$C_6H_3O_6N_3$	121	d.
2-Nitrotoluene	$C_7H_7O_2N$	− 4.1	221.7
3-Nitrotoluene	$C_7H_7O_2N$	16	232.6
4-Nitrotoluene	$C_7H_7O_2N$	51.9	237.7
2,4-Dinitrotoluene	$C_7H_6O_4N_2$	70	300
2,6-Dinitrotoluene	$C_7H_6O_4N_2$	64.3	–
2,4,6-Trinitrotoluene	$C_7H_5O_6N_3$	80.8	expl.
1-Nitronaphthalene	$C_{10}H_7O_2N$	61.5	304
2-Nitronaphthalene	$C_{10}H_7O_2N$	79	165 (20 mbar)

	Summation formula	Melting point (°C)	Boiling point (°C/1,013 mbar)
1,5-Dinitronaphthalene	$C_{10}H_6O_4N_2$	216	subl.
1,8-Dinitronaphthalene	$C_{10}H_6O_4N_2$	171.5	d.
2-Nitrophenol	$C_6H_5O_3N$	45.1	214.5
3-Nitrophenol	$C_6H_5O_3N$	97	194 (93 mbar)
4-Nitrophenol	$C_6H_5O_3N$	114	subl.
2,3-Dinitrophenol	$C_6H_4O_5N_2$	145.1	–
2,4-Dinitrophenol	$C_6H_4O_5N_2$	115	subl.
2,6-Dinitrophenol	$C_6H_4O_5N_2$	64	subl.
2,4,6-Trinitrophenol	$C_6H_3O_7N_3$	121.8	expl.
2-Nitroaniline	$C_6H_6O_2N_2$	71.5	284.1
3-Nitroaniline	$C_6H_6O_2N_2$	114	306.4
4-Nitroaniline	$C_6H_6O_2N_2$	147.8	331.7
11. *Sulfonic acids*			
Benzenesulfonic acid	$C_6H_6O_3S$	51	d.
4-Toluenesulfonic acid	$C_7H_8O_3S$	38	186 (0.2 mbar)
12. *Quinones*			
Naphthoquinone	$C_{10}H_6O_2$	125.5	subl.
Anthraquinone	$C_{14}H_8O_2$	286	379.8

expl.: explodes
subl.: sublimes
d.: decomposes

Bibliography

Chapter 1

1. Bäumler, E.: Ein Jahrhundert Chemie; Econ-Verlag, Düsseldorf (1963)
2. Haber, L.F.: The Chemical Industry during the Nineteenth Century; Oxford University Press, London (1958)
3. Hardie, D.W.F., Pratt, J.D.: A History of the Modern British Chemical Industry; Pergamon Press, Oxford (1966)
4. Krätz, O.: Zur Geschichte der Farben-Chemie; Chem. Lab. Betr., *30*, 525 (1979)
5. Menzi, K.: Die Basler Chemie im Wandel der Zeit; Swiss Chem., *5*, Nr.5a, 15 (1983)
6. Ress, F.M.: Geschichte der Kokereitechnik; Verlag Glückauf, Essen (1957)
7. Roggersdorf, W., Steinert, O.: Im Reiche der Chemie; Econ-Verlag, Düsseldorf (1965)
8. Schäfer, H.-G.: Zur Entwicklung der thermischen und chemischen Kohleveredelung; Techn. Mitteilungen, *75*, Nr.2/3, 82 (1982)
9. Schultz, G.: Die Chemie des Steinkohlenteers; F.Vieweg & Sohn, Braunschweig (1926)
10. Welham, R.D.: The Early History of the Synthetic Dye Industry I - The Chemical History; J.Soc. Dyers Colour., *79*, 98 (1963)
11. Wojtkowiak, B.: Histoire de la Chimie de l'Antiquité à 1950; Technique & Documentation - Lavoisier, Paris (1984)

Chapter 2

1. Allinger, N.L., Cava, M.P., de Jongh, D.C., Johnson, C.R., Lebel, N.A., Stevens, C.L.: Organische Chemie; Verlag Walter de Gruyter, Berlin (1980)
2. Balaban, A.T.: Is Aromaticity Outmoded? Pure Appl. Chem., *52*, 1409 (1980)
3. Bernasconi, C.F.: Mechanisms of Nucleophilic Aromatic and Hetero-aromatic Substitution; Chimia, *34*, 1 (1980)
4. Buddrus, J.: Grundlagen der Organischen Chemie; Verlag Walter de Gruyter, Berlin (1980)
5. Effenberger, F., Reisinger, F., Schönwälder, K.H., Bäuerle, P., Stezowski, J.J., Jogun, K.H., Schöllkopf, K., Stohrer, W.-D.: Structure and Reactivity of Aromatic σ-Complexes (Cyclohexadienylium Ions): A Correlated Experimental and Theoretical Study; J.Am. Chem. Soc., *109*, 882 (1987)
6. Effenberger, F.: Neues über die elektrophile Aromatensubstitution; Chem. i.u. Zeit, *13*, 87 (1979)
7. Garratt, P.J.: Aromaticity; McGraw-Hill, London (1986)
8. Griffiths, J.: Modern dye chemistry; Chemistry in Britain, Nr.11, 997 (1986)
9. Hafner, K.: August Kekulé dem Baumeister der Chemie zum 150.Geburtstag; Justus von Liebig Verlag, Darmstadt (1980)
10. Klessinger, M.: Konstitution und Lichtabsorption organischer Farbstoffe; Chem. i.u. Zeit, *12*, 1 (1978)
11. Lloyd, D.: Carbocyclic Non-Benzenoid Aromatic Compounds; Elsevier, Amsterdam/London/New York (1966)

12. Lloyd, D., Marshall, D.R.: An Alternative Approach to the Nomenclature of Cyclic Conjugated Polyolefins, together with some Observations on the Use of the Term "Aromatic"; Angew. Chem. Int. Ed., *11*, 5, 404 (1972)
13. Lowry, T.H., Schueller Richardson, K.: Mechanismen und Theorie in der Organischen Chemie; Verlag Chemie, Weinheim (1980)
14. March, J.: Advanced Organic Chemistry; 3.Ed., John Wiley, New York (1985)
15. Olah, G.A.: Friedel-Crafts and Related Reactions, I-General Aspects; Intersci. Publ., New York/London (1963)
16. Olah, G.A.: Friedel-Crafts and Related Reactions, II-Alkylation and Related Reactions; Intersci. Publ., New York/London/Sydney (1964)
17. Olah, G.A.: Friedel-Crafts and Related Reactions, III-Acylation and Related Reactions; Intersci. Publ., New York/London/Sydney (1964)
18. Olah, G.A.: Friedel-Crafts and Related Reactions, IV-Miscellaneous Reactions – Cumulative Indexes; Intersci. Publ., New York/London/Sydney (1965)
19. Olah, G.A.: Mechanism of Electrophilic Aromatic Substitutions; Acc. Chem. Res., *4*, 240 (1971)
20. Pines, H.: The Chemistry of Catalytic Hydrocarbon Conversions; Academic Press, New York (1981)
21. Sondheimer, F.: Non-benzenoid Aromatic π-Electron Systems; Chimia, *28*, 163 (1974)
22. Zander, M.: Aspekte der Physik und Chemie polycyclischer aromatischer Kohlenwasserstoffe; Naturwissenschaften, *69*, 436 (1982)
23. Zander, M.: Physical and Chemical Properties of Polycyclic Aromatic Hydrocarbons; In: Bjørseth, A., Ed.: Handbook of Polycyclic Aromatic Hydrocarbons; Marcel Dekker, New York/Basel (1983)
24. Zoltewicz, J.A.: New Directions in Aromatic Nucleophilic Substitution; Topics in current Chemistry, *59*, 33 (1975)

Chapter 3

1. Ahland, E., Friedrich, F., Romey, I., Strobel, B., Weber, H.: Verfahrensentwicklung in der Kohlehydrieranlage der Bergbau-Forschung; Erdöl, Erdgas, Kohle, *102*, 148 (1986)
2. Aiba, T., Kaji, H.: Residue Thermal Cracking By the Eureka Process; Chem. Eng. Progr., Nr.2, 37 (1981)
3. Anderson, R.F., Johnson, J.A., Mowry, J.R.: Cyclar – One Step Processing of LPG to aromatics and hydrogen; Am. Inst. Chem. Eng., Spring National Meeting, Houston, Texas, March 24-28, 1985
4. Anderson, R.P.: Recycling Solvent Techniques for the SRC Process; Coal Process. Technol., Nr.2, 130 (1975)
5. Berry, R.I.: Gasoline or olefins from an alcohol feed; In: Greene, R. et al., Ed.: Process Technology and Flowsheets, *II*, Chemical Engineering, 140; McGraw-Hill, London (1983)
6. Bertling, H., Nashan, G.: Energiewirtschaft des Verkokungsprozesses; Erdöl, Kohle, Erdgas, Petrochem., *34*, 397 (1981)
7. Bockrath, B.C.: Chemistry of Hydrogen Donor Solvents; Coal Sci., *2*, 65 (1983)
8. Chauvel, A., Lefebvre, G., Castex, L.: Procédés de pétrochimie, Caractéristiques techniques et économiques, 2.Ed., Editions Technip, Paris (1985)
9. Clements, L.D., Beck, S.R., Heintz, C.: Chemicals from Biomass Feedstocks; Chem. Eng. Progr., Nr.11, 59 (1983)
10. Collin, G.: Steinkohlenteerchemie: Bedeutung, Produkte und Verfahren; Erdöl, Kohle, Erdgas, Petrochemie, *38*, 489 (1985)
11. Collin, G., Zander, M.: Steinkohlenteerchemie – Aktuelle Entwicklungen aus Technik und Forschung; Erdöl, Erdgas, Kohle, *102*, 517 (1986)
12. Delannoy, G.: Les principes généraux de la gazéification du charbon; Industrie Minérale, *60*, 125 (1978)
13. Derbyshire, F.J., Varghese, P., Whitehurst, D.D.: Synergistic effects between light and heavy solvent components during coal liquefaction; Fuel, *61*, 859 (1982)

14. Dry, M.E.: The Fischer-Tropsch Synthesis; In: Anderson, J.R., Boudart, M., Ed.: Catalysis, Science and Technology, *1*, 159, Springer-Verlag, Berlin/Heidelberg/New York (1981)
15. v. Eberan-Eberhorst, C.C.A., Geldern, L., Pischinger, F., Seidel, G.: Mineralölprodukte und deren Anwendung; Erdöl, Kohle, Erdgas, Petrochem., *37*, 69 (1984)
16. Edgar, M.D.: Catalytic Reforming of Naphtha in Petroleum Refineries; In: Leach, B.E., Ed.: Applied Industrial Catalysis, Vol.1, 124, Academic Press New York (1983)
17. Eickermann, R.: Thermische Crackverfahren; Chem. Ing. Tech., *55*, 134 (1983)
18. Falbe, J.: Chemierohstoffe aus Kohle; Georg Thieme Verlag, Stuttgart (1977)
19. Fiedler, J.: Umweltverträglichkeit der Kohlenhydrierung am Beispiel der Kohleölanlage Bottrop; Glückauf, *121*, 48 (1985)
20. Franck, H.-G.: Die Kohle als Rohstoff für die Chemie; Chem. Ind., *30*, 185 (1978)
21. Franck, H.-G., Collin, G.: Steinkohlenteer – Chemie, Technologie und Verwendung; Springer-Verlag, Berlin/Heidelberg/New York (1968)
22. Franck, H.-G., Knop, A.: Kohleveredlung – Chemie und Technologie; Springer-Verlag, Berlin/Heidelberg/New York (1979)
23. Fricke, J.: Biomasse; Physik i.u. Zeit, *15*, 121 (1984)
24. Goossens, A.G., Westerduin, R.F., Mol, A.: New Developments in Ethylene Technology; Erdöl, Kohle, Erdgas, Petrochem., *28*, 471 (1975)
25. Griesbaum, K., Swodenk, W.: Entwicklungen und Entwicklungstendenzen in der Petrochemie; Erdöl, Kohle, Erdgas, Petrochem., *33*, 34 (1980)
26. Grosskinsky, O.: Handbuch des Kokereiwesens, Bd. I und Bd. II; Karl Knapp Verlag, Düsseldorf (1954)
27. Gundermann, K.-D.: Die chemische Konstitution der Steinkohle als Grundlage für die Erforschung ihres Reaktionsverhaltens; Erdöl, Erdgas, Kohle, *102*, 100 (1986)
28. Hatch, L.F., Matar, S.: From Hydrocarbons to Petrochemicals – Part 8: Production of olefins; Hydrocarb. Proc., *57*, Nr.1, 135 (1978)
29. Hatch, L.F., Matar, S.: From Hydrocarbons to Petrochemicals – Part 9: Production of olefins; Hydrocarb. Proc., *57*, Nr.3, 129 (1978)
30. Hedden, K., Weitkamp, J.: Thermisches Hydrocracken von Kohlenwasserstoffen; Chem. Ing. Tech., *55*, 907 (1983)
31. Hellwig, K.C., Alpert, S.B., Johanson, E.S., Wolk, R.H.: H-Oil- und H-Coal-Verfahren; Brennstoff-Chemie, *50*, 263 (1969)
32. Hiller, H.: Modern Coal Upgrading Processes; Chem. Econ. Eng. Rev., *16*, Nr.10, 10 (1984)
33. Hirotani, Y., Takeuchi, T., Miyabuchi, Y., Aiba, T., Shigeta, M.: Successful performance of a refinery with Eureka unit; ACS Div. Petrol. Chem., *26*, Nr.2, 465 (1981)
34. Hobson, G.D.: Modern Petroleum Technology, 4.Ed.; Appl. Sci. Publ., Barking/Ess. (1973)
35. Hölderich, W., Gallei, E.: Industrielle Anwendung zeolithischer Katalysatoren bei petrochemischen Prozessen; Chem. Ing. Tech., *56*, 908 (1984)
36. Hollerbach, A.: Grundlagen der organischen Geochemie; Springer-Verlag, Berlin/Heidelberg/New York (1985)
37. Hosoi, T., Keister, H.G.: Ethylene From Crude Oil; Chem. Eng. Progr., *71*, Nr.11, 63 (1975)
38. Hus, M.: Visbreaking process has strong revival; Oil Gas J., 109 (13.4.1981)
39. Jäckh, W.: Probleme der Hydrierung von Kohle; Erdöl, Kohle, Erdgas, Petrochem., *23*, 334 (1970)
40. Jahnig, C.E., Martin, H.Z., Campbell, D.L.: The development of fluid catalytic cracking; ChemTech, Nr.2, 106 (1984)
41. Jenkins, J.H., Stephens, T.W.: Kinetics of cat reforming; Hydrocarb. Proc., *59*, Nr.11, 163 (1980)
42. Jüntgen, H.: Zum Mechanismus von Pyrolyse und Hydropyrolyse; Erdöl, Kohle, Erdgas, Petrochem., *38*, 448 (1985)
43. Keim, K.H., Maziuk, J., Tönnesmann, A.: The Methanol-to-Gasoline (MTG) Process; Erdöl, Kohle, Erdgas, Petrochem., *37*, 558 (1984)
44. McKillip, W.J., Sherman, E.: Furan Derivatives; In: Kirk-Othmer, Encycl. Chem. Tech., 3.Ed., *11*, 499 (1980)
45. Knab, H., Mielicke, C., Rosum, E.: Kennzeichnende Verfahrens- und Einflußgrößen beim katalytischen Fließbett-Cracken; Chem. Ing. Tech., *54*, 79 (1982)

46. Kölling, G., Langhoff, J., Collin, G.: Kohlenwertstoffe und Verflüssigung von Kohle; Erdöl, Kohle, Erdgas, Petrochem., *37*, 394 (1984)
47. v. Krevelen, D. W.: Coal, Typology-Chemistry-Physics-Constitution; Elsevier, Amsterdam/London/New York/Princeton (1961)
48. Kronseder, J. G., Bogart, M. J. P.: Coal, Liquefaction, South Africa's Sasol II; Encycl. Chem. Proc. Des., *9*, 299 (1979)
49. Kürten, H.: Verfahrenstechnik der Kohlehydrierung in Sumpfphasen-Reaktoren; Chem. Ing. Tech., *54*, 409 (1982)
50. Kubo, H., Masamune, S., Sako, R.: Make BTX from cracker gasoline; Hydrocarb. Proc., *49*, Nr. 7, 111 (1970)
51. Langhoff, J., Dürrfeld, R., Wolowski, E.: Neue Technologien zur Steinkohlenveredlung; Erdöl, Kohle, Erdgas, Petrochem., *34*, 379 (1981)
52. Larsen, J. W., Lee, D., Shawver, S. E.: Coal macromolecular structure and reactivity; Fuel Process. Technol., *12*, 51 (1986)
53. Marcilly, Ch.: Progrès apportés par l'utilisation des zéolithes en cracking catalytique; Rev. Inst. Franc. Pétrole, *30*, 969 (1975)
54. Mikulla, K. D., Bölt, H., Richter, H.: Einsatzflexibilität in Olefinanlagen (Teil I); Erdöl, Kohle, Erdgas, Petrochem., *33*, 309 (1980)
55. Mochida, I., Shiraki, A., Korai, Y., Okuhara, T.: Carbonization of coals into anisotropic cokes; Fuel, *64*, 45 (1985)
56. Nashan, G.: Zur Entwicklung der Kokereitechnik und der Kokereiwirtschaft; Glückauf, *118*, 721 (1982)
57. v. Ness, J. H.: Vanillin; In: Kirk-Othmer Encycl. Chem. Tech., 3. Ed., *23*, 704 (1983)
58. Oberkobusch, R.: Neuere Entwicklung in der technischen Gewinnung reiner Steinkohlenteerbasen; Brennstoff-Chemie, *40*, 145 (1959)
59. Oelert, H.-H., Severin, D., Windhager, H.-J.: Charakterisierung des Aufbaus nichtsiedender Erdölanteile, I. Gesättigte Kohlenwasserstoffe. Erdöl, Kohle, Erdgas, Petrochem., *26*, 397 (1973)
60. Powell, T. G., Snowdon, L. R.: A Composite Hydrocarbon Generation Model; Erdöl, Kohle, Erdgas, Petrochem., *36*, 163 (1983)
61. Puppe, L.: Zeolithe – Eigenschaften und technische Anwendungen; Chem. i. u. Zeit, *20*, 117 (1986)
62. Rapp, L. M., v. Driesen, R. P.: H-Oil-Process Gives Product Flexibility; Hydrocarb. Proc., *44*, Nr. 12, 103 (1965)
63. Rhoé, A.: Aspects technologiques de la pyrolyse des charges lourdes; Rev. Inst. Franc. Petr., *36*, 191 (1981)
64. Riediger, B.: Die Verarbeitung des Erdöles; Springer-Verlag, Berlin/Heidelberg/New York (1971)
65. Ritzer, H.: Auswirkungen einer veränderten Rohstoffbasis auf die Produktion in Olefin-Anlagen; Erdöl, Kohle, Erdgas, Petrochem., *35*, 124 (1982)
66. Romey, I.: Stand der Kohlehydrierung in Europa; Erdöl, Kohle, Erdgas, Petrochem., *33*, 314 (1980)
67. Ruf, H.: Kleine Technologie des Erdöls; Birkhäuser-Verlag, Basel/Stuttgart (1963)
68. Schicketanz, W.: Ein Rückblick auf Forschungs- und Entwicklungsarbeiten zur Kohlehydrierung nach dem IG-Verfahren; Erdöl, Kohle, Erdgas, Petrochem., *35*, 287 (1982)
69. Schütze, B.: Thermische Crackprozesse zur Destillatoptimierung; Erdöl, Kohle, Erdgas, Petrochem., *37*, 156 (1984)
70. Semel, J., Steiner, R.: Nachwachsende Rohstoffe in der chemischen Industrie; Chem. Ind., *35*, 489 (1983)
71. Sfihi, H., Quinton, M. F., Legrand, A., Pregermain, S., Carson, D., Chiche, P.: Evaluation of the aromaticity in French coals by ^{13}C-^{1}H cross polarization, magic angle spinning and dipolar dephasing nuclear magnetic resonance spectroscopy; Fuel, *65*, 1006 (1986)
72. Singh, V. D.: Visbreaking Technology; Erdöl, Kohle, Erdgas, Petrochem., *39*, 19 (1986)
73. Specks, R., Klusmann, A.: German hard coal conversion projects; Energy Prog., *2*, 60 (1982)
74. Stadelhofer, J. W., Gerhards, R.: ^{13}C n.m.r. study on hydroaromatic compounds in anthracene oil; Fuel, *60*, 367 (1981)
75. Stadelhofer, J. W., Zander, M., Gerhards, R.: ^{13}C n.m.r. study on the hydrogen transfer during the distillation of crude coal tar; Fuel, *59*, 604 (1980)

76. Steinhofer, A., Frey, O.: Die Erzeugung von Äthylen aus Rohöl in der Wirbelschicht; Chem. Ing. Tech., *32*, 782 (1960)
77. Stolfa, F.: New roles for thermal cracking; Hydrocarb. Proc., *59*, Nr. 5, 101 (1980)
78. Teggers, H., Jüntgen, H.: Stand der Kohlevergasung zur Erzeugung von Brenngas und Synthesegas; Erdöl, Kohle, Erdgas, Petrochem., *37*, 163 (1984)
79. Tissot, B.: La genèse du pétrole; La Recherche, *8*, 326 (1977)
80. VandeVen, J., Mackey, J.: Progress report: continuous reforming; Oil Gas J., 116 (24.9.1973)
81. Weisz, P. B.: Molecular shape-selective catalysis - the personal adventure; Chem. Ind. (London), 392 (1985)
82. Weitkamp, J.: Hydrocracken, Cracken und Isomerisieren von Kohlenwasserstoffen; Erdöl, Kohle, Erdgas, Petrochem., *31*, 13 (1978)
83. Weitkamp, J.: Gewinnung leichter Kohlenwasserstoffe aus schweren Ölen - Verfahren und Entwicklungen; Chem. Ing. Tech., *54*, 101 (1982)
84. Welte, D. H.: Neue Wege in der Kohlenwasserstoffexploration; Erdöl, Kohle, Erdgas, Petrochem., *35*, 503 (1982)
85. Welte, D. H.: Erdöl und Kohle - Fossile Energierohstoffe und Zeugnisse vergangenen Lebens; Erdöl, Kohle, Erdgas, Petrochem., *31*, 139 (1978)
86. White, P. J.: How Cracker Feed Influences Yield; Hydrocarb. Proc., *47*, Nr. 5, 103 (1968)
87. Würfel, H.: Pyrosol - Das neue Kohleverflüssigungsverfahren der Saarbergwerke AG; Erdöl, Erdgas, Kohle, *102*, 45 (1986)
88. Zdonik, S. B., Bassler, E. J., Hallee, L. P.: How feedstocks affect ethylene; Hydrocarb. Proc., *53*, Nr. 2, 73 (1974)
89. Zürn, G., Kohlhase, K., Hedden, K., Weitkamp, J.: Entwicklungen der Raffinerietechnik; Erdöl, Kohle, Erdgas, Petrochem., *37*, 62 (1984)

Chapter 4

1. Craig, R. G., Doelp, L. C., Logwinuk, A. K.: Benzene by Hydrodealkylation Using the Detol Process; Erdöl, Kohle, Erdgas, Petrochem., *18*, 527 (1965)
2. Dermietzel, J., Bauer, F., Wienhold, C., Jockisch, W., Klempin, J., Barz, H.-J., Franke, H., Becker, K., John, H.: Die hydrokatalytische Isomerisierung technischer C_8-Aromatenfraktionen; Chem. Techn. (Leipzig), 30, 626 (1978)
3. Eisenlohr, K.-H., Wirth, J.: Verfahren zur Gewinnung von Reinaromaten aus Hydrierraffinaten und Reformaten; Chem. Ing. Tech., *32*, 789 (1960)
4. Eisenlohr, K.-H., Jäckh, R.: Toluol; In: Ullmanns Enzykl. Tech. Chem., 4. Aufl., *23*, 301 (1983)
5. Eisenlohr, K.-H., Jäckh, R.: Xylole; In: Ullmanns Enzykl. Tech. Chem., 4. Aufl., *24*, 525 (1983)
6. Feigelman, S., Lehman, L. M., Aristoff, E., Pitts, P. M.: Lowest Cost Route to Benzene: Thermally Dealkylate Toluene; Hydrocarb. Proc., *44*, No. 12, 147 (1965)
7. Fisher, J., Niclaes, H. J.: Aromatiques: une triple couronne; Rev. Assoc. Franc. Tech. Pétrole, Nr. 3/4, 57, (1974)
8. Folkins, H. O.: Benzene; In: Ullmann's Enzycl. Ind. Chem., 5. Ed., *A3*, 475 (1985)
9. Grandio, P., Schneider, F. H., Schwartz, A. B., Wise, J. J.: Toluene for benzene and xylenes; Hydrocarb. Proc., *51*, Nr. 8, 85 (1972)
10. Helms, G., John, P.: Récupération des aromatiques par distillation extractive; Inf. Chimie, Nr. 161, 105 (1976)
11. McKay, D. L., Dale, G. H., Tabler, D. C.: Para-xylene via fractional crystallization; Chem. Eng. Progr., *62*, Nr. 11, 104 (1966)
12. König, G.: Die Herstellung von p- und o-Xylol durch Isomerisierung und Adsorption; Erdöl, Kohle, Erdgas, Petrochem., *26*, 323 (1973)
13. Krönig, W., Halcour, K.: Hydrierende Aufarbeitung von Pyrolysebenzin; Brennstoff-Chemie, *50*, 258 (1969)
14. Lackner, K.: Aromatengewinnung mit Hilfe von N-Formylmorpholin (NFM); Chem. Tech., *11*, 3 (1982)
15. Logwinuk, A. K., Friedman, L., Weiss, A. H.: The Houdry Litol Process; Erdöl, Kohle, Erdgas, Petrochem., *17*, 532 (1964)

16. Lohr, B., Schliebener, C., Sohns, D.: Pyrolysebenzin, ein wertvolles Produkt bei der Olefin-Erzeugung; Erdöl, Kohle, Erdgas, Petrochem., *33*, 126 (1980)
17. Lorz, W., Craig, R.G., Cross, W.J.: Die Hydrodealkylierung von Aromaten; Erdöl, Kohle, Erdgas, Petrochem., *21*, 610 (1968)
18. Müller, E.: Use of N-methylpyrrolidone for aromatics extraction; Chem. Ind. (London), *518* (1973)
19. Müller, E.: Gewinnung und Abtrennung von Aromaten durch Extraktion und Extraktivdestillation; Verfahrenstechn., *8*, 88 (1974)
20. Nonnenmacher, H., Reitz, O., Schmidt, P.: Das BASF-Scholven-Verfahren zur Druckraffination von Rohbenzol; Erdöl, Kohle, Erdgas, Petrochem., *8*, 407 (1955)
21. Ockerbloom, N.E.: Xylenes and higher aromatics – Part 2: Metaxylene; Hydrocarb. Proc., *50*, Nr. 8, 113 (1971)
22. Ockerbloom, N.E.: Xylenes and higher aromatics – Part 4: Orthoxylene; Hydrocarb. Proc., *50*, Nr. 10, 101 (1971)
23. Ockerbloom, N.E.: Xylenes and higher aromatics – Part 6: Paraxylene; Hydrocarb. Proc., *51*, Nr. 1, 93 (1972)
24. Otani, S., Kanaoka, M., Matsumura, K., Akita, S., Sonoda, T.: Aromax Isolene – New Paraxylene Recovery and Xylene Isomerization Processes Developed by Toray; Chem. Econ. Eng. Rev., *3*, Nr. 6, 56 (1971)
25. Preusser, G., Emmrich, G.: Recovery of High-Purity Aromatics: New Process Technologies Improve Economy; Chem. Age India, *35*, 169 (1984)
26. Ransley, D.L.: Xylenes and Ethylbenzene; In: Kirk-Othmer, Encycl. Chem. Tech., *24*, 709 (1984)
27. Rittner, S., Steiner, R.: Die Schmelzkristallisation von organischen Stoffen und ihre großtechnische Anwendung; Chem. Ing. Tech., *57*, 91 (1985)
28. Seko, M., Miyake, T., Inada, K.: Sieves for mixed xylenes separation; Hydrocarb. Proc., *59*, Nr. 1, 133 (1980)
29. Sinfelt, J.H.: Catalytic Reforming of Hydrocarbons; In: Anderson, J.R., Boudart, M., Ed.: Catalysis, Science and Technology, *1*, 257, Springer Verlag, Berlin/Heidelberg/New York (1981)
30. Ueno, T.: MGC Xylene Extraction Process by use of HF-BF_3; In: Lo, T.C. et al., Ed.: Handbook of Solvent Extraction, 575, John Wiley, New York (1983)
31. Urban, W.: Die katalytische Druckraffination von Benzol; Erdöl, Kohle, Erdgas, Petrochem., *4*, 279 (1951)
32. Verdol, J.A.: Here's a new way to more xylenes; Oil Gas J., 63 (9.6.1969)
33. Ward, T.: Wasserstoffraffination von Kokerei-Rohbenzol nach dem Litol-Verfahren; Erdöl, Kohle, Erdgas, Petrochem., *26*, 440 (1973)

Chapter 5

1. Agnello, L.A.: Synthetic Phenol; Ind. Eng. Chem., *52*, 894 (1960)
2. Baier, E.: Buntpigmente – Grundbegriffe und Eigenschaften; Defazet, *29*, 54 (1975)
3. Bökelmann, F.: Cyclohexanol; In: Ullmanns Enzykl. Tech. Chem., 4. Aufl., *9*, 689 (1975)
4. Bonacci, J.C., Heck, R.M., Mahendroo, R.K., Patel, G.R., Allan, E.D.: Hydrogenate AMS to cumene; Hydrocarb. Proc., *59*, Nr. 11, 179 (1980)
5. Brownstein, A.M.: Trends in Petrochemical Technology; Petroleum Publ., Tulsa/Okl. (1976)
6. Budi, F., Neri, A., Stefani, G.: Future MA keys to butane; Hydrocarb. Proc., *61*, 159 (1982)
7. Büchel, K.H.: Pflanzenschutz und Schädlingsbekämpfung; Georg Thieme Verlag, Stuttgart (1977)
8. Canfield, R.C., Cox, R.P., McCarthy, D.M.: The New Cumene Process: Efficient and Economical; Chem. Eng. Prog., Nr. 8, 36 (1986)
9. Canfield, R.C., Unruh, T.L.: Improving Cumene yields via selective catalysis; Chem. Eng., *90*, 32 (1983)
10. DeMaio, D.A.: Will butane replace benzene as a feedstock for maleic anhydride; Chem. Eng., *87*, 104 (1980)

11. Dunlap, K. L.: Nitrobenzene and Nitrotoluenes; In: Kirk-Othmer, Encycl. Chem. Tech., 3. Ed., *15*, 916 (1981)
12. Dwyer, F. G., Lewis, P. J., Schneider, F. H.: Nouveau procédé pour l'éthylbenzène; Inform. Chim., Nr. 155, Spec. Mai, 141 (1976)
13. Erickson, S. H.: Salicylic Acid and Related Compounds; In: Kirk-Othmer, Encycl. Chem. Tech., 3. Ed., *20*, 500 (1982)
14. Ewers, J., Voges, H. W., Maleck, G.: Verfahren zur Herstellung von Hydrochinon; 24. Haupttagung der Deutschen Gesellschaft für Mineralölwissenschaft und Kohlechemie e. V. (DGMK), 30.9.- 3.10.1974, Hamburg, 487 (1974)
15. Fiege, H., Wedemeyer, K., Bauer, K. A., Krempel, A., Mölleken, R. G.: Further development of a classical process for the synthesis of aromatic hydroxyaldehydes; In: Croteau, R., Ed.: Fragrance Flavor Subst., Proc. Int. Haarmann Reimer Symp., 2nd 1979 (Pub. 1980), 63, D&PS Verlag, Pattensen
16. Fiege, H.: Cresols and Xylenols; In: Ullmann's Enzycl. Ind. Chem., 5. Ed., *A8*, 25 (1987)
17. Fisher, W. B., van Peppen, J. F.: Cyclohexanol and Cyclohexanone; In: Kirk-Othmer, Encycl. Chem. Tech., 3. Ed., *7* (1979)
18. Gans, M.: Which route to aniline; Hydrocarb. Proc., *55*, Nr. 11, 145 (1976)
19. Gelbein, A. P., Nislick, A. S.: Make phenol from benzoic acid; Hydrocarb. Proc., *57*, Nr. 11, 125 (1978)
20. Grolig, J., Swodenk, W., Blaschke, H.-G., Rauleder, G.: Chemierohstoffe aus Erdöl und Erdgas; In: Winnacker-Küchler, Chem. Technol., 4. Aufl., *5*, Org. Technol. I, 164 (1981)
21. Gupta, H.: Nitrosamine in der Gummiindustrie - Gefahr und Möglichkeiten zur Vermeidung; Gummi, Asbest, Kunstst., *39*, 6 (1986)
22. Hancock, E. G.: Benzene and its Industrial Derivatives; Ernest Benn, London (1975)
23. Hatch, L. F., Matar, S.: From hydrocarbons to petrochemicals - Part 13: Chemicals from benzene; Hydrocarb. Proc., *57*, Nr. 11, 291 (1978)
24. Hock, H., Kropf, H.: Autoxydation von Kohlenwasserstoffen und die Cumol-Phenol-Synthese; Angew. Chem., *69*, 313 (1957)
25. Hutzinger, O., Fink, M., Thoma, H.: PCDD und PCDF: Gefahr für Mensch und Umwelt?; Chem. i. u. Zeit, *20*, 165 (1986)
26. Innes, R. A., Occelli, M. L.: p-Methylstyrene from Toluene and Acetaldehyde; J. mol. Catal., *32*, 259 (1985)
27. Innes, R. A., Swift, H. E.: Toluene to styrene - a difficult goal; ChemTech, Nr. 4, 244 (1981)
28. Ito, K.: Make cresols from propylene; Hydrocarb. Proc., *51*, Nr. 8, 89 (1973)
29. Jordan, W., v. Barneveld, H., Gerlich, O., Ullrich, J., Bunge, W.: Phenol; In: Ullmanns Enzykl. Tech. Chem., 4. Aufl., *18*, 177 (1979)
30. Kaeding, W. W., Young, L. B., Prapas, A. G.: Para-methylstyrene; ChemTech, Nr. 9, 556 (1982)
31. Kosswig, K.: Tenside; In: Ullmanns Enzykl. Tech. Chem., 4. Aufl., *22*, 455 (1982)
32. Lewis, P. J., Hagopian, C., Koch, P.: Styrene; In: Kirk-Othmer, Encycl. Chem. Tech., 3. Ed., *21*, 770 (1983)
33. Lieb, M., Hildebrand, B.: Styrol; In: Ullmanns Enzykl. Tech. Chem., 4. Aufl., *22*, 293 (1982)
34. Lohwasser, H.: Nitrosamine in der Gummiindustrie; Gummi, Asbest, Kunstst., *39*, 385 (1986)
35. Malow, M.: Benzene or butane for MAN; Hydrocarb. Proc., *59*, Nr. 11, 149 (1980)
36. Mildenberger, H., Trösken, J.: Herstellung von Pflanzenschutzmitteln; In: Winnacker-Küchler, Chem. Technol., 4. Aufl., *7*, Org. Technol. III, 277 (1986)
37. Moyers, C. G.: Industrial Crystallization for Ultrapure Products; Chem. Eng. Progr., Nr. 5, 42 (1986)
38. Neuzil, R. W., Rosback, D. H., Jensen, R. H., Teague, J. R., deRosset, A. J.: Separation of Cresols by Continuous Countercurrent Adsorption; ACS/JCS Chem. Congr., Honolulu, Hawai, April 1-6 (1979)
39. Ohlinger, H., Stadelmann, S.: Die Entwicklung der Äthylbenzol-Dehydrierung zu Styrol in der BASF; Chem. Ing. Tech., *37*, 361 (1965)
40. Olzinger, A. H.: New Route to Hydroquinone; In: Cavaseno, V., Ed.: Process Technology and Flowsheets, 169, McGraw-Hill, New York (1979)
41. Pujado, P. R., Salazar, J. R., Berger, C. V.: Cheapest route to phenol; Hydrocarb. Proc., *55*, Nr. 3, 91 (1976)
42. Reed, H. W. B.: Alkylphenols; In: Kirk-Othmer, Encycl. Chem. Tech., 3. Ed., *2*, 72 (1978)

43. Rittner, S., Warning, K.: Herstellung von monocyclischen Aromaten; In: Winnacker-Küchler, Chem. Technol., 4. Aufl., *6*, Org. Technol. II, 150 (1982)
44. Robinson, W. D., Mount, R. A.: Maleic anhydride, maleic acid and fumaric acid; In: Kirk-Othmer, Encyl. Chem. Tech., 3. Ed., *14*, 770 (1981)
45. Roth, H. J., Kleemann, A.: Pharmazeutische Chemie I, Arzneistoffsynthese; Georg Thieme, Stuttgart/New York (1982)
46. Schaffel, G. S., Chem, S. S., Graham, J. J.: Maleic Anhydride from Butane; Erdöl, Kohle, Erdgas, Petrochem., *36*, 85 (1983)
47. Schultz, O.-E., Schnekenburger, J.: Einführung in die Pharmazeutische Chemie; 2. Aufl., Verlag Chemie, Weinheim (1984)
48. Schweter, W., Heimlich, G.: Alkylbenzole; In: Ullmanns Enzykl. Tech. Chem., 4. Aufl., *14*, 672 (1977)
49. Stevens, J. J.: What happened at Seveso; Chem. Ind. (London), 564 (1980)
50. Stobaugh, R. B.: Phenol: How, Where, Who – Future; Hydrocarb. Proc., *45*, Nr. 1, 143 (1966)
51. Thiem, K.-W., Sewekow, B., Kiel, W., Handschuh, V., Freese, H., Schimpf, R., Vagt, H., Bunge, W.: Nitroverbindungen, aromatische; In: Ullmanns Enzykl. Tech. Chem., 4. Aufl., *17*, 383 (1979)
52. Varagnat, J.: Hydroquinone and Pyrocatechol Production by Direct Oxidation of Phenol; Ind. Eng. Chem. Prod. Res. Dev., *15*, Nr. 3, 212 (1976)
53. Waldmann, H.: Phenol-Derivate; In: Ullmanns Enzykl. Tech. Chem., 4. Aufl., *18*, 219 (1979)
54. Waldmann, H., Jupe, C., Baumert, J., Seifert, H., Schümmer, G.: Brenzcatechin und Hydrochinon aus Phenol und Percarbonsäure – eine Verfahrensentwicklung; Chem. Ing. Tech., *53*, 664 (1981)
55. Ward, D. J.: Cumene; In: Kirk-Othmer, Encycl. Chem. Tech., 3. Ed., *7*, 286 (1979)
56. Weissermel, K., Arpe, H.-J.: Industrielle Organische Chemie – Bedeutende Vor- und Zwischenprodukte; 2. Aufl., Verlag Chemie, Weinheim (1978)
57. Wett, T.: Monsanto/Lummus styrene process is efficient; Oil Gas J., 76 (20.7.1981)
58. Weyens, E.: Recover maleic anhydride; Hydrocarb. Proc., *53*, Nr. 11, 132 (1974)
59. Wohlfarth, K., Emig, G.: Compare maleic anhydride routes; Hydrocarb. Proc., *59*, Nr. 6, 83 (1980)
60. Yamada, J., Shimizu, C., Saitoh, S.: Purification of organic chemicals by Kureha continuous crystal purifier; In: Jancic, S. J., de Jong, E. J., Ed.: Ind. Cryst. Proc. Symp. 8th 1981 (Pub. 1982), 265, North-Holland Publishing Comp., Amsterdam

Chapter 6

1. Brühne, F., Wright, E.: Benzyl Alcohol; In: Ullmann's Enzycl. Ind. Chem., 5. Ed., *A4*, 1 (1985)
2. Chadwick, D. H., Cleveland, T. H.: Isocyanates, organic; In: Kirk-Othmer, Encycl. Chem. Tech., 3. Ed., *13*, 789 (1981)
3. Cox, P. R., Strachan, A. N.: Two phase nitration of toluene – I; Chem. Eng. Sci., *27*, 457 (1972)
4. Fujiyama, S., Kasahara, T.: Make PTAL from CO and toluene; Hydrocarb. Proc., *57*, Nr. 11, 147 (1978)
5. Hancock, E. G.: Toluene, the Xylenes and their Industrial Derivatives; Elsevier, Amsterdam/Oxford/New York (1982)
6. Hatch, L. F., Mater, S.: From Hydrocarbons to Petrochemicals – Part 14: Chemicals from methylbenzenes; Hydrocarb. Proc., *58*, Nr. 1, 189 (1979)
7. Hersbach, G. J. M., Van der Beek, C. P., Van Dijck, P. W. M.: The Penicillins: Properties, Biosynthesis and Fermentation; In: Vandamme, E. J., Ed.: Biotechnology of Industrial Antibiotics; 45, Marcel Dekker, New York (1984)
8. Hoff, M. C.: Toluene; In: Kirk-Othmer, Encycl. Chem. Tech., 3. Ed., *23*, 246 (1983)
9. Leuenberger, H. G. W., Kieslich, K.: Biotransformationen; In: Präve, P., Faust, U., Sittig, W., Sukatsch, D. A., Ed.: Handbuch der Biotechnologie; 467, R. Oldenbourg-Verlag, München/Wien (1987)
10. Lindner, O.: Benzolsulfonsäuren und Derivate; In: Ullmanns Enzykl. Tech. Chem., 4. Aufl., *8*, 412 (1974)

11. Lipper, K.-A.: Chlorkohlenwasserstoffe, aromatische, seitenkettenchlorierte; In: Ullmanns Enzykl. Tech. Chem., 4. Aufl., *9*, 525 (1975)
12. Maki, T., Suzuki, Y.: Benzoic Acid and Derivatives; In: Ullmann's Enzycl. Ind. Chem., 5. Ed., *A3*, 555 (1985)
13. Melloh, W., Bloch, M., Inglis, R. P., Koebner, A., Leu, A.: Über ein neues Verfahren zur Sulfonierung von flüchtigen Aromaten; Tenside Detergents, *13*, 15 (1976)
14. Milligan, B., Gilbert, K. E.: Diaminotoluenes; In: Kirk-Othmer, Encycl. Chem. Tech., 3. Ed., *2*, 321 (1978)
15. Rehm, H.-J.: Industrielle Mikrobiologie; 2. Aufl., Springer-Verlag, Berlin/Heidelberg/New York (1980)
16. Ringk, W., Theimer, E. T.: Benzyl Alcohol and β-Phenethyl Alcohol; In: Kirk-Othmer, Encycl. Chem. Tech., 3. Ed., *3*, 793 (1978)
17. Roberts, S. M.: Beta-lactams – past and present; Chem. Ind. (London), 162 (1984)
18. Taverna, M., Chiti, M.: Compare Routes to Caprolactam; Hydrocarb. Proc., *49*, Nr. 11, 137 (1970)
19. Twitchett, H. J.: Chemistry of the Production of Organic Isocyanates; Chem. Soc. Rev., *3*, 209 (1974)
20. Williams, A. E.: Benzoic Acid; In: Kirk-Othmer, Encycl. Chem. Tech., 3. Ed., *3*, 778 (1978)

Chapter 7

1. Amemiya, T., Hayashi, K.: Industrial Development of Terephthalic Acid Production in Japan; Sixth World Petroleum Congress, Frankfurt/Main, June 19–26 (1963), Verein zur Förderung des 6. Welt-Erdöl-Kongresses, Hamburg
2. Ariki, S., Ohira, A.: Prospect for Metaxylene Production and Utilization; Chem. Econ. Eng. Rev., *5*, Nr. 7, 39 (1973)
3. Bartholomé, E., Hetzel, E., Horn, H. Ch., Molzahn, M., Rotermund, G. W., Vogel, L.: Verfahrenstechnische Probleme bei der Verfahrensentwicklung; Chem. Ing. Tech., *48*, 355 (1978)
4. Bemis, A. G., Dindorf, J. A., Horwood, B., Samans, C.: Phthalic Acids and other Benzenepolycarboxylic Acids; In: Kirk-Othmer, Encycl. Chem. Tech., 3. Ed., *17*, 732 (1982)
5. Hizikata, M.: New Process for Fiber-grade High-purity Terephthalic Acid (HTA); Chem. Econ. Eng. Rev., *9*, Nr. 11, 32 (1977)
6. Hoffmann, G., Mayer, D.: Terephthalsäure mit Isophthalsäure; In: Ullmanns Enzykl. Tech. Chem., 4. Aufl., *22*, 519 (1982)
7. Ibing, G.: Entwicklungstendenzen bei der großtechnischen Herstellung von Polycarbonsäuren und deren Anhydriden aus aromatischen Kohlenwasserstoffen; Erdöl, Kohle, Erdgas, Petrochem., *21*, 79 (1968)
8. Ichikawa, Y., Yamashita, G., Tokashiki, M., Yamaji, T.: New Oxidation Process for Production of Terephthalic Acid from p-Xylene; Ind. Eng. Chem., *62*, Nr. 4, 38 (1970)
9. Ichikawa, Y., Takeuchi, Y.: Compare pure TPA Processes; Hydrocarb. Proc., *51*, Nr. 11, 103 (1972)
10. Katzschmann, E.: Ein Verfahren zur Oxidation von Alkylaromaten; Chem. Ing. Techn., *38*, 1 (1966)
11. Matsuzawa, K.: Technological Development of Purified Terephthalic Acid; Chem. Econ. Eng. Rev., *8*, Nr. 9, 25 (1976)
12. Ockerbloom, N. E.: Xylenes and higher aromatics – Part 3: Phthalic anhydride; Hydrocarb. Proc., *50*, Nr. 9, 162 (1971)
13. Ockerbloom, N. E.: Xylenes and higher aromatics – Part 7: New uses for xylenes and derivatives; Hydrocarb. Proc., *51*, Nr. 2, 101 (1972)
14. Oga, T.: How to Make Xylylene Diamine From Xylenes; Hydrocarb. Proc., *45*, Nr. 11, 174 (1966)
15. Schoengen, T.: Liquid phase oxidation technology for aromatic hydrocarbons; Pet. Chem. Ind. Dev. Annu., 173 (1979)
16. Schwab, R. F., Doyle, W. H.: Hazards in phthalic anhydride plants; Chem. Eng. Progr., *66*, Nr. 9, 49 (1970)
17. Suter, H.: Phthalsäureanhydrid und seine Verwendung; Steinkopff-Verlag, Darmstadt (1972)

18. Ueda, T.: Hi-yield DMT by Mitsui; Hydrocarb. Proc., *59*, Nr. 11, 143 (1980)
19. Verde, L., Neri, A.: Make phthalic anhydride with low air ratio process; Hydrocarb. Proc., *63*, Nr. 11, 83 (1984)
20. Verde, L., Neri, A.: Un nouveau catalyseur ambivalent pour la production d'anhydride phtalique; Inf. Chimie, Nr. 247, 141 (1984)
21. Wirth, F., Franzischka, W., Bipp, H., Gelbke, H.-P.: Phthalsäure und Derivate; In: Ullmanns Enzykl. Tech. Chem., 4. Aufl., *18*, 521 (1979)

Chapter 8

1. Drayer, D. E.: Pyromellitic acid: oxidize durene; Hydrocarb. Proc., *50*, Nr. 6, 143 (1971)
2. Heering, R., Lobitz, P., Gehrke, K.: Zur Gewinnung von reinem Inden aus Pyrolyseölen; Plaste Kautschuk, *31*, 412 (1984)
3. Mitsutani, A., Maruyama, K.: Total Utilization of Aromatic Hydrocarbons and the Production of Polymethylbenzenes; Chem. Econ. Eng. Rev., *6*, Nr. 9, 36 (1974)
4. Ockerbloom, N. E.: Xylenes and higher aromatics – Part 8: Polymethylbenzenes; Hydrocarb. Proc., *51*, Nr. 4, 114 (1972)
5. Röhrscheid, F.: Carboxylic Acids, Aromatic; In: Ullmann's Enzycl. Ind. Chem., 5. Ed., *A5*, 249 (1986)
6. Szebényi, I., Széchy, G.: Die Bildung der C_9- und C_{10}-Polymethyl-Aromaten bei der katalytischen Benzinreformierung; Erdöl, Kohle, Erdgas, Petrochem., *37*, 262 (1984)

Chapter 9

1. Asselin, G. F., Erickson, R. A.: Benzene and naphthalene from petroleum by the Hydeal process; Chem. Eng. Progr., *58*, Nr. 4, 47 (1962)
2. Barbor, R. P.: Naphthalene from catalytic gas oil; Oil Gas J., 123 (21.5.1962)
3. Blank, U.: Herstellung von Naphthalin-Derivaten; In: Winnacker-Küchler, Chem. Technol., 4. Aufl., *6*, Org. Technol. II, 245 (1982)
4. Bretscher, H., Eigenmann, G., Plattner, E.: Integrierte Verfahrensentwicklung am Beispiel der Naphthalinsulfonsäurederivate; Chimia, *32*, 180 (1978)
5. Collin, G., Bunge, W.: Naphthalin und Naphthalinhydroverbindungen; In: Ullmanns Enzykl. Tech. Chem., 4. Aufl., *17*, 77 (1979)
6. Donaldson, N.: The Chemistry and Technology of Naphthalene Compounds; Edward Arnold, London (1958)
7. Dressler, H.: Naphthalene Derivatives; In: Kirk-Othmer, Encycl. Chem. Tech., 3. Ed., *15*, 719 (1981)
8. Engel, D., Schulz, R. C.: Anionische Homopolymerisation von 2-Isopropenylnaphthalin; Makromol. Chem., *182*, 3279 (1981)
9. Foster, N. R., Wainwright, M. S.: Recent Advances in Phthalic Anhydride Manufacture; Pace, Nr. 3, 28 (1978)
10. Gamburg, E. Y., Berents, A. D., Mukhina, T. N., Belyaeva, Z. G., Vinyukova, N. I., Vol'-Epshtein, A. B.: Production of Naphthalene from Liquid Products of the Pyrolysis of Hydrocarbon Feedstock; Soviet Chem. Ind., *13*, Nr. 9, 1119 (1981)
11. Gaydos, R. M.: Naphthalene; In: Kirk-Othmer, Encycl. Chem. Tech., 3. Ed., *15*, 698 (1981)
12. Graham, J. J.: The Fluidized Bed Phthalic Anhydride Process; Chem. Eng. Progr., *66*, Nr. 9, 2 (1970)
13. Graham, J. J., Young, B. J.: Coal Tar Naphthalene Unionfining; Erdöl, Kohle, Erdgas, Petrochem., *26*, 331 (1973)
14. Herbst, W., Hunger, K.: Industrielle Organische Pigmente – Herstellung, Eigenschaften, Anwendung; VCH Verlagsgesellschaft, Weinheim (1987)
15. Hörmeyer, H.: Calculation of Crystallization Equilibria; Ger. Chem. Eng., *6*, 277 (1983)
16. Hunger, K., Mischke, P., Rieper, W., Raue, R.: Azo Dyes; In: Ullmann's Enzycl. Ind. Chem., 5. Ed., *A3*, 245 (1986)
17. Kay, J.: Fractional crystallizer gives high purity with greater efficiency; Processing, Nr. 12, 25 (1978)

18. Molinari, J.G.D., Dodgson, B.V.: Theory and Practice related to the Brodie Purifier; The Chem. Engineer, Nr. 7/8, 460 (1974)
19. Ockerbloom, N.E.: Xylenes and higher aromatics – Part 5: Naphthalenes; Hydrocarb. Proc., 50, Nr.12, 101 (1971)
20. Pawellek, D., Behre, H., Bonse, G., Grolig, J., Bunge, W.: Naphthalin-Derivate; In: Ullmanns Enzykl. Tech. Chem., 4. Aufl., 17, 83 (1979)
21. Saxer, K., Papp, A.: Analysis of Multicomponent Separation in a Novel Process of Crystallization from the Melt; AIChE Meeting, November 25–29, 1979
22. Stobaugh, R.B: Naphthalene: How, Where, Who – Future; Hydrocarb. Proc., 45, Nr.3, 149 (1966)
23. Swodenk, W.: Umweltfreundlichere Produktionsverfahren in der chemischen Industrie; Chem. Ing. Tech., 56, 1 (1984)
24. Taglieri, G.: Il processo Brodie per la purificazione industriale dei prodotti organici; Ing. Chim. Ital., 15, Nr.11–12, 130 (1979)

Chapter 10

1. Bhattacharya, R.N.: Isomerisation of Methylnaphthalenes over Acidic Chalcide Catalysts; J. Indian Chem. Soc., 58, 682 (1981)
2. Collin, G.: Acenaphthen; In: Ullmanns Enzykl. Tech. Chem., 4. Aufl., 7, 10 (1974)
3. Gel'pern, N.I., Nosov, G.A., Zimnitskii, P.V.: Separation of β-Methylnaphthalene from its industrial fractions by the method of contact crystallization; Coke Chem. USSR, Nr.1, 44 (1982)
4. Solinas, V., Monaci, R., Marongiu, B., Forni, L.: Isomerisation of 1-Methylnaphthalene over Ω-Zeolithe; Appl. Catalysis, 5, Nr.2, 171 (1983)

Chapter 11

1. Chung, R.H.: Anthraquinone; In: Kirk-Othmer, Encycl. Chem. Tech., 3. Ed., 2, 700 (1978)
2. Dauter, R., Fuchs, H.: Herstellung von Farbstoffen und Pigmenten; In: Winnacker-Küchler, Chem. Technol., 4. Aufl., 7, Org. Technol. III, 6 (1986)
3. Düsing, G., Kleinschmit, P., Knippschild, G., Kunkel, W., Habersang, S.: Peroxoverbindungen, anorganische; In: Ullmanns Enzykl. Tech. Chem., 4. Aufl., 17, 691 (1979)
4. Greenhalgh, C.W.: Aspekte der Anthrachinon-Farbstoffchemie; Endeavour, 126, 134 (1976)
5. Haggin, J.: Electrochemical Theories Open Up Chemistry of Wood Digestion; Chem. Eng. News, 20 (15.10.1984)
6. Hohmann, W.: Herstellung von Anthracen-Derivaten; In: Winnacker-Küchler, Chem. Technol., 4. Aufl., 6, Org. Technol. II, 267 (1982)
7. Kirchner, J.R: Hydrogen Peroxide; In: Kirk-Othmer, Encycl. Chem. Tech., 3. Ed., 13, 12 (1981)
8. Lodh, S.B.: Introduction of Anthraquinone as a Pulping Catalytic Agent in Indian Paper Industry; Inst. Eng. (India), Part CH, 64, Nr.3, 73 (1984)
9. Meyer, F., Hausigk, D., Kölling, G.: Reaktion von Diphenylmethan in wasserfreiem HF/BF$_3$ zu Anthracen; Liebigs Ann. Chem., 736, 141 (1970)

Chapter 12

1. Collin, G.: Kohlenwasserstoffe; In: Ullmanns Enzykl. Tech. Chem., 4. Aufl., 14, 684 (1977)

Chapter 13

1. Alberts, B., Zwartbol, D.P.: Needle coke unit linked to ethylene plant; Oil Gas J., 137 (4.6.1979)

2. Alscher, A., Gemmeke, W., Alsmeier, F., Marrett, R.: Ageing and Rheological Properties of Binder Pitches; In: Miller, R.E., Ed.: Light Metals 1986, Proceedings of the 115th Annual Meeting, The Metallurgical Society, Inc., New Orleans, Louisiana, March 2-6, 605 (1986)
3. Bode, R.: Pigmentruße für Kunststoffe; Kautsch., Gummi, Kunstst., 36, 660 (1983)
4. Brown, G.H., Crooker, P.P.: Liquid crystals; Chem. Eng. News, 24 (31.1.1983)
5. Collin, G., Gemmeke, W.: Die Bedeutung des Steinkohlenteerpechs für die Aluminiumgewinnung; Erdöl, Kohle, Erdgas, Petrochem., 30, 25 (1977)
6. Davidson, H.W., Wiggs, P.K.C., Churchouse, A.H., Maggs, F.A.P., Bradley, R.S.: Manufactured Carbon; Pergamon Press, London (1968)
7. Eckert, A., Marrett, R., Stadelhofer, J.W.: Filtration von Steinkohlenteerpech; Erdöl, Kohle, Erdgas, Petrochem., 38, 510 (1985)
8. Eidenschink, R: Flüssige Kristalle; Chem. i.u. Zeit, 18, 168 (1984)
9. Franck, H.-G.: Die wahre Natur des Steinkohlenteerpechs; Brennst.-Chem., 36, 12 (1955)
10. Gambro, A.J., Shedd, D.T., Wang, H.W.: Delayed coking of coal tar pitch; Chem. Eng. Progr., 65, Nr.5, 75 (1969)
11. Gasparoux, H.: Cristaux Liquides et Mesophase Carbonée; J. Chimie Physique, 81, 759 (1984)
12. Ishikawa, T., Nagaoki, T.: Recent Carbon Technology; JEC Press, Cleveland (1983)
13. Kleinschmit, P., Kühner, G.: Industrieruße: Synthetischer Kohlenstoff für vielfältige Anwendungen; Chem. Ind., 37, 565 (1985)
14. Lewis, I.C.: Thermal Polymerization of Aromatic Hydrocarbons; Carbon, 18, 191 (1980)
15. Lewis, I.C.: Chemistry and Development of Mesophase in Pitch; J.Chimie Physique, 81, 751 (1984)
16. Lewis, J.E.: Mechanism of carbon-black formation in relation to compounded-rubber properties; In: Bacha, J.D., Newman, J.W., White, J.L., Ed.: Petroleum Derived Carbons, ACS Symposium, Ser.303, 269 (1986)
17. Marrett, R., Stadelhofer, J.W., Marsh, H.: Die Verkokung von flüssigen Kohlenwasserstoffen zur Herstellung von Kohlenstoff-Produkten; Chem. Ing. Tech., 55, 1 (1983)
18. Marsh, H., Calvert, C., Bacha, J.: Structure and formation of shot coke - a microscopy study; J.Mat. Sci., 20, 289 (1985)
19. Mochida, I., Nesumi, Y., Korai, Y.: A tube bomb as a model for a delayed coker; Ext. Abstract. Amer. Carb. Soc./Univ. of Kentucky, 17th Biennial Conference on Carbon, Lexington, Kentucky, June 16-21, 255 (1985)
20. Müller, R.: Der westeuropäische Rußmarkt - Situation und Trends; Kautsch., Gummi, Kunstst., 36, 303 (1983)
21. Newman, J.W. : What is petroleum pitch?; In: Deviney, M.L., O'Grady, T.M., Ed.: Petroleum Derived Carbons; ACS Symp., Ser.21, 52, Washington (1976)
22. Otani, S.: On the carbon fiber from the molten pyrolysis products; Carbon, 3, 31 (1965)
23. Reis, T.: About coke - and where the sulfur went; ChemTech., 7, 366 (1977)
24. Remirez, R.: Novel Process Makes Coke From Coal Tar Pitch; Chem. Eng., Nr.2, 74 (1969)
25. Sibal, A.K.: Carbon Fiber Development and Fabric Markets; J. Coated Fabrics, 13, 206 (1984)
26. Singer, L.S.: The Mesophase in Carbonaceous Pitches; Faraday Discuss. Chem. Soc., 79, 265 (1985)
27. Singer, L.S., Lewis, I.C., Greinke, R.A.: Characterization of the Phases in Centrifuged Mesophase Pitches; Mol. Cryst. Liq. Cryst., 132, 65 (1986)
28. Sprenger, K.-H.: Hochleistungsverbundwerkstoffe - Verstärkung mit Aramid- und Kohlenstoffasergewebe; Sprechsaal, 119, 453 (1986)
29. Stadelhofer, J.W., Marrett, R., Gemmeke, W.: The manufacture of high-value carbon from coal-tar pitch; Fuel, 60, 877 (1981)
30. Stadelhofer, J.W., Alsmeier, F.G., Diefendorf, R.J.: Mesophasen aus hochkondensierten Aromaten; Erdöl, Kohle, Erdgas, Petrochem., 38, 503, (1985)
31. Stadelhofer, J.W.: Industrielle Herstellung von Koks aus Mineralölprodukten und kohlestämmigen Rohstoffen; Erdöl, Kohle, Erdgas, Petrochem., 39, 541 (1986)
32. Stokes, C.A., Guercio, V.J.: Feedstocks for Carbon Black, Needle Coke and Electrode Pitch; Erdöl, Kohle, Erdgas, Petrochem., 38, 31 (1985)
33. Vohler, O., v.Sturm, F., Wege, E., v.Kienle, H., Voll, M., Kleinschmit, P.: Carbon; In: Ullmann's Enzycl. Ind. Chem., 5.Ed., A5, 95 (1986)

34. Vohler, O.J.: Carbon and Graphite in Future Markets; Erdöl, Kohle, Erdgas, Petrochem., *39*, 561 (1986)
35. Vohwinkel, K.: Augenblicklicher Stand und Entwicklungsrichtungen des Einsatzes von Ruß für Kautschuk; Kautschuk, Gummi, Kunstst., *39*, 810 (1986)
36. Wilkening, S.: One hundred years of carbon for the production of aluminium; Erdöl, Kohle, Erdgas, Petrochem., *39*, 551 (1986)
37. Wlochowicz, A.: Kohlenstoffasern aus Pech, ihre Herstellung und Eigenschaften; Textiltechnik, *34*, 595 (1984)
38. Zimmer, K.: Rußherstellung nach dem Oil-Furnace-Verfahren aus Schweröl einer thermischen Krackanlage; Erdöl, Kohle, Erdgas, Petrochem., *22*, 742 (1969)

Chapter 14

1. Adey, K.A.: Vinylverbindungen; In: Ullmanns Enzykl. Tech. Chem., 4. Aufl., *23*, 612 (1983)
2. Aiba, S., Tsunekawa, H., Imanaka, T.: New Approach to Tryptophan Production by Escherichia coli: Genetic Manipulation of Composite Plasmids in Vitro; Appl. Environm. Microbiol., *43*, 289 (1982)
3. Bang, W.-G., Behrendt, U., Lang, S., Wagner, F.: Continuous Production of L-Tryptophan from Indole and L-Serine by Immobilized Escherichia Coli Cells; Biotech. Bioeng., *25*, 1013 (1983)
4. Beschke, H., Friedrich, H.: Acrolein in der Gasphasensynthese von Pyridinderivaten; Chem.-Zeitung, *101*, 377 (1977)
5. Beschke, H., Friedrich, H., Schaefer, H., Schreyer, G.: Nicotinsäureamid aus β-Picolin; Chem.-Zeitung, *101*, 384 (1977)
6. Beschke, H., Kleemann, A., Clauss, W., Kurze, W., Mathes, K., Habersang, S.: Pyridin und Pyridin-Derivate; In: Ullmanns Enzykl. Tech. Chem., 4. Aufl., *19*, 591 (1980)
7. Bhattacharya, R.N., Roy, M.B., Banerjee, S.N., Nandi, G.C.: Recovery, Purification and Utilisation of Pyridine Bases from Coke Oven By-Products; URJA, *11*, Nr.4, 215 (1982)
8. Buchholz, B.: Thiophene and Thiophene Derivatives; In: Kirk-Othmer, Encycl. Chem. Tech., 3. Ed., *22*, 965 (1983)
9. Burakevich, J.V.: Cyanuric and Isocyanuric Acids; In: Kirk-Othmer, Encycl. Chem. Tech., 3. Ed., *7*, 397 (1979)
10. Collin, G.: Carbazol; In: Ullmanns Enzykl. Tech. Chem., 4. Aufl., *9*, 120 (1975)
11. Collin, G.: Chinolin; In: Ullmanns Enzykl. Tech. Chem., 4. Aufl., *9*, 311 (1975)
12. Collin, G., Kleffner, H.W.: Thiophen und Benzothiophen; In: Ullmanns Enzykl. Tech. Chem., 4. Aufl., *23*, 217 (1983)
13. Goe, G.L.: Pyridine and Pyridine Derivatives; In: Kirk-Othmer, Encycl. Chem. Tech., 3. Ed., *19*, 454 (1982)
14. Härtner, H.: Vitamin B_1 (Thiamin); In: Ullmanns Enzykl. Tech. Chem., 4. Aufl., *23*, 656 (1983)
15. Kleemann, A., Leuchtenberger, W., Hoppe, B., Tanner, H.: Amino Acids; In: Ullmann's Enzycl. Ind. Chem., 5. Ed., *A2*, 57 (1985)
16. Kriebitzsch, N., Klenk, H.: Cyanuric Acid and Cyanuric Chloride; In: Ullmann's Enzycl. Ind. Chem., 5. Ed., *A8*, 191 (1987)
17. Kusunoki, Y., Okazaki, H.: Make pyridines direct; Hydrocarb. Proc., *53*, Nr.11, 129 (1974)
18. Mensch, F.: Hydrodealkylierung von Pyridinbasen bei Normaldruck; Erdöl, Kohle, Erdgas, Petrochem., *22*, 67 (1969)
19. Nenz, A., Pieroni, M.: Commercial Synthetic Pyridine Bases; Hydrocarb. Proc., *47*, Nr.11, 139 (1968)
20. Nenz, A., Pieroni, M.: Commercial Synthetic Pyridine Bases; Hydrocarb. Proc., *47*, Nr.12, 103 (1968)
21. Paustian, J.E., Puzio, J.F., Stavropoulos, N., Sze, M.C.: A lesson in flow sheet design: nicotinamide and acid; ChemTech, Nr.3, 174 (1981)
22. Pollak, P., Romeder, G.: N-heterocycles provide performance options; Perform. Chem., *3*, Nr. 1, 34 (1988)
23. Reitter, L.. Lihotzky, R.: Melamin; In: Ullmanns Enzykl. Tech. Chem., 4. Aufl., *16*, 503 (1978)

24. Rittner, S., Warning, K.: Herstellung von heteroaromatischen Zwischenprodukten; In: Winnacker-Küchler, Chem. Technol., 4. Aufl., *6*, Org. Technol. II, 281 (1982)
25. Skogman, G. S., Sjöström, J.-E.: Factors Affecting the Biosynthesis of L-Tryptophan by Genetically Modified Strains of Escherichia coli; J. Gen. Microbiol., *130*, 3091 (1984)
26. Suter, Ch.: Vitamine; In: Ullmanns Enzykl. Tech. Chem., 4. Aufl., *23*, 706 (1983)
27. Sze, M. C., Gelbein, A. P.: Make aromatic nitriles this way; Hydrocarb. Proc., *55*, Nr. 2, 103 (1976)
28. Takahashi, N.: Novel Route to Synthesis of Alkylpyridines; Chem. Econ. Eng. Rev., *7*, Nr. 6, 34 (1975)
29. Terui, G.: Tryptophan; In: Yamada, K., Ed.: Microbial Prod. Amino Acids; 515, Kodansha, Tokyo (1972)
30. Uebel, H.-J., Moll, K.-K., Mühlstädt, M.: Untersuchungen zur Gewinnung von 3-Methylpyridin; Chem. Tech. (Leipzig), *22*, Nr. 12, 745 (1970)
31. Ujimaru, T., Kakimoto, T., Chibata, I.: L-Tryptophan Production by Achromobacter liquidum; Appl. Environm. Microbiol., *46*, 1 (1983)
32. Woodward, R. B., Doering, W. E.: The Total Synthesis of Quinine; J. Am. Chem. Soc., *67*, 860 (1945)

Chapter 15

1. Alexander, M.: Biodegradation of Chemicals of Environmental Concern; Science, *211*, 132 (1981)
2. Ames, B. N., Dietary Carcinogens and Anticarcinogens; Science, *221*, 1256 (1983)
3. Blackburn, G. M., Kellard, B.: Chemical carcinogens; Chem. Ind. (London), 607 (1986)
4. Blackburn, G. M., Kellard, B.: Chemical carcinogens - II; Chem. Ind. (London), 687 (1986)
5. Blackburn, G. M., Kellard, B.: Chemical carcinogens - III; Chem. Ind. (London), 770 (1986)
6. Daune, M.: La Cancerogenese Chimique; La Recherche, *11*, 1066 (19. 10. 1980)
7. Ghosal, D., You, I.-S., Chatterjee, D. K., Chakrabarty, A. M.: Microbial Degradation of Halogenated Compounds; Science, *228*, 135 (1985)
8. Lowe, J. P., Silverman, B. D.: Predicting Carcinogenicity of Polycyclic Aromatic Hydrocarbons; Acc. Chem. Res., *17*, 332 (1984)
9. Müller, R., Lingens, F.: Microbial Degradation of Halogenated Hydrocarbons: A Biological Solution to Pollution Problems?; Angew. Chem. Int. Ed., *25*, Nr. 9, 779 (1986)
10. Predel, H.: Angewandte Mikrobiologie der Kohlenwasserstoffe; Erdöl, Kohle, Erdgas, Petrochem., *37*, 502 (1984)
11. Sax, N. I.: Dangerous Properties of Industrial Materials; 6. Ed., Van Nostrand Reinhold, New York (1984)
12. Sax, N. I.: Cancer Causing Chemicals; Van Nostrand Reinhold, New York (1981)
13. Schäfer, H. K.: Sicherheit in der Chemie; Hanser-Verlag, München/Wien (1979)
14. Schäfer-Ridder, M.: Carcinogenese durch polycyclische aromatische Kohlenwasserstoffe; Nachr. Chem. Tech. Lab., *27*, 4 (1979)
15. Sprenger, B., Ebner, H. G., Klein, J.: Biologischer Abbau nach Wunsch; Chem. Ind., *109*, 596 (1986)
16. Zander, M.: Polycyclic Aromatic and Heteroaromatic Hydrocarbons; In: Hutzinger, O., Ed.: The Handbook of Environm. Chem., *3/A*, Springer-Verlag, Berlin/Heidelberg (1980)

Chapter 16

1. Lenz, R. W.: Structure-order relationships in liquid crystalline polyesters; Pure Appl. Chem., *57*, 1537 (1985)
2. Moseley, J. D., Nowak, R. M.: Engineering Thermoplastics: Materials for the Future; Chem. Eng. Progr., Nr. 6, 49 (1986)
3. Rose, J. B.: Improved engineering thermoplastics from aromatic polymers; Shell Polym., *8*, 88 (1984)

4. Scott, C.D., Strandberg, G.W., Lewis, S.N.: Microbial Solubilization of Coal; Biotech. Progr., *2*, 131 (1986)
5. Stütz, A.: Allylamine Derivatives – a New Class of Active Substances in Antifungal Chemotherapy; Angew. Chem. Int. Ed. *26*, Nr. 4, 320 (1987)
6. Wendorff, J.H.: Flüssig-kristalline Kunststoffe – Struktur und Eigenschaften; Kunststoffe, *73*, 524 (1983)

Subject index

acenaphthene 38, 340
- oxidation 340
acenaphthylene 340, 342
acetanilide 205
5-acetoacetylaminobenzimidazolone 225
N-acetoacetyl-o-toluidide 239
N-acetoacetyl-2,4-xylidide 282
acetone 149, 166
acetophenone 150
acetophenone synthetic resins 150
2-acetylaminofluorene
- carcinogenicity 428
p-acetylaminophenol 180
acetylsalicylic acid 6, 176
acid black 1 326
acid black 26 319
acid blue 25 353
acid blue 62 352
acid green 25 356
acid orange 7 208, 317
acid orange 19 315
acid red 18 321
acid red 26 321
acid red 114 240
acid red 337 326
acid yellow 23 208
acid yellow 24 298
acid yellow 151 209
acifluorfen 250
acridine 26, 387
adamantane 33
additives for lubricants 171, 174
adipic acid 193
aflatoxin B_1 444
alachlor 206
alizarin 4, 343, 351
alizarine brilliant blue R 352
alizarine cyanine green G 356
alizarine saphirol A 353
alkyd resins 213
alkylanthraquinone 359
- hydrogenation 360
alkylbenzene

- linear 212
- production capacities 213
- sulfonation 212
alkylbenzene sulfonate 210, 212
- production 211
alkylnaphthalene 329, 331, 334
- hydrodealkylation 307
alkylphenol 163
- higher 174
alkylsalicylates 174
N-allyltoluidine 2
amines
- aromatic
- - carcinogenicity 428
1-aminoanthraquinone 352
2-aminoanthraquinone 357
o-aminoazotoluene
- MAK value 430
1-aminobenzene-3-β-hydroxyethylsulfone 354
3-aminobenzenesulfonic acid 187
3-aminobenzotrifluoride 258
p-aminobenzylaniline 199
4-aminobiphenyl
- MAK value 430
1-amino-4-bromo-anthraquinone-2-sulfonic acid 352
3-amino-9-ethylcarbazole 424
- MAK value 430
1-amino-8-hydroxynaphthalene-3,6-disulfonic acid 324
2-amino-5-hydroxynaphthalene-7-sulfonic acid 324
2-amino-8-hydroxynaphthalene-6-sulfonic acid 324
aminonaphthalene 327
1-aminonaphthalene-4-sulfonic acid 315
2-aminonaphthalene-1-sulfonic acid 320
1-aminonaphthalene-3,6,8-trisulfonic acid 324
2-amino-4-nitrotoluene
- MAK value 430
6-aminopenicillanic acid 254

aminophenol 180, 187, 194, 195
2-aminopyridine
- MAK value 430
4-aminosalicylic acid 187
4-aminotoluene-3-sulfonic acid 242
4-amino-1,2,4-triazole 391
amitrole 393
amoxycillin 254
ampicillin 254
aniline 2, 196, 207, 209
- alkylation 206
- N-alkylderivatives 203
- derivatives 199
- diazotization 206, 209
- hydrogenation 200
- MAK value 430
- production capacities 199
- production
- - from nitrobenzene 196
- - from phenol 198
aniline dyes 3
aniline yellow 6
4-anilino-1-methylpiperidine 389
[16]-annulene 12
[18]-annulene 12
anodes 379
anthracene 24, 38, 343
- magnesium complex 346
- maleic anhydride adduct 363
- oxidation 346
- production 343
- synthesis 344
anthracene oil 343
anthranilic acid 263, 276, 417
anthraquinone 4, 15, 311
- alkylated 359
- derivatives 350
- production 346
- - from anthracene 347
- - from benzophenone 349
- - from benzoquinone 349
- - from naphthoquinone 313, 348
- - from phthalic anhydride 348
- - from styrene 349
- sulfonation 351
- uses
- - for hydrogen peroxide production 359
- - for wood pulping 360
anthraquinone dyes 350
anthraquinone-1-sulfonic acid 351
anthrone 357
anti-aromatic systems 11
antioxidants 170, 174, 184, 190, 199, 204, 226, 239, 321, 420
antipyrine 6, 207
API-gravity 58
aramid fibers 196, 228, 280, 382

aromaticity 12, 32, 97
aromatics
- heterocyclic 387
- nitration
- - influence of substituents 18
- polycyclic 39
- production
- - from alcohols 86
- - by extractive distillation 112
- - from propane/butane 88
- production processes
- - review 96
- solvent power for coal 48
- substitution
- - directing effect of functional groups 17
aromatics extraction 107
- by diethylene glycol 108
- by dimethylsulfoxide 111
- by N-formylmorpholine 110
- by N-methylpyrrolidone 110
- by sulfolane 109
- process conditions 111
aryne 19
aryne mechanism 165
aspartame 255
aspirin 6, 176
atrazine 414
- MAK value 430
auramine 428
- MAK value 430
azine dyes 227
azinphos-methyl 277
- MAK value 431
azo-coupling 16
azo dyes 5, 206, 315, 317, 326, 350
azulene 11

4B acid 242
Bakelite resins 158
barbituric acid 412
basic green 4 203
basic violet 3 203
basic violet 14 3
benomyl 224
bentazone 276
benzal chloride 256
benzaldehyde 154, 256
benzaldehyde-2-sulfonic acid 261
benz[a]anthracene 40
benzanthrone 357
benzene 2, 8, 10, 444
- azeotropic mixtures 106
- chlorination 218, 229
- crude, see benzole
- dehydrogenation 334
- derivatives
- - process review 234

Subject index   473

- higher alkylated 210
- - production 211
- hydrogenation 99, 191
- - energy diagram 9
- MAK value 431
- nitration 194
- - relative reaction rates 16
- oxidation 214
- oxychlorination 152, 220
- polyalkylated 291
- production 99
- - by dealkylation 122
- - from aromatic mixtures 114
- production figures 129
- propylation 147
- quality standards 128
- refining 122
- sulfonation 152
- uses 99, 132
benzene-1,3-disulfonic acid 186
1,3,5-benzenetricarboxylic acid 294
benzidine 194, 205, 428
- MAK value 431
benzidine rearrangement 23
benzil 257
benzilic acid rearrangement 23
benzimidazole polymers 223
benzimidazolone 243
benzimidazolone pigments 225
benzoctamine 345
benzoic acid 154, 247
- production 248
- uses 248
benzoin 257
benzole 34
- composition 103
- hydrogenation 104
- recovery 35
- sulfuric acid refining 104
- typical data 105
benzonitrile 248
benzo[a]phenalenyl 40
benzo[c]phenanthrene 40
benzophenone 249
benzo[a]pyrene 25, 427, 444
- MAK value 431
benzo[e]pyrene 25, 427
benzoquinone 213
- MAK value 431
benzothiazole 418
- derivatives 200
2,3-benzothiophene 416
benzotriazole 418
benzotrichloride 248, 258
benzotrifluoride 258
o-benzoyl benzoic acid 348
benzoylchloride 249, 258

benzyl alcohol 154, 251
benzyl benzoate 249
benzyl chloride 251
benzyl cyanide 252
benzyl penicillin 252
BHT 170
bioengineering
- decomposition of aromatics 444
- production
- - of ephedrine 258
- - of D-lactic acid 178
- - of penicillin 253
- - of tryptophan 418
biomass 89
- uses 90
biphenyl 334
- MAK value 431
2,2'-bipyridyl 401
4,4'-bipyridyl 401
bisphenol A 158
- crystallization 159
- uses 160
bisphenol F 158
bisphenol S 158
bitertanol 393
blown bitumen 83
bromamine acid 352
bromoxynil 182
BTX aromatics 100
- physical data 106
- production 105
- - by azeotropic distillation 106, 112
- - by dealkylation 122
- - by disproportionation 126
- - by extractive distillation 106, 112
- - by isomerization 125
- - by liquid-liquid extraction 106
- - review of processes 130
- quality standards 128
butane
- oxidation 215
butter yellow 428
2-tert-butylanthraquinone 359
4-tert-butylbenzaldehyde 264
4-tert-butylbenzoic acid 264
butyl benzyl phthalate 251
4-tert-butylcatechol 184
butylphenol 163, 174
- MAK value 431
4-tert-butyltoluene 264
- MAK value 431
butyrolactone 217

C_9-aromatics 291
calcium cyclohexylsulfamate 203
camphor 272
ε-caprolactam 163, 191, 193, 248

captan 217
carbaryl 314
- MAK value 432
carbazole 26, 38, 198, 344, 387, 423
N-(3-carbazole)-1,4-benzoquinonimine 425
carbendazim 224
carbofuran 184
carbolic oil
- composition 156
carbon black 369
- production 382, 383
- production capacities 385
- spectrum of grain sizes 384
- types 384
carbon black feedstocks 382
- characteristic data 383
carbon bonds
- free bond enthalpie 77
carbon fibers 368, 380
- production figures 382
- properties 381
- raw materials 381
carbon products 368
carbonium ion 14
carbonization 34, 35
carbonless copy papers 335
4-carboxybenzaldehyde 284
carminic acid 350
carvacrol 168
cat-cracker 63
- flow diagram 64
cat-cracker naphtha 67
catechol 183
- production 184
CBS 202
cefalotin 389
cellulose 30, 91, 360
cetylpyridinium chloride 401
chemical raw materials
- renewable 90
chemicals law 428
chloramine T 264
chloramphenicol 256
chlorinated biphenyls
- MAK value 432
chloroacetic acid 168
1-chloro-2-aminoanthraquinone 357
2-chloro-5-aminotoluene-4-sulfonic acid 262
chloroaniline 195, 223, 226
2-chloroanthraquinone 274
- production 356
chlorobenzene 218
- ammonolysis 198
- MAK value 432
- nitration 222
- production 218
- production capacities 220

4-chlorobenzotrichloride 260
4-chlorobenzotrifluoride 260
2-(4'-chlorobenzoyl)-benzoic acid 356
4-chloro-o-cresol 168
4-chloro-3,5-dimethylphenol 173
4-chloro-3,5-dinitrobenzotrifluoride 260
chlorodioxins 179
3-chloro-4-fluoronitrobenzene 230
chloroflurenol-methylester 364
3-chloro-4-methylaniline 242
4-chloro-2-methylphenoxy-acetic acid 168
4-chloro-2-methylphenoxy-carboxylic acids
- production figures 168
2-(4-chloro-2-methylphenoxy)-propionic
  acid 168
1-chloronaphthalene 313
chloronitrobenzene 180, 194, 195, 222
- ammonolysis 224
- MAK value 432
- production figures 222
- products 223
2-chloro-5-nitrotoluene-4-sulfonic acid 262
chlorophenol, see phenol - chlorinated
chlorophyll 28
2-chloropropionic acid 168
2-chloropyridine 402
chloroquine 420
chlorothalonil 281
chlorotoluene 165, 260
- MAK value 432
2-chlorotoluene-4-sulfonic acid 262
chloro-o-toluidine
- MAK value 432
2-chloro-triphenylchloromethane 392
chlorpyrifos 403
chlorsulfuron 448
chlortoluron 242
cholestane 29
cholesterol 29
chrysene 24, 40
- MAK value 432
cimetidine 392
cinnamaldehyde 256, 257
cinnamic alcohol 256
clotrimazole 392
CLT acid 262
coal 27, 31
- aromatic extracts 33
- ^{13}C-NMR-spectra 34
- degasification 35
- model of structure 32
- origin 27
- producers 32
- production figures 31
- reserves 31
- structure 32
coal conversion

- thermal 35
coal extraction 46, 47
coal gasification 34, 43
- tar 45
coal hydrogenation 34, 46
- *Exxon*-Donor-Solvent-process 52
- H-coal-process 53
- *IG*-process 47
- *Ruhrkohle/Veba Oel*-process 54
- *SRC*-process 52
coal liquefaction 46
- mechanisms 47
coalification 30, 90
coke 375
- anisotropic 369
- calcined
- - composition 378
coke oven 36
- flow chart of quantities 35
coke oven benzole 100
coke oven gas
- purification 37
coking 375
- of pitch 380
colchicine 444
collidine 410
congo red 6
conversion processes 62
coronene 24
corrin 391
coumarine 183
coumarone 297
cracking 62
- catalytic 63
- - coke formation 66
- - composition of products 64
- - naphtha yield 65
- thermal 76, 85
- - Dubbs process 85
cresol 163, 164
- alkylation 168
- chlorination 168
- crystallization 165
- MAK value 433
- phosphate esters 171
- production
- - from chlorotoluene 165
- - from cymene 166
- - from isoprene/vinylacetate 168
- - from toluene 168
- production figures 171
- sources 164
- uses 168
crude benzene, see benzole
crude oil 27, 55
- aromatics content 59
- basic composition 57

- characteristic data 58, 61
- classification 58
- cracking 81
- distillation 60
- geological age 60
- origin 27
- production figures 56
- refining 60
- reserves 31, 55
crystal violet 203
crystallization
- of bisphenol A 160
- of m/p-cresol 165
- of dichlorobenzene 229
- of naphthalene 302
cumene 146
- conversion 149
- nitration 296
- oxidation
- - kinetics 22
- production
- - from benzene 146
- production figures 148
- specifications 148
cumene hydroperoxide 149
cumenesulfonic acid 296
cumidine 296
cyanazine 415
cyanuric chloride 26, 412
cyclobutadiene 11
1,3-cyclohexadiene 9
cyclohexane 191
- oxidation 155
- production 191
- production figures 193
cyclohexane carboxylic acid 248
cyclohexanol 161
- ammonolysis 200
- production
- - from phenol 162
cyclohexanone 161
- production
- - from phenol 162
cyclohexene 9
cyclohexylamine 199, 200
cyclohexylbenzothiazolesulfenamide 202
cyclohexylchloride
- ammonolysis 200
cyclohexylsulfamate 203
cyclopentadiene
- ammoxidation 400
cyclopentadienyl anion 10
cyclopropenyl cation 10
cymene 166
cypermethrin 170

2,4-D 177
- MAK value 433
dangerous substances
- aromatic 429
Dangerous Substances Act 429
DCBS 202
DDT 218, 221
- MAK value 433
dealkylation
- kinetics 20
- reaction diagram 21
decant oils 382
decomposition
- of aromatics
- - biochemical 444
dehydrobenzene 19
delayed coking process 375
- flow diagram 376
- yields 378
delocalization
- of π-electrons 9
deltamethrin 170
detergents 262
dexchlorpheniramine 402
2,4-diaminoanisole
- MAK value 433
diaminoanthraquinone 259
1,4-diaminoanthraquinone dyestuffs 356
3,3'-diaminobenzidine 223
4,4'-diaminobiphenyl 205
3,4'-diaminodiphenyl ether 196
4,4'-diaminodiphenyl ether 295
- MAK value 433
4,4'-diaminodiphenyl sulfide
- MAK value 433
4,4'-diaminodiphenylmethane 199, 294
- MAK value 433
- production 200
diaminostilbenedisulfonic acid 415
diaminotoluene 242
dian 158
diazepam 228
diazinon 412
- MAK value 433
diazotization 327
dibenz[a,h]anthracene 427, 444
4,4'-dibenzanthronyl 358
dibenzofuran 180, 425
dibenzo[c,g]phenanthrene 25
dibenzothiazyl disulfide 202
dibenzothiophene 363
dibenzoyl peroxide
- MAK value 433
dibenzyl ether 252
2,6-di-tert-butyl-p-cresol 170
2,6-di-tert-butylphenol 174
3,4-dichloroaniline 229

dichlorobenzene 229
- crystallization 229
- MAK value 434
- nitration 231
- production 229
- uses 229
3,3'-dichlorobenzidine 206, 444
- MAK value 434
2,4-dichloro-3,5-dimethylphenol 173
4,4'-dichlorodiphenylsulfone 161
dichloronitrobenzene 230, 231
2,4-dichlorophenol 177
- production 177
2,4-dichlorophenoxy-acetic acid 177
2,4-dichlorophenoxy-propionic acid 177
4,7-dichloroquinoline 195, 420
2,5-dichlorothioindoxyl 231
α,α-dichlorotoluene
- MAK value 434
diclobutrazol 393
diclofop-methyl 178
dicyclohexylbenzothiazolesulfenamide 202
N,N'-di[2-(4,6-dichlorotriazino)]-4,4'-diamino-
   stilbene-2,2'-disulfonic acid 241
2,6-diethylaniline 206
diethylbenzene
- dehydrogenation 145
diethyleneglycol 111
4,4'-difluorodiphenylketone 190
9,10-dihydroanthracene 344
1,4-dihydroxyanthraquinone 354
o-dihydroxybenzene 183
m-dihydroxybenzene 185
p-dihydroxybenzene 187
- MAK value 434
2,4-dihydroxybenzophenone 187
2,2'-dihydroxybiphenyl 425
1,2-dihydroxydihydrobenzene 444
2,6-dihydroxynaphthalene 320
2,5-dihydroxyterephthalic acid 190
16,17-dihydroxyviolanthrone 358
diisooctylphthalate 272
- production 273
p-diisopropylbenzene 189
diisopropylnaphthalene 329
3,3'-dimethoxybenzidine
- MAK value 434
dimethylacetamide 138
N,N-dimethylaniline 203
- MAK value 434
3,4-dimethylaniline 278
3,5-dimethylaniline 173
3,3'-dimethylbenzidine 206
- MAK value 434
α,α-dimethylbenzylhydroperoxide
- MAK value 435
dimethylbiphenyl 336

3,3'-dimethyl-4,4'-diaminodiphenylmethane
- MAK value 435
2,6-dimethylnaphthalene 339
dimethylnitrobenzene 277, 282
3,5-dimethylphenol 172
- production
- - from m-xylene 281
dimethylsulfoxide 111
dimethylterephthalate 286
- production figures 289
dinitrobenzene
- MAK value 435
4,6-dinitro-o-cresol
- MAK value 435
dinitronaphthalene
- MAK value 435
4,4'-dinitrostilbene-2,2'-disulfonic acid 241
dinitrotoluene 238, 424
- MAK value 435
- reduction 244
- uses 242
dinonylnaphthalene 331
di-sec-octylphthalate
- MAK value 435
dioxins 180
dipentamethylenethiuram tetrasulfide 403
2,2'-diphenic acid 364
diphenyl ether 336
- MAK value 435
diphenylamine 198, 204
- production 23
diphenylene oxide 425
diphenylene sulfide 363
diphenyl oxide 336
N,N'-diphenylguanidine 208
diphenylmethane 345
4,4'-diphenylmethanediisocyanate 199
- MAK value 436
N,N'-diphenyl-p-phenylenediamine 239
diquat 401
direct blue 15 326
direct blue 71 321
disperse red 65 240
disperse yellow 3 227
dithiosodiumsalicylate 263
1,1-ditolylethane 146
N,N'-ditolyl-p-phenylenediamine 239
diuron 230
divinylbenzene 145
- production 145
dodecylbenzene 211
dodecylphenol 163
- production 174
DOP 272
2,4-DP 177
DPG 208
DPPD 239

DPTT 403
DTPD 239
Dubbs process 85
Durene 291, 295

electrode binders 379
π-electron system 9
electrophilic substitution 13
- reaction-energy diagram 14
emissions
- of carcinogenic substances 444
emissions law 444
eosine 265
L-(−)-ephedrine 258
epoxy resins 160
1,2-epoxybenzene 444
p-ethoxyacetanilide 6
5-ethoxymethyl-4-hydroxy-2-methyl-
  pyrimidine 411
N-ethyl-2-aminonaphthalene 209
2-ethylanthraquinone 359
ethylbenzene 133
- dehydrogenation 139
- - adiabatic process 140
- - isothermal process 139
- MAK value 436
- production 134
- production capacities 137
- recovery
- - by distillation 133
- synthesis 133
- thermodynamic equilibrium diagram 134
- uses 137
ethylene 77, 81
2-ethyl-6-methylaniline 238
5-ethyl-2-methylpyridine 403
- production 404
O-ethyl-O-(4-nitrophenyl)-phenylthio-
  phosphonate
- MAK value 436
2-ethyl-5,6,7,8-tetrahydroanthraquinone
  epoxide 359
eugenol 95

fenitrothion 169
fenthion
- MAK value 436
fenvalerate 169, 260
Fischer-Tropsch synthesis 43
flamprop-methyl 230
flavanthrone 357
fluazifop-butyl 407
fluid catalytic cracking
- flow diagram 64
fluometuron 258
fluoranthene 38, 365
fluorene 364

fluorene-9-hydroxy-9-carboxylic acid 23
fluorenone 364
fluorescein 265
fluorescent brightener 28 415
fluorobenzene 209
flurenol-n-butyl ester 363
folpet 276
N-formylmorpholine 111
fossils
- geochemical 28
Friedel-Crafts acylation 15
Friedel-Crafts alkylation 14
fuchsin 3, 428
fumaric acid 217
furan 26, 387
furfural 92, 388
- MAK value 436
- refining 93
furfuryl alcohol
- MAK value 436
furnace carbon black 384

G-acid 321, 324
γ-acid 324, 326
gasoline 68, 76
- production
- - from methanol 87
gas-works tar 2
glutarimide 403
graphite 368
- structure 369
graphite electrodes 379
green coke 378
- composition 378
Griesheim red 319
guaiacol 96, 183

H-acid 324
- production 325
HCH 232
heavy naphtha 291, 297
hemicellulose 92
hemimellitene 291
heterocyclics 387
- condensed 416
hexachlorobenzene 180
hexachlorocyclohexane 232
- production 233
n-hexadecylsalicylate 175
hexamethylbenzene 291
hexaphenyl 371
hippuric acid 247
homogentisic acid 445
homoveratric acid 422
homoveratrylamine 422
Hückel's rule 10
humic acids 30

hydantoin 232
hydrazobenzene 205
hydrocracking 73
- ideal 73
- product specifications 76
- relative reactions rates 74
- yields 76
hydro-dealkylation 20
hydron blue R 424
hydropyrolysis 55
hydroquinone 183
- production 189
- - from benzene 190
- - from p-diisopropylbenzene 188
- - from phenol 189
- production figures 190
p-hydroxybenzaldehyde 182
3-hydroxybenzoic acid 250
4-hydroxybenzoic acid 181
p-hydroxybenzoic acid propyl ester 182
p-hydroxybenzonitrile 182
5-(2'-hydroxyethyl)-4-methyl thiazole 411
9-hydroxyfluorene-9-carboxylic acid 363
1-hydroxy-2-isopropoxybenzene 185
7-hydroxyisoquinoline 422
2-hydroxynaphthalene-3,6-disulfonic acid
  321, 324
2-hydroxynaphthalene-6,8-disulfonic acid
  321, 324
1-hydroxynaphthalene-4-sulfonic acid 315
2-hydroxynaphthalene-1-sulfonic acid 320
2-hydroxy-3-naphthoic acid 318
5-(2'-hydroxy-3'-naphthoyl)-aminobenz-
  imidazolone 225
2-hydroxy-4-octyloxybenzophenone 187
(−)-1-hydroxy-1-phenylacetone 257
8-hydroxyquinoline 419
hydroxysulfone blue 354

igepal NA 210
imazalil 392
imazapyr 420
imazaquin 448
imidazole 391
impregnation oil 2
indamine 227
indan 297
indanthrene 357
indanthrene brilliant green B 358
indene 297
indene/coumarone resins 297
indigo 4, 207
indole 387, 417
indoxyl 207
indra red 319
ioxynil 182
IPPD 226

iprodione 232
isodurene 291, 295
isoeugenol 95
isoniazid 409
isophorone 172
isophthalic acid 265, 279
isophthalodinitrile 280
isophthaloyl chloride 280
isopropenylbenzene 146
- MAK value 440
isopropenylnaphthalene 331
p-isopropenylphenol 160
4-isopropylaniline 296
isopropylbenzene
- MAK value 440
2-isopropyl-(4-chlorophenyl)-acetic acid chloride 169, 260
isopropylnaphthalene 317
- oxidation 317
- production 329
4-isopropylphenol
- ammonolysis 296
4-isopropylphenylisocyanate 296
isoproturon 296
isoquinoline 26, 38, 421

J-acid 321, 324

kapton 295
keoxifene 417
kermesic acid 351
kerogen 28
kevlar 228, 290
Koch-acid 324

D-lactic acid 178
LAS 211
lethal dose 426
- of aromatics 427
letter acids 324
lignin 30, 91, 93, 360
- hydrogenation 94
lignite 30
lindane 232
linuron 229
liquid crystals 371
lubricant additives 217, 331
luminal 412
lutidine 26, 410

MAK value 429
- of aromatics 430
malachite green 203
maleic anhydride 213, 270
- production
- - from benzene 214
- - from butane 215

- production figures 216
martius yellow 298
mauveine 3
MBS 202
MBT 202
MBTS 202
MCPA 168, 363
MCPP 168
MDI 199, 200
Meisenheimer complex 17
melamine 412
- production
- - from urea 413
menadiol 337
1-menadiol monoacetate 337
menadione 337
menthane 167
menthol 169
MEP 403
2-mercaptobenzothiazole 199, 202
mesidine 295
mesitylene 291, 294
mesogens 370
mesophase 368
mesophase pitch 374
metalaxyl 171, 282
metamitron 249, 416
metanilic acid 187
metathesis 143
metazachlor 282
methabenzthiazuron 203
methiocarb 173
methoxyaniline
- MAK value 436
methoxychlor
- MAK value 436
2-methoxynaphthalene 322
methyl parathion 181
2-methyl-6-acetyl-naphthalene 338
N-methylaniline 203
- MAK value 437
p-methylbenzoic acid methyl ester 287
methylbiphenyl 154, 336
methyldopa 96
4,4'-methylene-bis-(2-chloroaniline)
- MAK value 437
4,4'-methylene-bis-(N,N-dimethylaniline)
- MAK value 437
3,4-methylenedioxy-benzaldehyde 184
5-methylimidazole-4-carboxylic acid ethyl ester 392
2-methyl-5-isopropylphenol 168
4-methylmercapto-3,5-xylenol 173
methylnaphthalene 336
- oxidation 337
1-methyl-3-phenylindane 349
2-methylpyridine 394, 408

3-methylpyridine 394
4-methylpyridine 409
1-methylpyrrole 390
N-methylpyrrolidone 109, 111
α-methylstyrene 146, 148, 149
p-methylstyrene 146
methylstyrenes
- MAK value 437
N-methyl-2,4,6-N-tetranitroaniline
- MAK value 437
metolachlor 238
metribuzin 416
michlers ketone 203
- MAK value 437
morphine 362
2-morpholinobenzothiazylsulfenamide 202
mucochloric acid 259
munjistin 350

NA 341
NAA 337
nafcillin 322
naftifine 448
naphtha
- steam cracking 101
naphthacene 24, 40
naphthalene 5, 24, 38, 298
- alkylated 329
- alkylation 317
- condensation products 373
- conversion processes
- - review 332
- crystallization 302
- - Brodie-process 304
- - *Sulzer-MWB*-process 303
- derivatives 308
- explosive limits 309
- hydrogenation 328
- hydrorefining 300
- MAK value 437
- oxidation 5, 248, 266, 309, 310
- - kinetics 21
- production figures 307
- recovery 299
- - from coal tar 299
- - from petroleum-derived raw materials 305
- - from reforming residues 307
- sulfonation 16, 322
- sulfonic acid derivatives 322
- uses 308
2,6-naphthalenedicarboxylic acid 331, 338
1,5-naphthalene-diisocyanate
- MAK value 437
naphthalene oil
- distillation 300
naphthalenesulfonic acids 210, 313, 317

1,4,5,8-naphthalenetetracarboxylic acid 366
naphthalic anhydride 340
1-naphthol 313
- production 313
2-naphthol 226, 316
- production 316
naphthol AS 318
naphtho[8,1,2-bcd]perylene 25
naphthoquinone 310
naphthyl chloride 337
1-naphthylacetic acid 337
naphthylamine 327
1-naphthylamine 314
2-naphthylamine 428, 444
- MAK value 437
1-naphthalenylthiourea
- MAK value 430
naproxen 322
naptalam 327
natural gas 27
- deposits 31
- origin 27
needle coke 378
Nevile-Winther acid 315
niacin 403
nicotinamide 403
nicotine 403
- MAK value 438
nicotinic acid 403
- production 406
nitration 15
5-nitroacenaphthene
- MAK value 438
2-nitro-4-aminophenol
- MAK value 438
3-nitro-4-aminotoluene 241
nitroaniline 224, 226
- MAK value 438
1-nitroanthraquinone 352
nitrobenzene
- hydrogenation 196
- MAK value 438
- production 194
- production capacities 195
- reduction 196, 205
nitrobenzenesulfonic acid 194, 195
4-nitrobiphenyl
- MAK value 438
o-nitrocinnamic acid 5
nitrocresol 169, 237
nitrocumene 148, 296
nitrocyclohexane
- reduction 200
nitrodiphenyl ether 226
4-nitrodiphenylamine 226
nitrofen 178, 226
nitrofurantoin 388

nitrofurazone 388
5-nitrofurfural diacetate 388
nitronaphthalene 327
- MAK value 438
1-nitronaphthalene-3,6,8-trisulfonic acid 326
nitrophenol 180, 226, 237
2-nitro-p-phenylenediamine
- MAK value 438
nitropyrenes
- MAK value 438
N-nitrosodiphenylamine 204
N-nitrosoethylphenylamine
- MAK value 439
N-nitrosomethylphenylamine
- MAK value 439
nitrotoluene 237, 241
- MAK value 439
4-nitrotoluene-2-sulfonic acid 241
NMR-spectroscopy 34
nomenclature
- of aromatics 23
nomex 280
nonylphenol 163
- production 174
norflurazon 259
nucleophilic substitution 17
nylon 99, 191

octahydroanthracene 345
octahydrophenanthrene 345
octane number 68
4'-octylbiphenyl-4-carbonitrile 370
octylphenol
- production 174
optical brighteners 241
oxazole 391
oxidation reactions 21
oxine 419
N-(oxydiethylene)-2-benzothiazylsulfen-
  amide 202

papaverine 185, 422
para red 226
paracetamol 180
paraquat 401
- MAK value 439
parathion
- MAK value 439
PBI 223
PCB 336
peat 30
PEEK 190
penicillin G 252
- production 253
penicillin V 182
penicillins
- semi-synthetic 254, 257, 322

pentachlorophenol 177, 179
- MAK value 439
pentamethylbenzene 291
2-pentylanthraquinone 359
perinone pigments 366
persistence 426, 445
perylenetetracarboxylic acid 341
perylenetetracarboxylic-diimide 341
phenacetin 6
phenanthrene 24, 38, 334, 362
phenanthrene polymer 373
9,10-phenanthrenequinone 363
phenmedipham 240
phenol 2, 38, 148, 221, 248
- alkylation 164, 167, 174
- ammonolysis 198
- chlorinated 176
- derivatives 158
- hydrogenation 162
- MAK value 439
- nitration 180
- polyhydric 183
- production
- - from benzene 151, 220
- - from chlorobenzene 153
- - from coal pyrolysis products 155
- - from cumene 149
- - from cyclohexane 155
- - from lignin 94
- - from toluene 154
- production figures 157
- sources 148
- specifications 157
- uses 157
phenol/formaldehyde resins 158
phenolphthalein 265, 274
phenothiazine 204
phenoxyacetic acid 182
m-phenoxybenzaldehyde 169
m-phenoxybenzoic acid methyl ester 169
m-phenoxytoluene 169
phenylacetic acid 252
phenylacetylcarbinol 258
phenylalanine 255
1-phenyl-2,3-dimethylpyrazolin-5-one 6, 207
phenylenediamine 224, 227, 280, 290
- arylated 239
- MAK value 439
phenylglycidyl ether
- MAK value 439
D-phenylglycine 254, 257
N-phenylglycine 207
phenylglycine-o-carboxylic acid 5
phenylhydrazine 206
- MAK value 439
phenylhydroxylamine 180
1-phenyl-3-methylpyrazolin-(5)-one 206

482 Subject index

N-phenyl-2-naphthylamine 321
- MAK value 440
phenylnitromethane 237
phloroglucinol 190
phosalone 180
photosynthesis 27
phthalic acid
- decarboxylation 248
phthalic anhydride 5, 265
- MAK value 440
- other products 274
- production
- - from naphthalene 266, 309
- - from o-xylene 265
- production figures 271
- uses 272
phthalic esters
- characteristic data 273
- production 272
phthalimide 275
phthalocyanine 275
phthalodinitrile 275
C_{20}-phytane 28
phytol 337
phytomenadione 337
phytoplankton 28
picloram 409
α-picoline 26, 394, 408
- production 408
β-picoline 394
- ammoxidation 407
γ-picoline 394, 409
picramic acid 181
picric acid 181, 218
pigment orange 5 317
pigment orange 13 207
pigment orange 36 225
pigment orange 43 366
pigment red 1 226
pigment red 3 241, 317
pigment red 7 319
pigment red 49: 1 320
pigment red 52: 1 262
pigment red 53: 1 262
pigment red 57: 1 242
pigment red 63: 1 320
pigment red 88 231
pigment red 112 319
pigment red 149 341
pigment red 171 225
pigment red 179 341
pigment red 194 366
pigment violet 19 290
pigment violet 23 424
pigment yellow 1 209, 241
pigment yellow 13 209, 282
pigment yellow 14 239

piperidine 403
piperonal 184
pirimicarb 412
piroxicam 263
pitch
- characteristic data 379
- viscosity pattern 375
pitch coke 380
pitch coking 380
pitch fibers 380
plasticizers 249, 251, 264, 272, 274, 294
platform process 70
PMP 206
PNDPA 226
polyadamantane 33
polycarbonate 161
polychlorinated biphenyl 336
polydimethylphenylene oxide 171
polyester 213, 217, 246, 288
polyester fibers 283
polyetheretherketone 190
polyimides 294, 295
poly-1,3-phenylene-2,2'-(5,5'-bis-benz-imidazole) 223
polyphenylenesulfide 231
polystyrene 138
polysulfone plastics 161
polyurethane 200, 209, 246
6PPD 226
PPS 231
praziquantel 421
prehnitene 291
premium coke 375
prochloraz 392
procion brilliant orange GS 416
profenofos 176
promazin 204
prometon 415
prometryn 415
propanil 230
propiconazole 393
propoxur 185
- MAK value 440
propranolol 314
propylene 77
1,3-propyleneglycol dibenzoate 249
protocatechnic acid 445
pseudocumene 291
- oxidation 293
pseudopurpurin 350
purpurin 350
pyrazine 411
pyrazon 207, 412
pyrene 24, 38, 40, 365
pyrethroid insecticides 169, 170, 260
pyridazine 411

pyridine 26, 387, 394
- alkylated 403
- chlorination 402
- derivatives 401
- MAK value 440
- synthesis 396
- - from acrolein 399
- - from crotonaldehyde 400
- - from cyclopentadiene 400
- - from formaldehyde/acetaldehyde 397
pyridine bases 394
- composition 394
- refining 395
2,3-pyridine dicarboxylic acid 420
pyridine-thiones 402
pyridoxine 410
pyridoxol 410
pyrimidine 411
pyrocatechol 445
pyrolysis 20, 368, 382
- of biomass 90
- of naphthalene
- - condensation products 373
pyrolysis gasoline 100, 101
- C_9-aromatics 292
- composition 101
- - after hydrogenation 102
- desulfurization 101
- hydrogenation 101
- properties 101
pyrolysis naphtha 78
- characteristic data 80
pyrolysis oil 382
pyrolysis tar 78, 306
- composition 81
pyromellitic dianhydride 295
pyrrole 26, 387, 390

quaterphenyl 334
quinacridone pigments 190, 290
quinaldine 274
quinic acid 187
quinine 2, 422
quinizarin 354
quinoline 26, 38, 387, 419
- alkylated 421
- sulfonation 419
quinoline-8-sulfonic acid 419
quinolinic acid 420
quinophthalone 274

R-acid 321, 324
radical reactions 20
ranitidine 388
reactive blue 19 353
reactive orange 1 416
rearrangement reactions 22

reformer gasoline 100
- composition 100
reforming
- catalytic 68
- - C_9-aromatics 292
- - thermodynamic data 69
- - typical data 71
renewable raw materials 89
resonance 9
resonance energy 9
- of aromatic hydrocarbons 10
resorcinol 183, 185
- production 186
- production figures 187
resorcinol resins 187
riboflavin 278
rotenone 222
- MAK value 440
rubiadin 350

saccharin 263
safranine B extra 227
salicylaldehyde 175, 183
salicylic acid 175
- production 175
- production figures 176
Sandmeyer reaction 20
Schäffer acid 319
sodium benzoate 249
sodium cyclamate 203
sodium cyclohexylsulfamate 203
solvent
- for carbonless copy papers 329
solvent red 19 209
solvent red 23 209
solvent yellow 1 6
solvent yellow 2 428
special coke 368
steam cracking 77
- of naphtha 78
stearamido-methylpyridinium chloride 402
styrene 137
- distillation 141
- MAK value 441
- production
- - from acetophenone 143
- - from butadiene 144
- - from chloroethylbenzene 142
- - from ethylbenzene 138
- - from ethylbenzene hydroperoxide 143
- - from toluene 143
- production figures 144
- specifications 142
- substituted 144
sulfamethoxazole 205
sulfanilic acid 208
sulfolane 111

sulfonation 16, 212
surface-active substances 210
surfactants 174, 213
syncrude 46
syntans 331

2,4,5-T 179
T-acid 324
tanning substances 331
tar 1, 34, 299
- characteristic data 40
- components 38
- dyes 3
- production figures 38
- quinoline insolubles 40
- recovery 35
- refining 37, 41
- toluene insolubles 40
tartrazine 208
TDI 242
technora 196
terephthalic acid 265, 283
- production
- - from phthalic anhydride 287
- - from toluene 287, 290
- - from p-xylene 284
- production figures 289
- refining 284
terephthalic acid diglycol ester 288
terephthalic acid dimethyl ester
- production 286
terephthalic acid dinitrile 290
terephthaloyl dichloride 228, 290
terphenyl 334
- hydrogenated 335
3,3',4,4'-tetraaminobiphenyl 223
1,3,6,8-tetrabromopyrene 366
1,2,4,5-tetrachlorobenzene 179
2,3,7,8-tetrachloro-dibenzo-p-dioxin 179
- MAK value 441
tetrachloroisophthalodinitrile 281
tetrahydroanthraquinone 311, 361
4,4'-tetrahydrobipyridyl 401
tetrahydrofurfuryl alcohol 400
1,2,3,4-tetrahydronaphthalene 301
1,2,3,6-tetrahydrophthalic anhydride 217
tetralin 301, 328
- oxidation 314
tetralol 328
tetralone 314, 328
tetranitromethane 237
thalidomide 277
thenalidine 389
thiabendazole 393
thiamine 411
thianaphthene 38, 302, 387, 416
thiazole 391

thiazolidine-β-lactam 254
2-thienylmethyl chloride 389
thiophanate-methyl 224
thiophene 26, 387, 389
thymol 168
timber impregnation 4
TNT 99, 247
Tobias acid 320
α-tocopherol 173
2-tolidine 238, 239
tolmetin 390
tolnaphthate 322
toluene 99
- alkylation 264
- chlorine derivatives 250
- dealkylation 20, 122, 335
- derivatives 236
- dinitration 242
- disproportionation 126
- MAK value 441
- nitration 237
- nuclear chlorination 259
- oxidation 154, 248
- oxychlorination 168
- production
- - from aromatic mixtures 114
- production figures 129
- propylation 166
- quality standards 128
- side-chain chlorination 250
- sulfonation 261
- sulfonic acid derivatives 261
- uses 236
toluenediamine 243
- MAK value 441
toluene diisocyanate 199, 229, 238, 241, 242
- MAK value 441
- production 245
- production capacities 246
toluenesulfonic acid 261
toluenesulfonyl chloride 263
2-toluidine 238
- MAK value 441
4-toluidine 241
3-tolyl isocyanate 240
N-o-tolyl-N'-phenyl-p-phenylenediamine 239
torlan 294
toxicity 426
- of aromatics 427
toxicology 426
TPS 211
triadimefon 176, 393
2,4,6-triamino-1,3,5-triazine 412
1,2,4-triazine 416
1,3,5-triazine 412

1,2,4-triazole 176, 391
1,2,4-trichlorobenzene
- MAK value 441
1,1,1-trichloro-2,2-bis(4-chlorophenyl)-ethane 218
1,3-bis-(trichloromethyl)-benzene 280
N-trichloromethylthiophthalimide 276
2,4,5-trichlorophenol 177, 233
- production 179
2,4,5-trichlorophenoxyacetic acid 179
- MAK value 441
α,α,α-trichlorotoluene
- MAK value 442
2,4,6-trichloro-1,3,5-triazine 241, 412
- production 414
trifluoromethylaniline 258, 326
3-trifluoromethylnitrobenzene 258
trifluralin 260
1,3,5-triisopropylbenzene 190
trimellitic anhydride
- MAK value 442
trimesic acid 294
trimethoprim 205
2,3,5-trimethylaniline 173, 294
2,4,5-trimethylaniline
- MAK value 442
2,4,6-trimethylaniline 295
trimethylbenzene 291
trimethyl-1,2-dihydroquinoline 420
2,4,7-trinitrofluorenone
- MAK value 442
2,4,6-trinitrophenol 181
    MAK value 442
trinitrotoluene 99, 238, 247
- MAK value 442
triphenylamine 198
triphenylene 24, 40, 334
triphenylmethane dyes 203
triptycene 19
TRK value 429
α-tropolone 11
tryptophan 417
twaron 228, 290

udel 161

vanillin 95, 185
- synthesis 96
vat black 27 424
vat blue 4 357
vat blue 43 424
vat green 1 358
vat orange 15 424
vat yellow 1 357
vat yellow 2 259
veratrole 185

veronal 412
N-vinylcarbazole 423
2-vinylpyridine 409
vinyltoluene 145
violanthrone 358
16,17-violanthronequinone 358
visbreaker process 84
vitamin B_1 411
vitamin B_2 278
vitamin B_3 403
vitamin B_6 410
vitamin B_{12} 391
vitamin E 173, 294
vitamin K_1 337
vitamin K_3 337
vitamin PP 403
vulcanization accelerator 202, 208, 403
vulcanization retarder 204

warfarin
- MAK value 442
water resources law 446
wetting agents 331
wood 91, 93
wood preservatives 176
wood processing 93

xylenes 100
- dealkylation 122
- derivatives 283
- equilibrium concentrations 125
- isomerization 121, 125
- - reaction diagram 126
- MAK value 442
- production
- - by adsorption 119
- - by crystallization 116
- - by distillation 116
- - by formation of complexes 121
- production figures 129
- quality standards 128
- separation 115
- uses 100
o-xylene 265
- explosive limits 309
- nitration 277
- oxidation 265, 267, 270
- - reaction diagram 267
m-xylene
- ammoxidation 281
- derivatives 279
- nitration 282
- oxidation 279
p-xylene
- ammoxidation 281, 290
- oxidation 283
- - kinetics 22

xylenol 163, 171
- ammonolysis 282
- production 171
- - from phenol 167
- uses 171
xylidine 277, 282

- MAK value 443
m-xylylenediamine 280

zeolite 63, 67, 75, 86, 88, 126, 127, 136, 339
zooplankton 28

M. B. Hocking
Modern Chemical Technology and Emission Control

1985. 152 figures. XVI, 460 pages.
ISBN 3-540-13466-2

Contents: Background and Technical Aspects of the Chemical Industry. – Air Quality and Emission Control. – Water Quality and Emission Control. – Natural and Derived Sodium and Potassium Salts. – Industrial Bases by Chemical Routes. – Electrolytic Sodium Hydroxide and Chlorine and Related Commodities. – Sulfur and Sulfuric Acid. – Phosphorus and Phosphoric Acid. – Ammonia, Nitric Acid and their Derivatives. – Aluminium and Compounds. – Ore Enrichment and Smelting of Copper. – Production of Iron and Steel. – Production of Pulp and Paper. – Fermentation Processes. – Petroleum Production and Transport. – Petroleum Refining. – Formulae and Conversion Factors. – Subject Index.

Springer-Verlag
Berlin Heidelberg New York
London Paris Tokyo

Catalysis Science and Technology

Editors: J.R. Anderson, M. Boudart

Volume 1
1981. 107 figures. X, 309 pages. ISBN 3-540-10353-8*

Contents: *H. Heinemann:* History of Industrial Catalysis. – *J. C. R. Turner:* An Introduction to the Theory of Catalytic Reactors. – *A. Ozaki, K. Aika:* Catalytic Activation of Dinitrogen. – *M. E. Dry:* The Fischer-Tropsch Synthesis. – *J. H. Sinfelt:* Catalytic Reforming of Hydrocarbons.

Volume 2
1981. 145 figures. X, 282 pages. ISBN 3-540-10593-X*

Contents: *G.-M. Schwab:* History of Concepts in Catalysis. – *J. Haber:* Crystallography of Catalyst Types. – *G. Froment, L. Hosten:* Catalytic Kinetics: Modelling. – *A. J. Lecloux:* Texture of Catalysts. – *K. Tanabe:* Solid Acid and Base Catalysts.

Volume 3
1982. 91 figures. X, 289 pages. ISBN 3-540-11634-6*

Contents: *E. E. Donath:* History of Catalysis in Coal Liquefaction. – *G. K. Boreskov:* Catalytic Activation of Dioxygen. – *M. A. Vannice:* Catalytic Activation of Carbon Monoxide on Metal Surfaces. – *S. R. Morrison:* Chemisorption on Nonmetallic Surfaces. – *Z. Knor:* Chemisorption of Dihydrogen.

Volume 4
1983. 106 figures. X, 289 pages. ISBN 3-540-11855-1*

Contents: *P. N. Rylander:* Catalytic Processes in Organic Conversions. – *H.-P. Boehm, H. Knözinger:* Nature and Estimation of Functional Groups on Solid Surfaces. – *G. Ertl:* Kinetics of Chemical Processes on Well-defined Surfaces.

Volume 5
1984. 122 figures. X, 281 pages. ISBN 3-540-12665-1*

Contents: *J. R. Rostrup-Nielsen:* Catalytic Steam Reforming. – *K. Taylor:* Automobile Catalytic Converters. – *J. B. Peri:* Infrared Spectroscopy in Catalytic Research. – *P. Gallezot:* X-Ray Techniques in Catalysis.

Springer-Verlag
Berlin Heidelberg New York
London Paris Tokyo

Volume 6
1984. 142 figures. X, 313 pages. ISBN 3-540-12815-8*

Contents: *J. B. Butt:* Catalyst Deactivation and Regeneration. – *I. Pasquon, U. Giannini:* Catalytic Olefin Polymerization. – *G. Maire, F. Garin:* Metal Catalysed Skeletal Reactions of Hydrocarbons on Metal Catalysts. – *K. Foger:* Dispersed Metal Catalysts.

Volume 7
1985. 94 figures. XII, 223 pages. ISBN 3-540-15035-8*

Contents: *S. A. Topham:* The History of the Catalytic Synthesis of Ammonia. – *J. V. Sanders:* The Electron Microscopy of Catalysts. – *B. E. Koel, G. A. Somorjai:* Surface Structural Chemistry. – Subject Index. – Author Index Volumes 1–6.

Volume 8
1987. 60 figures. XII, 262 pages. ISBN 3-540-15034-X*

G. Chinchen, P. Davies, R. J. Sampson: The Historical Development of Catalytic Oxidation Processes

The present chapter outlines the history of this oxidation technology, indicating that, in the cases of many products, earlier routes which did not involve direct oxidation have been replaced by lower-cost oxidation routes.

J. C. Mol, J. A. Moulijn: Catalytic Metathesis of Alkenes

The metathesis of cyclic and acyclic alkenes is reviewed with respect to various hetero- and homogeneous catalytical systems.

J. J. Carberry: Physico-Chemical Aspects of Mass and Heat Transfer in Heterogeneous Catalysis.

This chapter shows how long and short range gradients of concentrations and temperature may persist in heterogenerously catalyzed reaction(s)-reactor networks.

K. C. Pratt: Small Scale Laboratory Reactors

Central to the evaluation presented here is the assessment of the catalyst's reaction performance, carried out using one or more of the many forms of laboratory reactor available.

J. H. Lundsford: EPR Methods in Heterogeneous Catalysis

In this review the application of EPR spectroscopy to eleven types of catalytic reactions is described.

*Distribution rights for all socialist countries: Akademie-Verlag, Berlin